Encyclopedia of Archaea

Volume I

Encyclopedia of Archaea
Volume I

Edited by **Giles Watkins**

R CALLISTO REFERENCE

New York

Published by Callisto Reference,
106 Park Avenue, Suite 200,
New York, NY 10016, USA
www.callistoreference.com

Encyclopedia of Archaea: Volume I
Edited by Giles Watkins

International Standard Book Number: 978-1-63239-207-7 (Hardback)

Printed in the United States of America.

Contents

Permissions

List of Contributors

Preface

While earlier biologists used to classify all living organisms into two broad categories namely, animal and plant; by the 1970s a new category was discovered called archaea. Even these were initially considered a form of bacteria but further studies proved them to be quite distinct.

Archea are one of the earliest forms of life that appeared on the earth billion of years ago. They do share some similar features with bacteria but differences are equally prominent. Archaea can thrive in the most extreme of environment while they do exist in normal living conditions too.

The study of archaea opens up a new path of understanding and investigating the principles of survival. But such studies and experimental testing are tough, costly and labour intensive. More so, when worldwide mining and milling activities have introduced dangerous levels of radionuclide and heavy metals in the soil and aquatic environments.

These microorganisms are used in the mining of precious metal like gold and area. Due to its ability to survive in extreme conditions, archaea combines with various chemical in extracting gold, copper, cobalt and silver from their ores. Certain other form of archaea brought a revolution in microbiology by triggering a chain reaction that could be used for the technique of DNA cloning. In the food processing industry, these archaea enable the production of whey and low lactose milk. Another form of archaea, known as methanogenic archaea, is pivotal in sewage treatment. These same microorganisms are also useful in producing one of the most efficient renewable fuels, biogas.

Also in the field of medicine, archaea plays an important role in developing useful antibiotics. Although the entire power and scope of archaea is far from being explored; but whatever has been developed is fruitful and beneficial for the humanity. This book delves deep into these minute looking powerhouses to come up with more information about them.

I would like to thank all the contributing authors, the publishing team and my family for making this endeavor a success.

Editor

Deposition of Biogenic Iron Minerals in a Methane Oxidizing Microbial Mat

Christoph Wrede,[1,2] Sebastian Kokoschka,[1] Anne Dreier,[1,3] Christina Heller,[4,5] Joachim Reitner,[3,4] and Michael Hoppert[1,4]

[1] *Institute of Microbiology and Genetics, Georg-August-University, Grisebachstr. 8, 37077 Göttingen, Germany*

[2] *Hannover Medical School, Institute of Functional and Applied Anatomy, Carl-Neuberg-Str. 1, 30625 Hannover, Germany*

[3] *Courant Centre Geobiology, Georg-August-University, Goldschmidtstr. 3, 37077 Göttingen, Germany*

[4] *Geoscience Centre Göttingen, Georg-August-University, Goldschmidtstr. 3, 37077 Göttingen, Germany*

[5] *Federal Institute for Geosciences and Natural Resources, Stilleweg 2, 30655 Hannover, Germany*

Correspondence should be addressed to Christoph Wrede; cwrede@gwdg.de

Academic Editor: Charles Cockell

The syntrophic community between anaerobic methanotrophic archaea and sulfate reducing bacteria forms thick, black layers within multi-layered microbial mats in chimney-like carbonate concretions of methane seeps located in the Black Sea Crimean shelf. The microbial consortium conducts anaerobic oxidation of methane, which leads to the formation of mainly two biomineral by-products, calcium carbonates and iron sulfides, building up these chimneys. Iron sulfides are generated by the microbial reduction of oxidized sulfur compounds in the microbial mats. Here we show that sulfate reducing bacteria deposit biogenic iron sulfides extra- and intracellularly, the latter in magnetosome-like chains. These chains appear to be stable after cell lysis and tend to attach to cell debris within the microbial mat. The particles may be important nuclei for larger iron sulfide mineral aggregates.

1. Introduction

Frequently, biofilm formation in marine and freshwater systems is accompanied by precipitation of minerals. These minerals are also structurally integrative parts of the microbial biofilm [1]. In most cases, mineral precipitates are deposited in close contact to and in interaction with organic macromolecules, that is, carbohydrates and/or proteins [2]. Formation of a biomineral in a microbial biofilm may be detrimental to the organisms which is mainly due to the enclosure of the living biomass by mineral precipitates. However, also positive effects, for example, when lithified precipitates provide a matrix or scaffold for the microbial biomass, may be expected. It has also been considered that beneficial effects predominate, for example, when biominerals act as chemical filters or shield UV radiation [3]. It is known that, in certain cases, biological macromolecules influence solubility of minerals (e.g., by buffering the aqueous environment or by chelating ions) and may direct the formation of a mineral matrix in a more or less specific way. As a consequence, the shape of biomineral deposits varies considerably at narrow scales and seemingly similar environmental conditions [4].

Mineral deposits caused by the activity of microorganisms are mostly based on either carbonates or silicates [4]. These mineral phases are regularly intermixed with other organic or mineralic compounds (overviews in [2, 5]). A special case of these organomineral precipitations is microbialite formation during anaerobic oxidation of methane (AOM). AOM is conducted by various groups of archaea in a metabolic pathway reverting methanogenesis [6]. Mostly, sulfate reducing bacteria (SRB) participate in AOM [7–9]. The role of SRB is still not fully understood, though is generally accepted that, along with the oxidation of methane, sulfate is reduced: $CH_4 + SO_4^{2-} \rightarrow HCO_3^- + HS^- + H_2O$ [8, 10]. As a result, carbonate phases (calcite and aragonite) and iron sulfides are generated as byproducts of the metabolic process.

It is is known that AOM occurs worldwide in anoxic sediments when methane and electrons acceptors are available (e.g., [11]). The formation of large (several centimeters and bigger) carbonate concretions depends on high methane concentrations under hydrostatic pressure and on the presence of sulfate [8, 9, 12]. In the anaerobic water column of the Black Sea, huge carbonate concretions have been observed at the Crimean shelf [8]. The carbonate buildups may be considered as highly porous "fixed bed" bioreactors, allowing the percolation of methane and the exchange of sea water. The outer and inner surfaces of these carbonate buildups are covered by complex microbial mats, primarily formed by the organisms involved in AOM.

In previous investigations, distinct layers in these microbial mats were discriminated. On the surface, exposed to the sea water, a black layer consists mainly of aggregates between methane-oxidizing archaea of the ANME-2 group and sulfate reducing bacteria (SRB). SRB of this mat type often exhibit intracytoplasmic magnetosome-like chains of greigite precipitations [8, 13, 14]. Our results imply that greigite magnetosomes are one sink for (otherwise toxic) sulfides. These particles were found inside SRB but were also present in the extracellular matrix of the biofilm.

2. Materials and Methods

Microbial mat samples were collected in 2001 during a cruise with the Russian R/V Professor Logachev in the methane seep area located in the GHOSTDABS field (Black Sea north east the Crimean shelf). These samples have already been subjected to extended geochemical and structural analyses [9, 14]. Specific antibodies, directed against methyl-coenzyme M reductase (MCR), the key enzyme of (reverse) methanogenesis, were generated after purification of MCR as essentially described according to [15] by immunization of rabbits following established protocols (e. g., [16] and, the references therein). Specificity of the antibody was extensively studied for methanogenic archaea and reverse methanogens as already described [14, 16, 17].

For microscopic analyses, the samples were chemically fixed in a 4.0% (v/v) aqueous formaldehyde solution (from a 10%, w/v, stock solution, pH 8.0, freshly prepared from paraformaldehyde) and stored in 100 mM PBS (phosphate buffered saline, pH 7.0) at 4°C until further use. The material was then washed several times in PBS and cut to small fragments of about 200 μL volume. Samples were then chemically fixed in a 0.5% (v/v) glutardialdehyde solution (in 100 mM PBS) for 2 h. The samples were then processed as described [18], for electron and light microscopy, and finally cut in ultrathin or semithin sections of either 100–300 nm or 1 μm in thickness. Semithin sections were transferred, with the aid of a transfer loop, on microscope slides, and ultrathin sections were picked up with Formvar-coated grids. For light microscopy, the sections were treated either with an anti-MCR antibody or with the lectin concanavalin A (ConA) coupled to fluorescent marker molecules (Sigma-Aldrich, Deisenhofen, Germany) as described [18]. The lectin ConA IV coupled to Alexa Fluor 546 as fluorescence marker

(Molecular Probes, Eugene, OR, USA) was used in 1/1000 working concentration dilutions in PBS supplemented with 1 mM $CaCl_2$ and $MnCl_2$ (lectin buffer). The sections were mounted on glass slides by heat fixation at 60°C for 15 min and then incubated for 30 min at room temperature. After this, the lectin dilution was soaked off, and the sections were briefly rinsed in pure lectin buffer and covered with coverslips. For immunofluorescence microscopy, the heat-fixed semithin sections were incubated with the antiserum (dil. 1/1000 with PBS, pH 7.5) for 2 h. The sections were rinsed three times in 100 μL drops of PBS (supplemented with 0.01% Tween 20) and incubated with a secondary goat anti-rabbit antibody, coupled to Alexa Fluor 546 fluorescent dye (Molecular Probes, Eugene, OR, USA), diluted 1 : 250 [14]. The rinsing steps were repeated. Fluorescence microscopy was performed with an Axio Scope light microscope using filter set 43 (BP: 545/25, FT 570, LP: 605/70) and the AxioVision software package (Zeiss, Göttingen, Germany). For comparison, phase contrast images were taken and were digitally merged with fluorescence images.

For transmission electron microscopy, ultrathin sections obtained from five distinct samples were mounted on Formvar-coated 300 mesh specimen grids. Immunolocalization with antibodies directed against MCR was performed as described. Mounted sections were stained with phosphotungstic acid (3%, w/v), if not stated otherwise [14]. Electron microscopy was performed in a Zeiss EM 902 transmission electron microscope (Zeiss, Oberkochen, Germany), equipped with a eucentric goniometer stage. Images were recorded with a 1 KB digital camera. Detection and enhancement of colloidal gold markers in digitized electron micrographs were performed as described [19]. Electron energy loss spectroscopy (EELS) of iron-containing particles was performed essentially as described in [20], with the aid of the analysis V software package (Olympus-SIS, Münster, Germany). Electron energy loss was measured between 655 eV and 751 eV. The L2 (720.6 eV) and L3 (708.0 eV) edges, observed for pure iron minerals, were represented by one broad peak at approximately 715 eV of the deposits in embedded and ultrathin sectioned biofilms.

Goniometry was performed by tilting 300 nm sections ±60 degrees with 1-degree increments. Tomograms were performed with the EM3D 2.0 software package (Department of Neurobiology, Stanford University [21]).

3. Results and Discussion

Two layers of microbial mats retrieved from the the Black Sea Crimean shelf have been identified as important for AOM: the orange (or pink) layer and the black layer. The orange layer consists of various cell morphotypes [8, 22]. Most of them were identified as ANME-1 archaea. ANME-1 cells are morphologically similar to the filamentous methanogens *Methanospirillum* and *Methanosaeta*; these cells are covered by a tight and very rigid protein sheath [23]. Sulfate reducing bacteria were present in large clusters, but not in direct contact with ANME-1 cells [8]. Visually, the black layer could be clearly distinguished from other layers. In contrast

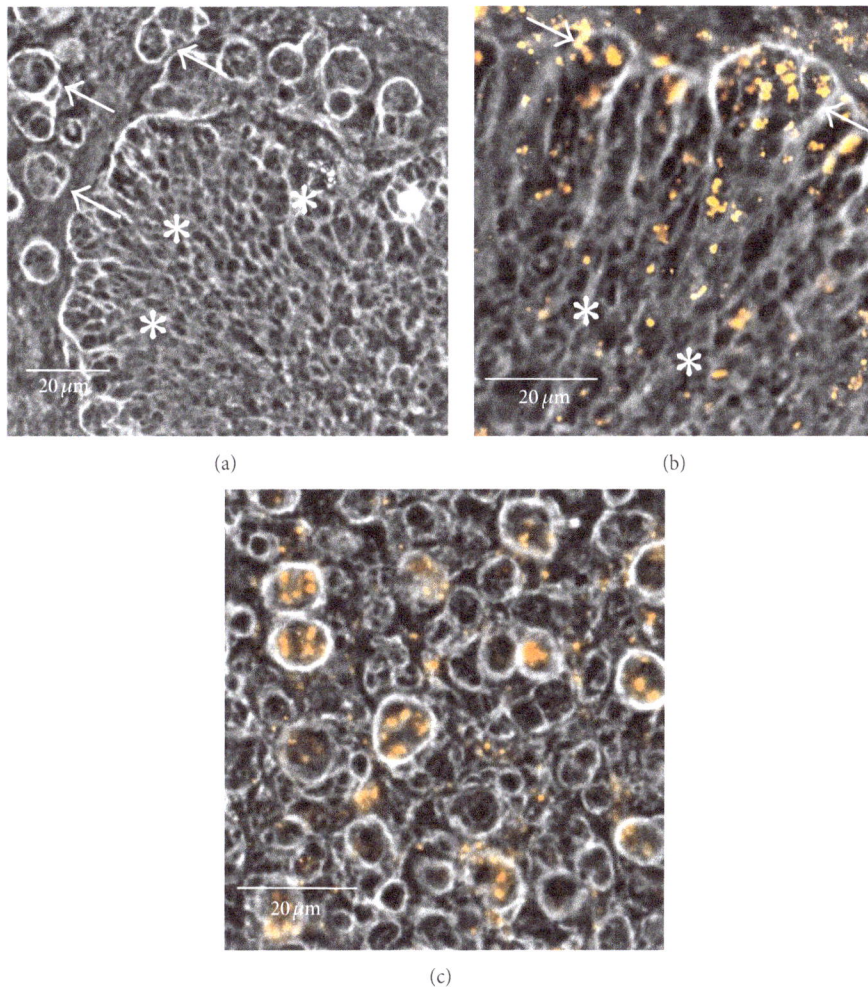

FIGURE 1: Light micrographs of semithin sections after immunofluorescence staining with anti-MCR antibodies (merged fluorescence/phase contrast images). (a) One large, cauliflower-like aggregate (lower right, the whole aggregate marked by three asterisks) surrounded by small globular aggregates (arrows) in an unstained semithin section (phase contrast microscopy). (b) Periphery of a large aggregate. The cell density at the periphery of the aggregate (arrows) is higher than in the central area (asterisks). (c) Small aggregates after fluorescence staining.

to the orange layer, the black layer consists of aggregates formed by ANME-2 and SRB of the DSS group [8, 24, 25]. Immunofluorescence labelling of MCR, performed on resin sections, marks the position of the MCR-expressing ANME-2 inside large cauliflower-shaped aggregates, consisting of thousands of cells visible in a section (Figure 1(a), asterisks, Figure 1(b)). Similar labelling experiments have been also performed on the same samples with antibodies directed against the dissimilatory adenosine-5′-phosphosulfate (APS) reductase, along with the identification of the respective gene in the microbial mat samples [14]. APS reductase is a key enzyme of sulfate reduction and could be localized in the magnetosome-bearing cell type (see below), identified as SRB.

In addition to these large cauliflower-shaped aggregates, also a smaller globular-shaped aggregate type of 5–20 labeled ANME-2 cells was identified (Figure 1(c)). This aggregate type is located in the surrounding (Figure 1(a), arrows) and in the center of the cauliflower-shaped features. All aggregates are separated by areas of low cell densities. It has to be noted that not all cells visible in the depicted sections are labelled, since markers do only bind to cells with their cytoplasm exposed to the section surface. In the large type of aggregates, cells are not completely randomly distributed; higher cell densities are observable at the periphery, separated in irregular lobes (Figure 1(b), arrows). Both types of aggregates consist of ANME-2/SRB consortia [8, 14, 22]. The electron micrographs in Figure 2 show ANME-2 (immunogold labelled) and SRB (nearly unlabeled) in a globular aggregate, surrounded by multiple layers of extracellular material. The gaps between the large aggregates are filled with EPS [18]. These gaps show a distinct lectin labelling (Figure 3(a), arrows), in contrast to the unlabeled extracellular surrounding of the ANME-2/SRB consortia. Various morphotypes of prokaryotic cells could be detected in these empty spaces (Figure 3(b), cf. [14]), including thin filaments of several μm in length and 200 nm in diameter (arrows in Figure 3(b)). Some filaments still contain a dark stained

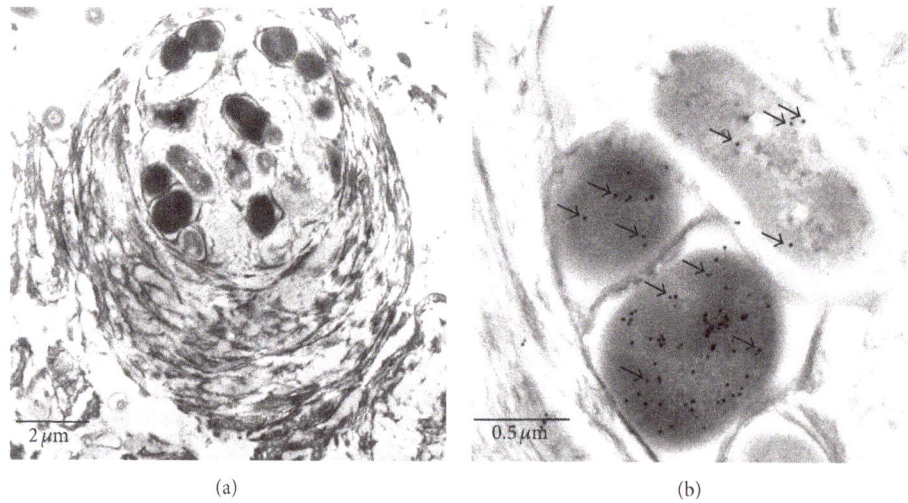

(a)   (b)

FIGURE 2: Immunoelectron microscopy. (a) Electron microscopy of a typical small aggregate (cf. Figure 1(c)), consisting of ANME-2/SRB. Several single cells are surrounded by a thick multilayered mucilage. (b) Detail of the aggregate as depicted in (a). The MCR expressing ANME-2 cells are labelled with small gold dots (black arrows point to some dots); the SRB (upper right cell) show a low background labelling (black arrows).

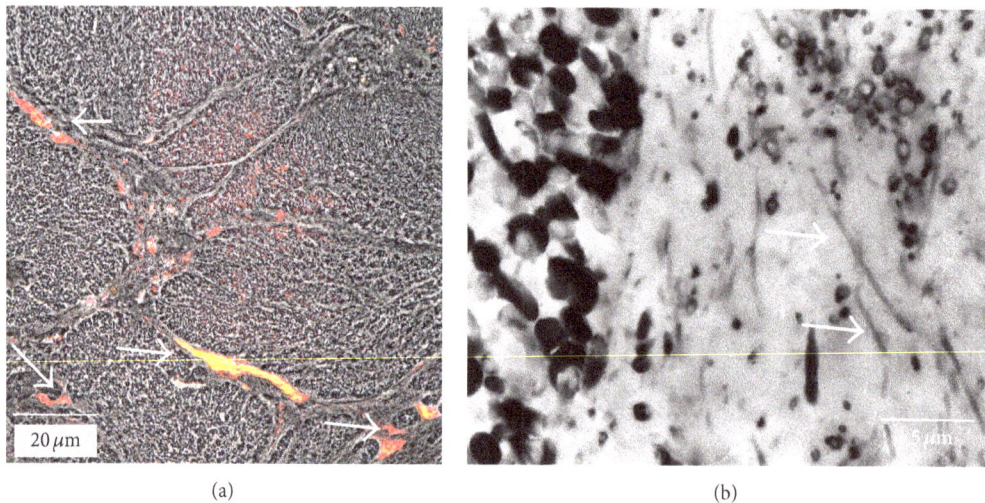

(a)   (b)

FIGURE 3: Appearance of gaps between large aggregates. (a) Overview: large patches outside aggregates show intense fluorescence (arrows) after staining with fluorescently labeled ConA lectin. (b) Electron micrograph of a 300 nm thick section from the aggregate periphery with intact cells (left) and cell debris embedded in EPS (right). Arrows point to long filaments.

matrix, putatively cytoplasmic contents. Shallowly stained filaments likely represent empty cell envelopes (Figures 4(b), 5(a), and 5(b)).

Within these large, cauliflower-shaped aggregates, SRB exhibit peculiar cytological features. Apart from occasionally observed intracytoplasmic membranes, intracellular magnetosome-like particles, arranged in straight rows and composed of greigite, were observed frequently [8, 14]. Here, we show that these chains appear to be stable after cell death and cell lysis, but exhibit structural modifications. In intact cells, chains are arranged in straight rows, mostly as parallel pairs (Figure 4(b), inset, Figure 4(c)). Figure 4(c) shows

magnetosome chains in unstained sections; that is, the cells, though still present and morphologically intact, are invisible here. Basically, three variants of magnetosome-like chains could be observed. These variants may represent three stages of development. In a (putatively) early stage, chains appear to be absent (Figure 4(a)). During aggregate development, the organisms deposit intracellular chains (Figures 4(b) and 4(c)). Finally, the organisms get lysed, and in stained sections, just the magnetosomes and some cell debris are still present; the free chains still mark the positions of the SRB in the aggregate (Figure 4(d)); distances between these deposits are similar to the distances between magnetosome chains in

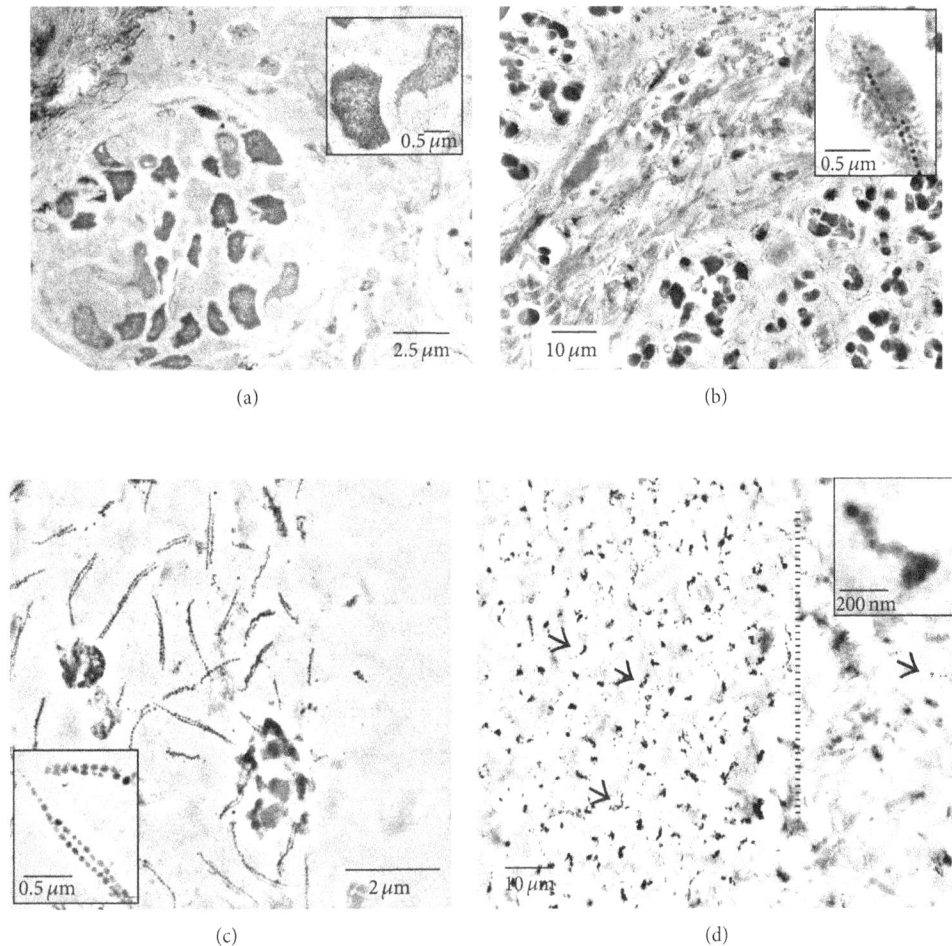

FIGURE 4: Aggregates in different stages of magnetosome-like chain formation. (a) ANME-2/SRB aggregate without visible precipitates. (b) Periphery of a large aggregate (asterisk), with gap between aggregates (arrows). Cells with multiple magnetosome-like chains (inset; see also (c)). Note cell debris, mainly consisting of envelopes from filamentous cells (arrows), outside the aggregates. (c) Periphery of an intact aggregate as depicted in (b) (unstained section; cells are invisible), showing the position of straight magnetosome chains inside cells. (d) ANME-2/SRB aggregate after cell lysis (cells are absent in spite of staining, compare (c)), with magnetosome chains still in place (some chains are marked by arrows). The inset shows a single chain. The dotted line marks the border of the aggregate (right of the dotted line). Some chains are found outside the area of the aggregate.

neighboured intact cells (Figure 4(c)). Chains of the same size are also intermixed with the cell debris in the gaps between the ANME-2/SRB aggregates; here they appear to be attached to cell envelopes of filamentous morphotypes (Figure 5(a)). Mostly, the chains exhibit curves and wrinkles, perhaps due to the loss of their intracellular scaffolding structure (cf. [26]), though also straight chains are observable. Figures 5(b) and 5(c) show particles of increased size, from 30 nm to up to 100 nm in diameter. The particles appear to be attached to the filamentous morphotype (Figure 5(c)). Larger agglomerations of these particles are found near 0.5 μm sized microcrystals (arrows in Figure 5(d)), similar to particles forming pyrite framboids (cf. [7]). It is unclear if the greigite magnetosomes contribute, in the end, to the formation of these crystals and/or framboidal pyrite

(e.g., [7, 27, 28]), but the contribution of free magnetosome chains to the iron sulfide minerals in the black layer appears to be obvious. In particular, the "reactive" surfaces of prokaryotic cell envelopes are involved in binding and accumulation of these particles. Figure 6 summarizes our observations and proposes a schematic sequence of the observed features. Active consortia may not contain magnetosome-like chains in the beginning (a), but in most of the aggregates, well-developed chains are present ((b) and (c)). The involvement of these magnetosomes in chemotaxis appears to be doubtful, since the SRB are immotile during all stages of biofilm development. It may be speculated that the particles are, in this case, an intracellular "dead end" storage granule, accumulating iron sulfides as waste product from sulfate reduction (cf. [29]). It has to be expected that not all reduced

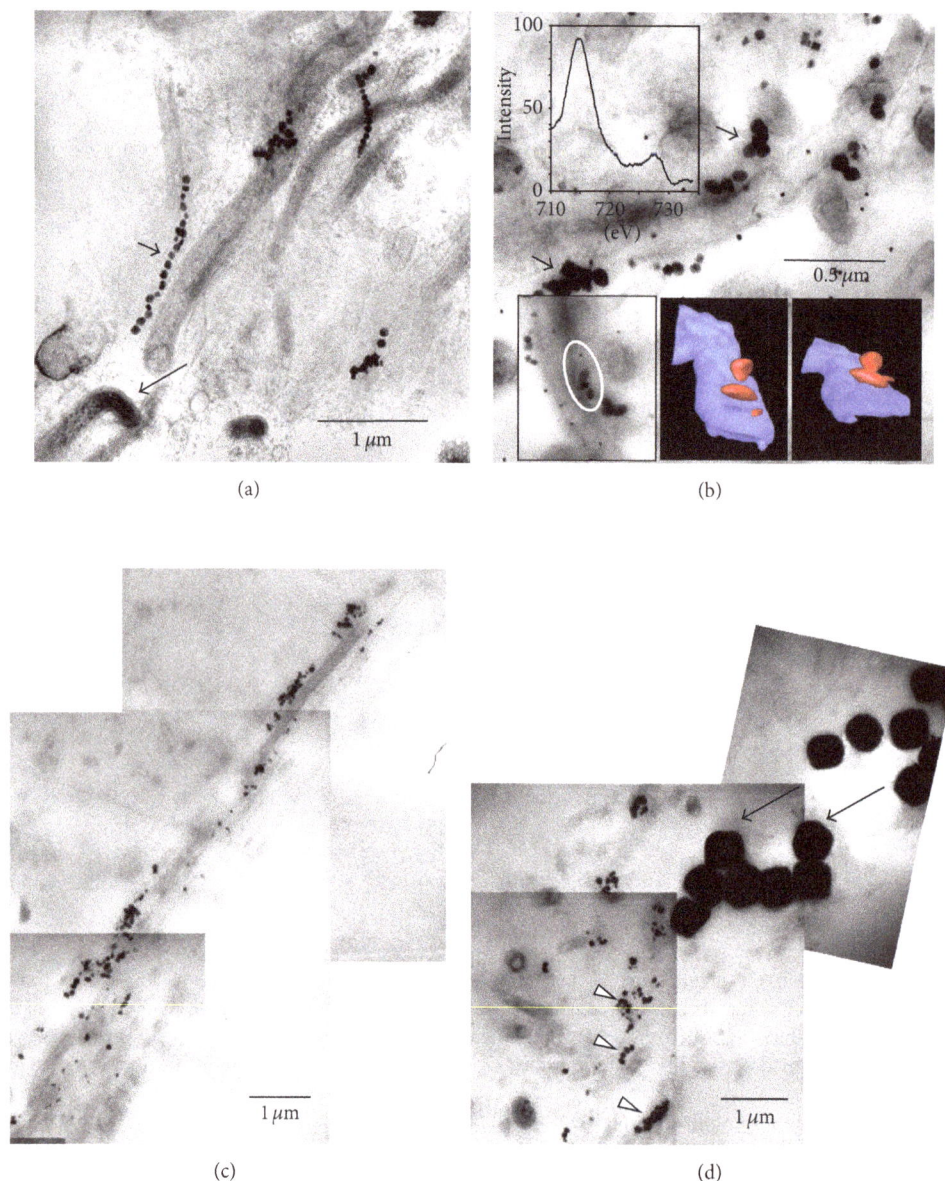

FIGURE 5: Extracellular magnetosome-like features. (a) Extracellular magnetosome chains (short arrow) bound to filamentous cell envelopes. Occasionally, also intact organisms, filled with dark cytoplasm, are visible (long arrow). (b) Particles (arrows), double in size of magnetosomes close or attached to filaments. The EELS spectrum (upper left inset) shows the energy loss at the Fe L2/L3 edges. The lower right insets show a tomogram of a small section (encircled) as depicted in the lower left inset (blue: cell envelope, red: particles). The small dots represent randomly distributed colloidal gold particles (no markers) necessary for image alignment of the tilted sections. (c) Overview image showing aggregates of the particles adjacent to filamentous envelopes. (d) Aggregates of particles (arrowheads) and typical microcrystals from framboidal pyrite (arrows).

sulfur compounds end up inside cells and, as known from other sulfate reducing bacteria, sulfides are also deposited outside cells. However, intracellular deposition may be also a rapid way to keep the concentration of sulfides as low as possible. It is obvious that both syntrophic partners die and lyse (possibly all cells at the same time) leaving the magnetosome chains as still visible remains (d) inside the aggregates. Free chains (e) migrate, by diffusion in the matrix outside the aggregate bind to a specific type of cell envelope (f), and lose their chain-like appearance and regular size (g).

Conflict of Interests

The authors declare that they have no conflict of interests.

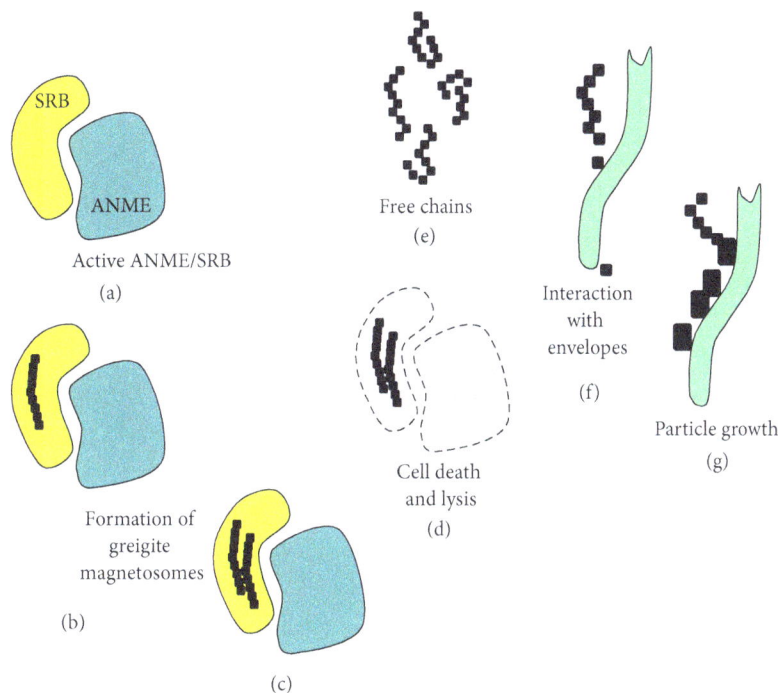

FIGURE 6: Turnover of magnetosome-like chains in the black layer. See Section 3 for further explanation.

Acknowledgments

The authors thank the crew of the R/V Professor Logachev, the Hamburg research group of Professor W. Michaelis, and the Jago-Submersible Team (J. Schauer and K. Hissmann) for collaboration during the sampling cruise for deep sea sediments. The authors thank also Jörn Peckmann (RCOM-Bremen) for the helpful discussions. Part of this study received financial support by the Bundesministerium für Bildung und Forschung (BMBF-GEOTECHNOLOGIEN-Program GHOSTDABS 03G0559A) and the Deutsche Forschungsgemeinschaft (DFG-Research Unit 571—Geobiology of Organo—and Biofilms, DFG Grants Re 665/31-1 and Ho 1830/2-1). Anne Dreier is supported by a fellowship of the Studienstiftung des Deutschen Volkes. The authors also acknowledge the support by the Open Access Publication Funds of the Göttingen University. This is a Courant Research Centre Geobiology publication.

References

[1] R. V. Burne and L. S. Moore, "Microbialites: organosedimentary deposits of benthic microbial communities," *Palaios*, vol. 2, no. 3, pp. 241–254, 1987.

[2] A. W. Decho, "Microbial biofilms in intertidal systems: an overview," *Continental Shelf Research*, vol. 20, no. 10-11, pp. 1257–1273, 2000.

[3] V. R. Phoenix and K. O. Konhauser, "Benefits of bacterial biomineralization," *Geobiology*, vol. 6, no. 3, pp. 303–308, 2008.

[4] D. S. S. Lim, B. E. Laval, G. Slater et al., "Limnology of Pavilion Lake, B. C., Canada: characterization of a microbialite forming environment," *Fundamental and Applied Limnology*, vol. 173, no. 4, pp. 329–351, 2009.

[5] P. U. P. A. Gilbert, M. Abrecht, and B. H. Frazer, "The organic-mineral interface in biominerals," *Reviews in Mineralogy and Geochemistry*, vol. 59, pp. 157–185, 2005.

[6] S. J. Hallam, N. Putnam, C. M. Preston et al., "Reverse methanogenesis: testing the hypothesis with environmental genomics," *Science*, vol. 305, no. 5689, pp. 1457–1462, 2004.

[7] J. Peckmann, A. Reimer, U. Luth et al., "Methane-derived carbonates and authigenic pyrite from the northwestern Black Sea," *Marine Geology*, vol. 177, no. 1-2, pp. 129–150, 2001.

[8] J. Reitner, J. Peckmann, M. Blumenberg, W. Michaelis, A. Reimer, and V. Thiel, "Concretionary methane-seep carbonates and associated microbial communities in Black Sea sediments," *Palaeogeography, Palaeoclimatology, Palaeoecology*, vol. 227, no. 1–3, pp. 18–30, 2005.

[9] J. Reitner, J. Peckmann, A. Reimer, G. Schumann, and V. Thiel, "Methane-derived carbonate build-ups and associated microbial communities at cold seeps on the lower Crimean shelf (Black Sea)," *Facies*, vol. 51, no. 1–4, pp. 66–79, 2005.

[10] J. T. Milucka, T. G. Ferdelman, L. Polereck et al., "Zero-valent sulphur is a key intermediate in marine methane oxidation," *Nature*, vol. 491, no. 7425, pp. 541–546, 2012.

[11] K. U. Hinrichs and A. Boetius, "The anaerobic oxidation of methane: new insights in microbial ecology and biogeochemistry," in *Ocean Margin Systems*, G. Wefer, D. Billet, D. Hebbeln, B. B. Jørgensen, M. Schlüter, and T. van Weering, Eds., pp. 457–477, Springer, Heidelberg, Germany, 2002.

[12] W. Michaelis, R. Seifert, K. Nauhaus et al., "Microbial reefs in the black sea fueled by anaerobic oxidation of methane," *Science*, vol. 297, no. 5583, pp. 1013–1015, 2002.

[13] C. T. Lefèvre, N. Menguy, F. Abreu et al., "A cultured greigite-producing magnetotactic bacterium in a novel group of sulfate-reducing bacteria," *Science*, vol. 334, no. 6063, pp. 1720–1723, 2011.

[14] C. Wrede, V. Krukenberg, A. Dreier, J. Reitner, C. Heller, and M. Hoppert, "Detection of metabolic key enzymes of methane turnover processes in cold seep microbial biofilms," *Geomicrobiology Journal*, vol. 30, no. 3, pp. 214–227, 2013.

[15] M. Hoppert and F. Mayer, "Electron microscopy of native and artificial methylreductase high-molecular-weight complexes in strain Go 1 and Methanococcus voltae," *FEBS Letters*, vol. 267, no. 1, pp. 33–37, 1990.

[16] I. J. Braks, M. Hoppert, S. Roge, and F. Mayer, "Structural aspects and immunolocalization of the F420-reducing and non-F420-reducing hydrogenases," *Journal of Bacteriology*, vol. 176, no. 24, pp. 7677–7687, 1994.

[17] M. Krüger, A. Meyerdierks, F. O. Glöckner et al., "A conspicuous nickel protein in microbial mats that oxidize methane anaerobically," *Nature*, vol. 426, no. 6968, pp. 878–881, 2003.

[18] C. Wrede, C. Heller, J. Reitner, and M. Hoppert, "Correlative light/electron microscopy for the investigation of microbial mats from Black Sea Cold Seeps," *Journal of Microbiology Methods*, vol. 73, no. 2, pp. 85–91, 2008.

[19] M. Kämper, S. Vetterkind, R. Berker, and M. Hoppert, "Methods for in situ detection and characterization of extracellular polymers in biofilms by electron microscopy," *Journal of Microbiological Methods*, vol. 57, no. 1, pp. 55–64, 2004.

[20] R. Bauer, "Electron spectroscopic imaging: an advanced technique for imaging and analysis in transmission electron microscopy," *Methods in Microbiology*, vol. 20, pp. 113–146, 1988.

[21] D. Ress, M. L. Harlow, M. Schwarz, R. M. Marshall, and U. J. McMahan, "Automatic acquisition of fiducial markers and alignment of images in tilt series for electron tomography," *Journal of Electron Microscopy*, vol. 48, no. 3, pp. 277–287, 1999.

[22] C. Heller, M. Hoppert, and J. Reitner, "Immunological localization of coenzyme M reductase in anaerobic methane-oxidizing archaea of ANME 1 and ANME 2 type," *Geomicrobiology Journal*, vol. 25, no. 3-4, pp. 149–156, 2008.

[23] T. J. Beveridge, G. D. Sprott, and P. Whippey, "Ultrastructure, inferred porosity, and gram-staining character of Methanospirillum hungatei filament termini describe a unique cell permeability for this archaeobacterium," *Journal of Bacteriology*, vol. 173, no. 1, pp. 130–140, 1991.

[24] A. Boetius, K. Ravenschlag, C. J. Schubert et al., "A marine microbial consortium apparently mediating anaerobic oxidation methane," *Nature*, vol. 407, no. 6804, pp. 623–626, 2000.

[25] L. Schreiber, T. Holler, K. Knittel, A. Meyerdierks, and R. Amann, "Identification of the dominant sulfate-reducing bacterial partner of anaerobic methanotrophs of the ANME-2 clade," *Environmental Microbiology*, vol. 12, no. 8, pp. 2327–2340, 2010.

[26] A. Scheffel, M. Gruska, D. Faivre, A. Linaroudis, J. M. Plitzko, and D. Schüler, "An acidic protein aligns magnetosomes along a filamentous structure in magnetotactic bacteria," *Nature*, vol. 440, no. 7080, pp. 110–114, 2006.

[27] J. Schieber, "Sedimentary pyrite: a window into the microbial past," *Geology*, vol. 30, no. 6, pp. 531–534, 2002.

[28] L. C. W. MacLean, T. Tyliszczak, P. U. P. A. Gilbert et al., "A high-resolution chemical and structural study of framboidal pyrite formed within a low-temperature bacterial biofilm," *Geobiology*, vol. 6, no. 5, pp. 471–480, 2008.

[29] M. Pósfai, B. M. Moskowitz, B. Arató et al., "Properties of intracellular magnetite crystals produced by *Desulfovibrio magneticus* strain RS-1," *Earth and Planetary Science Letters*, vol. 249, no. 3-4, pp. 444–455, 2006.

The *Nitrosopumilus maritimus* CdvB, but Not FtsZ, Assembles into Polymers

Kian-Hong Ng,[1] Vinayaka Srinivas,[1,2] Ramanujam Srinivasan,[3] and Mohan Balasubramanian[1,2,3]

[1] *Cell Division Laboratory, Temasek Life Sciences Laboratory, 1 Research Link, National University of Singapore, Singapore 117604*
[2] *Department of Biological Sciences, National University of Singapore, 14 Science Drive 4, Singapore 117543*
[3] *Mechanobiology Institute, National University of Singapore, 5A Engineering Drive 1, Singapore 117411*

Correspondence should be addressed to Kian-Hong Ng; kianhong@tll.org.sg

Academic Editor: Jerry Eichler

Euryarchaeota and Crenarchaeota are two major phyla of archaea which use distinct molecular apparatuses for cell division. Euryarchaea make use of the tubulin-related protein FtsZ, while Crenarchaea, which appear to lack functional FtsZ, employ the Cdv (cell division) components to divide. Ammonia oxidizing archaeon (AOA) *Nitrosopumilus maritimus* belongs to another archaeal phylum, the Thaumarchaeota, which has both FtsZ and Cdv genes in the genome. Here, we used a heterologous expression system to characterize FtsZ and Cdv proteins from *N. maritimus* by investigating the ability of these proteins to form polymers. We show that one of the Cdv proteins in *N. maritimus*, the CdvB (Nmar_0816), is capable of forming stable polymers when expressed in fission yeast. The *N. maritimus* CdvB is also capable of assembling into filaments in mammalian cells. However, *N. maritimus* FtsZ does not assemble into polymers in our system. The ability of CdvB, but not FtsZ, to polymerize is consistent with a recent finding showing that several Cdv proteins, but not FtsZ, localize to the mid-cell site in the dividing *N. maritimus*. Thus, we propose that it is Cdv proteins, rather than FtsZ, that function as the cell division apparatus in *N. maritimus*.

1. Introduction

Cell division mechanisms in archaea, the third domain of life, have been relatively less elucidated until recent years. As opposed to eukarya and bacteria, which use actomyosin ring and FtsZ ring, respectively, for cell division, archaea appear to be more diverse in terms of their use of cell division machineries. It appears that FtsZ acts as a primary cell division apparatus in nearly all members of Euryarchaeota [1–4]. Nevertheless, FtsZ is notably absent from the other major phylum of archaea, the Crenarcheota, which consists of extremophiles that survive at extremely harsh conditions like high temperatures and high acidic environments. Recent findings strongly suggest that crenarchaeon *Sulfolobus acidocaldarius* utilizes the Cdv components (also known as endosomal sorting complex required for transport (ESCRT) in eukaryotes) for cell division [5–7]. ESCRT apparatus in eukaryotes is made up of several complexes that play important roles in different cellular processes, for instance, multivesicular body formation, membrane abscission during cytokinesis, and virus egression [8–11]. In *S. acidocaldarius*, the ESCRT-III-like CdvB (Saci_1373), Vps4-like CdvC (Saci_1372), and another gene that encodes for a coiled-coil domain protein, CdvA (Saci_1374), are arranged in an operon-like structure [5, 6]. *S. acidocaldarius* CdvB and CdcC localize to the mid cell during cell division, and their localization corresponds to the membrane ingression site between two segregated nucleoids. Overexpression of a dominant negative form of CdvC has been shown to result in enlarged cells with elevated DNA content and also cells devoid of DNA, a strong indication of cell division defects [6]. In a recent work reported by Samson et al., CdvB and CdvA were shown to cooperatively deform membranes in vitro [7], a feature that is consistent with their roles in membrane attachment, force generation, and execution of binary fission in *Sulfolobus* cells.

N. maritimus belongs to a phylum of archaea known as Thaumarchaeota [12, 13]. It is an ammonia-oxidizing archaeon (AOA) that contributes to the nitrification process in marine nitrogen cycle [14–16]. Interestingly, in the genome of the *N. maritimus*, there are genes that encode for both Cdv proteins and FtsZ, raising the question of which of the two components is used for cell division. A recent report by Pelve et al. showed that the *N. maritimus* Cdv proteins, but not FtsZ, localized to the mid-cell region during cell division [17], suggesting that Cdv proteins rather than FtsZ function in cytokinesis in this organism.

One of the important characteristics for cell division apparatus is the ability of one or more proteins to form polymeric structures. Actin and FtsZ have been shown to polymerize both in vivo and in vitro, and their polymerization activities are essential for cell division [18–23]. We have shown in our previous studies that tubulin-like FtsZ and actin-like MreB in bacteria form elaborate filaments in a yeast expression system [24, 25]. In this study, we seek to further understand thaumarchaeal cell division by identifying *N. maritimus* proteins that are capable of forming filament-like structures. We have focused our study on Cdv proteins and the FtsZ-like protein. We show that one of the *N. maritimus* CdvB proteins, Nmar_0816, is able to polymerize and form filament-like structures in both yeast and mammalian cells. By contrast, the FtsZ homolog in *N. maritimus*, Nmar_1262, does not polymerize or form any higher-order structure. Our findings are in agreement with the conclusions of Pelve et al., suggesting that the *N. maritimus* is likely to use Cdv proteins for cell division.

2. Results and Discussion

2.1. Expression of N. maritimus CdvB and CdvC in Fission Yeast.
CdvB (Saci_1373) from *S. acidocaldarius* has been shown to play a central role in crenarchaeal cell division [5, 6]. In eukaryotes, ESCRT-III proteins are shown to form polymeric structures in vivo and in vitro [26–34]. In addition, several Cdv proteins from the crenarchaeon *Metallosphaera sedula* were first demonstrated to form filament-like structures in vitro in a study done by Moriscot et al. [35]. The authors showed that *M. sedula* CdvA formed helical filaments in association with DNA. Interestingly, they also demonstrated that a C-terminally deleted CdvB was capable of forming polymers even though its full-length form did not. These findings have suggested an intricate link between cell constriction/membrane deformation and the polymerizing activity of proteins involved in cell division. Since both the *N. maritimus* and the *S. acidocaldarius* CdvB proteins share substantial sequence similarity (see Figure S1 in Supplementary Material available online at http://dx.doi.org/10.1155/2013/104147), we addressed if any of the *N. maritimus* CdvB proteins could potentially polymerize into filamentous structures, an important feature that would further lend support to the claim that thaumarchaea use Cdv proteins for cell division. Since genetic manipulation techniques are yet to be developed for *N. maritimus*, we sought to answer the question using an established green fluorescence protein- (GFP-) tagging system in yeast for the examination of

nonnative cytoskeletal elements. We expressed all of the three *N. maritimus* CdvB paralogs (Nmar_0029, Nmar_0061, and Nmar_0816) and the CdvC (Nmar_1088) in fission yeast with a GFP fusion at their C-terminus. Interestingly, one of the CdvB paralogs, the Nmar_0816, was found to readily form distinct polymeric structures upon expression in fission yeast (Figure 1(a)). All of the other CdvB paralogs and the CdvC examined showed only diffuse GFP signals throughout the cells, without discernible polymer formation (Figure 1(a)). It is still unclear to us why the other two CdvB paralogs (Nmar_0029 and Nmar_0061) did not form filament-like structure despite their close similarity with Nmar_0816 (Figure S1). One possibility is that fusion of GFP to the proteins might have altered the protein conformation and hence inhibited their polymerizing activity. It is also likely that Nmar_0029 and Nmar_0061 represent a distinct group of CdvB from Nmar_0816, as both Nmar_0061 and Nmar_0029 share ~50% in protein sequence identity with each other, but they share ~30% sequence identity with Nmar_0816. It is thus possible that both groups of CdvB would have distinctive properties and roles in *N. maritimus*.

Next, we took a closer look at the polymers formed by the Nmar_0816 and found that polymerized Nmar_0816 could exist in various forms ranging from simple elongated structures to closed circular and intertwined structures (Figure S2). Since these higher-order structures closely resemble those formed by cytoskeletal proteins like tubulin, actin, MreB, and FtsZ, we tested if drugs inhibiting polymerization of these cytoskeletal elements would affect the Nmar_0816 polymer formation. We found that the Nmar_0816 polymers, though sharing substantial morphological similarity to filaments formed by cytoskeletal proteins, were not affected by treatments of Latrunculin A (inhibiting actin polymerization), A22 (affecting MreB polymerization), TBZ, and MBC (inhibiting microtubule polymerization) (data not shown). These observations also established that the host cytoskeleton was dispensable for Nmar_0816 polymer formation and stability.

2.2. Formation of the Nmar_01816 Polymers in Fission Yeast.
To understand how the Nmar_0816 polymers were formed, we took time-lapse movies of Nmar_0816-GFP in growing fission yeast cells. We found that Nmar_0816-GFP first formed an aggregate in cells with strong fluorescent signals (Figure 1(b) and Video S1). Later, elongated filament-like polymers started to emerge from the aggregate in a unidirectional manner. In cells with elaborate filaments, these polymers were also curved, circularized, and intertwined, thus forming various forms of the Nmar_0816-GFP polymers (Figure S2 and data not shown). Interestingly, in some cases, the cytokinetic apparatus of the fission yeast cells failed to cut through the Nmar_0816-GFP filaments, which resulted in extensive cytosolic vacuolization (Figure S3A and Video S2), a strong indicator of cell death in yeast (Figure S3B). Low-level expression of the Nmar_0816 did not result in polymer formation, as initial induction of Nmar_0816-GFP (first 16 hours) only resulted in cells with fluorescent signal but no filament-like structures. A further 4–6 hours induction was

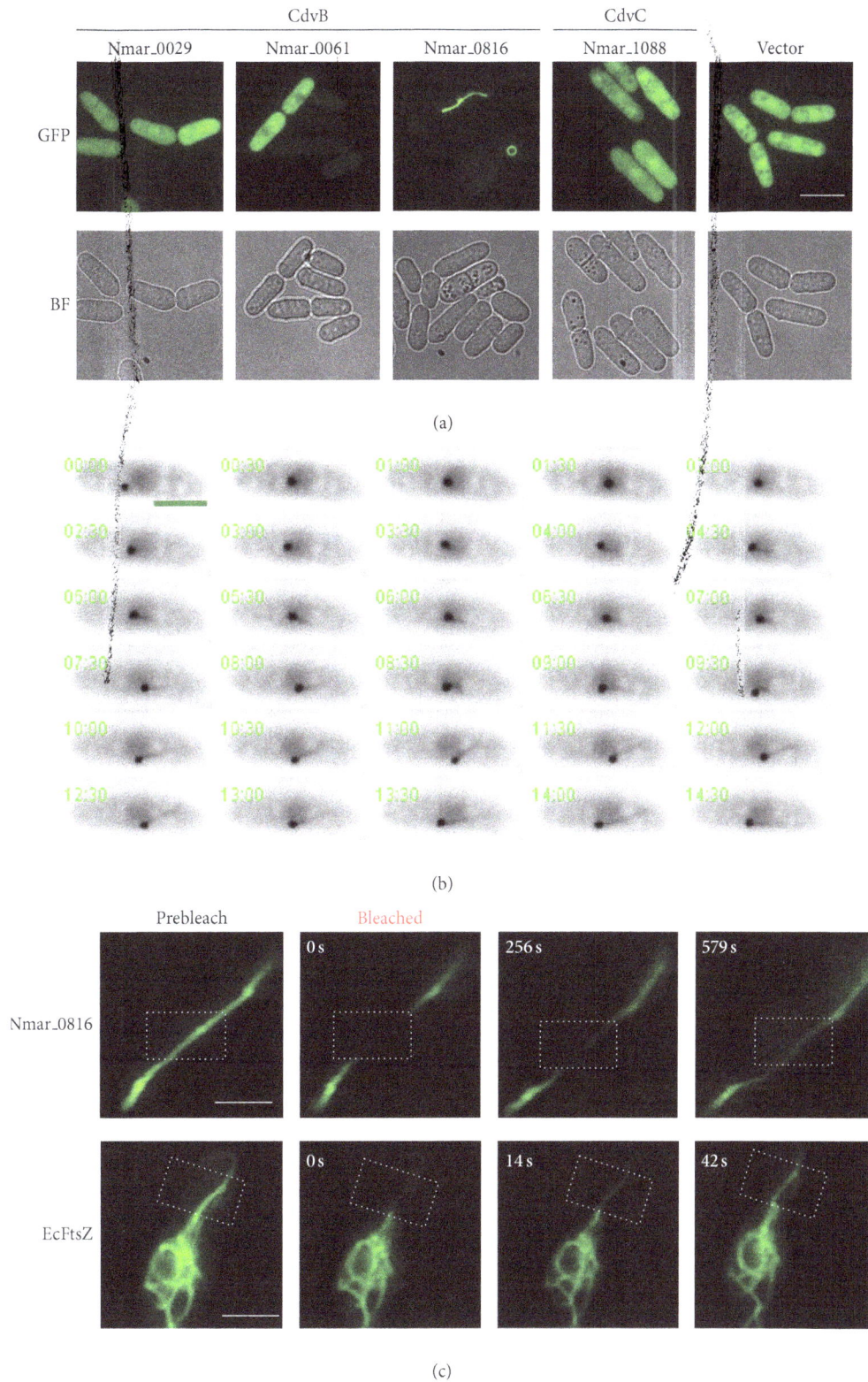

FIGURE 1: *N. maritimus* CdvB (Nmar_0816) forms filament-like structures in yeast. (a) Images of fission yeast cells expressing *N. maritimus* CdvB paralogs and CdvC fused with GFP. BF: bright field; Scale bar: 10 μm. (b) Time-lapse images of Nmar_0816 polymer formation in fission yeast. Cells carrying pREP42-Nmar_0816-GFP were cultured in the absence of thiamine for 24 h at 24°C and monitored for GFP signals. Frames were captured at 15 s intervals. (Video S1). Scale bar: 5 μm. (c) Time-lapse images of the Nmar_0816-GFP polymers versus the *E. coli* FtsZ-GFP polymers upon fluorescence recovery after photobleaching (FRAP). Dotted rectangle indicates bleached region. Scale bar: 3 μm.

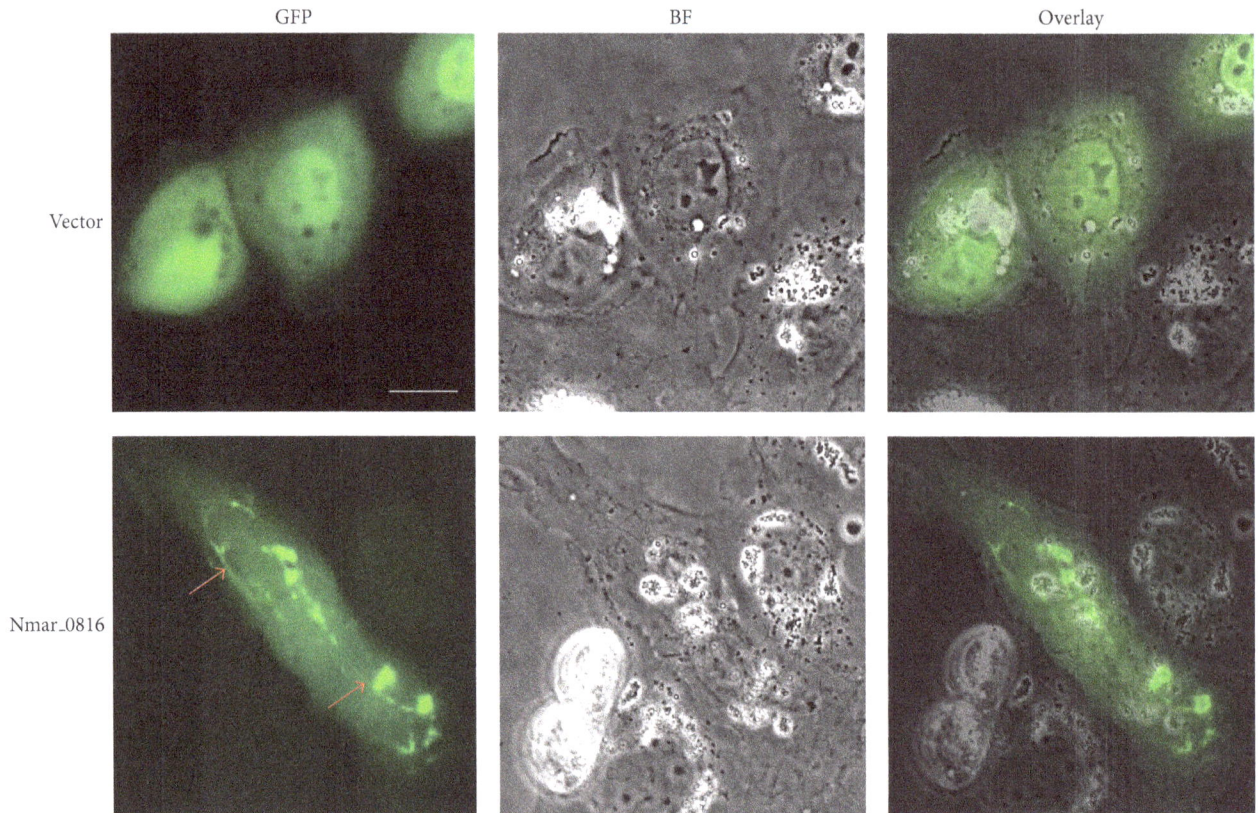

FIGURE 2: Nmar_0816 forms filament-like structures in mammalian cells. Images of mammalian NRK cells expressing GFP-Nmar_0816. Red arrows point to filament-like structures. BF: bright field; Scale bar: 7 μm.

needed for the polymer formation (data not shown). Our experiments suggest that Nmar_0816 protein has to reach a likely threshold level before aggregation and polymerization can be initiated.

Next, we sought to understand the protein turnover property of the Nmar_0816 polymers by fluorescence recovery after photobleaching (FRAP). FRAP analysis revealed a very slow signal recovery for Nmar_0816 polymers upon photobleaching as compared to the *E. coli* FtsZ (~2% signal recovery for Nmar_0816 polymers at 56 seconds postbleaching as compared to ~37% signal recovery for FtsZ filaments at 52 seconds postbleaching), indicating that the higher-order structures formed by the Nmar_0816 protein were stable (Figure 1(c)). Collectively, our data suggest that the Nmar_0816 is capable of forming stable higher-order polymers in fission yeast cells. In contrast to rapid turnover of the FtsZ protein that ensures proper control of *E. coli* division, a stable CdvB might be needed in *N. maritimus* for division. It is also likely that the *N. maritimus* CdvB requires other factors like CdvC to regulate its turnover.

2.3. The N. maritimus CdvB (Nmar_0816) Polymerizes into Higher-Order Structures in Mammalian NRK Cells. To determine if the formation of the Nmar_0816 polymers was limited to the cellular context of fission yeast, we transiently transfected mammalian NRK cells with a construct expressing the Nmar_0816 with a GFP fusion at its N-terminus. The transfected cells with GFP-Nmar_0816 showed filament-like

polymers (Figure 2). By contrast, in the control cells transfected with GFP-containing vector, only diffuse GFP signals were seen. Intriguingly, in some transfected cells, the GFP-Nmar_0816 seemed to localize to the rim of enlarged endosomes (Figure S4), a feature similar to that observed upon overexpression of truncated hVps2-1, hVps24, and hSnf7-1 in mammalian cells [36]. Vps20, a component of ESCRT-III, contains a myristoylation site that potentially facilitates its direct association with the membrane. It has been shown that Vps20 mediates the recruitment of Vps24 and Vps2 to the membrane [8]. Interestingly, Nmar_0816 lacks the putative lipid modification site, and it shares 12% and 5% sequence identity with mammalian Vps24 (hVps24) and hVps20, respectively (Figure S5). As ESCRT-III members have been shown to interact with each other, it is possible that Nmar_0816 is targeted to the rim of endosomes through its potential interaction with the mammalian Vps20. Nevertheless, further experiments are needed to demonstrate if the targeting of CdvB to the rim of endosomes is specific. Our findings suggest an interesting link between the *N. maritimus* CdvB and the mammalian ESCRT-IIIs. Nevertheless, we have not observed endosome association of the Nmar_0816-GFP in yeast. Different cellular factors may have contributed to distinct forms and localization of the Nmar_0816 in fission yeast and in mammalian cells. Taken together, our data showed that the Nmar_0816 was able to assemble into polymeric structures not only in fission yeast, but also in mammalian cells.

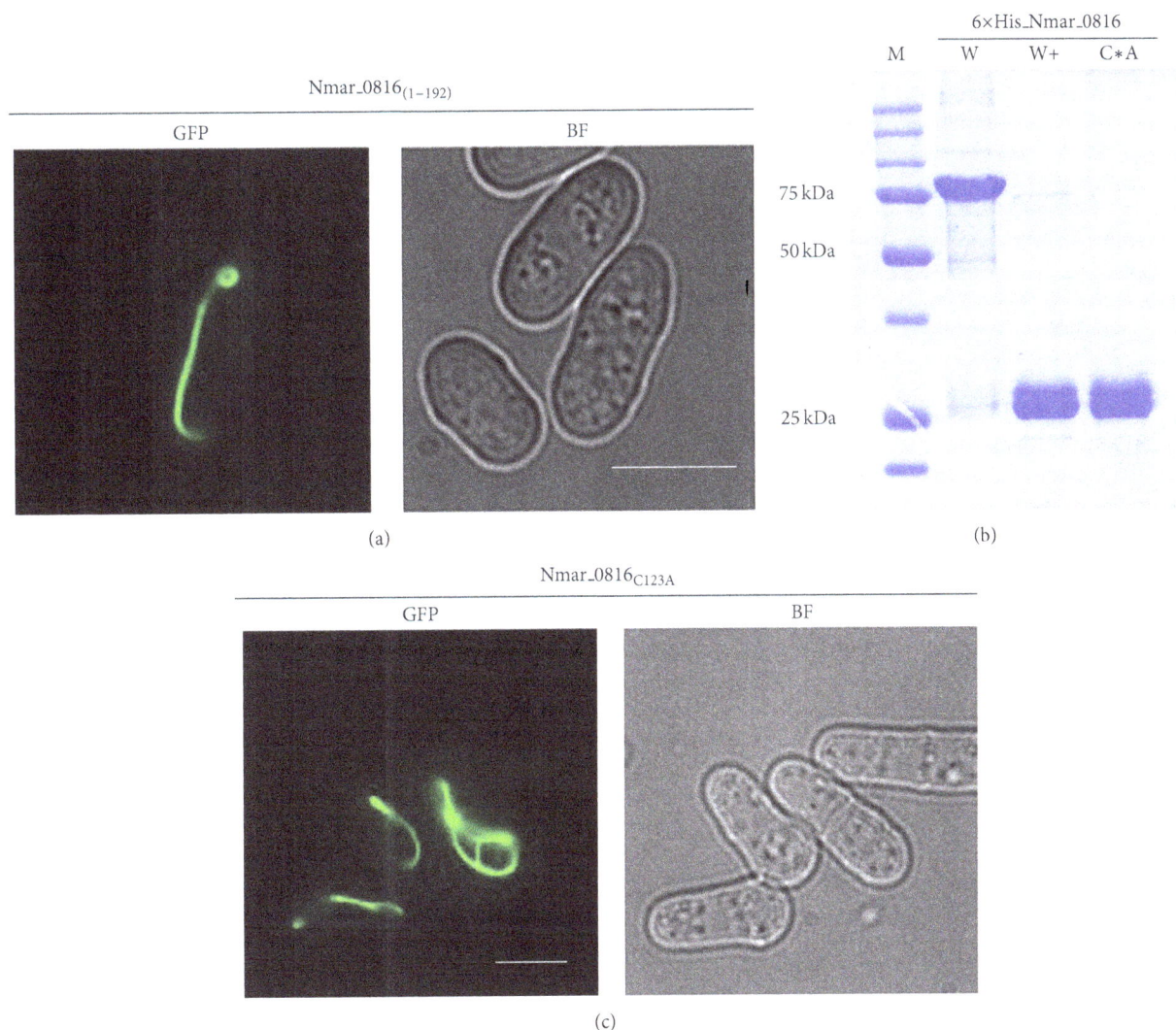

FIGURE 3: The core domain of the CdvB is necessary for Nmar_0816 polymer formation in yeast. (a) Images of yeast cells expressing Nmar_0816$_{(1-192)}$-GFP. Scale bar: 7 μm. (b) Expression and purification of Nmar_0816 in *E. coli*. W: 6 × His-Nmar_0816, nonreducing condition. W+: 6 × His-Nmar_0816, in the presence of 5% β-mecaptoethanol. C*A: 6 × His-Nmar_0186$_{C123A}$, nonreducing condition. (c) Images of yeast cells expressing Nmar_0186$_{C123A}$-GFP. Scale bar: 5 μm.

2.4. Evaluation of the N. maritimus CdvB (Nmar_0816) Filament-Forming Property.

Previous studies have shown that the core domain of the eukaryotic ESCRT-III was essential and sufficient for polymerization [33, 36]. We sought to understand if this filament-forming property was also conserved in the *N. maritimus* CdvB. We generated N-terminal (1–108 aa) and C-terminal (109–206 aa) Nmar_0816-GFP, in which the core domain was disrupted in both cases. Not surprisingly, expression of both GFP-fusion proteins resulted in diffuse GFP signals without discernible polymeric structure (data not shown). By contrast, a small C-terminal deletion of the Nmar_0816 that retained its core domain (1–192 aa) formed elaborate filaments similar to the full length Nmar_0816 (Figure 3(a)). As the C-terminal deletion also removed the putative MIT-interacting motif 2 (MIM2, for the interaction with MIT domain of the CdvC protein) of the Nmar_0816, it is also likely that the formation and stability of the Nmar_0816 polymers are independent of its interaction

with the CdvC protein. Since the putative MIM2 motif in Nmar_0816 is not highly conserved with those of Saci_1373 and mammalian CHMP6 (Figures S1 and S5), it is not clear if it is capable of interacting with CdvC. Taken together, we showed that the core domain of the CdvB was sufficient for Nmar_0816 polymerization (Table 1 and Figure 3(a)). Interestingly, it is still not known to us why the other *N. maritimus* CdvB paralogs (the Nmar_0029 and the Nmar_0061) did not polymerize as both of them possess similar core domains for such a function. It is possible that interference from the GFP fusion might have negative impact on the polymerizing activity of the other two CdvB paralogs. In addition, sequence variation in the C-terminal regions (Figure S1) might have contributed to distinctive regulation of the CdvB paralogs for polymerization. It is also possible that specific and individualistic differences of protein folding may have contributed to such distinction.

TABLE 1: Summary on the domain analysis of the Nmar_0816 (1–216 aa) protein for its polymerization activity in yeast.

	Polymerization activity
Nmar_0816$_{(1-216)}$	Yes
Nmar_0816$_{(1-108)}$	No
Nmar_0816$_{(109-216)}$	No
Nmar_0816$_{(1-192)}$	Yes
Nmar_0816$_{C123A}$	Yes

In Nmar_0816$_{(1-108)}$ and Nmar_0816$_{(109-216)}$, the core domain (1–181 aa) was disrupted. In Nmar_0816$_{C123A}$, the cysteine residue was mutated to alanine. In Nmar_0816$_{(1-192)}$, the putative MIM2 motif was removed.

To further investigate the protein properties of the Nmar_0816, we expressed recombinant Nmar_0816 in *E. coli* and found that purified Nmar_0816 existed as both monomer and dimer/multimer under nonreducing condition (Figure 3(b)). The dimer/multimer forms were readily resolved into monomeric form of ~25 kDa in the presence of reducing agents. To examine if the dimeric/multimeric forms of the Nmar_0816 were due to the presence of a disulfide bond, we mutated the codon coding for cysteine into alanine and found that the mutated Nmar_0816 was resolved only as a monomer in nonreducing SDS-PAGE. As polymerization could be initiated from monomeric or dimeric/multimeric forms of a protein, we were interested to know if the dimeric/multimeric forms of the CdvB protein served as protein nucleators for the polymerization. To address this issue, we expressed the mutated form of the Nmar_0816 in yeast to examine if its polymerization activity would be disrupted if there is only the monomeric form present. As shown in Figure 3(c), the cysteine to alanine mutated Nmar_0816 effectively polymerized into filament-like structures that were indistinguishable from the wild type suggesting that dimeric/multimeric forms of Nmar_0816 are not required for its polymerization. It is noteworthy that the Nmar_0816 is the only CdvB paralog that contains a cysteine residue (Figure S1, C123). Even though we have ruled out that polymerization requires dimeric/multimeric forms of the Nmar_0816, it will be interesting to know if the ability of the Nmar_0816 to dimerize/multimerize has any significant role in its protein folding and stability, or function in vivo.

2.5. The N. maritimus FtsZ Is Substantially Different from the E. coli FtsZ and Does Not Polymerize into Filament. N. maritimus genome contains a gene for FtsZ. Since FtsZ is used in bacteria and euryarchaea as the cell division apparatus and is capable of polymerization, we asked if the *N. maritimus* FtsZ (Nmar_1262) has similar polymerization activity. We aligned and compared the sequence of the *Nitrosopumilus* FtsZ with the *E. coli* FtsZ. As shown in the sequence alignment, the *N. maritimus* FtsZ does not share substantial sequence similarity with the *E. coli* FtsZ even at the most conserved domains for GTP binding and hydrolysis (T7-loop) (Figure 4(a), highlighted in yellow and green, resp.). We have previously shown that expression of the *E. coli* FtsZ in yeast resulted in elaborate FtsZ filaments [25]. However,

the *N. maritimus* FtsZ did not show any polymeric structure when expressed in yeast (Figure 4(b)). Since the *N. maritimus* FtsZ does not possess a conserved motif for GTP binding [37], we replaced the GTP-binding motif of *N. maritimus* FtsZ with the conserved GTP-binding motif (GGGTGTG) of *E. coli* FtsZ. The mutant *Nitrosopumilus* FtsZ (Nmar_1262**) formed small aggregate-like structures in yeast but still did not polymerize into filaments (Figure 4(b)). Interestingly, *E. coli* FtsZ carrying a mutation in the T7 hydrolysis loop (D209A, Figure 4(a)) formed similar aggregate-like structures when expressed in fission yeast (Figure S6). Mutations in the T7 loop are known to disrupt FtsZ polymerization but not GTP binding [38, 39]. *N. maritimus* FtsZ seems to lack the conserved T7 loop (NVDFAD, Figure 4(a)). It is likely that the altered localization and morphology of fluorescent foci in Nmar_1262** might be due to GTP binding in the absence of an active T7 loop. Collectively, our result suggests that the *N. maritimus* FtsZ might have undergone multiple nucleotide changes following loss of GTP-binding activity such that even replacement of GTP-binding motif is not sufficient to restore its ability to polymerize into filaments.

Our findings and those from the study of Pelve et al. have both suggested that *N. maritimus* is likely to use Cdv proteins for cell division even though there is also a gene encoding for FtsZ-like protein. This raises another intriguing question on what is the function of the FtsZ in *N. maritimus* if not for cell division. However, from the sequence alignment with *E. coli* FtsZ, it is clear to us that *N. maritimus* FtsZ does not have a conserved GTP-binding motif (hence, might not bind to GTP) and is completely lacking the T7-loop for GTP hydrolysis. Thus, it is likely that the *N. maritimus* FtsZ has lost its cell division function. Interestingly, Thaumarchaeota has been suggested to have diverged before the speciation of Euryarchaeota and Crenarchaeota [12]. It is not impossible that an ancestral lineage of archaea with both FtsZ-like and Cdv proteins had evolved to give rise to two distinct archaeal lineages where one uses Cdv proteins for cell division and loses the FtsZ, while the other relies on FtsZ and does away with Cdv proteins. In that way, thaumarchaea might represent a "living fossil" in which Cdv proteins are being used for cell division, while FtsZ is losing its functional role in the evolution of cell division machinery.

An interesting aspect of CdvB is that it often exists as multiple paralogs in the genomes (three in *N. maritimus* and four in *S. acidocaldarius*). It would be intriguing to know if these paralogs have overlapping functions. In *S. acidocaldarius*, only one of the CdvB proteins (Saci_1373) has been shown to be involved in cell division [5, 6]. Two other CdvB paralogs (Saci_0451 and Saci_1416) have been suggested to have originated from a more recent gene duplication event [40]. Interestingly, these two paralogs were found to be in secreted membrane vesicles [41], indicating that they might play a distinct role from Saci_1373 in *S. acidocaldarius*. In *N. maritimus*, based on the sequence identities, it is likely that the polymer-forming Nmar_0816 and the other two paralogs (Nmar_0029 and Nmar_0061) form distinct groups of CdvB that would have played different roles. Nevertheless, more experiments need to be done to reveal their respective functions in *N. maritimus*. It would also be interesting to

```
EcFtsZ       MFEPMELTNDAVIKVIGVGGGGGNAVEHMVRERIEGVEFFAVNTDAQALRKTAVGQTIQI    60
Nmar_1262    ----MSFQVKEPVLVVGLGG-AGSKLALKAKDSLN-SDCLLISNDSKDFAGDVPSVHVST    54
             * :  : :: *:*:*** .*.:   :: .: .::::   ::*:: .:  : *

EcFtsZ       GSGITKGLGAGANPEVGRNAADEDRDALRAALEGADMVFIAAGMGGGTGTGAAPVVAEVA   120
Nmar_1262    DSVVNP------SMQLIRGSTYNASEEIKSKISGYSTIVMMSNLAGKAGSAMAPVVSEMC   108
             .* ::.     .  :: *.::   :  : :: *  :. .  : .:   * :*:. ****:*:.

EcFtsZ       KDLGILTVAVVTKPFNFEGKKRMAFAEQGITELSKHVDSLITIPNDKLLKVLGRGISLLD   180
Nmar_1262    KESDIGLVSFAIMPFKYE-KDRIFNSGVSLKRVRENSECTVVLDNDSLLESN-PDLTPKA   166
             *:  .*  *:.. **.:*  *.*:   . .*   :  :..: ::  :  .:: **.**:  .::

EcFtsZ       AFGAANDVLKGAVQGIAELITRPGLMNVDFADVRTVMSEMGYAMMGSGVASGEDRAEEAA   240
Nmar_1262    CYDIANSAIMHVVESLG-------------------TSEMSHDTN-ILTTSKEGQDIEDS   206
             .  :.  **:: . :   .*:.:.             *** .:    .:: *.: * :

EcFtsZ       EMAISSPLLEDIDLSGARGVLVNITAGFDLRLDEFETVGNTIRAFASDNATVVIGTSLDP   300
Nmar_1262    LRDSLKMLYENAPPNAVKRSMLYVVGGSNIPVGVLNSITNLTSGILGESNSQIDMTSEHE   266
             . * .*     ...:  :    *  ..:  :*:.    :  : .   *  .  .

EcFtsZ       DMNDELRVTVVATGIGMDKRPEITLVTNKQVQQPVMDRYQQHGMAPLTQEQKPVAKVVND   360
Nmar_1262    ES----KVVMLSSIQGMTK---------------FDNYDPLGMIPQED---TLDWSTPD   303
             :     :*.:.::: **  :              .:*.*:  ** ::  :  .:

EcFtsZ       NAPQTAKEPDYLDIPAFLRKQAD   383
Nmar_1262    CSIDCELDLYQLE----------   316
             :  :    :   : *:
```

(a)

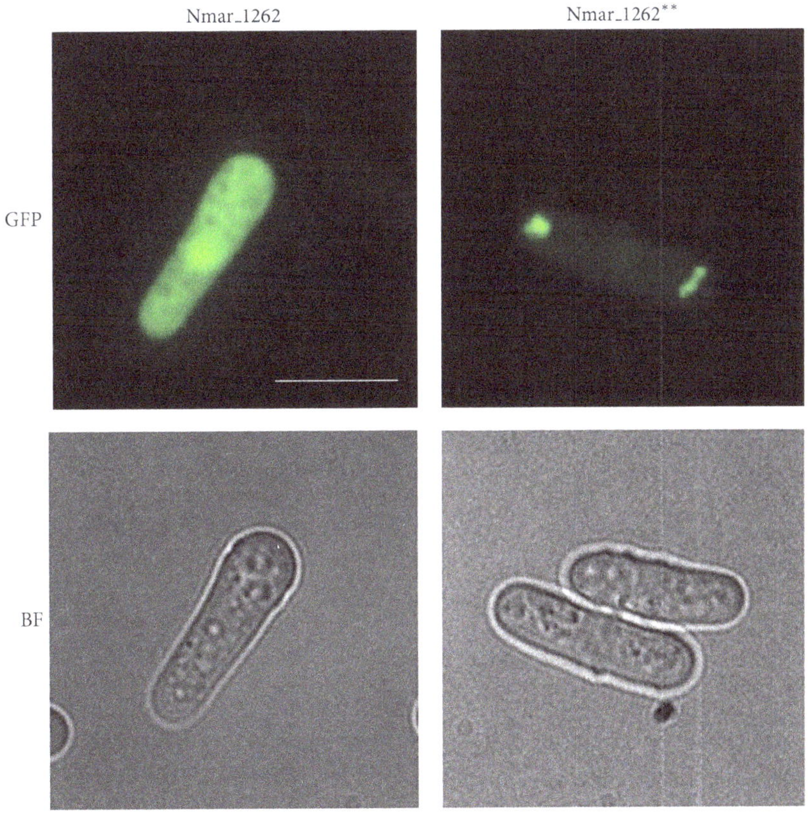

(b)

FIGURE 4: *N. maritimus* FtsZ-GFP (Nmar_1262) does not form filament-like structures in yeast. (a) Sequence alignment of the Nmar_1262 with the *E. coli* FtsZ (EcFtsZ). Regions highlighted in yellow and green in the EcFtsZ sequence are two conserved motifs for tubulin/FtsZ, GGGTGTG, and NVDFAD (T7-loop) for GTP binding and hydrolysis, respectively. Sequence alignment was performed using ClustalW2 program available from the website http://www.ebi.ac.uk/Tools/msa/clustalw2/. (b) Images of yeast cells expressing Nmar_1262-GFP and Nmar_1262**-GFP, in which the presumptive GTP-binding motif AGKAGSA was replaced with the conventional GGGTGTG. Scale bar: 10 μm.

understand which of the CdvB paralogs are essential and which are not for the organisms. Gene deletion analysis would provide a quick glimpse on this point.

In summary, we have shown that one of the *N. maritimus* CdvB paralogs, Nmar_0816, was capable of assembling into filaments. By contrast, the FtsZ did not polymerize in our assay. As cytoskeletal polymers are involved in cell division, we conclude that *N. maritimus* likely uses CdvB as its cell division machinery.

3. Experimental Procedures

3.1. Plasmid Preparation. Coding sequences for the *N. maritimus* Cdv components (Nmar_0029, Nmar_0061, Nmar_0816, and Nmar_1088) and the FtsZ (Nmar_1262) were codon optimized, synthesized, and cloned into pUC57 plasmids by commercial gene synthesis service (GenScript Inc. USA). *S. pombe* GFP-tagging expression vector pREP42-GFP containing a uracil biosynthesis gene for auxotroph selection was from our lab collection. Expression of the GFP fusion proteins with pREP42-GFP vector was under the control of a mid-strength thiamine-repressible nmt promoter [42].

3.2. Gap-Repair Cloning and Yeast Transformation. Gap-repair cloning was performed as described [43]. Gene-specific fragments (see Table S1 for the primer list) from *N. maritimus* with ~50nts overlapping regions from pREP42-GFP at N- and C-terminus were obtained by PCR amplification using high fidelity taq polymerase (Roche) according to manufacturer's instruction. Different pUC57 plasmids carrying codon-optimized *N. maritimus* genes were used as templates. PCR fragments were analyzed using agarose gel electrophoresis. The pREP42-GFP vector was linearized by *BamHI* and *NdeI* double digestion (New England Biolabs) and purified using column purification kit (QIAGEN). PCR fragments and linearized pREP42-GFP were mixed in 10 : 1 ratio and transformed into MBY192 strain (*ura4*-D18, *leu1*-32, h-) using lithium acetate method as previously described [44]. Briefly, yeast cells were grown to an optical density of 0.5. Cells were rinsed once with sterile water and washed once with 1x LiAc/TE solution (100 mM lithium acetate, 10 mM Tris-HCl, 1 mM EDTA, pH 7.5). Cells were resuspended in 100 μL of 1x LiAc/TE solution and incubated with respective plasmids for 10 min at room temperature. The cells were further incubated with 240 μL of PEG/LiAc/TE solution (LiAc/TE solution with 40% polyethylene glycol 4000) for 30 min at 30°C. 42 μL of dimethyl sulfoxide (DMSO) was added to the cells prior to heat shock at 42°C for 5 min. After heat shock, cells were washed once with sterile water and plated on Edinburgh minimal medium (EMM) supplemented with amino acids and 15 μM of thiamine. After 4-5 days of growth, at least 20 colonies from each of the transformants were picked and re-streaked on fresh supplemented EMM plates in the absence of thiamine and incubated at 30°C for 4-5 days. Colonies were then examined for fluorescence under fluorescence microscope. Transformants with GFP signals were grown in liquid cultures for detailed

examination. To further verify filament-forming Nmar_0816-GFP, cells were transformed with pREP42-Nmar_0816-GFP plasmid constructed by conventional restriction-based cloning method. The transformants of pREP42-Nmar_0816-GFP showed filament-forming properties that were indistinguishable from transformants obtained by gap-repair cloning. To construct pREP42-Nmar_1262**-GFP (replacement of variant GTP binding motif AGKAGSA with conventional GGGTGTG), two independent N-terminal and C-terminal fragments were obtained by PCR amplification using primers incorporating the corresponding changes of nucleotide sequences (see Table S1 for primer sequences). Both fragments were used for an overlapping PCR to get final PCR fragment of Nmar_1262 with a swap of GTP binding motif (Nmar_1262**). The PCR fragment of Nmar_1262** was then used for subsequent gap repair cloning.

3.3. Cell Culture, Transfection, and Expression of GFP-Nmar_0816 in NRK Cells. For the expression of GFP-Nmar_0816 in mammalian cells, full length Nmar_0816 was amplified and cloned into pEGFP-C1 vector (Clontech). NRK cells were maintained in Kaighn's modified F12 (F12K, Sigma) medium supplemented with 10% fetal bovine serum (FBS), 1 mM L-glutamine, 100 U/mL penicillin, and 100 μg/mL streptomycin (GIBCO) at 37°C and 5% CO_2. For transfection, cells were grown on a cover slip chamber to 60%–70% confluency. Cells were rinsed once with Opti-MEM I medium (LifeTechnologies) immediately before transfection. The rinsed cells were then transfected with 1 μg each of the pEGFP-C1-Nmar_0816 and the pEGFP-C1 using Lipofectamine 2000 reagent (Invitrogen) according to manufacturer's instructions. After 4 h of incubation, the medium containing DNA-Lipofectamine was replaced with the F12K medium containing 10% FBS, and the cells were cultured for an additional 14–16 h before imaging.

3.4. Protein Expression and SDS-PAGE Analysis. Full length Nmar_0816 was PCR amplified (see Table S1 for primers information) and cloned into pQE30 vector (QIAGEN) for expression as 6 × His-Nmar_0816 recombinant protein in *E. coli* M15. The recombinant 6 × His-Nmar_0816 was induced with 1 mM IPTG and purified on nickel column (QIAGEN) according to manufacturer's instruction. To express the 6 × His-Nmar_0816$_{C123A}$ (cysteine to alanine mutation) in *E. coli*, two independent N-terminal and C-terminal PCR fragments were obtained using primers incorporating the cysteine to alanine mutation in nucleotide sequences (see Table S1 for primer sequences). Overlapping PCR was performed using both N- and C-terminal PCR fragments to get a final PCR fragment of the Nmar_0816 with cysteine to alanine mutation. The Nmar_0816$_{C123A}$ fragment was then cloned into pQE30 vector for expression and purification. The purified recombinant proteins were analyzed by 10% SDS-PAGE.

3.5. Microscopy and FRAP Analysis. For the epifluorescence imaging of yeast, images were acquired on an Olympus IX71 inverted microscope equipped with a charge-coupled device (CCD) camera (CoolSNAP ES, Photometrics), a 100x/1.45

NA Plan Apo objective lens (Olympus), and Metamorph (v.7.6) software. Fluorescence recovery after photobleaching (FRAP) was performed using Zeiss Meta 510 inverted confocal microscope with a 100x/1.25 NA Apochromat objective lens. For fluorescence imaging in mammalian NRK cells, the images were taken using Axiovert 200 M inverted microscope (Carl Zeiss) with a 100x/1.30 NA Plan-Neofluar lens or Zeiss LSM Meta 510 inverted confocal microscope with a 100x/1.4 NA Plan-Apochromat lens. All images from Axiovert 200 M were acquired using a cooled CCD camera (CoolSNAP$_{HQ}$, Roper Scientific) and MetaView imaging software (Universal Imaging). Images were processed with ImageJ software (http://rsb.info.nih.gov/ij/) for presentation.

Acknowledgments

The authors would like to thank Drs. Mithilesh Mishra, Xie Tang, Phing-Chian Chai, Singh Nongmaithem Sadananda, and Sreepathy Sachin Seshadri for critical reading of the paper and all members of Cell Division Laboratory for technical advices. Special thanks to Dr Snezhana Oliferenko for valuable comments on the paper. This work was supported by research funds from Temasek Life Sciences Laboratory and Mechanobiology Institute.

References

[1] P. Baumann and S. P. Jackson, "An archaebacterial homologue of the essential eubacterial cell division protein FtsZ," *Proceedings of the National Academy of Sciences of the United States of America*, vol. 93, no. 13, pp. 6726–6730, 1996.

[2] W. Margolin, R. Wang, and M. Kumar, "Isolation of an ftsZ homolog from the archaebacterium *Halobacterium salinarium*: implications for the evolution of FtsZ and tubulin," *Journal of Bacteriology*, vol. 178, no. 5, pp. 1320–1327, 1996.

[3] A. Poplawski, B. Gullbrand, and R. Bernander, "The *ftsZ* gene of *Haloferax mediterranei*: sequence, conserved gene order, and visualization of the FtsZ ring," *Gene*, vol. 242, no. 1-2, pp. 357–367, 2000.

[4] X. Wang and J. Lutkenhaus, "FtsZ ring: the eubacterial division apparatus conserved in archaebacteria," *Molecular Microbiology*, vol. 21, no. 2, pp. 313–319, 1996.

[5] A. C. Lindås, E. A. Karlsson, M. T. Lindgren, T. J. G. Ettema, and R. Bernander, "A unique cell division machinery in the Archaea," *Proceedings of the National Academy of Sciences of the United States of America*, vol. 105, no. 48, pp. 18942–18946, 2008.

[6] R. Y. Samson, T. Obita, S. M. Freund, R. L. Williams, and S. D. Bell, "A role for the ESCRT system in cell division in archaea," *Science*, vol. 322, no. 5908, pp. 1710–1713, 2008.

[7] R. Y. Samson, T. Obita, B. Hodgson et al., "Molecular and structural basis of ESCRT-III recruitment to membranes during Archaeal cell division," *Molecular Cell*, vol. 41, no. 2, pp. 186–196, 2011.

[8] M. Babst, D. J. Katzmann, E. J. Estepa-Sabal, T. Meerloo, and S. D. Emr, "ESCRT-III: an endosome-associated heterooligomeric protein complex required for MVB sorting," *Developmental Cell*, vol. 3, no. 2, pp. 271–282, 2002.

[9] J. E. Garrus, U. K. Von Schwedler, O. W. Pornillos et al., "Tsg101 and the vacuolar protein sorting pathway are essential for HIV-1 budding," *Cell*, vol. 107, no. 1, pp. 55–65, 2001.

[10] H. L. Hyung, N. Elia, R. Ghirlando, J. Lippincott-Schwartz, and J. H. Hurley, "Midbody targeting of the ESCRT machinery by a noncanonical coiled coil in CEP55," *Science*, vol. 322, no. 5901, pp. 576–580, 2008.

[11] E. Morita, V. Sandrin, H. Y. Chung et al., "Human ESCRT and ALIX proteins interact with proteins of the midbody and function in cytokinesis," *EMBO Journal*, vol. 26, no. 19, pp. 4215–4227, 2007.

[12] C. Brochier-Armanet, B. Boussau, S. Gribaldo, and P. Forterre, "Mesophilic crenarchaeota: proposal for a third archaeal phylum, the Thaumarchaeota," *Nature Reviews Microbiology*, vol. 6, no. 3, pp. 245–252, 2008.

[13] A. Spang, R. Hatzenpichler, C. Brochier-Armanet et al., "Distinct gene set in two different lineages of ammonia-oxidizing archaea supports the phylum Thaumarchaeota," *Trends in Microbiology*, vol. 18, no. 8, pp. 331–340, 2010.

[14] M. Könneke, A. E. Bernhard, J. R. De La Torre, C. B. Walker, J. B. Waterbury, and D. A. Stahl, "Isolation of an autotrophic ammonia-oxidizing marine archaeon," *Nature*, vol. 437, no. 7058, pp. 543–546, 2005.

[15] W. Martens-Habbena, P. M. Berube, H. Urakawa, J. R. De La Torre, and D. A. Stahl, "Ammonia oxidation kinetics determine niche separation of nitrifying Archaea and Bacteria," *Nature*, vol. 461, no. 7266, pp. 976–979, 2009.

[16] C. B. Walker, J. R. De La Torre, M. G. Klotz et al., "Nitrosopumilus maritimus genome reveals unique mechanisms for nitrification and autotrophy in globally distributed marine crenarchaea," *Proceedings of the National Academy of Sciences of the United States of America*, vol. 107, no. 19, pp. 8818–8823, 2010.

[17] E. A. Pelve, A. C. Lindas, W. Martens-Habbena, J. R. de la Torre, D. A. Stahl et al., "Cdv-based cell division and cell cycle organization in the thaumarchaeon Nitrosopumilus maritimus," *Molecular Microbiology*, vol. 82, pp. 555–566, 2011.

[18] M. K. Balasubramanian, R. Srinivasan, Y. Huang, and K. H. Ng, "Comparing contractile apparatus-driven cytokinesis mechanisms across kingdoms," *Cytoskeleton*, vol. 69, no. 11, pp. 942–956, 2012.

[19] H. P. Erickson, D. E. Anderson, and M. Osawa, "FtsZ in bacterial cytokinesis: cytoskeleton and force generator all in one," *Microbiology and Molecular Biology Reviews*, vol. 74, no. 4, pp. 504–528, 2010.

[20] W. Margolin, "FtsZ and the division of prokaryotic cells and organelles," *Nature Reviews Molecular Cell Biology*, vol. 6, no. 11, pp. 862–871, 2005.

[21] M. Osawa, D. E. Anderson, and H. P. Erickson, "Reconstitution of contractile FtsZ rings in liposomes," *Science*, vol. 320, no. 5877, pp. 792–794, 2008.

[22] T. D. Pollard, "Mechanics of cytokinesis in eukaryotes," *Current Opinion in Cell Biology*, vol. 22, no. 1, pp. 50–56, 2010.

[23] B. Wickstead and K. Gull, "The evolution of the cytoskeleton," *The Journal of Cell Biology*, vol. 194, no. 4, pp. 513–525, 2011.

[24] R. Srinivasan, M. Mishra, M. Murata-Hori, and M. K. Balasubramanian, "Filament formation of the *Escherichia coli* actin-related protein, MreB, in fission yeast," *Current Biology*, vol. 17, no. 3, pp. 266–272, 2007.

[25] R. Srinivasan, M. Mishra, L. Wu, Z. Yin, and M. K. Balasubramanian, "The bacterial cell division protein FtsZ assembles into cytoplasmic rings in fission yeast," *Genes and Development*, vol. 22, no. 13, pp. 1741–1746, 2008.

[26] G. Bodon, R. Chassefeyre, K. Pernet-Gallay, N. Martinelli, G. Effantin et al., "Charged multivesicular body protein 2B (CHMP2B) of the endosomal sorting complex required for transport-III (ESCRT-III) polymerizes into helical structures deforming the plasma membrane," *The Journal of Biological Chemistry*, vol. 286, pp. 40276–40286, 2011.

[27] I. Fyfe, A. L. Schuh, J. M. Edwardson, and A. Audhya, "Association of the endosomal sorting complex ESCRT-II with the Vps20 subunit of ESCRT-III generates a curvature-sensitive complex capable of nucleating ESCRT-III filaments," *The Journal of Biological Chemistry*, vol. 286, pp. 34262–34270, 2011.

[28] S. Ghazi-Tabatabai, S. Saksena, J. M. Short et al., "Structure and disassembly of filaments formed by the ESCRT-III subunit Vps24," *Structure*, vol. 16, no. 9, pp. 1345–1356, 2008.

[29] J. Guizetti, L. Schermelleh, J. Mäntler et al., "Cortical constriction during abscission involves helices of ESCRT-III-dependent filaments," *Science*, vol. 331, no. 6024, pp. 1616–1620, 2011.

[30] P. I. Hanson, R. Roth, Y. Lin, and J. E. Heuser, "Plasma membrane deformation by circular arrays of ESCRT-III protein filaments," *The Journal of Cell Biology*, vol. 180, no. 2, pp. 389–402, 2008.

[31] W. M. Henne, N. J. Buchkovich, Y. Zhao, and S. D. Emr, "The endosomal sorting complex ESCRT-II mediates the assembly and architecture of ESCRT-III helices," *Cell*, vol. 151, pp. 356–371, 2012.

[32] S. Lata, G. Schoehn, A. Jain et al., "Helical structures of ESCRT-III are disassembled by VPS4," *Science*, vol. 321, no. 5894, pp. 1354–1357, 2008.

[33] Y. Lin, L. A. Kimpler, T. V. Naismith, J. M. Lauer, and P. I. Hanson, "Interaction of the mammalian endosomal sorting complex required for transport (ESCRT) III protein hSnf7-1 with itself, membranes, and the AAA$^+$ ATPase SKD1," *The Journal of Biological Chemistry*, vol. 281, no. 50, p. 38966, 2006.

[34] S. Saksena, J. Wahlman, D. Teis, A. E. Johnson, and S. D. Emr, "Functional Reconstitution of ESCRT-III assembly and disassembly," *Cell*, vol. 136, no. 1, pp. 97–109, 2009.

[35] C. Moriscot, S. Gribaldo, J. M. Jault et al., "Crenarchaeal CdvA forms double-helical filaments containing DNA and interacts with ESCRT-III-like CdvB," *PLoS ONE*, vol. 6, no. 7, Article ID e21921, 2011.

[36] S. Shim, L. A. Kimpler, and P. I. Hanson, "Structure/function analysis of four core ESCRT-III proteins reveals common regulatory role for extreme C-terminal domain," *Traffic*, vol. 8, no. 8, pp. 1068–1079, 2007.

[37] K. K. Busiek and W. Margolin, "Split decision: a thaumarchaeon encoding both FtsZ and Cdv cell division proteins chooses Cdv for cytokinesis," *Molecular Microbiology*, vol. 82, no. 3, pp. 535–538, 2011.

[38] A. Mukherjee, C. Saez, and J. Lutkenhaus, "Assembly of an FtsZ mutant deficient in GTpase activity has implications for FtsZ assembly and the role of the Z ring in cell division," *Journal of Bacteriology*, vol. 183, no. 24, pp. 7190–7197, 2001.

[39] D. J. Scheffers, J. G. De Wit, T. Den Blaauwen, and A. J. M. Driessen, "GTP hydrolysis of cell division protein FtsZ: evidence that the active site is formed by the association of monomers," *Biochemistry*, vol. 41, no. 2, pp. 521–529, 2002.

[40] T. J. G. Ettema and R. Bernander, "Cell division and the ESCRT complex: a surprise from the Archaea," *Communitative and Integrative Biology*, vol. 2, no. 2, pp. 86–88, 2009.

[41] A. F. Ellen, S. V. Albers, W. Huibers et al., "Proteomic analysis of secreted membrane vesicles of archaeal *Sulfolobus* species reveals the presence of endosome sorting complex components," *Extremophiles*, vol. 13, no. 1, pp. 67–79, 2009.

[42] G. Basi, E. Schmid, and K. Maundrell, "TATA box mutations in the *Schizosaccharomyces pombe nmt1* promoter affect transcription efficiency but not the transcription start point or thiamine repressibility," *Gene*, vol. 123, no. 1, pp. 131–136, 1993.

[43] A. Chino, K. Watanabe, and H. Moriya, "Plasmid construction using recombination activity in the fission yeast Schizosaccharomyces pombe," *PLoS ONE*, vol. 5, no. 3, Article ID e9652, 2010.

[44] S. Moreno, A. Klar, and P. Nurse, "Molecular genetic analysis of fission yeast *Schizosaccharomyces pombe*," *Methods in Enzymology*, vol. 194, pp. 795–823, 1991.

3

The Effect of Saturated Fatty Acids on Methanogenesis and Cell Viability of *Methanobrevibacter ruminantium*

Xuan Zhou,[1] **Leo Meile,**[2] **Michael Kreuzer,**[1] **and Johanna O. Zeitz**[1]

[1] *ETH Zurich, Institute of Agricultural Sciences, Universitaetstrasse 2, 8092 Zurich, Switzerland*
[2] *ETH Zurich, Institute of Food, Nutrition and Health, Schmelzbergstrasse 7, 8092 Zurich, Switzerland*

Correspondence should be addressed to Johanna O. Zeitz; j.zeitz@gmx.de

Academic Editor: Yoshizumi Ishino

Saturated fatty acids (SFAs) are known to suppress ruminal methanogenesis, but the underlying mechanisms are not well known. In the present study, inhibition of methane formation, cell membrane permeability (potassium efflux), and survival rate (LIVE/DEAD staining) of pure ruminal *Methanobrevibacter ruminantium* (DSM 1093) cell suspensions were tested for a number of SFAs. Methane production rate was not influenced by low concentrations of lauric (C_{12}; 1 μg/mL), myristic (C_{14}; 1 and 5 μg/mL), or palmitic (C_{16}; 3 and 5 μg/mL) acids, while higher concentrations were inhibitory. C_{12} and C_{14} were most inhibitory. Stearic acid (C_{18}), tested at 10–80 μg/mL and ineffective at 37°C, decreased methane production rate by half or more at 50°C and ≥50 μg/mL. Potassium efflux was triggered by SFAs ($C_{12} = C_{14} > C_{16} > C_{18}$ = control), corroborating data on methane inhibition. Moreover, the exposure to C_{12} and C_{14} decreased cell viability to close to zero, while 40% of control cells remained alive after 24 h. Generally, tested SFAs inhibited methanogenesis, increased cell membrane permeability, and decreased survival of *M. ruminantium* in a dose- and time-dependent way. These results give new insights into how the methane suppressing effect of SFAs could be mediated in methanogens.

1. Introduction

Methane (CH_4) as a potent greenhouse gas is among the most important drivers of compositional changes of atmospheric gas and thus global warming [1]. Agricultural CH_4 emissions account for about 50% of total CH_4 from anthropogenic sources, where the single largest one is from enteric fermentation in ruminant livestock [2]. Methane is generated by a subgroup of the Archaea, the methanogens, which are, in the ruminant's fore-stomach (rumen), dominated by *Methanobrevibacter* [3]. At undisturbed rumen function, proteins and polymeric carbohydrates as main components of the diet are degraded by microorganisms and fermented mainly to volatile fatty acids (VFAs), ammonia, hydrogen (H_2), and carbon dioxide (CO_2). Ruminal methanogens primarily utilize H_2 as energy source to reduce CO_2 to CH_4 in a series of reactions that are coupled to ATP synthesis [4, 5]. As CH_4 cannot be utilized in the metabolism of the animal, ruminal methanogenesis also impairs feed conversion efficiency and represents a significant waste of energy (2% to 12% of energy intake; [6]).

Therefore, inhibition of ruminal methanogenesis should be approached by various interventions. Among the most effective are dietary medium- and long-chain saturated fatty acids (SFAs). Nonesterified lauric acid (C_{12}) was reported to have a particularly high potential in suppressing ruminal methanogenesis, followed by myristic acid (C_{14}) [7–9]. By contrast, long-chain SFAs (LCFAs) such as palmitic acid (C_{16}) and stearic acid (C_{18}) were not effective in suppressing ruminal methanogenesis *in vitro* [7, 10]. The production of CH_4 by pure, growing, cultures of *M. ruminantium*, a dominant ruminal methanogen [3], was found to be inhibited by the addition of unsaturated [11, 12] and saturated medium-chain (C_{12}–C_{16}; [12]) fatty acids. When testing the nonruminal methanogens *Methanothermobacter thermoautotrophicus* and *Methanococcus voltae*, C_{12} and C_{14} were found to inhibit methanogenesis as well [13]. However, systematic studies on dose-response relationships with SFAs on methanogenesis in pure ruminal methanogen cultures are missing. Besides, it is unclear why long-chain SFAs do not inhibit methanogenesis and if this is related to the low solubility of these long-chain SFAs at temperatures below 40°C [13, 14]. Furthermore, although fatty acids (FAs) are known to have antimicrobial and cytotoxic properties [15] and are used by a wide range of organisms like humans [16], molluscs [17], and brown algae [18] to defend against pathogens, the mechanisms which lead to the inhibition effect are still not definitely known. Several mechanisms have been proposed [15]. The primary target of the action seems to be the microbial cell membrane and various essential processes that occur within and at the membrane [15]. Fatty acids, including C_{12}, C_{14}, C_{16}, and C_{18}, have been shown to pass protein-free phospholipid bilayers in their unionized form [19]. Saturated and unsaturated fatty acids may be adsorbed by bacterial cell membranes [20], damage the bacterial cell membrane as determined by loss of potassium (K^+) [21], ATP, and proteins [16] and by electron microscopy [22, 23], and play a role in cell death [22, 24, 25]. As the composition of the cell envelope of methanogens is fundamentally different from the bacterial cell envelope, and the methanogens are phylogenetically and physiologically distinct from all other cell types [26], the mechanisms of FA action on methanogens may differ from that valid for other organisms. However, since the methanogen cell envelope normally acts as a diffusion barrier between the cytoplasm and the extracellular medium, it might also represent a key point for the identification of inhibitor targets. Therefore, we hypothesized that membrane integrity is disturbed and leakage of cell metabolites including inorganic ions such as K^+ occurs through the interaction of the SFA with the cell membrane lipids and that this results in an impaired cell survival. Like in most prokaryotes, K^+ is accumulated in the cytoplasm of methanogens in exchange for Na^+ [27].

In the present study, pure cultures of *M. ruminantium* were treated with pure nonesterified SFAs in order to exclude all confounding factors such as interactions between feed, minerals, and microbes occurring *in vivo* or with rumen fluid *in vitro*. The aims of the present study were (i) to investigate the relationship between SFA type and dosage and the inhibition of methanogenesis in nongrowing cells, that is, cell suspensions, and (ii) to get first insights into the modes of action underlying in this process. In detail, K^+ efflux was used as an indicator of membrane integrity. Finally, cell survival was monitored using the LIVE/DEAD BacLight Kit which has been successfully used in Archaea before [28, 29].

2. Materials and Methods

2.1. Strain and Growth Conditions. A pure culture of *M. ruminantium* M1 (DSM 1093) was obtained from the "Deutsche Sammlung von Mikroorganismen und Zellkulturen," Braunschweig, Germany. It was anaerobically cultivated in the strain-specific cultivation medium 119 prepared according to DSMZ (http://www.dsmz.de) in 120 mL serum bottles, which were sealed with butyl rubber stoppers (20 mm size; 2048-11800, Bellco, Vineland, USA) and aluminum seals (2048-11020, Bellco). Reagents for the media were dissolved in boiled oxygen-deprived distilled water and stirred on a magnetic stirrer overnight in an anaerobic chamber (Coy Laboratory Products, Grasslake, USA). Heat stable solutions of media ingredients were sterilized in a batch autoclave (Sauter, Belimed Sauter AG, Sulgen, Switzerland) for 20 min at 121°C.

Heat susceptible solutions, that is, vitamins, sodium formate, and SFA, were filtrated through a 0.2 μm Minisart-plus filter (Sartorius AG, Göttingen, Germany). Ruminal fluid was obtained from a rumen-cannulated cow, filtered through four layers of medicinal gauze (REF 200137, Novamed, Jerusalem, Israel) and then centrifuged twice for 15 min at 4,000 ×g (Varifuge K, Heraeus, Osterode, Germany). The supernatant was adjusted to pH 7.0 with HCl and NaOH, gassed with N_2 to an atmospheric pressure of 150 kPa, autoclaved, and stored at −20°C for up to 6 months, before being used to prepare the media. Aliquots of the prepared medium were filled into 250-mL bottles, closed with rubber septa, gassed with N_2 to atmospheric pressure of 250 kPa, autoclaved, and stored either at 4°C for 8 weeks or at −20°C for up to 6 months before being used. *M. ruminantium* was grown under atmospheric pressure of 250 kPa of a CO_2/H_2 mixture (20 : 80) (Pangas AG, Dagmarsellen, Switzerland). The gas mixture in the headspace was renewed every 24 h and 3 mL precultures were transferred to 27 mL fresh medium every four days. The culture bottles were incubated in horizontal position in an incubation shaker (Incu Shaker 10 L, Benchmark, Korea) at 37°C with a shaking speed of 150 rpm. Growth of the cultures was monitored by recording CH_4 production, gas consumption, and optical density. A volume of 0.15 mL of gas was collected from the headspace of the cultivation bottle with a gas-tight syringe (Hamilton, model 1725/RN 250 mL, Fisher Scientific AG, Wohlen, Switzerland), and its CH_4 concentration was analyzed with a gas chromatograph (model 6890N, Agilent Technologies, Santa Clara, CA, USA) equipped with a flame ionization detector operated at 250°C and a 234 mm × 23 mm column (80/100; 166 mesh; Porapak Q, Fluka Chemie AG, Buchs, Switzerland). Overpressure in the cultivation bottles was detected with a manometer (GDH 200-13, Greisinger Electronic GmbH, Regenstauf, Germany). One milliliter of culture liquid was collected in acrylic absorption cuvettes (1 cm path length; (VWR, Leuven, Germany)), and its optical density was measured at 600 nm (OD_{600}) with a UV-160A recording spectrophotometer (Shimadzu, Kyoto, Japan). The growth phases distinguished were lag, exponential, stationary, and death phase. Prior to each experiment, methanogens were inoculated into fresh medium with 3 mL of pre-culture in their early to mid-exponential growth phase.

2.2. Experiment 1. Lauric acid, C_{14}, C_{16}, and C_{18} (≥97% purity) were obtained from Sigma-Aldrich, Buchs, Switzerland, to be used as experimental supplements. Stock solutions were prepared by dissolving the SFA in the sterile-filtered solvent dimethyl sulfoxide (DMSO) (Sigma-Aldrich) to reach concentrations of 1, 3, 5, 10, and 30 mg/mL (C_{12} to C_{16}) as well as 50 and 80 mg/mL (C_{18}). They were stored at room temperature before supplementation. The C_{18} solution to be applied later at 50°C was heated to 50°C before use.

As OD_{600} was used to estimate cell dry matter (DM) concentration in growing cultures prior to harvesting, a regression line between OD_{600} and cell DM concentration was established before the start of the experiment. Seventeen bottles of medium were prepared and inoculated with *M. ruminantium* as described before. From three bottles each, 21 mL of culture liquid were collected after 24, 48, 53,

72, 77, and 96 h covering the development from the early exponential growth phase to the stationary phase. Thereof, 1 mL was used for measurement of OD_{600}, and 20 mL was dried at 70°C to constant weight in a 50 mL Falcon tube after the wet weight had been recorded in order to calculate culture DM content. The regression curve established from in total 17 OD/DM pairs (OD range: 0.348 to 0.986) was linear and reads DM (mg/mL) = $7.6092 \times OD_{600} + 0.4754$ ($R^2 = 0.95$). This relationship was used to adjust and equalize cell DM concentration in cell suspensions.

In order to prepare the experimental cell suspensions in an anaerobic chamber, always 20 mL of culture were harvested in the mid-exponential growth phase and transferred to two 50 mL sterilized Falcon tubes and centrifuged for 10 min at 3,000 ×g. The supernatant was discarded and the pellet was washed twice with an autoclaved phosphate buffer of pH 6.8 containing 0.025 M KH_2PO_4, 0.025 M K_2HPO_4, 0.5 mM titanium citrate, 0.1 M NaCl, and 1 mM $MgCl_2$ [30]. Titanium citrate was prepared according to Jones and Pickard [31], by anaerobically adding 5 mL of a 15% titanium(III) chloride solution (Merck Millipore, Darmstadt, Deutschland) to 50 mL of 0.2 M sodium citrate solution, adjusting with a saturated sodium carbonate solution to pH 7, gassing the bottle with N_2, followed by autoclaving. Syringes were used for all withdrawals. After washing, the cell pellet was then resuspended in the same buffer to a final concentration of 6 mg cell DM/mL adjusted with the help of the regression line relating OD and culture DM concentration. Under anaerobic condition, 1 μL of the differently concentrated SFA stock solutions was added to 999 μL cell suspensions in 25 mL serum bottles to reach concentrations of 1, 3, 5, 10, and 30 μg/mL of C_{12}, C_{14}, and C_{16} as well as 50 and 80 μg/mL of C_{18}. The bottles were sealed with rubber stoppers (size 18D, 203018; Glasgerätebau Ochs, Lenglern, Germany), gassed to atmospheric pressure of 250 kPa with a CO_2/H_2 mixture (20 : 80) and stored on ice waiting for incubation start by putting into a waterbath (Julabo shake Temp, Merck, Switzerland) at set intervals due to time needed for GC measurement (3.2 min/sample). Cell suspensions were incubated at 37°C and 50°C (only C_{18}) shaking suspensions at 150 rpm. Finally, suspensions where no SFA had been added were supplemented with either 1 μL/mL DMSO, equal to the DMSO concentration in treatment groups (control group) or with 1 μL/mL of the buffer (blank group). The CH_4 concentration (mol %) was determined by gas chromatography after 1, 2, 5, and 24 h had passed. The CH_4 production rate (μmol CH_4/mg cell DM per min) was calculated from bottle head space gas volume and the volume of CH_4 produced. The amount of gas present in the bottles at the start of the experiment were set to 0.0023 mol as calculated from using the ideal gas law ($n = p \times V/R \times T$, where p is the sum of the overpressure of the gas in the bottle (150000 Pa) and the standard air pressure (96600 Pa for Zurich), V is the volume of the gas = 24×10^{-6} m^3, n is the amount of gas in the bottle in mol, T is the temperature of the gas = 309.15 K, and R is the ideal gas constant = 8.314 J K^{-1} mol^{-1}). The amounts of CH_4 produced in each bottle (Y; in mol) were calculated considering the stoichiometry of methanogenesis from H_2 and CO_2, that is, that 5 mol of gas are consumed to produce 1 mol of CH_4 meaning $Y/(0.0023 - 4 \times Y)$ = mol% CH_4 ($X/100$) and therefore $Y = 0.0023X/(100 + 4X)$.

For each SFA, a minimum of two independent cell suspension incubations were performed with freshly grown $M.$ $ruminantium$ culture, each performed at least in triplicate.

2.3. Experiments 2 and 3. Cells were harvested as described before and resuspended to a final concentration of 6 mg cell DM/mL in K^+-free buffer containing 0.025 M $(NH_4)_2HPO_4$, 0.025 M $NH_4H_2PO_4$, 0.01 M NaCl, 1 mM $MgCl_2$, and 0.5 mM titanium citrate. Two resting cell suspension experiments were performed at 37°C as described before, and in Experiment 2, C_{12} was supplemented to final concentrations of 10, 15 and 30 μg/mL, and in Experiment 3, C_{12}, C_{14}, C_{16} and C_{18} were added to reach a final concentration of 10 μg/mL. After 3 h and 24 h of incubation, 300 μL of cell suspension were transferred to a 2 mL centrifuge tube inside the anaerobic chamber, centrifuged at 10,000 ×g for 10 min, and the K^+ concentration in the supernatant was analyzed by Inductively Coupled Plasma-Optical Emission Spectrometer (715-ES Radial ICP OES, Varian, Canada). A stock solution containing 1 mg/L KNO_3 (Merck, Darmstadt, Germany) and 1% HNO_3 in distilled water was used to prepare a calibration curve with concentrations of 0, 25, 50, 75, and 100 μL/L. Samples were diluted 50-fold by using a diluter (Microlab 1000, Hanmilton, Martinsried, Germany) in 5 mL of total volume. The survival rate of $M.$ $ruminantium$ in cell suspensions after 3 and 24 h was assessed by using the LIVE/DEAD BacLight Bacterial Viability Kit for microscopy and quantitative assays (Kit L7012; Invitrogen GmbH, Darmstadt, Germany). The kit applied contained two fluorescent dyes: propidium iodide with red fluorescence penetrates cells with damaged membranes; SYTO 9 with green fluorescence accumulates only in living cells. Thus, undestroyed archaeal cells with intact membranes have green fluorescence, while cells with damaged membranes display red fluorescence. Occasionally, an intermediate ambiguous yellowish color has been observed which has been observed also in studies of others [28]. Cells showing this color have been categorized as living cells with damaged membrane but were not included into the category of living cells in the tables. Staining was performed according to the manufacturer's protocol with several modifications. An amount of 0.5 μL of a 1 : 1 mixture of SYTO 9 and propidium iodide dyes was added to 100 μL of cell suspension under aerobic conditions, mixed thoroughly and incubated at room temperature in the dark for 10 min. No washing was required before staining because background fluorescence was low in this experimental system, and oxygen exposure was minimized by this way. An amount of 5 μL of the stained cell suspension was trapped between a microscope slide and an 18 mm square cover glass. All samples were examined at 600 and 1000 times magnification using a fluorescence microscope (BX60; Olympus GmbH, Voketswil, Switzerland) and a digital camera (FView; adapter U-CMAD, Olympus, Switzerland). Three locations on each sample were chosen and captured at random. Fluorescent micrographs (exposure time: 50 ms) of the very same sample section were taken applying appropriate filter sets for propidium iodide (wavelengths: excitation 530–545 nm, emission >610 nm) and SYTO 9 (excitation 440–470 nm, emission 525–550 nm) and using the digital image analysis software Analysis (Soft Image System GmbH, Münster, Germany). The two false-colored images of one sample section were combined using the same software, and dead and live cells were counted with Adobe Photoshop CS5 (Adobe, San Jose, USA). Postacquisition processing involved adjustments of the brightness/contrast to optimize the visualization of live and dead cells within the images. Viability was calculated as viability = $N/N_0 \times 100$, where N_0 are the total fluorescence counts and N are the green fluorescence counts after 3 h and 24 h of reaction. Experiments 2 and 3 were performed in triplicate with three samples per treatment group and additionally, three samples for LIVE/DEAD staining and K^+ leakage determination after 3 h.

2.4. Statistical Analysis. For Experiment 1, analysis of variance was performed using the MIXED procedure of SAS (version 9.1 of 2003; SAS Institute Inc., Cary, NC) with treatment group and time point and its interaction as fixed factors and the repeated statement to compare control and SFA-supplemented cultures at each time-point. For Experiments 2 and 3, treatment group was considered as fixed and replicate as random factor to compare CH_4 inhibition rate, K^+ leakage and cell viability both at 3 and 24 h. The Bonferroni correction was used for multiple comparisons among means. Differences were declared statistically significant at $P < 0.05$. The results are presented as means ± standard errors.

3. Results

3.1. Inhibition of Methane Production of Methanobrevibacter ruminantium by Saturated Fatty Acids as Depending on Dose in Experiment 1. All SFAs investigated influenced CH_4 production by *M. ruminantium* in a dose-dependent way, but the extent of the effect differed (Figure 1). In Figure 1, only one of the two incubations performed per SFA is shown (the other is given as Supplementary Figure 1), but values were similar between incubations. For C_{12}, the CH_4 production rate was inhibited in a dose-dependent way with (μg/mL) 30 > 10 = 5 ≥ control ≥ 1 (incubation 1; Figure 1(a)) and 30 > 10 = 5 > 1 = control (incubation 2; Supplementary Figure 1). For C_{14}, the sequence was 30 > 10 > 1 = control > 5 (incubation 1; Figure 1(b)) and 30 = 10 > 1 ≥ 5 ≥ control (incubation 2). The inhibitory pattern of C_{16} was different from C_{12} and C_{14}; C_{16} needed more time to exert its influence: dosages of 10 and 30 μg/mL inhibited the CH_4 production rate at 24 h completely (incubation 1; Figure 1(c)) or by half (incubation 2) but not at earlier time points. Lower concentrations did not inhibit CH_4 production during the measurement period. C_{18} was not effective at 37°C (Figure 1(d)) but at 50°C, a temperature closer to the melting point of C_{18} of 69°C. At 50°C, C_{18} decreased the CH_4 production rate in a dose-dependent way after 5 h by 55% and 68% at 10 and 30 μg/mL, respectively (incubation 1; Figure 1(e)). At 50 μg/mL, the CH_4 production rates started to decline even earlier and were

decreased by 63% and 99% at 5 h and 24 h, and at 80 μg/mL, by 52%, 94%, and 100% at 1 h, 3 h, and 5 h, respectively.

3.2. Influence of Lauric Acid on Methane Production, K^+ Leakage and Cell Viability in Experiment 2. In K^+-free buffer, the CH_4 inhibitory pattern of C_{12} (Table 1) was similar as compared to Experiment 1 in K^+-containing buffer; concentrations of ≥10 μg/mL decreased the CH_4 production rate very fast and, with 30 μg/mL, stopped it completely after already 3 h. A quick increase in extracellular K^+ concentration occurred in C_{12}-treated groups after 3 h of incubation (Table 1). Especially in groups where 15 and 30 μg/mL was added, extracellular K^+ concentration reached its peak already at 3 h and did not increase as reaction time progressed. The viability of the *M. ruminantium* cells as verified using LIVE/DEAD staining at 3 h and 24 h after supplementation of 10, 15, and 30 μg C_{12}/mL is shown in Table 1. Although methanogenesis was completely inhibited and marked K^+ leakage occurred in groups supplemented with 15 and 30 μg C_{12}/mL at 3 h, cell viability was still 27% and 29%, respectively, instead of being zero. Within 24 h, C_{12} caused more cell death.

3.3. Influence of Saturated Fatty Acids on Methane Production, K^+ Leakage, and Cell Viability in Experiment 3. All SFAs were supplemented in the same concentration (10 μg/mL) in a single incubation to allow a direct comparison between SFAs (Table 2). C_{12} and C_{14} had a similar inhibitory effect on methanogenesis. Both immediately started displaying their influence. C_{16} needed more time and its effect was weaker than that of the former two SFAs, while C_{18} showed no effect at 37°C, which was consistent with the results of Experiment 1. The patterns of methanogenesis inhibition and K^+ efflux were similar (Table 2). C_{12} and C_{14} also had the strongest effect of all SFAs tested in triggering K^+ leakage, while C_{16} caused lower K^+ efflux compared to C_{12} and C_{14}, but the extracellular K^+ concentration was higher ($P < 0.05$) than in control (Table 2). In summary, the K^+ efflux was (in decreasing order): $C_{12} = C_{14} > C_{16} > C_{18} >$ control. Interestingly, C_{18} showed no inhibitory effect on CH_4 production rate but did cause K^+ efflux (+23% as compared to the control after 3 h). C_{12} and C_{14} had the strongest effect on cell viability, as 57% and 64% of the cells were categorized as dead after 3 h, while in the C_{16} group only 32% of cells were dead or, as part of the cells were not red but yellow, damaged (Figure 2). At 24 h, nearly all cells treated with C_{12} and C_{14} were dead, compared to 60% of dead cells found in the control (Table 2 and Figure 2). Also in the C_{16} treatment 88% of cells were dead after 24 h, which implies that the inhibition of methanogenesis and K^+ efflux are somehow correlated. C_{18} did not cause significant extra cell death when compared to the control group.

4. Discussion

The antifungal and bactericidal properties of FA have been extensively investigated, and, as a generalization, the cell membrane seems to be the prime target to explain the effects

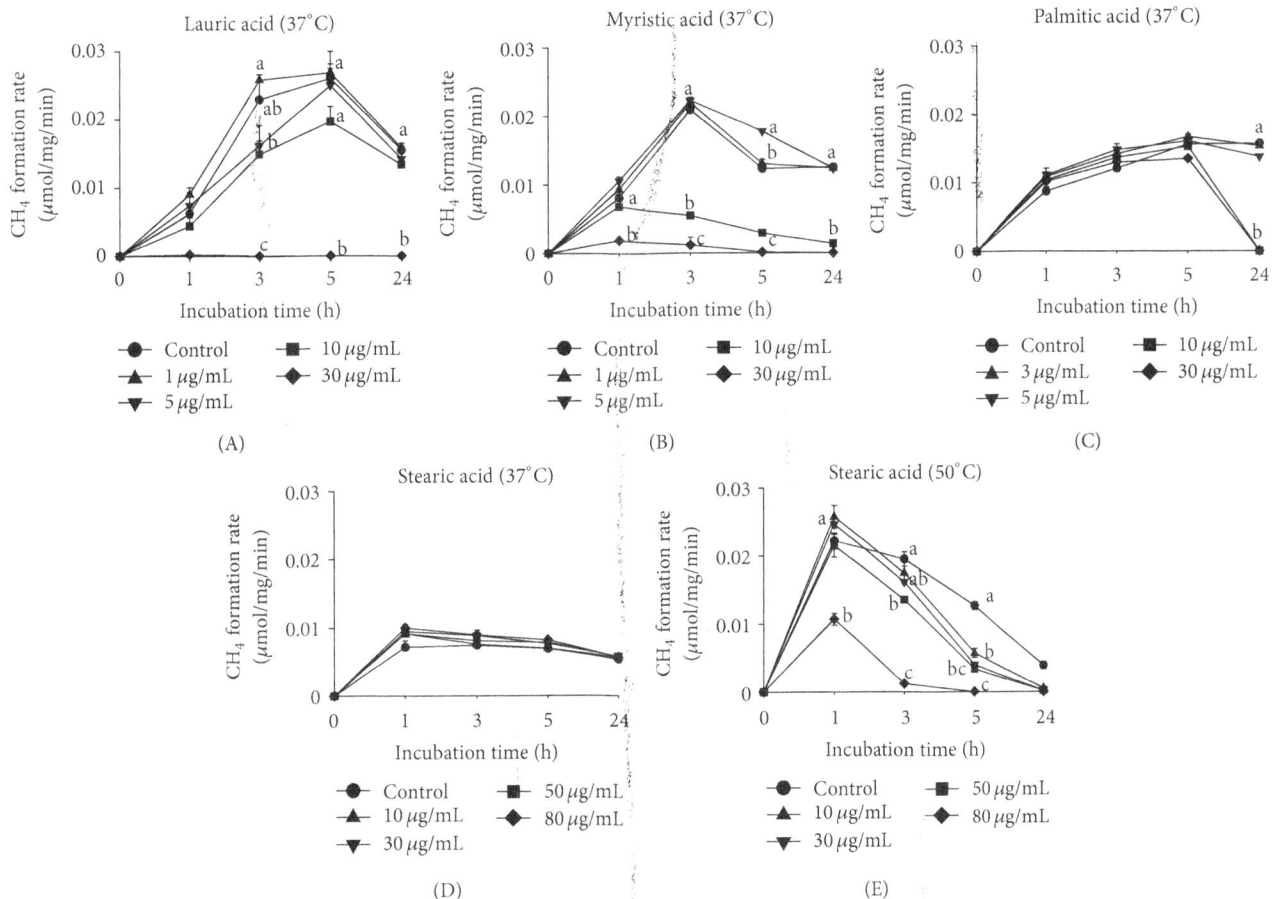

FIGURE 1: Methane production rate (μmol/mg cell DM/min) in cell suspensions of *M. ruminantium* in K^+-containing buffer ($n = 3$) in response to supplementation of different concentrations of lauric acid (A), myristic acid (B), palmitic acid (C), and stearic acid (D) at 37°C and of stearic acid at 50°C (E) (Experiment 1). Means within time point with unequal letters (a, b) are different at $P < 0.05$. Bars represent standard errors.

TABLE 1: Methane inhibition rate, K^+ efflux, and cell viability in cell suspensions treated with C_{12} in different concentrations in Experiment 2 ($n = 3$; means ± standard error).

Time		3 h			24 h	
Treatment	K^+ (mg/L)	CH_4 inhibition (%)[1]	Cell viability (%)[2]	K^+ (mg/L)	CH_4 inhibition (%)[1]	Cell viability (%)[2]
Blank	12.8 ± 1.7[b]	20.7 ± 8.1[b]	75 ± 3[a]	16.0 ± 0.2[b]	13.5 ± 16.3[b]	56 ± 2[ab]
Control	11.1 ± 0.2[b]	—[b]	79 ± 2[a]	16.3 ± 0.4[b]	—[b]	61 ± 5[a]
10 μg/mL	12.8 ± 0.3[b]	89.6 ± 5.1[a]	24 ± 5[b]	18.5 ± 0.1[a]	95.1 ± 2.5[a]	53 ± 7[ab]
15 μg/mL	18.4 ± 0.2[a]	99.8 ± 0.1[a]	27 ± 4[b]	19.3 ± 0.2[a]	99.8 ± 0.1[a]	35 ± 4[bc]
30 μg/mL	19.0 ± 0.3[a]	100.0 ± 0.1[a]	29 ± 6[b]	19.1 ± 0.4[a]	100 ± 0.0[a]	13 ± 3[c]
P values	0.0003	<0.0001	<0.0001	<0.0001	0.0018	0.0004

[a–c]Treatment means with unequal superscripts are different at $P < 0.05$.
[1]Calculated from methane production rate (μmol/mg cell DM/min) in percent of the value of the control group after 3 and 24 h, respectively.
[2]Percentage of live cells (green) of total cells (green, yellow, and red) as determined with the LIVE/DEAD BacLight Kit.

of SFAs on the activity of cells and microorganisms [15]. However, studies on the effects of SFAs on pure cultures of ruminal methanogens are limited [12]. Although, finally, potential inhibitors of ruminal methanogenesis have to be evaluated with the mixed microbial community and in the presence of feeds, elucidating the SFA effects on individual methanogen species in the absence of further influencing factors is very important to differentiate direct and indirect SFA effects on methanogens and to identify the mechanisms which lead to the inhibition of methanogenesis by SFAs.

4.1. Efficiency of Saturated Fatty Acids to Inhibit Methanogenesis in Methanobrevibacter ruminantium. In the present

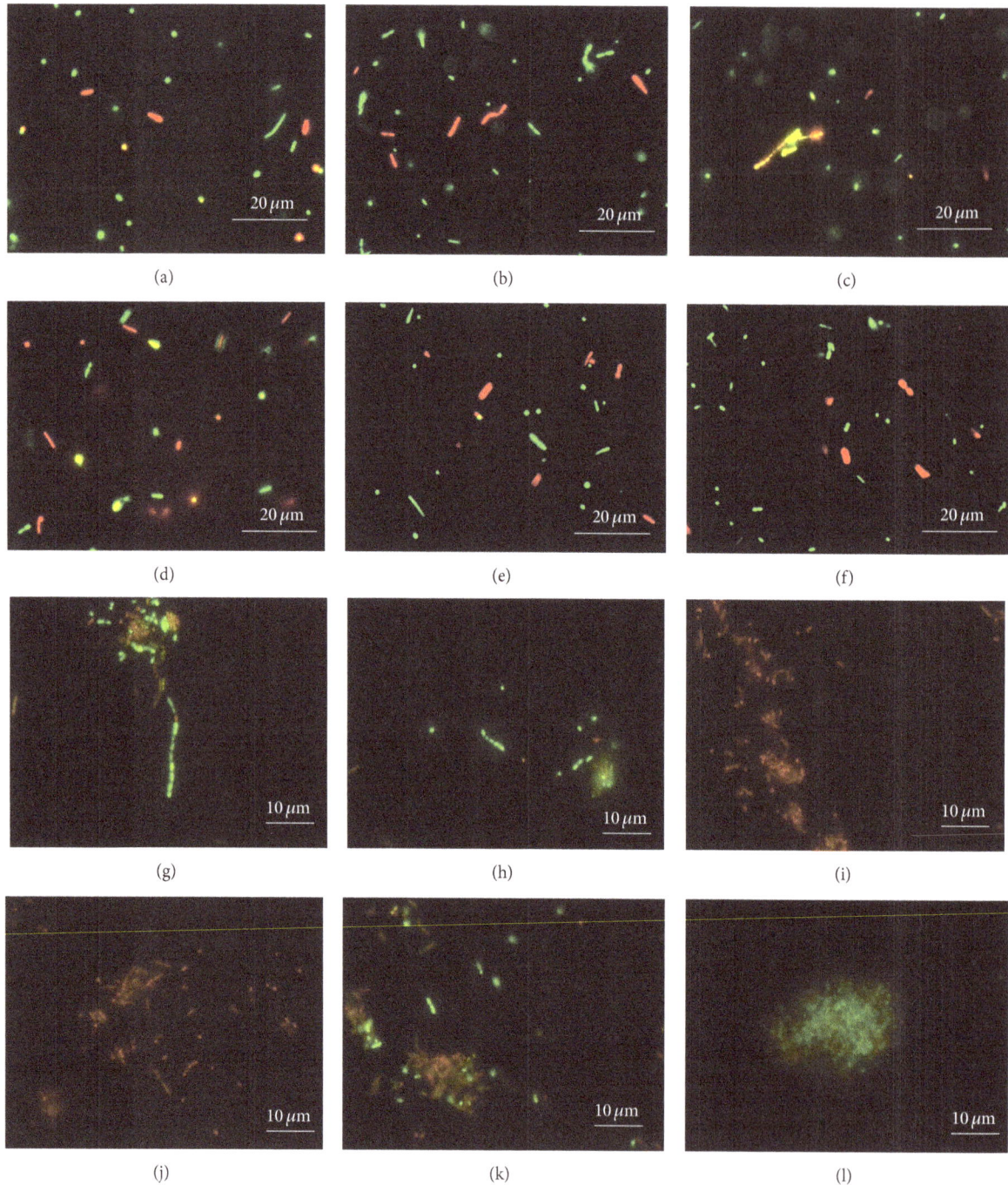

FIGURE 2: Fluorescence images illustrating cell viability of *M. ruminantium* cell suspensions exposed to different saturated fatty acids provided in a concentration of 10 μg/mL in K^+-free buffer and stained with the LIVE/DEAD BacLight Kit (Experiment 3). Green and red cells represent living and dead cells, respectively. Yellow cells were not categorized as living but included in total cell counts. (a–f) Images taken 3 h after SFA supplementation; (g–l) Images taken after 24 h. The images selected are representative for blank (a, g), control (b, h), C_{12} (c, i), C_{14} (d, j), C_{16} (e, k), and C_{18} (f, l).

study, at first the effect of SFAs on CH_4 production by cell suspensions of a major ruminal methanogen, *M. ruminantium,* was examined. The inhibition of methanogenesis in washed cell suspensions of *M. ruminantium* was getting more pronounced with decreasing chain length (C_{12} = C_{14} > C_{16} > C_{18}) and increasing SFA concentration (1 to 80 μg/mL

suspension) or SFA/cell DM ratio (0.2 to 13 μg/mg cell DM). Although cell inoculum each time was always applied by transferring the same volume using the microbes at almost the same growth phase and the cell suspensions were prepared by following the same protocol in each incubation, it seems that cell susceptibility varied between incubations,

TABLE 2: Methane inhibition rate, K^+ efflux, and cell viability in cell suspensions treated with different saturated fatty acids at $10\,\mu g/mL$ in Experiment 3 ($n = 3$; means ± standard error).

Time		3 h			24 h	
Treatment	K^+ (mg/L)	CH_4 inhibition rate (%)[1]	Cell viability (%)[2]	K^+ (mg/L)	CH_4 inhibition rate (%)[1]	Cell viability (%)[2]
Blank	5.1 ± 0.1^d	-6.8 ± 3.5^b	79 ± 7^a	14.2 ± 0.3^b	-2.5 ± 11.1^b	50 ± 8^a
Control	5.4 ± 0.4^d	$-^b$	81 ± 2^a	12.9 ± 0.3^b	$-^b$	40 ± 6^a
C_{12}	13.8 ± 0.2^a	99.9 ± 0.0^a	43 ± 2^b	15.7 ± 0.2^a	100 ± 0.1^a	1 ± 0^b
C_{14}	13.9 ± 0.2^a	99.8 ± 0.2^a	36 ± 5^b	15.7 ± 0.3^a	100 ± 0.0^a	3 ± 1^b
C_{16}	8.7 ± 0.2^b	85.4 ± 1.8^a	68 ± 6^a	13.5 ± 0.1^b	100 ± 0.1^a	12 ± 5^b
C_{18}	7.0 ± 0.1^c	7.9 ± 17.6^b	78 ± 5^a	13.9 ± 0.2^b	44 ± 16.2^{ab}	38 ± 5^a
P values	<0.0001	<0.0001	<0.0001	<0.0001	0.0015	<0.0001

$^{a-d}$Treatment means with unequal superscripts are different at $P < 0.05$.
[1]Calculated from methane production rate ($\mu mol/mg$ cell DM/min) in percent of the value of the control group after 3 and 24 h, respectively.
[2]Percentage of live cells (green) of total cells (green, yellow, and red) as determined with the LIVE/DEAD BacLight Kit.

which also caused variability in the CH_4 production patterns of the control groups (Figure 1 and Supplementary Figure 1). In agreement with studies performed at 35–38°C and neutral pH in cultures of ruminal and nonruminal methanogens and bacteria [12, 13, 32] and in sheep *in vivo* [33], the present data also indicate that C_{12} and C_{14} are the most effective SFAs. C_{12} had also been the most inhibitory representative of the SFAs against 12 Gram-positive microorganisms [32]. In the present study, the hydrophobic SFAs were dissolved in DMSO to guarantee distribution of SFA in the hydrophilic *M. ruminantium* cell suspension. Nevertheless, despite using DMSO, the SFA solubility was visually observed to decrease as SFA chain length increased. Solubility was especially weak when using C_{18} at 37°C where also no CH_4 inhibition occurred. C_{18} was only inhibitory at 50°C, which corresponds with its increased solubility at this temperature. This supports the hypothesis that SFAs need to be at least partly dissolved in the buffer or medium to be able to exert an effect [13]. Further experiments have to investigate if the SFAs state (protonated versus dissociated) plays a role in *M. ruminantium*. Lowering the pH of the incubation medium has been shown to increase adsorption of SFAs onto bacteria and also their sensitivity against SFAs [20, 34]. The SFA concentrations needed to achieve a 50% reduction in CH_4 formation rates were much lower in the present study than those required in the study of Henderson [12], where 0.5 g/L of C_{12} and C_{14} were necessary to reduce the growth rate of *M. ruminantium* by 50% compared to the control. This might have resulted either from the difference in metabolic state between cell cultures and cell suspensions or from differences in growth states before SFA supplementation and harvesting or both. Still, the SFA concentrations where a significant inhibition of methanogenesis occurred in the present study (10 to 80 $\mu g/mL$) were in the same order of magnitude than those reported earlier (30 to 1000 $\mu g/mL$) in growing methanogen cultures [12, 13, 20, 32]. This indicates that in cell cultures and cell suspensions generally the same type of effect occurs. Presumably, no cell growth occurred in the washed cell suspensions used due to the absence of nutrients needed for growth of *M. ruminantium*, like acetate and coenzyme M [35], and, in case of K-containing buffer, also nitrogen. Therefore, only CH_4 production, that is, energy metabolism, was performed which

indicates that the SFAs directly affect the process of CH_4 formation. Each dose-response test had been repeated at least once and in both incubations in three replicates each to allow robust conclusions. Although the extent of the inhibition of methanogenesis by the different SFA concentrations was not exactly the same in the two incubations, the ranking and inhibition extent of the treatments with regard to the level of effect were coinciding. Slight variations in CH_4 formation rates and peak times as well as in the SFA effects might be due to slight differences in growth phase between incubations when the cells being in their mid-exponential phase were harvested.

4.2. Indications for Modes of Action of Saturated Fatty Acids.
In the present study, the findings on K^+ leakage, an indicator of a damaged membrane [21], indicate that the cell membrane permeability increases after SFA exposure. The integrity of the archaeal membrane is fundamental to maintain the chemiosmotic balance, which is essential for the membrane-associated energetic metabolism of cells [5, 26]. The K^+ leakage also occurred concomitantly to the inhibition of methanogenesis which seems to have been followed by increasing occurrence of cell death. The K^+ efflux in *M. ruminantium* responded to different SFAs and to different C_{12} concentrations similarly as the CH_4 production rate. Accordingly, C_{12} and C_{14} triggered the largest K^+ efflux and had the strongest inhibitory effects of all SFAs tested, and increasing C_{12} concentrations increasingly inhibited methanogenesis and promoted K^+ efflux compared to the lower dosages. The LIVE/DEAD BacLight bacterial viability kit has been already shown to be a useful tool to indicate cell viability in Archaea [28, 29]. Although the CH_4 production rate declined to zero, corroborated by heavy K^+ leakage, in treatment groups supplemented with C_{12} and C_{14} at 3 h, and the percentage of cells with damaged membrane was significantly different to all other groups, it was not zero. It seems that cell death does not occur immediately but is delayed in time because after 24 h, and the cells in these two groups were nearly all dead.

4.3. Conclusion.
The inhibitory effect of SFAs on the production of the important greenhouse gas methane by

M. ruminantium was demonstrated to be dependent on SFA concentration, SFA type, and incubation temperature (37°C versus 50°C). The present study showed for the first time with a ruminal methanogen, *M. ruminantium*, that supplementation of SFAs can also damage the cell membrane and trigger K^+ efflux. The identification of the detailed mechanism on how SFAs are detrimental to the methanogens needs further studies.

Conflict of Interests

The authors declare that there is no conflict of interests.

Acknowledgments

The authors are very grateful to R. Thauer for his advice and helpful discussions and to B. Studer for potassium analysis. This study was supported by the China Scholarship Council.

References

[1] D. J. Wuebbles and K. Hayhoe, "Atmospheric methane and global change," *Earth-Science Reviews*, vol. 57, no. 3-4, pp. 177–210, 2002.

[2] E. A. Scheehle and D. Kruger, "Global anthropogenic methane and nitrous oxide emissions," *The Energy Journal*, vol. 27, pp. 33–44, 2006.

[3] P. H. Janssen and M. Kirs, "Structure of the archaeal community of the rumen," *Applied Microbiology and Biotechnology*, vol. 74, no. 12, pp. 3619–3625, 2008.

[4] T. A. McAllister and C. J. Newbold, "Redirecting rumen fermentation to reduce methanogenesis," *Australian Journal of Experimental Agriculture*, vol. 48, no. 2, pp. 7–13, 2008.

[5] R. K. Thauer, A. K. Kaster, H. Seedorf, W. Buckel, and R. Hedderich, "Methanogenic archaea: ecologically relevant differences in energy conservation," *Nature Reviews Microbiology*, vol. 6, no. 8, pp. 579–591, 2008.

[6] K. A. Johnson and D. E. Johnson, "Methane emissions from cattle," *Journal of Animal Science*, vol. 73, no. 8, pp. 2483–2492, 1995.

[7] F. Dohme, A. Machmüller, A. Wasserfallen, and M. Kreuzer, "Ruminal methanogenesis as influenced by individual fatty acids supplemented to complete ruminant diets," *Letters in Applied Microbiology*, vol. 32, no. 1, pp. 47–51, 2001.

[8] A. Machmüller and M. Kreuzer, "Methane suppression by coconut oil and associated effects on nutrient and energy balance in sheep," *Canadian Journal of Animal Science*, vol. 79, no. 1, pp. 65–72, 1999.

[9] C. R. Soliva, L. Meile, I. K. Hindrichsen, M. Kreuzer, and A. Machmüller, "Myristic acid supports the immediate inhibitory effect of lauric acid on ruminal methanogens and methane release," *Anaerobe*, vol. 10, no. 5, pp. 269–276, 2004.

[10] C. M. Zhang, Y. Q. Guo, Z. P. Yuan et al., "Effect of octadeca carbon fatty acids on microbial fermentation, methanogenesis and microbial flora *in vitro*," *Animal Feed Science and Technology*, vol. 146, no. 3-4, pp. 259–269, 2008.

[11] R. A. Prins, C. J. Van Nevel, and D. I. Demeyer, "Pure culture studies of inhibitors for methanogenic bacteria," *Antonie van Leeuwenhoek*, vol. 38, no. 1, pp. 281–287, 1972.

[12] C. Henderson, "The effects of fatty acids on pure cultures of rumen bacteria," *Journal of Agricultural Science*, vol. 81, no. 1, pp. 107–112, 1973.

[13] J. O. Zeitz, S. Bucher, X. Zhou, L. Meile, M. Kreuzer, and C. R. Soliva, "Inhibitory effects of saturated fatty acids on methane production by methanogenic Archaea," *Journal of Animal and Feed Science*, vol. 22, no. 1, pp. 44–49, 2013.

[14] F. Dohme, A. Machmüller, A. Wasserfallen, and M. Kreuzer, "Comparative efficiency of various fats rich in medium-chain fatty acids to suppress ruminal methanogenesis as measured with RUSITEC," *Canadian Journal of Animal Science*, vol. 80, no. 3, pp. 473–482, 2000.

[15] A. P. Desbois and V. J. Smith, "Antibacterial free fatty acids: activities, mechanisms of action and biotechnological potential," *Applied Microbiology and Biotechnology*, vol. 85, no. 6, pp. 1629–1642, 2010.

[16] J. B. Parsons, J. Yao, M. W. Frank, P. Jackson, and C. O. Rock, "Membrane disruption by antimicrobial fatty acids releases low-molecular-weight proteins from *Staphylococcus aureus*," *Journal of Bacteriology*, vol. 194, no. 19, pp. 5294–5304, 2012.

[17] K. Benkendorff, A. R. Davis, C. N. Rogers, and J. B. Bremner, "Free fatty acids and sterols in the benthic spawn of aquatic molluscs, and their associated antimicrobial properties," *Journal of Experimental Marine Biology and Ecology*, vol. 316, no. 1, pp. 29–44, 2005.

[18] F. C. Küpper, E. Gaquerel, E. M. Boneberg, S. Morath, J. P. Salaün, and P. Potin, "Early events in the perception of lipopolysaccharides in the brown alga *Laminaria digitata* include an oxidative burst and activation of fatty acid oxidation cascades," *Journal of Experimental Botany*, vol. 57, no. 9, pp. 1991–1999, 2006.

[19] F. Kamp, J. A. Hamilton, and H. V. Westerhoff, "Movement of fatty acids, fatty acid analogues, and bile acids across phospholipid bilayers," *Biochemistry*, vol. 32, no. 41, pp. 11074–11086, 1993.

[20] H. Galbraith and T. B. Miller, "Effect of long chain fatty acids on bacterial respiration and amino acid uptake," *Journal of Applied Bacteriology*, vol. 36, no. 4, pp. 659–675, 1973.

[21] P. Boyaval, C. Corre, C. Dupuis, and E. Roussel, "Effects of free fatty acids on propionic acid bacteria," *Lait*, vol. 75, no. 1, pp. 17–29, 1995.

[22] L. L. Wang and E. A. Johnson, "Inhibition of *Listeria monocytogenes* by fatty acids and monoglycerides," *Applied and Environmental Microbiology*, vol. 58, no. 2, pp. 624–629, 1992.

[23] P. Tangwatcharin and P. Khopaibool, "Activity of virgin coconut oil, lauric acid or monolaurin in combination with lactic acid against *Staphylococcus aureus*," *The Southeast Asian Journal of Tropical Medicine and Public Health*, vol. 43, no. 4, pp. 969–985, 2012.

[24] G. Bergsson, J. Arnfinnsson, O. Steingrímsson, and H. Thormar, "Killing of Gram-positive cocci by fatty acids and monoglycerides," *APMIS*, vol. 109, no. 10, pp. 670–678, 2001.

[25] C. L. Fischer, D. R. Drake, D. V. Dawson, D. R. Blanchette, K. A. Brogden, and P. W. Wertz, "Antibacterial activity of sphingoid bases and fatty acids against Gram-positive and Gram-negative bacteria," *Antimicrobial Agents and Chemotherapy*, vol. 56, no. 3, pp. 1157–1161, 2012.

[26] S. V. Albers and B. H. Meyer, "The archaeal cell envelope," *Nature Review Microbiology*, vol. 9, no. 6, pp. 414–426, 2011.

[27] G. D. Sprott and K. F. Jarrell, "K^+, Na^+, and Mg^{2+} content and permeability of *Methanospirillum hungatei* and *Methanobacterium thermoautotrophicum*," *Canadian Journal of Microbiology*, vol. 27, no. 4, pp. 444–451, 1981.

[28] S. Leuko, A. Legat, S. Fendrihan, and H. Stan-Lotter, "Evaluation of the LIVE/DEAD BacLight kit for detection of

extremophilic archaea and visualization of microorganisms in environmental hypersaline samples," *Applied and Environmental Microbiology*, vol. 70, no. 11, pp. 6884–6886, 2004.

[29] C. Bang, A. Schilhabel, K. Weidenbach et al., "Effects of antimicrobial peptides on methanogenic Archaea," *Antimicrobial Agents and Chemotherapy*, vol. 56, no. 8, pp. 4123–4130, 2012.

[30] H. J. Perski, J. Moll, and R. K. Thauer, "Sodium dependence of growth and methane formation in *Methanobacterium thermoautotrophicum*," *Archives of Microbiology*, vol. 130, no. 4, pp. 319–321, 1981.

[31] G. A. Jones and M. D. Pickard, "Effect of titanium (III) citrate as reducing agent on growth of rumen bacteria," *Applied and Environmental Microbiology*, vol. 39, no. 6, pp. 1144–1147, 1980.

[32] J. J. Kabara, D. M. Swieczkowski, A. J. Conley, and J. P. Truant, "Fatty acids and derivatives as antimicrobial agents," *Antimicrobial Agents and Chemotherapy*, vol. 2, no. 1, pp. 23–28, 1972.

[33] K. L. Blaxter and J. Czerkawski, "Modifications of the methane production of the sheep by supplementation of its diet," *Journal of the Science of Food and Agriculture*, vol. 17, no. 9, pp. 529–540, 1966.

[34] J. Yang, X. Hou, P. S. Mir, and T. A. McAllister, "Anti-Escherichia coli O157:H7 activity of free fatty acids under varying pH," *Canadian Journal of Microbiology*, vol. 56, no. 3, pp. 263–267, 2010.

[35] C. D. Taylor, B. C. McBride, R. S. Wolfe, and M. P. Bryant, "Coenzyme M, essential for growth of a rumen strain of *Methanobacterium ruminantium*," *Journal of Bacteriology*, vol. 120, no. 2, pp. 974–975, 1974.

4

Microbial Diversity and Biochemical Potential Encoded by Thermal Spring Metagenomes Derived from the Kamchatka Peninsula

Bernd Wemheuer,[1] Robert Taube,[1] Pinar Akyol,[1] Franziska Wemheuer,[2] and Rolf Daniel[1]

[1] *Department of Genomic and Applied Microbiology and Goettingen Genomics Laboratory, Institute of Microbiology and Genetics, Georg-August-University Goettingen, Grisebachstraße 8, 37077 Goettingen, Germany*
[2] *Section of Agricultural Entomology, Department for Crop Sciences, Georg-August-University Goettingen, Grisebachstraße 6, 37077 Goettingen, Germany*

Correspondence should be addressed to Rolf Daniel; rdaniel@gwdg.de

Academic Editor: Michael Hoppert

Volcanic regions contain a variety of environments suitable for extremophiles. This study was focused on assessing and exploiting the prokaryotic diversity of two microbial communities derived from different Kamchatkian thermal springs by metagenomic approaches. Samples were taken from a thermoacidophilic spring near the Mutnovsky Volcano and from a thermophilic spring in the Uzon Caldera. Environmental DNA for metagenomic analysis was isolated from collected sediment samples by direct cell lysis. The prokaryotic community composition was examined by analysis of archaeal and bacterial 16S rRNA genes. A total number of 1235 16S rRNA gene sequences were obtained and used for taxonomic classification. Most abundant in the samples were members of *Thaumarchaeota*, *Thermotogae*, and *Proteobacteria*. The Mutnovsky hot spring was dominated by the Terrestrial Hot Spring Group, *Kosmotoga*, and *Acidithiobacillus*. The Uzon Caldera was dominated by uncultured members of the Miscellaneous Crenarchaeotic Group and *Enterobacteriaceae*. The remaining 16S rRNA gene sequences belonged to the *Aquificae*, *Dictyoglomi*, *Euryarchaeota*, *Korarchaeota*, *Thermodesulfobacteria*, *Firmicutes*, and some potential new phyla. In addition, the recovered DNA was used for generation of metagenomic libraries, which were subsequently mined for genes encoding lipolytic and proteolytic enzymes. Three novel genes conferring lipolytic and one gene conferring proteolytic activity were identified.

1. Introduction

Sites of volcanic activity can be found all over the world and even under the sea. Volcanic regions provide a variety of different environments for extremophilic archaeal and bacterial microorganisms. Well-known examples of such extreme environments are terrestrial surface hot springs. With respect to geographical, physical, environmental, and chemical characteristics, hot springs are unique sites for extremophilic microorganisms [1–3]. Extremophiles inhabiting hot springs are considered to be the closest living descendants of the earliest life forms on Earth [4, 5]. Therefore, these springs provide insights into the origin and evolution of life. In addition, thermophiles and hyperthermophiles produce a variety of hydrolytic enzymes such as lipases, glycosidases, peptidases and other biomolecules, which are of industrial interest [6–8]. For example, Hotta et al. [9] found an extremely stable carboxylesterase in the hyperthermophilic archaeon *Pyrobaculum calidifontis* VA1, and Arpigny et al. [10] identified a novel heat-stable lipolytic enzyme in *Sulfolobus acidocaldarius* DSM 639.

Especially in extreme environments, most microorganisms are reluctant to cultivation-based approaches [11, 12]. Therefore, culture-independent metagenomic strategies are promising approaches to assess the phylogenetic composition and functional potential of microbial communities living in extreme environments [7, 13, 14]. For example, Simon et al. studied the prokaryotic community in glacier ice and found a

Microbial Diversity and Biochemical Potential Encoded by Thermal Spring Metagenomes Derived from the Kamchatka Peninsula

29

highly diverse bacterial community [15]. In 1998, Hugenholtz et al. [1] investigated the bacterial diversity in the Obsidian Pool in Yellowstone National Park and identified several new bacterial candidate divisions. The same pool and two others were studied later by Meyer-Dombard et al. [16]. They encountered diverse bacterial and archaeal communities in all three hot springs.

In the present study, we investigated the phylogenetic composition and metabolic potential of two microbial communities derived from two extreme sites of the Kamchatka peninsula, which is located in the Far East of Russia. The Kamchatka peninsula comprises an area of approximately $472,300 \, \text{km}^2$ and is described as the *land of fire* by its first explorers due to the high density of volcanoes and associated volcanic phenomena. For example, the largest active volcano of the northern hemisphere, the Klyuchevskaya Sopka, is located on the Kamchatka Peninsula. Sediment samples analyzed in this study were taken from two hot springs providing a thermoacidophilic (70°C, pH 3.5–4) or a thermophilic (81°C, pH 7.2–7.4) environment. The composition of the prokaryotic communities of the two Kamchatkian hot springs was assessed by 16S rRNA gene analysis. In addition, metagenomic libraries were generated and screened for novel biocatalysts.

2. Materials and Methods

2.1. Sampling and DNA Extraction. Two sediment samples were taken from the hot springs located on the Kamchatka peninsula in summer 2001. The first sample was collected from a thermoacidophilic spring (70°C, pH 3.5–4) at the Mutnovsky volcano (52.453 N, 158.195 E). The second sample was taken from a thermophilic spring (81°C, pH 7.2–7.4) in the Uzon Caldera (54.5 N, 159.967 E). The chemical analysis of both sediment samples is shown in Table 1.

DNA was extracted as described by Zhou et al., 1996 [17]. The concentration of the recovered DNA was quantified using a NanoDrop ND-1000 spectrophotometer (PEQLAB, Erlangen, Germany).

2.2. Amplification of 16S rRNA Genes and Generation Clone Libraries. To assess the prokaryotic community structure, archaeal and bacterial 16S rRNA genes were amplified by PCR and analyzed. The PCR reaction mixture (50 μL) contained 2.5 μL of 10-fold Mg-free *Taq* polymerase buffer, 200 μM of each of the four deoxynucleoside triphosphates, 1.75 mM $MgCl_2$, 0.4 μM of each primer, 1 U of *Taq* DNA polymerase (Fermentas, St. Leon-Rot, Germany), and approximately 25 ng of recovered DNA as a template. Prokaryotic 16S rRNA genes were amplified with the following set of primers: 8F 5$'$-AGAGTTTGATCMTGGC-3$'$ [18] and 1114R 5$'$-GGG-TTGCGCTCGTTRC-3$'$ [19], A800F 5$'$-GTAGTCCYGGCY-GTAAAC-3$'$ [20] and A1530R 5$'$-GGAGGTGATCCAGCC-G-3$'$ [21], and Arch8F 5$'$-TCCGGTTGATCCTGCCGG-3$'$ [15] and Arch958R 5$'$-YCCGGCGTTGAMTCCAATT-3$'$ [22]. The following thermal cycling scheme was used: initial denaturation at 94°C for 2 min, 25 cycles of denaturation at 94°C for 1.5 min, annealing at 56°C (8F and 1114R), 51°C

TABLE 1: Chemical analysis of the two investigated sediment samples (DM: dry matter, DIN: in accordance with the DIN (German Institute for Standardization) norm, VDLUFA: Association of German Agricultural Analytic and Research Institutes).

Element	Method	Mutnovsky (mg/kg DM)	Uzon Caldera (mg/kg DM)
Aluminum	DIN ISO 22036	22000	13000
Arsenic	DIN ISO 22036	21	590
Barium	DIN ISO 22036	20	120
Beryllium	DIN ISO 22036	<0.20	<0.20
Boron	DIN ISO 22036	<3.0	8.7
Cadmium	DIN ISO 22036	<0.10	<0.10
Calcium	DIN ISO 22036	23000	17000
Chromium	DIN ISO 22036	580	320
Cobalt	DIN ISO 22036	45	11
Copper	DIN ISO 22036	240	22
Iron	DIN ISO 22036	97000	27000
Lead	DIN ISO 22036	5.7	<2.0
Magnesium	DIN ISO 22036	2800	4200
Manganese	DIN ISO 22036	150	440
Nickel	DIN ISO 22036	270	150
Phosphorus	DIN ISO 22036	230	130
Potassium	DIN ISO 22036	120	530
Sodium	DIN ISO 22036	110	1200
Strontium	DIN ISO 22036	24	40
Titanium	DIN ISO 22036	1100	870
Vanadium	DIN ISO 22036	82	57
Zinc	DIN ISO 22036	73	42
pH ($CaCl_2$)*	VDLUFA-Method A 5.1.1	6.4	6.8
TOC	DIN EN 13137	1.06	0.67

* pH values of the two sediments were measured in situ.

(A800F and A1530R), or 55°C (Arch8F and 958R), followed by extension at 72°C (1 min for 1 kb). The final extension was carried out at 72°C for 10 min. Negative controls were performed by using the reaction mixture without template. The obtained PCR products were purified using the Wizard SV Gel and PCR clean up system (Promega, Madison, USA) and subsequently cloned into pCR2.1-TOPO as recommended by the manufacturer (Invitrogen, Carlsbad, USA). The resulting recombinant plasmids were used to transform *Escherichia coli* TOP 10 cells. A total of 1271 insert-carrying plasmids were isolated from randomly selected *E. coli* clones. The insert sequences were determined by the Göttingen Genomics Laboratory (Göttingen, Germany).

2.3. Analysis of 16S rRNA Genes. To assess the prokaryotic community structure, the retrieved 16S rRNA gene sequences were analyzed using QIIME [23]. The obtained 16S gene sequences were edited using gap4 [24] and initially checked for the presence of chimeric sequences using Mallard [25], Bellerophon [26], and Chimera Check [27]. Remaining sequences were clustered employing the UCLUST algorithm [28] and the following QIIME scripts: pick_otus.py and pick_rep_set.py. The sequences were clustered in operational taxonomic units (OTUs) at 1, 3, and 20% genetic dissimilarity.

The phylogenetic composition of the prokaryotic communities in both samples was determined using the QIIME assign_taxonomy.py script. A BLAST alignment [29] against the most recent SILVA ARB database [30] was performed. Sequences were classified with respect to the taxonomy of their best hit in the ARB database. Finally, OTU tables were generated. Rarefaction curves, Shannon indices [31], and Chao1 indices [32] were calculated employing QIIME. In addition, the maximal number of OTUs (n_{max}) was estimated for each sample using the Michaelis-Menten-fit alpha diversity metrics included in the QIIME software package.

One sequence per OTU (1% genetic distance) was further used for the construction of phylogenetic trees. Sequences were imported into the most recent SSU Ref SILVA database of the ARB program package [33]. Multiple sequence alignments were checked manually and improved by employing the ARB editor tool. Phylogenetic trees were created by employing the maximum parsimony algorithm implemented in ARB. The robustness of obtained tree topologies was evaluated by bootstrap analysis with 100 resemblings.

2.4. Construction of Small-Insert Metagenomic Libraries. To exploit the biochemical potential, metagenomic small-insert libraries were generated. Due to the low DNA recovery, starting material for the generation of these libraries was obtained by multiple displacement amplification employing the GenomiPhi V2 DNA Amplification Kit (GE Healthcare, Munich, Germany). To improve cloning efficiency, hyperbranched structures were resolved and the DNA was inserted into pCR-XL-TOPO (Invitrogen) as described by Simon et al. [34]. In this way, two metagenomic libraries were generated.

2.5. Screening for Hydrolytic Activity and Identification of Corresponding Genes. The constructed metagenomic libraries were screened for genes conferring lipolytic or proteolytic activity using a function-driven approach. The constructed libraries were used to transform *E. coli* DH5α cells. Recombinant cells were plated on LB agar plates containing either 1% tributyrin (lipolytic activity) [35] or 2% skim milk (proteolytic activity) [36]. Plates were incubated at 37°C for up to two weeks. Hydrolytic activity is indicated by halo formation. To determine the substrate specificity of the protease-producing clones, skim milk was replaced by 0.3% (w/v) azocasein, azoalbumin, or elastin-Congo red. Insert sequences of recombinant plasmids derived from positive clones were determined by the Göttingen Genomics Laboratory. The retrieved insert sequences were edited employing gap4 [24], and putative ORFs were annotated using Artemis (version 11.0) [37].

2.6. Cloning of Genes Conferring Lipolytic or Proteolytic Activity into Expression Vectors and Purification of the Corresponding Gene Products. For enzyme production of

lipolytic proteins, identified genes were cloned into pET101-TOPO according to the Champion pET101 directional TOPO Expression Kit (Invitrogen). In this way, sequences encoding a His$_6$ tag and a V5 epitope provided by the vector were added to the 3′ end of the coding regions. Alternative start codons were replaced by ATG. *Escherichia coli* BL21 Star (DE3) (Invitrogen) was used as a host for enzyme production. The production was performed as recommended by the manufacturer. Subsequently, recombinant cells were harvested by centrifugation, washed with Tris buffer (30 mM, pH 8.0), and resuspended in 50 mmol L^{-1} of sodium phosphate buffer containing 0,3 mol L^{-1} NaCl (pH 8.0). The cells were disrupted employing a French Press (1.38×10^8 Pa). Subsequently, the extract was cleared by centrifugation at 14,000 g and 4°C for 45 min. The supernatant was used as a source for soluble proteins. Recombinant proteins were purified using the Protino Ni-TED 2000 packed columns (Macherey-Nagel, Düren, Germany) as recommended by the manufacturer. Protein preparations were dialyzed using sodium phosphate buffer (50 mM, pH 7.5) to remove residual imidazole. Protein concentration in the purified sample was determined with Roti-Quant (Carl Roth, Karlsruhe, Germany) as suggested by the manufacturer. The purity of the protein preparations was analyzed by SDS-polyacrylamide gel electrophoresis according to Laemmli [38].

For protein production of the putative protease, the corresponding gene (*pepBW1*) was cloned into pBAD/Myc-His (Invitrogen) using a modified fusion method [39]. The gene was amplified in two PCR reactions with the following two sets of primer pairs containing synthetic sites (underlined) for cloning into the vector: pair 1, 5′-CATGGTGTTCAAT-AAATATGTCTT-3′ and 5′-CTAGGGTAGACTTAACGC-3′ and pair 2, 5′-GTGTTCAATAAATATGTCTTATT-3′ and 5′-CGCTAGGGTAGACTTAACGC-3′. After mixing, denaturation, and hybridization, four different hybridization products are formed of which one contained overhangs complementary to vector digested with *Nco*I (Fermentas) and *Bsp*119I (Fermentas). The preligation mixture (10 μL) contained 1 μL O-buffer (Fermentas) and approximately 50 ng of each PCR product. To create the appropriate overhangs for cloning, the following thermal cycling scheme was used: denaturation at 95°C for 3 min, 4 cycles of reannealing at 65°C for 2 min, and reannealing at 25°C for 15 min. The ligation reaction mixture (20 μL) contained the preligation mixture, 1 μL of O-Buffer (Fermentas), 1 μL ATP (10 mM), 1 U of T4 DNA ligase, and 10 ng of pBAD/Myc-His digested with *Nco*I and *Bsp*119I. The reaction was incubated at 16°C overnight and inactivated by heating at 65°C for 10 min. The resulting recombinant plasmids were then used to transform *E. coli* top 10 cells.

2.7. Characterization of Lipolytic Activity. Determination of enzyme specificity against different triacylglycerides was determined by growing *E. coli* BL21 (DE3) harboring the recombinant plasmids on LB agar plates containing triacylglycerides with different chain length (C4 to C18). Plates were supplemented with IPTG to a final concentration of 0.1 mM to induce gene expression.

Microbial Diversity and Biochemical Potential Encoded by Thermal Spring Metagenomes Derived from the
Kamchatka Peninsula

31

For quantitative analysis, p-nitrophenyl esters with various chain lengths were used as described by Rashamuse et al. [40]. Routine esterase activity assays were performed by measuring the release of p-nitrophenol from a p-nitrophenyl (p-NP) ester at 410 nm using a Cary 100 UV-Vis spectrophotometer (Varian, Palo Alto, USA) with a Peltier temperature controller. Unless otherwise described, enzyme activity was measured at 50°C in Tris-HCl (50 mmol L^{-1}; pH 7.5) with p-NP caprylate (1 mmol L^{-1}; dissolved in 2-propanol) as a substrate. p-NP caprylate was used as a substrate in the standard assay because of its stability at high temperatures and alkaline pH values. Enzyme activity of EstBW2 was determined with p-NP butyrate as the enzyme showed no activity with the other substrates tested. Therefore, temperature and pH dependence of EstBW2 were not measured above 75°C and pH 8. All measurements were performed in triplicate.

To determine substrate specificity, enzyme activity was measured at standard assay conditions employing the following p-NP esters of various chain lengths: p-NP acetate (C2), p-NP butyrate (C4), p-NP caprylate (C8), p-NP caprate (C10), p-NP laurate (C12), and p-NP palmitate (C16).

The temperature dependence of enzyme activity was determined between 20 and 95°C under standard assay conditions. To compensate temperature effects on pH values, buffers were preheated to set-point temperature and adjusted using Tris buffer (50 mmol L^{-1}). Thermostability was measured by incubating the enzyme at different temperatures over various time periods. Enzyme activity was subsequently measured under standard assay conditions. Optimal pH values for enzyme activity were measured under standard assay conditions employing different overlapping buffer solutions (50 mM): sodium acetate buffer (pH 4 and 5), sodium phosphate buffer (pH 5, 6, and 7), Tris-HCl (pH 7, 8, and 9), CHES (pH 9 and 10), and CAPS (pH 10 and 11).

The effect of different detergents on enzyme activity was determined by under standard assay conditions in the presence of 1 mM $AgNO_3$, 1 mM $CaCl_2$, 1 mM $CoCl_2$, 1 mM cetyltrimethylammonium bromide (CTAB), 1 mM $CuCl_2$, 1 mM ethylenediaminetetraacetic acid (EDTA), 1 mM $FeCl_3$, 1 mM KCl, 1 mM $MgCl_2$, 1 mM $MnCl_2$, 1 mM NaCl, 1 mM $NiSO_4$, 1 mM Sodium Dodecyl Sulfate (SDS), 1 mM $ZnCl_2$, 0.01% (v/v) Tween 80, or 0.01% (v/v) 2-Mercaptoethanol. The serine dependence of the recovered lipolytic enzymes was validated under standard assay conditions by incubation in the presence of 1 mM phenylmethylsulfonyl fluoride (PMSF). In addition, we analyzed the significance of the determined effects on enzyme activity. As the enzyme tests were performed in triplicate, we assumed that all measured enzyme activities were normally distributed. The variance homogeneity was tested employing the F-test, and the significance of the detergent effects was subsequently tested either with the Student's t-test (homogenous variances) or the Welch's t test (heterogeneous variances). All statistical analyses were performed in R [41].

2.8. Nucleotide Sequence Accession Numbers. The 16S rRNA gene sequences have been deposited in GenBank under accession numbers HM149792–HM150618. Nucleotide

sequences of the four identified genes have been deposited in GenBank under the accession numbers HM063743 (plpBW1), HM063744 (estBW1), HM063745 (estBW2), and HM063746 (pepBW1).

3. Results

3.1. Sampling and Chemical Properties of the Investigated Sediments. Sediment samples were collected from two Kamchatkian hot springs. The springs were located near the Mutnovsky volcano (Mutnovsky sample) and in the Uzon Caldera (Uzon sample), which represent a thermoacidophilic (70°C, pH 3.5–4) and a thermophilic (81°C, pH 7.2–7.4) environment, respectively. Both investigated sediments were chemically distinct from each other (Table 1). The Mutnovsky sample contained higher Al, Ca, Co, Cu, Fe, Pb, and Zn concentrations than the Uzon sample. For As, B, Ba, K, Mn, and Na concentrations, the opposite was recorded. The concentrations of Cr, Mg, Ni, Sr, P, Ti, and V and the total organic carbon contents were almost identical in both samples.

3.2. Isolation of Metagenomic DNA and Construction of Metagenomic Libraries. To assess the prokaryotic diversity and metabolic potential by metagenomic approaches, environmental DNA was extracted from both samples. Approximately, 2.7 μg DNA per 10 g sediment was recovered from both samples. After removal of remaining salts, archaeal and bacterial 16S rRNA genes were amplified from the purified DNA. The resulting PCR products were used for the generation of 16S rRNA gene libraries. A total of 1271 clones were sequenced from these libraries. After quality filtering and removal of potential chimeric sequences, 1235 high-quality 16S rRNA gene sequences were obtained (536 for the Mutnovsky sample, 699 for the Uzon sample). The DNA from both samples was also used to construct metagenomic libraries. The Mutnovsky library comprised approximately 479,000 plasmids with an average insert size of 5.3 kb. The percentage of insert-carrying plasmids was 74%. The Uzon library consisted of approximately 117,000 plasmids with an average insert size of 4 kb. The percentage of insert-carrying plasmids was 85%. In summary, the generated small-insert metagenomic libraries harbored approximately 2.27 Gbp of cloned environmental DNA.

3.3. Archaeal Community Structures. We were able to assign 265 16S rRNA gene sequences of both samples to the domain *Archaea*. The classified sequences were affiliated to four different archaeal phyla (Figure 1). The *Thaumarchaeota* was the most abundant archaeal phylum in both samples (57% and 68% of all sequences, resp.). Most of the sequences were affiliated to Miscellaneous Crenarchaeotic Group nowadays belonging to the recently proposed *Thaumarchaeota* (37%). Another abundant thaumarchaeotic group was the Terrestrial Hot Spring Group (24.7%). The majority of the remaining sequences of the Mutnovsky sample were affiliated to uncultured members of the *Euryarchaeota* (34.7%), which were only detected in this sample. The majority of the

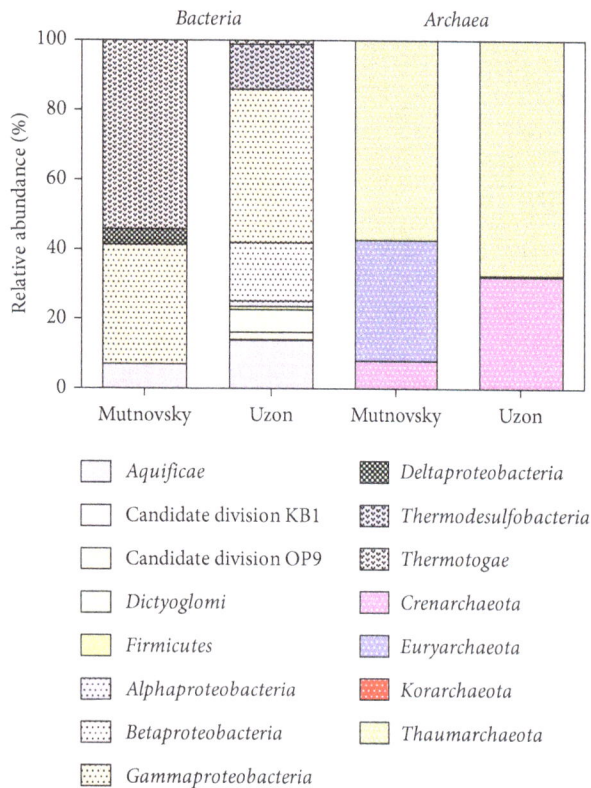

FIGURE 1: Relative sequence abundances of different archaeal and bacterial phyla and proteobacterial classes.

remaining Uzon sequences belonged to the *Crenarchaeota* (32.1%). Sequences were affiliated to known genera such as *Sulfophobococcus* (12.1%), *Thermofilum* (6.8%), *Ignisphaera* (5.3%), and to *Desulfurococcus kamchatkensis* (4.2%). The archaeal phylum *Korarchaeota* was only identified in the Uzon sample (1 sequence).

3.4. Diversity and Species Richness of Archaeal Communities.

To determine the archaeal diversity and richness, rarefaction analyses were performed with QIIME. The observed OTU numbers in the Mutnovsky sample and the Uzon sample were 33 and 13 (1% genetic distance), 25 and 11 (3% genetic distance), and 7 and 5 (20% genetic distance), respectively (Table 2). The maximal expectable number of clusters for both samples was determined based on the Michaelis-Menten_fit metrics. On average, more than 90% of the entire archaeal community was covered by the surveying effort. Shannon indices of the Mutnovsky and Uzon sample were 1.83 and 2.96 (1% genetic distance), 2.70 and 1.83 (3% genetic distance), and 0.86 and 1.87 (20% genetic distance), respectively. These indicated low archaeal diversity in the investigated samples. Comparison of the rarefaction analyses with the number of OTUs determined by Chao1 richness estimator revealed that, at 1 and 3% genetic distance, the rarefaction curves were almost saturated (Figure 4). Thus, the majority of the estimated richness was recovered by the surveying effort (Table 2).

3.5. Bacterial Community Structures.

We were able to assign 271 sequences for Mutnovsky and 434 sequences for Uzon sample to the domain *Bacteria*. The classified sequences were affiliated to three and eight different bacterial phyla and candidate divisions in the Mutnovsky sample and the Uzon sample, respectively (Figure 1). The *Thermotogae* was the most abundant bacterial phylum in the Mutnovsky sample (54%). This phylum was almost absent in the Uzon sample (1%). Interestingly, all sequences in the Mutnovsky sample were further affiliated to uncultured members of the genus *Kosmotoga*. This genus was completely absent in the Uzon Caldera. The *Proteobacteria* were the second most abundant phylum in the Mutnovsky sample (39%) and the most abundant one in the Uzon Caldera sample (62%). Most of these sequences were further assigned to *Acidithiobacillus caldus* ATCC 51756 (28%) in the Mutnovsky sample and different genera within the *Enterobacteriacaeae* (41%) in the Uzon sample. The *Aquificae* were the third most abundant phylum in the Uzon sample (13.4%). The corresponding sequences were assigned to *Sulfurihydrogenibium rodmanii* (9.7%) and *Thermosulfidibacter takaii* (2.8%) and uncultured members of the *Aquificae*. Another abundant bacterial phylum was *Thermodesulfobacteria* (12.9%). All the sequences belonged to the genus *Caldimicrobium*. The remaining sequences were affiliated to *Dictyoglomus thermophilum* and *Dictyoglomus turgidum* of the *Dictyoglomi* (6.5%), the Candiate division OP9 (2.1%), the *Firmicutes* (0.9%), and the Candidate division KB1 (0.2%).

3.6. Diversity and Species Richness of Bacterial Communities.

The observed OTU numbers in both hot springs were 17 and 50 (1% genetic distance), 12 and 42 (3% genetic distance), and 10 and 11 (20% genetic distance) in the Mutnovsky sample and the Uzon sample, respectively (Table 2). Analysis of the maximal expectable number of clusters indicated that more than 94% of the entire bacterial community was recovered by the surveying effort. Correspondingly, comparison of the rarefaction analyses with the number of OTUs determined by Chao1 richness estimator revealed that at 1%, 3%, and 20% genetic distance, the rarefaction curves were almost saturated (Table 2, Figure 4).

3.7. Screening of Metagenomic Libraries.

The two generated metagenomic libraries were employed in a function-based screening to identify novel lipolytic and proteolytic enzymes. Three novel genes encoding lipolytic enzymes (*plpBW1*, *estBW1*, and *estBW2*) and one gene encoding a proteolytic enzyme (*pepBW1*) were identified during the screening of the metagenomic library derived from the Uzon sample. No hydrolytic enzymes were identified within the metagenomic library derived from the Mutnovsky sample.

The closest relatives of all identified protein sequences originated from known thermophiles. They were similar to uncharacterized putative gene products derived from *Desulfurococcus kamchatkensis* (PepBW1), *Sulfurihydrogenibium azorense* (PlpBW1 and EstBW2), and *Thermobaculum terrenum* (EstBW1) (Table 3).

Microbial Diversity and Biochemical Potential Encoded by Thermal Spring Metagenomes Derived from the
Kamchatka Peninsula

33

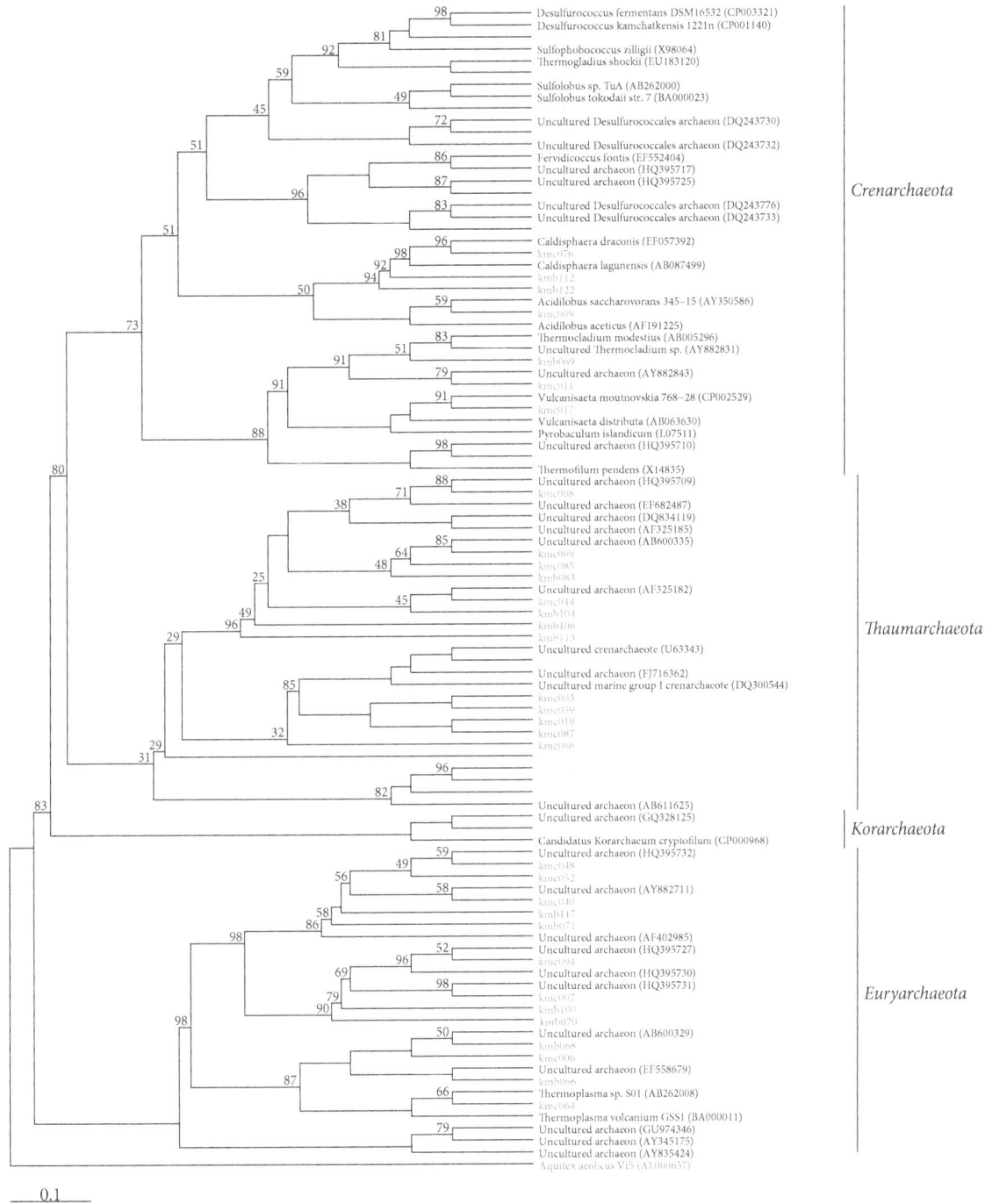

FIGURE 2: Maximum parsimony phylogenetic tree based on all archaeal 16S rRNA gene sequences. The tree was calculated using the ARB software package [33]. Sequences are characterized by sample designation (km, Kamchatka Mutnovsky; ku, Kamchatka Uzon Caldera), length of amplicon ((b), 730 bp; (c), 950 bp), number of sequence, and accession number. Sequences derived from the Mutnovsky sample are shown in red and those from the Uzon sample in blue. Numbers at branch nodes are bootstrap values (only values ≥25 are shown). The tree is rooted with the 16S rRNA gene sequence of *Aquifex aeolicus* Vf5 as an outgroup.

PlpBW1 was affiliated to the patatin-like proteins (PLPs). Four conserved domains are described for this enzyme type [42], which could all be identified within the amino acid sequence (data not shown). Interestingly, PLPs do not possess a catalytic triad. The lipolytic activity is conferred by a catalytic dyad formed by a serine residue and an aspartate residue [42]. EstBW1 and EstBW2 were affiliated to family V of lipolytic enzymes according to the classification system of Arpigny and Jaeger [43]. PepBW1 was classified employing the MEROPS database [44]. It was affiliated to the subtilisin family (family S8). Interestingly, the *pepBW1* gene sequence was almost identical to that of a putative gene encoding a

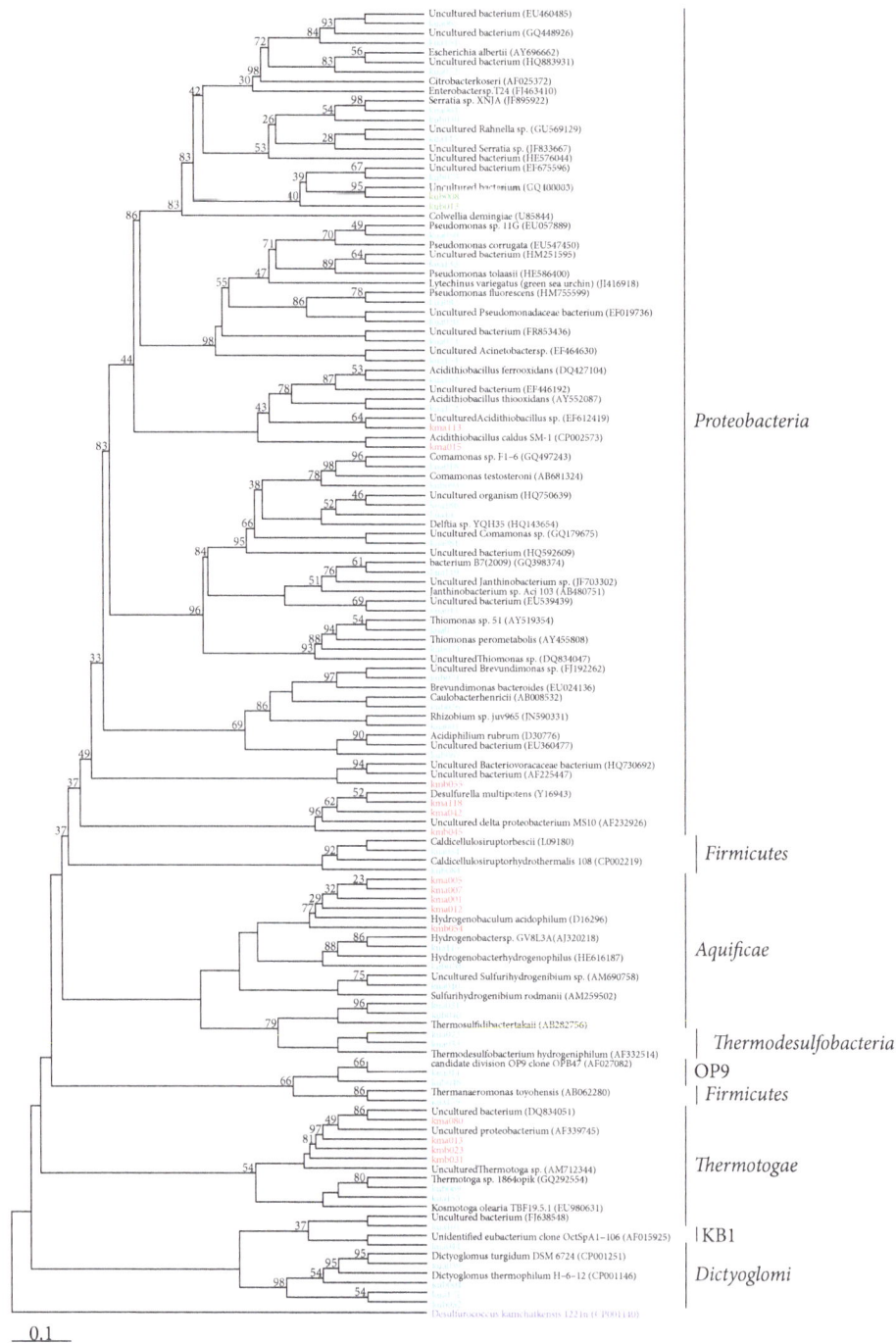

FIGURE 3: Maximum parsimony phylogenetic tree based on all bacterial 16S rRNA gene sequences. The tree was calculated using the ARB software package [33]. Sequences are characterized by sample designation (km, Kamchatka Mutnovsky; ku, Kamchatka Uzon Caldera), length of amplicon ((a), 1100 bp, (b), 730 bp), number of sequence, and accession number. Sequences derived from the Mutnovsky are shown in red and those from the Uzon Caldera in blue. OTUs shared by both sites are depicted in green. Numbers at branch nodes are bootstrap values (only values ≥25 are shown). The tree is rooted with the 16S rRNA gene sequence of *Desulfurococcus kamchatkensis* 1221n as an outgroup.

serine peptidase of *Desulfurococcus kamchatkiensis* (Table 3); *Desulfurococcus kamchatkiensis* belongs to the *Crenarchaeota* and was also isolated from a thermal spring within the Uzon Caldera [45]. In addition, the 16S rRNA gene sequence of this species was found in our 16S analysis of the Uzon sample.

3.8. Characterization of Recombinant Enzymes. To characterize all recombinant proteins, the genes conferring lipolytic and proteolytic activity were cloned into expression vectors. The recombinant *E. coli* strain containing PepBW1 was tested towards different proteins and showed proteolytic activity

Microbial Diversity and Biochemical Potential Encoded by Thermal Spring Metagenomes Derived from the Kamchatka Peninsula

35

TABLE 2: Prokaryotic diversity and richness at 1, 3, and 20% genetic distance. Numbers of observed OTUs as well as Shannon and Chao1 indices were calculated with QIIME [16]. The maximal number of OTUs (n_{max}) was calculated using the Michaelis-Menten-fit diversity metrics implemented in the QIIME package. Coverage was determined by dividing the observed number of OTUs with n_{max}.

Sample	Observed OTUs			Max. OTUs (n_{max})			Coverage (%)			Shannon index (H')			Chao1		
	1%	3%	20%	1%	3%	20%	1%	3%	20%	1%	3%	20%	1%	3%	20%
Archaea															
Mutnovsky	33	25	7	37.7	27.4	7.2	88	91.2	97.2	2.96	2.70	1.87	44.3	28.3	7
Uzon	13	11	5	14	11.6	5	93	94.8	100	1.83	1.67	0.86	13	11	5
Bacteria															
Mutnovsky	17	12	10	18.2	12.3	10	93.4	97.6	100	2	1.81	1.65	32	15	13
Uzon	50	42	11	62.9	52.5	11.3	95.8	80.0	97.3	2.9	2.60	1.63	69.3	56.4	11.5

TABLE 3: Novel lipolytic and proteolytic enzymes and their closest relatives in the NCBI database.

Protein	Length (amino acids)	Closest similar protein, accession no. of similar protein	Corresponding organism	*E*-value	Amino acid homology to the closest similar protein (% identity)
PlpBW1	250	Patatin, YP_002729059	*Sulfurihydrogenibium azorense* Az-Ful	9e − 111	1–250 (75%)
EstBW1	254	Alpha/beta hydrolase family protein, ZP_03857090	*Thermobaculum terrenum* ATCC BAA-798	4e − 61	2–251 (47%)
EstBW2	191	Hypothetical protein SULAZ_0137, YP_002728134	*Sulfurihydrogenibium azorense* Az-Ful	2e − 91	1–188 (85%)
PepBW1	411	Subtilisin-like serine protease, YP_002428837.1	*Desulfurococcus kamchatkensis*	0	1–411 (98%)

with skim milk and elastin-Congo red but not with azoalbumin or azocasein.

The activities of the recombinant lipolytic proteins were tested towards different triacylglycerides. All proteins showed activity with tributyrin as substrate. In addition, PlpBW1 showed activity with long-chain triacylglycerides, up to trimyristin (C14). Hydrolysis of different p-nitrophenyl esters was used to further analyze the substrate specificity (Figure 5(a)). PlpBW1 and EstBW1 showed highest activity with p-NP acetate and p-NP butyrate, respectively. Both enzymes exhibit activity towards all tested p-NP esters, except p-NP palmitate. The activity decreased with increasing chain length. In contrast, EstBW2 showed only activity towards p-NP butyrate. Specific activities under standard assay conditions using the optimal substrate were 2.6 ± 0.3 U/mg (PlpBW1), 2.33 ± 0.32 U/mg (EstBW1), and 1.89 ± 0.21 U/mg (EstBW2). Based on the results, all three lipolytic enzymes are most likely carboxylesterases and not lipases.

All lipolytic enzymes were active over a wide temperature range. PlpBW1, EstBW1, and EstBW2 retained a minimum of 50% activity from 60 to 90°C, 65 to 95°C, and 40 to 75°C, respectively (Figure 5(b)). Maximal activities were recorded for PlpBW1 at 85°C, for EstBW1 at 90°C, and for EstBW2 at 65°C. We further determined the stability of the three lipolytic enzymes with respect to different temperatures. The half-lives of PlpBW1 were 45 min at 70°C, 15 min at 80°C, and 5 min at 90°C. EstBW1 exhibited half-lives of 5 h at 70°C, 2.5 h at 80°C, and a remarkable half-life of 15 min at 90°C. EstBW2 was less stable at 90°C (7 min half-live), but the activity was almost unaffected by 5 h of incubation at 70°C and 80°C (data not shown).

The pH effect on enzyme activity was measured at pH values ranging from 4 to 11 (Figure 5(c)). All enzymes exhibited high activity at neutral or alkaline pH values. Maximal activities were determined at pH 10 (PlpBW1) and pH 7 (EstBW1 and EstBW2).

Addition of EDTA, KCl, or NaCl to the reaction mixture had no significant effect on enzyme activity ($P > 0.05$), whereas all other tested detergents exhibited an effect on the activity of at least one of the recovered enzymes (Table 4). CTAB, Tween 80, and $ZnCl_2$ impacted the activities of all three enzymes significantly ($P < 0.05$). PlpBW1 showed a more than 2.5-fold higher activity in presence of CTAB, whereas EstBW1 and EstBW2 displayed a loss in activity. Tween 80 increased enzyme activity of PlpBW1 and EstBW2 but not that of EstBW1. The addition of $ZnCl_2$ decreased the activity of all three recombinant enzymes. Lipolytic activity of EstBW1 and EstBW2 was completely inhibited by the phenylmethylsulfonyl fluoride, indicating the presence of a serine residue at the active site of both enzymes. Interestingly, the activity of PlpBW1 was not affected by PMSF.

4. Discussion

4.1. Prokaryotic Community Composition in the Kamchatkian Springs. The number of metagenomic studies has been rapidly increased over the past years. Metagenomics has been

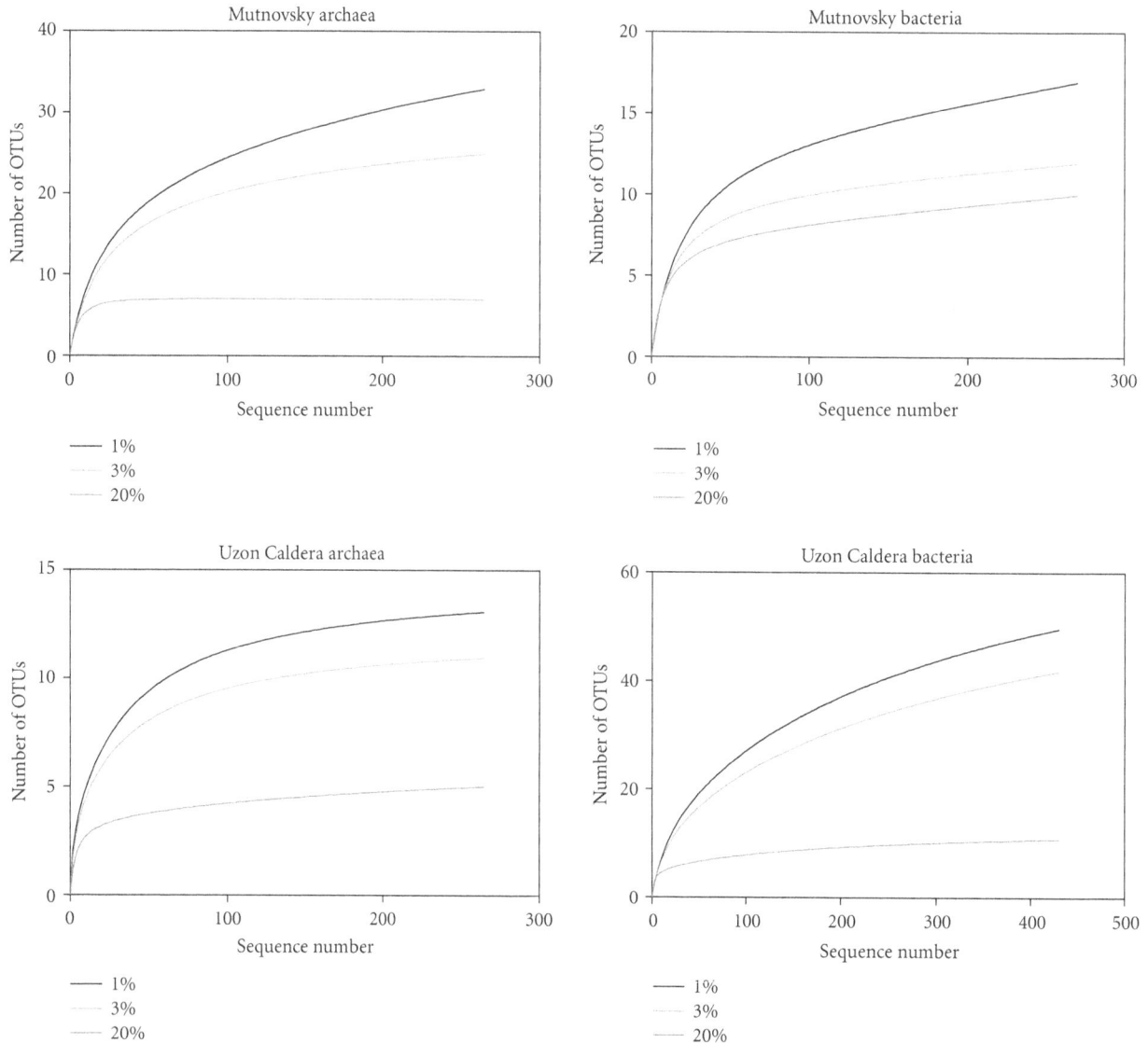

FIGURE 4: Rarefaction curves for both sampling sites. Curves were calculated at 1%, 3%, and 20% genetic distance level employing QIIME [23].

employed to assess and exploit the biodiversity of many habitats including environments of extremophiles [1, 15, 16, 46, 47]. In this study, we investigated the prokaryotic diversity of two hot springs located on the Kamchatka peninsula. We found different bacterial and archaeal communities at both sites, which were dominated by *Proteobacteria*, *Thermotogae*, and *Thaumarchaeota*.

Jackson et al. (2001) studied a mat derived from the Norris Geyser Basin, an acidic thermal spring in the Yellowstone National Park [46]. They found community pattern comparable to that in the Mutnovsky sample with one difference. They were not able to identify *Thaumarchaeota*, which is not surprising as this phylum was first proposed in 2008 [48]. Members of this phylum are not restricted to thermophilic habitats as they were originally described as mesophilic *Crenarchaeota* [48–50]. A study by Meyer-Dombard et al. (2005) investigated the prokaryotic community in three thermal

springs in the Yellowstone National Park (the Silvan Spring, the Bison Pool, and the Obsidian Pool) [16]. Whereas the other pools have a rather neutral milieu, the Silvan Spring has a low pH of 5. However, the prokaryotic community structure of this acidic spring was different to that found in the acidic Mutnovsky spring sample. Meyer-Dombard et al. identified the *Crenarchaeota* as the most abundant archaeal group, whereas *Thaumarchaeota* were the most abundant group in our study. A more recent study on prokaryotic community composition of hot springs on the Tibetan Plateau also found *Thaumarchaeota* as the dominant archaeal group [48].

Analysis of the Uzon sample revealed a more diverse prokaryotic community than in the Mutnovsky sample. Only two OTUs at 1% genetic distance were shared, whereas all the other OTUs were unique for each sample (Figures 2 and 3). The observed differences in community composition between the two sampling sites might be due to the different

(a)

(b)

(c)

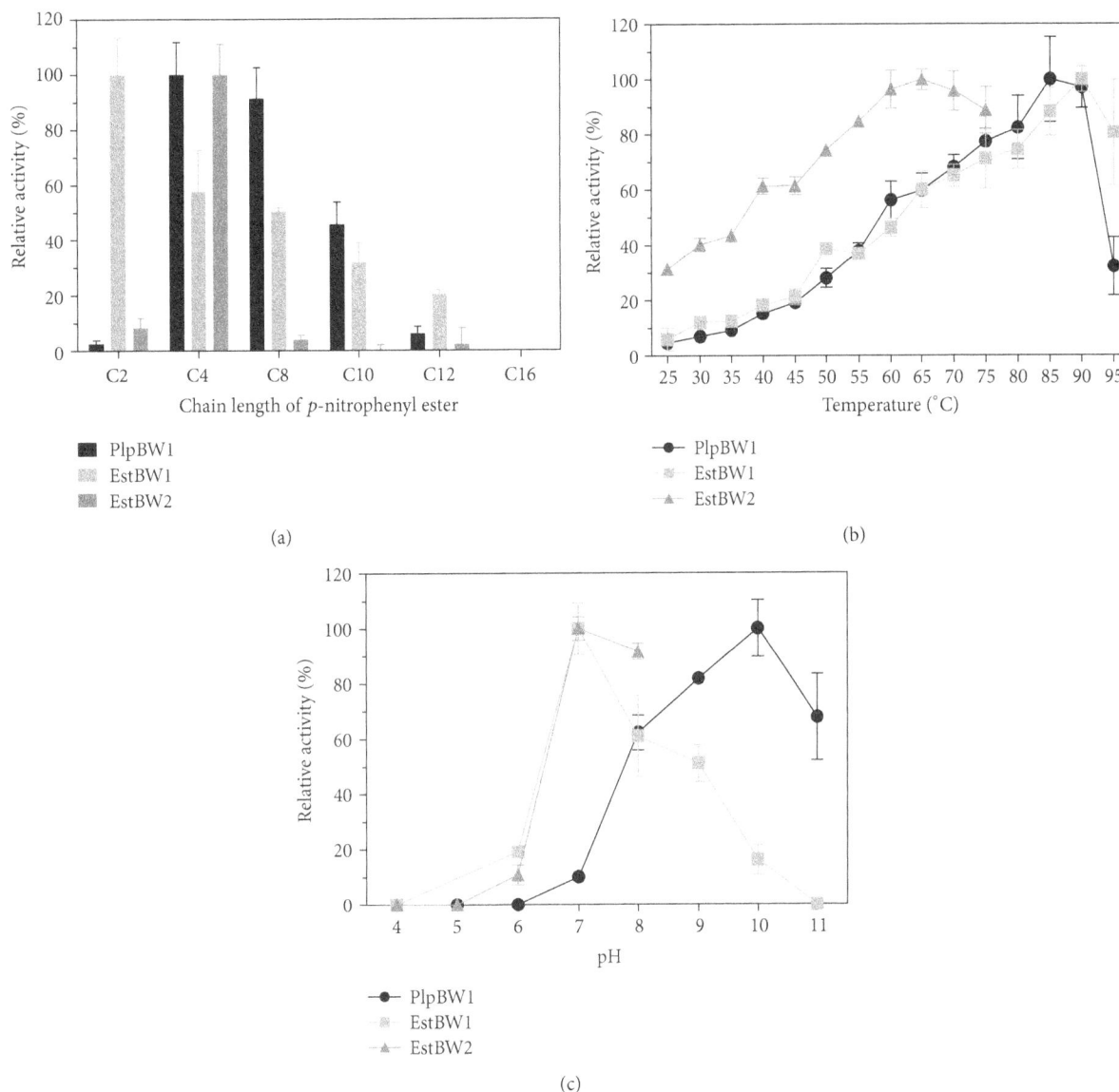

FIGURE 5: Relative activities of the three lipolytic enzymes towards different p-nitrophenyl esters with various chain lengths (a), at different temperatures (b), and pH values (c). The activity of estBW2 could not be measured above 75°C and pH 8 due to the instability of the p-NP butyrate under these conditions.

temperatures and pH values at the sites. However, Huang et al. (2011) found no statistical correlation between temperature and diversity [48].

Despite the geographical separation, the Obsidian Pool and the Uzon Caldera hot spring share a very similar community structure, as almost the same dominant archaeal and bacterial groups were identified [16]. In addition, some rare phyla were present in both samples, that is, the *Korarchaeota*. This phylum is a relatively new phylum first described by Barns et al. in 1996 [51]. Another rare bacterial group found in both samples was the Candidate division OP9. These results confirm also the presumption proposed for other extreme environments that similar environmental conditions result in similar microbial communities [15].

4.2. Hydrolytic Enzymes. In the present study, we were able to identify three novel lipolytic enzymes and one proteolytic enzyme. The determined optimal temperatures and pH values reflect the environmental conditions of the samples used for DNA isolation, indicating that the environment shapes the characteristics of the enzymes. Correspondingly, the characterized lipolytic enzymes (PlpBW1, EstBW1, and EstBW2) showed features similar to those of other metagenome-derived esterases, which were identified in thermophilic sites. Rhee et al. (2005) identified a thermophilic esterase in metagenomic libraries generated from hot spring and mud hole DNA [52]. The enzyme was active from 30 to 95°C and exhibited an optimal pH value of approximately 6.0. Tirawongsaroj et al. (2008) screened metagenomic libraries

TABLE 4: Relative activities of recombinant esterases in the presence of different chemical compounds. The effect of the additives was further tested for significance. Significant effects ($P < 0.05$) are written in bold type.

Detergent	Relative activity of PlpBW1 (%)	Relative activity of EstBW1 (%)	Relative activity of EstBW2 (%)
$AgNO_3$	106.80 ± 16.06	34.27 ± 1.66	163.17 ± 4.88
$CaCl_2$	$\mathbf{132.39 \pm 9.63}$	85.35 ± 6.30	97.10 ± 0.98
$CoCl_2$	$\mathbf{131.17 \pm 5.93}$	104.94 ± 7.29	$\mathbf{64.00 \pm 5.60}$
CTAB	$\mathbf{274.39 \pm 34.52}$	$\mathbf{40.66 \pm 6.67}$	$\mathbf{8.11 \pm 3.30}$
$CuCl_2$	125.48 ± 2.70	$\mathbf{81.03 \pm 3.98}$	94.20 ± 0.57
EDTA	119.45 ± 7.50	101.47 ± 10.75	96.03 ± 4.24
$FeCl_3$	106.67 ± 8.94	$\mathbf{53.56 \pm 12.56}$	99.47 ± 5.01
KCl	124.50 ± 19.51	99.29 ± 12.11	96.40 ± 5.55
2-Mercaptoethanol	84.82 ± 7.31	$\mathbf{0.00 \pm 0.00}$	78.74 ± 19.95
$MgCl_2$	$\mathbf{133.37 \pm 7.25}$	90.38 ± 13.06	101.54 ± 2.87
$MnCl_2$	72.07 ± 2.31	82.14 ± 14.16	$\mathbf{48.90 \pm 5.22}$
NaCl	131.24 ± 15.17	98.59 ± 20.07	92.42 ± 3.60
$NiSO_4$	121.86 ± 14.71	89.64 ± 4.60	45.65 ± 8.03
PMSF	77.49 ± 16.91	$\mathbf{0.08 \pm 16.43}$	0 ± 4.16
SDS	68.45 ± 17.38	$\mathbf{22.47 \pm 9.64}$	$\mathbf{39.49 \pm 3.14}$
Tween 80	$\mathbf{174.52 \pm 2.29}$	$\mathbf{60.70 \pm 12.74}$	$\mathbf{141.50 \pm 4.94}$
$ZnCl_2$	$\mathbf{72.01 \pm 1.77}$	59.26 ± 5.72	$\mathbf{8.64 \pm 1.71}$

Abbreviations: CTAB: cetyltrimethylammonium bromide; EDTA: ethylenediaminetetraacetic acid; SDS: sodium dodecyl sulfate; PMSF: phenylmethylsulfonyl fluoride.

derived from a Thailand hot spring and identified two novel lipolytic enzymes, of which one was also characterized as a patatin-like protein [53]. To our knowledge, PlpBW1 is the second reported patatin-like protein derived from a hot spring metagenomic library up to now [53]. In contrast to most other lipolytic enzymes containing a serine residue in the active site, PlpBW1 is not inactivated by the inhibitor PMSF [51, 54]. The effect of Zn^{2+} ions recorded for all recombinant enzymes investigated in this study was also mentioned for esterases studied by Chu et al. [54]. They also recorded a decrease of activity in presence of Zn^{2+} ions. The activity of the recovered lipolytic enzymes was positively or negatively influenced by addition of CTAB or Tween 80 (Table 4). It has been shown that these detergents can either promote or decrease activity of lipolytic enzymes by formation of micellar aggregates and monomers which then interact with hydrophobic parts of the enzymes [55].

In addition to the esterases, metagenomic libraries were mined for proteolytic activity. The identified serine peptidase, PepBW1, is the first metagenome-derived peptidase from a thermophilic environment. As PepBW1 is derived from an archaeal organism, it illustrates that screening in a heterologous host can be successful, even if the target gene originates from a different domain of life [56].

4.3. Ecology of Hot Springs.

As most studies on ecology of hot springs are targeting the prokaryotic diversity, for example, via 16S rRNA gene or other marker gene analyses, little is known on the global relevance of these extremophilic communities. Burgess et al. (2011) studied two thermal pools in the Uzon Caldera by 16S rRNA gene analysis and related some community members to different archaeal and bacterial groups, which might play a role in cycling of C, N, and S [57].

The vital role of *Archaea* in N_2 fixation and denitrification is well established [58]. The first step of nitrification, ammonium oxidation, was originally thought to be restricted to some *Proteobacteria* [59]. However, recent metagenomic studies provided evidence that *Archaea* are capable to oxidize ammonium to nitrate [58, 59]. Until recently, methanogenic *Euryarchaeota* were thought to be the only archaeal group of global relevance for element cycling. This presumption changed with the discovery of ammonia-oxidizing archaea [60], which are affiliated to the recently proposed phylum *Thaumarchaeota*. Members of this phylum contribute significantly to the global N cycle, as their high abundance and extremely low substrate threshold provides compelling evidence for a dominant role as ammonia oxidizers in open oceans [60]. In our study, we identified diverse thaumarchael groups in both investigated sediment samples. Thus, hot springs may also play a major role in the global N cycle.

Acknowledgments

The authors thank Dr. Joanna S. Potekhina from the Institute of Ecology of the Volga River Basin, (Russian Academy of Sciences, Togliatti, Russia) for providing the environmental samples. This work was supported by the Bundesministerium für Bildung und Forschung (BMBF).

References

[1] P. Hugenholtz, C. Pitulle, K. L. Hershberger, and N. R. Pace, "Novel division level bacterial diversity in a Yellowstone hot spring," *Journal of Bacteriology*, vol. 180, no. 2, pp. 366–376, 1998.

[2] V. T. Marteinsson, S. Hauksdóttir, C. F. V. Hobel, H. Kristmannsdóttir, G. O. Hreggvidsson, and J. K. Kristjánsson, "Phylogenetic diversity analysis of subterranean hot springs in Iceland," *Applied and Environmental Microbiology*, vol. 67, no. 9, pp. 4242–4248, 2001.

[3] T. Kvist, B. K. Ahring, and P. Westermann, "Archaeal diversity in Icelandic hot springs," *FEMS Microbiology Ecology*, vol. 59, no. 1, pp. 71–80, 2007.

[4] G. J. Olsen, C. R. Woese, and R. Overbeek, "The winds of (evolutionary) change: breathing new life into microbiology," *Journal of Bacteriology*, vol. 176, no. 1, pp. 1–6, 1994.

[5] C. R. Woese, O. Kandler, and M. L. Wheelis, "Towards a natural system of organisms: proposal for the domains Archaea, Bacteria, and Eucarya," *Proceedings of the National Academy of Sciences of the United States of America*, vol. 87, no. 12, pp. 4576–4579, 1990.

[6] J. Handelsman, "Metagenomics: application of genomics to uncultured microorganisms," *Microbiology and Molecular Biology Reviews*, vol. 68, no. 4, pp. 669–685, 2004.

Microbial Diversity and Biochemical Potential Encoded by Thermal Spring Metagenomes Derived from the Kamchatka Peninsula

39

[7] R. Daniel, "The soil metagenome—a rich resource for the discovery of novel natural products," *Current Opinion in Biotechnology*, vol. 15, no. 3, pp. 199–204, 2004.

[8] W. R. Streit, R. Daniel, and K. E. Jaeger, "Prospecting for biocatalysts and drugs in the genomes of non-cultured microorganisms," *Current Opinion in Biotechnology*, vol. 15, no. 4, pp. 285–290, 2004.

[9] Y. Hotta, S. Ezaki, H. Atomi, and T. Imanaka, "Extremely stable and versatile carboxylesterase from a hyperthermophilic archaeon," *Applied and Environmental Microbiology*, vol. 68, no. 8, pp. 3925–3931, 2002.

[10] J. L. Arpigny, D. Jendrossek, and K. E. Jaeger, "A novel heat-stable lipolytic enzyme from *Sulfolobus acidocaldarius* DSM 639 displaying similarity to polyhydroxyalkanoate depolymerases," *FEMS Microbiology Letters*, vol. 167, no. 1, pp. 69–73, 1998.

[11] R. I. Amann, W. Ludwig, and K. H. Schleifer, "Phylogenetic identification and in situ detection of individual microbial cells without cultivation," *Microbiological Reviews*, vol. 59, no. 1, pp. 143–169, 1995.

[12] P. Lorenz, K. Liebeton, F. Niehaus, and J. Eck, "Screening for novel enzymes for biocatalytic processes: accessing the metagenome as a resource of novel functional sequence space," *Current Opinion in Biotechnology*, vol. 13, no. 6, pp. 572–577, 2002.

[13] C. Elend, C. Schmeisser, C. Leggewie et al., "Isolation and biochemical characterization of two novel metagenome-derived esterases," *Applied and Environmental Microbiology*, vol. 72, no. 5, pp. 3637–3645, 2006.

[14] C. Elend, C. Schmeisser, H. Hoebenreich, H. L. Steele, and W. R. Streit, "Isolation and characterization of a metagenome-derived and cold-active lipase with high stereospecificity for (R)-ibuprofen esters," *Journal of Biotechnology*, vol. 130, no. 4, pp. 370–377, 2007.

[15] C. Simon, A. Wiezer, A. W. Strittmatter, and R. Daniel, "Phylogenetic diversity and metabolic potential revealed in a glacier ice metagenome," *Applied and Environmental Microbiology*, vol. 75, no. 23, pp. 7519–7526, 2009.

[16] D. R. Meyer-Dombard, E. L. Shock, and J. P. Amend, "Archaeal and bacterial communities in geochemically diverse hot springs of Yellowstone National Park, USA," *Geobiology*, vol. 3, no. 3, pp. 211–227, 2005.

[17] J. Zhou, M. A. Bruns, and J. M. Tiedje, "DNA recovery from soils of diverse composition," *Applied and Environmental Microbiology*, vol. 62, no. 2, pp. 316–322, 1996.

[18] G. Muyzer, A. Teske, C. O. Wirsen, and H. W. Jannasch, "Phylogenetic relationships of *Thiomicrospira* species and their identification in deep-sea hydrothermal vent samples by denaturing gradient gel electrophoresis of 16S rDNA fragments," *Archives of Microbiology*, vol. 164, no. 3, pp. 165–172, 1995.

[19] A. Wilmotte, G. Van Der Auwera, and R. De Wachter, "Structure of the 16 S ribosomal RNA of the thermophilic cyanobacterium *Chlorogloeopsis* HTF (*Mastigocladus laminosus HTF*) strain PCC7518, and phylogenetic analysis," *FEBS Letters*, vol. 317, no. 1-2, pp. 96–100, 1993.

[20] T. V. Kolganova, B. B. Kuznetsov, and T. P. Turova, "Designing and testing oligonucleotide primers for amplification and sequencing of archaeal 16S rRNA genes," *Mikrobiologiya*, vol. 71, no. 2, pp. 283–286, 2002.

[21] T. Itoh, K. I. Suzuki, and T. Nakase, "*Vulcanisaeta distributa* gen. nov., sp. nov., and *Vulcanisaeta souniana* sp. nov., novel hyperthermophilic, rod-shaped crenarchaeotes isolated from hot springs in Japan," *International Journal of Systematic and Evolutionary Microbiology*, vol. 52, no. 4, pp. 1097–1104, 2002.

[22] E. F. DeLong, "Archaea in coastal marine environments," *Proceedings of the National Academy of Sciences of the United States of America*, vol. 89, no. 12, pp. 5685–5689, 1992.

[23] J. G. Caporaso, J. Kuczynski, J. Stombaugh et al., "QIIME allows analysis of high-throughput community sequencing data," *Nature Methods*, vol. 7, no. 5, pp. 335–336, 2010.

[24] R. Staden, K. F. Beal, and J. K. Bonfield, "The Staden package, 1998," *Methods in Molecular Biology*, vol. 132, pp. 115–130, 2000.

[25] K. E. Ashelford, N. A. Chuzhanova, J. C. Fry, A. J. Jones, and A. J. Weightman, "New screening software shows that most recent large 16S rRNA gene clone libraries contain chimeras," *Applied and Environmental Microbiology*, vol. 72, no. 9, pp. 5734–5741, 2006.

[26] T. Huber, G. Faulkner, and P. Hugenholtz, "Bellerophon: a program to detect chimeric sequences in multiple sequence alignments," *Bioinformatics*, vol. 20, no. 14, pp. 2317–2319, 2004.

[27] J. R. Cole, B. Chai, T. L. Marsh et al., "The Ribosomal Database Project (RDP-II): previewing a new autoaligner that allows regular updates and the new prokaryotic taxonomy," *Nucleic Acids Research*, vol. 31, no. 1, pp. 442–443, 2003.

[28] R. C. Edgar, "Search and clustering orders of magnitude faster than BLAST," *Bioinformatics*, vol. 26, no. 19, Article ID btq461, pp. 2460–2461, 2010.

[29] C. Camacho, G. Coulouris, V. Avagyan et al., "BLAST+: architecture and applications," *BMC Bioinformatics*, vol. 10, article 421, 2009.

[30] E. Pruesse, C. Quast, K. Knittel et al., "SILVA: a comprehensive online resource for quality checked and aligned ribosomal RNA sequence data compatible with ARB," *Nucleic Acids Research*, vol. 35, no. 21, pp. 7188–7196, 2007.

[31] C. E. Shannon, "A mathematical theory of communication," *SIGMOBILE Mobile Computing and Communications Review*, vol. 5, no. 1, pp. 3–55, 2001.

[32] A. Chao and J. Bunge, "Estimating the number of species in a stochastic abundance model," *Biometrics*, vol. 58, no. 3, pp. 531–539, 2002.

[33] W. Ludwig, O. Strunk, R. Westram et al., "ARB: a software environment for sequence data," *Nucleic Acids Research*, vol. 32, no. 4, pp. 1363–1371, 2004.

[34] C. Simon, J. Herath, S. Rockstroh, and R. Daniel, "Rapid identification of genes encoding DNA polymerases by function-based screening of metagenomic libraries derived from glacial ice," *Applied and Environmental Microbiology*, vol. 75, no. 9, pp. 2964–2968, 2009.

[35] A. Henne, R. A. Schmitz, M. Bömeke, G. Gottschalk, and R. Daniel, "Screening of environmental DNA libraries for the presence of genes conferring lipolytic activity on *Escherichia coli*," *Applied and Environmental Microbiology*, vol. 66, no. 7, pp. 3113–3116, 2000.

[36] T. Waschkowitz, S. Rockstroh, and R. Daniel, "Isolation and characterization of metalloproteases with a novel domain structure by construction and screening of metagenomic libraries," *Applied and Environmental Microbiology*, vol. 75, no. 8, pp. 2506–2516, 2009.

[37] K. Rutherford, J. Parkhill, J. Crook et al., "Artemis: sequence visualization and annotation," *Bioinformatics*, vol. 16, no. 10, pp. 944–945, 2000.

[38] U. K. Laemmli, "Cleavage of structural proteins during the assembly of the head of bacteriophage T4," *Nature*, vol. 227, no. 5259, pp. 680–685, 1970.

[39] D. Tillett and B. A. Neilan, "Enzyme-free cloning: a rapid method to clone PCR products independent of vector restriction enzyme sites," *Nucleic Acids Research*, vol. 27, no. 19, pp. e26–e28, 1999.

[40] K. Rashamuse, T. Ronneburg, F. Hennessy et al., "Discovery of a novel carboxylesterase through functional screening of a pre-enriched environmental library," *Journal of Applied Microbiology*, vol. 106, no. 5, pp. 1532–1539, 2009.

[41] R Development Core Team, *R: a Language and Environment for Statistical Computing*, R Foundation for Statistical Computing, Vienna, Austria, 2005.

[42] S. Banerji and A. Flieger, "Patatin-like proteins: a new family of lipolytic enzymes present in *Bacteria*?" *Microbiology*, vol. 150, no. 3, pp. 522–525, 2004.

[43] J. L. Arpigny and K. E. Jaeger, "Bacterial lipolytic enzymes: classification and properties," *Biochemical Journal*, vol. 343, part 1, pp. 177–183, 1999.

[44] N. D. Rawlings, A. J. Barrett, and A. Bateman, "MEROPS: the peptidase database," *Nucleic Acids Research*, vol. 38, supplement 1, pp. D227–D233, 2010.

[45] I. V. Kublanov, S. K. Bidjieva, A. V. Mardanov, and E. A. Bonch-Osmolovskaya, "*Desulfurococcus kamchatkensis* sp. nov., a novel hyperthermophilic protein-degrading archaeon isolated from a Kamchatka hot spring," *International Journal of Systematic and Evolutionary Microbiology*, vol. 59, no. 7, pp. 1743–1747, 2009.

[46] C. R. Jackson, H. W. Langner, J. Donahoe-Christiansen, W. P. Inskeep, and T. R. McDermott, "Molecular analysis of microbial community structure in an arsenite-oxidizing acidic thermal spring," *Environmental Microbiology*, vol. 3, no. 8, pp. 532–542, 2001.

[47] N. Byrne, M. Strous, V. Crépeau et al., "Presence and activity of anaerobic ammonium-oxidizing *Bacteria* at deep-sea hydrothermal vents," *ISME Journal*, vol. 3, no. 1, pp. 117–123, 2009.

[48] Q. Huang, C. Dong, R. Dong et al., "Archaeal and bacterial diversity in hot springs on the Tibetan Plateau, China," *Extremophiles*, vol. 15, no. 5, pp. 549–563, 2011.

[49] C. Brochier-Armanet, S. Gribaldo, and P. Forterre, "Spotlight on the *Thaumarchaeota*," *The ISME Journal*, vol. 6, no. 2, pp. 227–230, 2012.

[50] A. Spang, R. Hatzenpichler, C. Brochier-Armanet et al., "Distinct gene set in two different lineages of ammonia-oxidizing archaea supports the phylum *Thaumarchaeota*," *Trends in Microbiology*, vol. 18, no. 8, pp. 331–340, 2010.

[51] S. M. Barns, C. F. Delwiche, J. D. Palmer, and N. R. Pace, "Perspectives on archaeal diversity, thermophily and monophyly from environmental rRNA sequences," *Proceedings of the National Academy of Sciences of the United States of America*, vol. 93, no. 17, pp. 9188–9193, 1996.

[52] J. K. Rhee, D. G. Ahn, Y. G. Kim, and J. W. Oh, "New thermophilic and thermostable esterase with sequence similarity to the hormone-sensitive lipase family, cloned from a metagenomic library," *Applied and Environmental Microbiology*, vol. 71, no. 2, pp. 817–825, 2005.

[53] P. Tirawongsaroj, R. Sriprang, P. Harnpicharnchai et al., "Novel thermophilic and thermostable lipolytic enzymes from a Thailand hot spring metagenomic library," *Journal of Biotechnology*, vol. 133, no. 1, pp. 42–49, 2008.

[54] X. Chu, H. He, C. Guo, and B. Sun, "Identification of two novel esterases from a marine metagenomic library derived from South China Sea," *Applied Microbiology and Biotechnology*, vol. 80, no. 4, pp. 615–625, 2008.

[55] V. Delorme, R. Dhouib, S. Canaan, F. Fotiadu, F. Carrière, and J. F. Cavalier, "Effects of surfactants on lipase structure, activity, and inhibition," *Pharmaceutical Research*, vol. 28, no. 8, pp. 1831–1842, 2011.

[56] S. Kocabiyik and B. Demirok, "Cloning and overexpression of a thermostable signal peptide peptidase (SppA) from *Thermoplasma volcanium* GSS1 in *E. coli*," *Biotechnology Journal*, vol. 4, no. 7, pp. 1055–1065, 2009.

[57] E. Burgess, J. Unrine, G. Mills, C. Romanek, and J. Wiegel, "Comparative geochemical and microbiological characterization of two thermal pools in the Uzon Caldera, Kamchatka, Russia," *Microbial Ecology*, vol. 63, no. 3, pp. 471–489, 2012.

[58] R. Cavicchioli, M. Z. DeMaere, and T. Thomas, "Metagenomic studies reveal the critical and wide-ranging ecological importance of uncultivated archaea: the role of ammonia oxidizers," *BioEssays*, vol. 29, no. 1, pp. 11–14, 2007.

[59] L. J. Reigstad, A. Richter, H. Daims, T. Urich, L. Schwark, and C. Schleper, "Nitrification in terrestrial hot springs of Iceland and Kamchatka," *FEMS Microbiology Ecology*, vol. 64, no. 2, pp. 167–174, 2008.

[60] M. Pester, C. Schleper, and M. Wagner, "The *Thaumarchaeota*: an emerging view of their phylogeny and ecophysiology," *Current Opinion in Microbiology*, vol. 14, no. 3, pp. 300–306, 2011.

5

On Physical Properties of Tetraether Lipid Membranes: Effects of Cyclopentane Rings

Parkson Lee-Gau Chong, Umme Ayesa, Varsha Prakash Daswani, and Ellah Chay Hur

Department of Biochemistry, Temple University School of Medicine, 3420 North Broad Street, Philadelphia, PA 19140, USA

Correspondence should be addressed to Parkson Lee-Gau Chong, pchong02@temple.edu

Academic Editor: Yosuke Koga

This paper reviews the recent findings related to the physical properties of tetraether lipid membranes, with special attention to the effects of the number, position, and configuration of cyclopentane rings on membrane properties. We discuss the findings obtained from liposomes and monolayers, composed of naturally occurring archaeal tetraether lipids and synthetic tetraethers as well as the results from computer simulations. It appears that the number, position, and stereochemistry of cyclopentane rings in the dibiphytanyl chains of tetraether lipids have significant influence on packing tightness, lipid conformation, membrane thickness and organization, and headgroup hydration/orientation.

1. Introduction

Archaea are subdivided into two kingdoms: euryarchaeota and crenarchaeota [1]. Euryarchaeota include methanogens and halophiles, whereas crenarchaeota are traditionally comprised of thermophilic or hyperthermophilic archaea [1]. Halophiles and some methanogens are found mostly in high salt water or hypersaline systems such as natural brines, alkaline salt lakes, and salt rocks; while thermophilic and hyperthermophilic archaea are found in very high temperature environments [2]. In recent years, crenarchaeota have also been found in nonextreme environments such as soil and pelagic areas [3, 4].

The plasma membranes of archaea are rich in tetraether lipids (TLs) and diphytanylglycerol diethers, also known as archaeols (reviewed in [11–13]). TLs are the dominating lipid species in crenarchaeota, particularly in thermoacidophilic archaea (~90–95%). They are also found in methanogens (0–50%) but are virtually absent in halophiles. Archaeal TLs contain either a caldarchaeol (GDGT) or a calditoglycerocaldarchaeol (GDNT) hydrophobic core (Figure 1) [13–17]. GDGT has two glycerols at both ends of the hydrophobic core. GDNT has a glycerol backbone at one end of the hydrophobic core and the calditol group at

the other end. Typically, TLs in methanogens contain only GDGT, but TLs in thermoacidophiles, particularly in the members of the order *Sulfolobales*, have both GDGT and GDNT components. The *Metallosphaera sedula TA-2* strain from hot springs in Japan, which has only GDGT-based lipids, is an exception [18]. TLs have been thought to play an important role in the thermoacidophile's high stability against extreme growth conditions such as high temperatures (e.g., 65-90°C) and acidic environments (e.g., pH 2-3) [19]. However, more recent studies showed that GDGT-based TLs are also abundant in nonextremophilic crenarchaeota present in marine environments, lakes, soils, peat bogs, and low temperature areas [20, 21]. The functional role of tetraether lipids in crenarchaeota is not fully understood.

The hydrophobic core of archaeal TLs is made of dibiphytanyl hydrocarbon chains, which may contain up to 8 cyclopentane rings per molecule (reviewed in [13]). The number of cyclopentane rings increases as growth temperature increases [22–25], but decreases with decreasing pH in growth media [26]. The presence of cyclopentane rings is a structural feature unique for archaeal tetraether lipids. Therefore, it is of great interest to unravel its biological roles.

Various polar headgroups can be attached to the glycerol and calditol backbones and yield either monopolar or bipolar

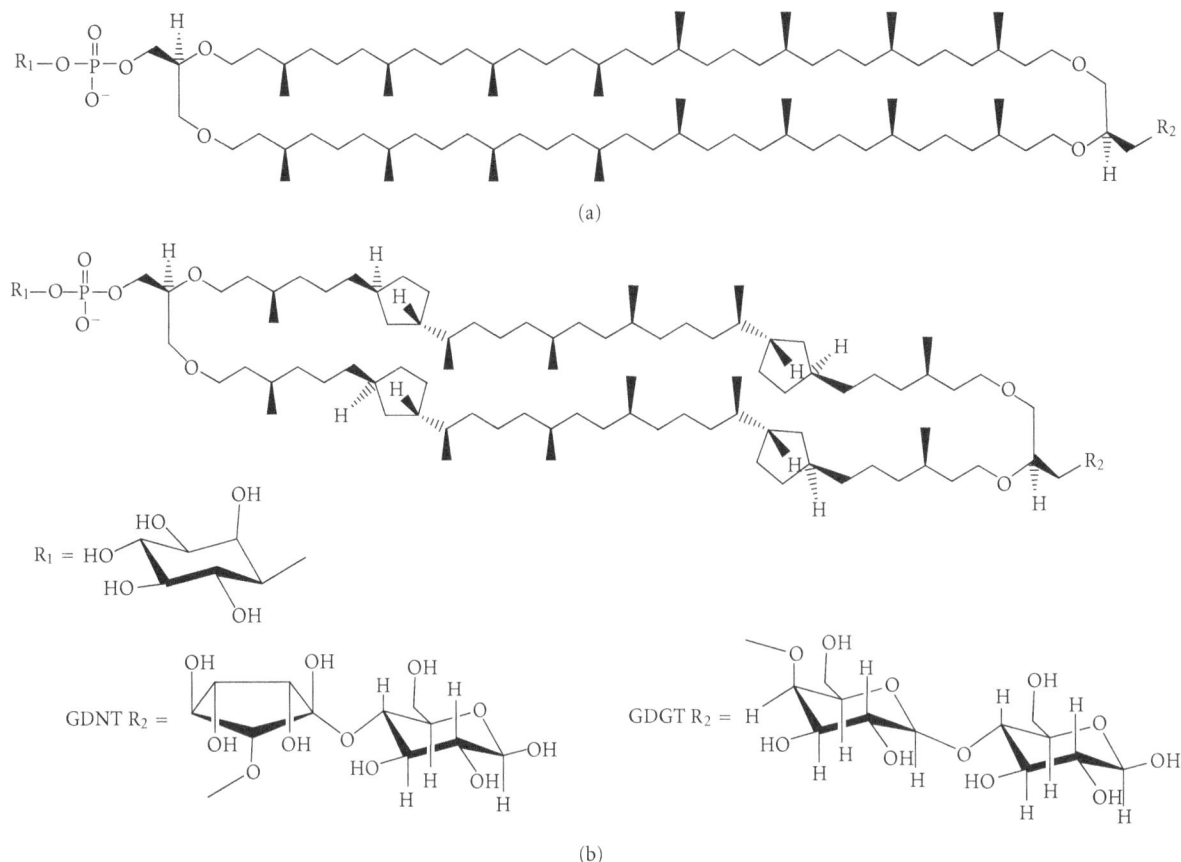

FIGURE 1: Illustrations of the molecular structures of the bipolar tetraether lipids in the polar lipid fraction E (PLFE) isolated from *S. acidocaldarius*. PLFE contains (a) GDGT (or caldarchaeol) and (b) GDNT (or calditolglycerocaldarchaeol). The number of cyclopentane rings in each biphytanyl chain can vary from 0 to 4. The different head groups of GDNT and GDGT are presented at the bottom. GDG(N)T-0 and GDG(N)T-4 contain 0 and 4 cyclopentane rings per molecule, respectively (taken from [5], reproduced with permission).

tetraether (BTL) lipids. Archaeal BTLs are glycolipids or phosphoglycolipids (illustrated in Figure 1). Liposomes that are made of BTLs containing two or more sugar moieties exhibit lower proton permeability than those containing only one sugar molecule [26]. It has been proposed that thermoacidophilic archaea cells adapt to low pH and high temperature by increasing the number of sugar moieties and cyclopentane rings [26, 27]. Increasing the number of cyclopentane rings tightens membrane packing (discussed later) [27]. Sugar moieties and the phosphate group in the BTL polar headgroup regions interact with each other to form a strong hydrogen bond network at the membrane surface [28].

BTLs are unique to archaea and cannot be biosynthesized by eukaryotic or bacterial cells. The ether formation from glycerol has been studied to a great extent ([29] and references cited therein). The calditol moiety of GDNTs can be synthesized via an aldol condensation between dihydroxyacetone and fructose [30]. Calditol is then reduced and alkylated to form GDNTs [30]. An *in vitro* study showed that with the aid of 1L-*myo*-inositol 1-phosphate synthase, archaetidylinositol phosphate (AIP)

synthase and AIP-phosphatase, archaeal inositol phospholipid (see Figure 1 e.g.) can be formed from CDP-archaeol and D-glucose-6-phosphate via *myo*-inositol-1-phosphate and AIP [31]. It has been proposed that the cyclopentane rings in BTLs of *Sulfolobus* are synthesized from glucose by a "cyclase" enzyme of the calditol carbocycle [32].

In this paper we focus on the recent findings related to the physical properties of tetraether lipid membranes, with special attention on the effects of the number, position, and configuration of cyclopentane rings on membrane properties. We discuss the findings obtained from model membranes composed of naturally occurring archaeal tetraether lipids and synthetic tetraethers as well as the results from computer simulations.

2. Physical Properties of Model Membranes Composed of Thermoacidophilic Tetraether Lipids

2.1. Membranes Made of Total Polar Lipid Extracts. The stability and physical properties of liposomes made from the total polar lipids (TPLs) extracted from archaea have been studied extensively (reviewed in [11, 12, 33, 34]). TPL extracts contain both diether and tetraether lipids. The general trend shows that membranes become more stable as the mole fraction of tetraether lipids increases. As an example, liposomes made of diether lipids such as *Methanosarcina mazei* TPL (0 wt% in caldarchaeols) were unstable against simulated human bile while those made of TPL from *Methanobacterium espanolae* (65% in caldarchaeols) and *Thermoplasma acidophilum* (90% in caldarchaeols) were relatively more stable [35]. Solute and water permeability also decrease as the content of tetraether lipids in membranes made with archaeal TPLs increases [36].

Sprott et al. [37] demonstrated that liposomes made with TPL from the archaeon *M. smithii AL1* can be highly fusogenic when exposed to low pH and α- and β-glucosidases. It was suggested that, at low pH (4.8), the positively charged glucosidases interact with the anionic phospholipids in *M. smithii* TPL, which in turn causes archaeosomes to rapidly aggregate [37]. Aggregation is a prerequisite for membrane fusion. This result is somewhat surprising because previous studies showed that tetraether liposomes are resistant to fusogenic compounds [38–40]. Since TPL of *M. smithii AL1* contains a significant amount of diethers, in addition to caldarchaeols (~40 wt %), it is possible that the strong fusogenic activity mentioned above comes from the diether component.

2.2. Membranes Made of Partially Purified Tetraether Lipid Fractions.

Since tetraethers are the dominating lipid species in thermoacidophiles, and the presence of diethers in the total polar lipid extracts makes the data interpretation more difficult, it is of biophysical interest to study membranes made only with tetraether lipids. The physical properties of lipid membranes made of partially purified polar lipid fractions from the archaeon *Sulfolobus solfataricus* have been reviewed [11, 34]. In this section, we focus on the recent studies of membranes made of partially purified polar lipid fractions isolated from the archaeon *Sulfolobus acidocaldarius*.

2.2.1. PLFE.

The polar lipid fraction E (PLFE) is one of the major bipolar tetraether lipids (BTLs) found in the thermoacidophilic archaeon *S. acidocaldarius* [41, 42]. PLFE is a mixture of GDNT and GDGT (Figure 1). The GDNT component (~90% of total PLFE) contains phospho-*myo*-inositol on the glycerol end and β-glucose on the calditol end, whereas the GDGT component (~10% of total PLFE) has phospho-*myo*-inositol attached to one glycerol and β-D-galactosyl-D-glucose to the other glycerol skeleton (Figure 1). The nonpolar regions of these lipids consist of a pair of 40-carbon biphytanyl chains, each of which may contain up to four cyclopentane rings [22].

2.2.2. PLFE Liposomes.

PLFE lipids can form stable unilamellar (~60–800 nm in diameter), multilamellar, and giant unilamellar (~10–150 μm) vesicles [40, 41, 43]. The lipids in these vesicles span the entire lamellar structure, forming a monomolecular thick membrane [44], which contrasts to the bilayer structure formed by monopolar diester (or diether) phospholipids. Compared to liposomes made of diester or diether lipids, PLFE liposomes exhibit extraordinary membrane properties (reviewed in [11, 12, 34]). PLFE liposomes exhibit low proton permeability and dye leakage [45, 46], high stability against autoclaving and Ca^{2+}-induced vesicle fusion [40, 47], tight and rigid membrane packing [43], and low enthalpy and volume changes associated with the phase transitions [48, 49].

It is known that a decrease in archaeal cell growth temperature (T_g) decreases the number of cyclopentane rings in archaeal TLs [22]. In the case of *S. acidocaldarius*, the average number of cyclopentane rings per tetraether lipid molecule decreases from 4.8 to 3.4 when T_g drops from 82°C to 65°C [23]. Recent experimental work (see below) has addressed the effect of T_g, inferentially the number of cyclopentane rings, on the physical properties of tetraether lipid membranes.

2.2.3. Effect of Cyclopentane Rings on Phase Behavior of PLFE Liposomes.

The phase behavior of PLFE liposomes has been characterized by small angle X-ray scattering, infrared and fluorescence spectroscopy, and differential scanning calorimetry (DSC). PLFE liposomes exhibit two thermally-induced lamellar-to-lamellar phase transitions at ~47–50°C and ~60°C [34, 43, 48, 49] and a lamellar-to-cubic phase transition at ~74–78°C [48, 49] all of which involve small or no volume changes as revealed by pressure perturbation calorimetry (PPC) [49]. The calorimetry experiments also suggested that the number of cyclopentane rings in the dibiphytanyl chains affect membrane packing in PLFE liposomes because the liposomes derived from different cell growth temperatures showed different thermodynamic properties [49]. DSC allows us to determine the enthalpy change (ΔH) of the phase transition. PPC, on the other hand, allows us to determine the relative volume change ($\Delta V/V$) at the phase transition and the thermal expansivity coefficient (α) at each temperature.

For PLFE liposomes derived from cells grown at 78°C, the DSC heating scan exhibited an endothermic transition at 46.7°C, which can be attributed to a lamellar-to-lamellar phase transition and has an unusually low ΔH (3.5 kJ/mol), when compared to that for the main phase transitions of saturated diacyl monopolar diester lipids (e.g., 1,2-dimyristoyl-*sn*-glycero-3-phosphocholine, DMPC). The PPC scan revealed that, at this same phase transition, the relative volume change ($\Delta V/V$) in the membrane is very small (~0.1%) and much lower than the $\Delta V/V$ value 2.8% for the main phase transition of DMPC. The low ΔH and $\Delta V/V$ values may arise from the restricted *transauche* conformational changes in the dibiphytanyl chain due to the presence of cyclopentane rings, branched methyl groups, and to the spanning of the lipid molecules over the whole membrane [49].

For PLFE liposomes derived from cells with growth temperature of 65°C, similar DSC and PPC profiles were obtained. However, the lower cell growth temperature

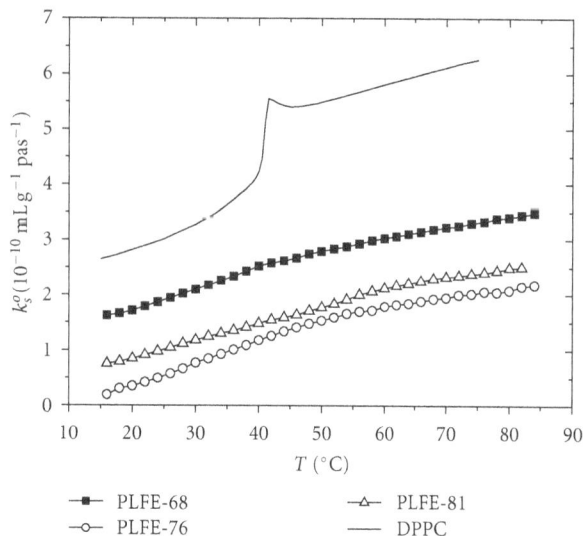

FIGURE 2: Adiabatic compressibilities (k_S^o) of PLFE liposomes derived from cells grown at three different temperatures: 68°C (dark squares), 76°C (open circles), and 81°C (open triangles). Solid line: DPPC liposomes for comparison (taken from [6], reproduced with permission).

yielded a higher $\Delta V/V$ (~0.25%) and ΔH (14 kJ/mol) value for the lamellar-to-lamellar phase transition measured at pH 2.1. The lower growth temperature also generated less negative temperature dependence of α. The changes in $\Delta V/V$, ΔH, and the temperature dependence of α can be attributed to the decrease in the number of cyclopentane rings in PLFE due to the lower growth temperature [49]. A decrease in the number of cyclopentane rings makes the membrane less tight and less rigid; thus, a higher $\Delta V/V$ value is shown through the phase transition.

2.2.4. Effect of Cyclopentane Rings on Compressibility and Membrane Volume Fluctuations of PLFE Liposomes.

The isothermal and adiabatic compressibility and relative volume fluctuations of PLFE liposomes have been determined by using calorimetry (DSC and PPC) and molecular acoustics (ultrasound velocimetry and densimetry) [50]. The compressibility values of PLFE liposomes were low, compared to those found in a gel state of 1,2-dipalmitoyl-sn-glycero-3-phosphocholine (DPPC) [50]. Relative volume fluctuations of PLFE liposomes at any given temperature examined were 1.6–2.2 times more damped than those found in DPPC liposomes [50]. Volume fluctuations are closely related to solute permeation across lipid membranes [51] and lateral motion of membrane components [52]. Thus, the low values of relative volume fluctuations explain why PLFE liposomes exhibit unusually low proton permeation and dye leakage [45, 46] as well as limited lateral mobility, especially at low temperatures (e.g., <26°C) [43, 53].

Zhai et al. [6] have used the growth temperature T_g to alter the structure of PLFE lipids. They determined the compressibilities and volume fluctuations of PLFE liposomes derived from different cell growth temperatures

(T_g = 68, 76, and 81°C). The compressibility and volume fluctuation values of PLFE liposomes exhibit small but significant differences with T_g. Figure 2 shows that adiabatic compressibility (k_S^o) of PLFE liposomes changes significantly with T_g: $k_S^o(T_g = 68°C) > k_S^o(T_g = 81°C) > k_S^o(T_g = 76°C)$. For isothermal compressibility (k_T^o), isothermal compressibility coefficient (β_T) and relative volume fluctuations, a similar, but somewhat different, trend is seen: (T_g = 68°C) > (T_g = 81°C) $\geq$ or $\approx$ (T_g = 76°C). These data indicate that, among the three employed growth temperatures, the growth temperature 76°C leads to the least compressible, and inferentially the most tightly packed PLFE lipid membranes. Note that 76°C is in the temperature range for optimal growth of *S. acidocaldarius* (75–80°C, [54, 55]). This finding suggests that membrane packing in PLFE liposomes may actually vary with the number of cyclopentane rings in a nonlinear manner, reaching maximal tightness when the tetraether lipids are derived from cells grown at the optimal growth temperatures [6].

2.2.5. Future Studies of Physical Properties of Tetraether Lipid Membranes.

PLFE is a mixture of GDNT- and GDGT-derived BTLs with varying numbers of cyclopentane rings. Furthermore, at any given growth temperature, there is always a broad distribution of the number of cyclopentane rings. In order to gain more insight into the effect of cyclopentane rings on compressibility and membrane volume fluctuations, it will be necessary to use purified archaeal BTLs with a well-determined number and location of cyclopentane rings. It has been reported that intact polar lipids (archaeols (diethers) and caldarchaeols (GDGT)) of the archaeon *Thermoplasma acidophilum* can be separated with single cyclopentane ring resolution by high-performance liquid chromatography (HPLC) as detected by evaporative light-scattering detection [26, 56]. However, the study by Shimada et al. on *T. acidophilum* was limited to GDGT-based BTLs. To separate intact archaea BTLs at single cyclopentane ring resolution when both GDNT- and GDGT-derived BTLs are present remains a major challenge.

Hydrolyzed BTLs can also be separated with single cyclopentane ring resolution using normal phase HPLC and positive ion atmospheric pressure chemical ionization mass spectrometry [7]. Figure 3 shows the structures of the cyclopentane-containing GDGT hydrophobic cores previously identified from the archaeon *Sulfolobus solfataricus*. These structures were determined by mass spectrometry. Compounds F′ and G′ (Figure 3) were reported as minor components in *S. solfataricus* [7]. The relative distribution of these GDGT structures varies from species to species. The GDGT fraction of *S. solfataricus* is dominated by those structures with one (Structures E and G, Figure 3) or two (F) biphytanyl chains with two cyclopentane rings. The distribution of GDGTs in the extract of the archaeon *M. sedula* is somewhat different. In this case, the distribution is dominated by structures containing one or two biphytanyls with one cyclopentane ring. Physical properties of liposomes made of hydrolyzed BTLs (without sugar and phosphate

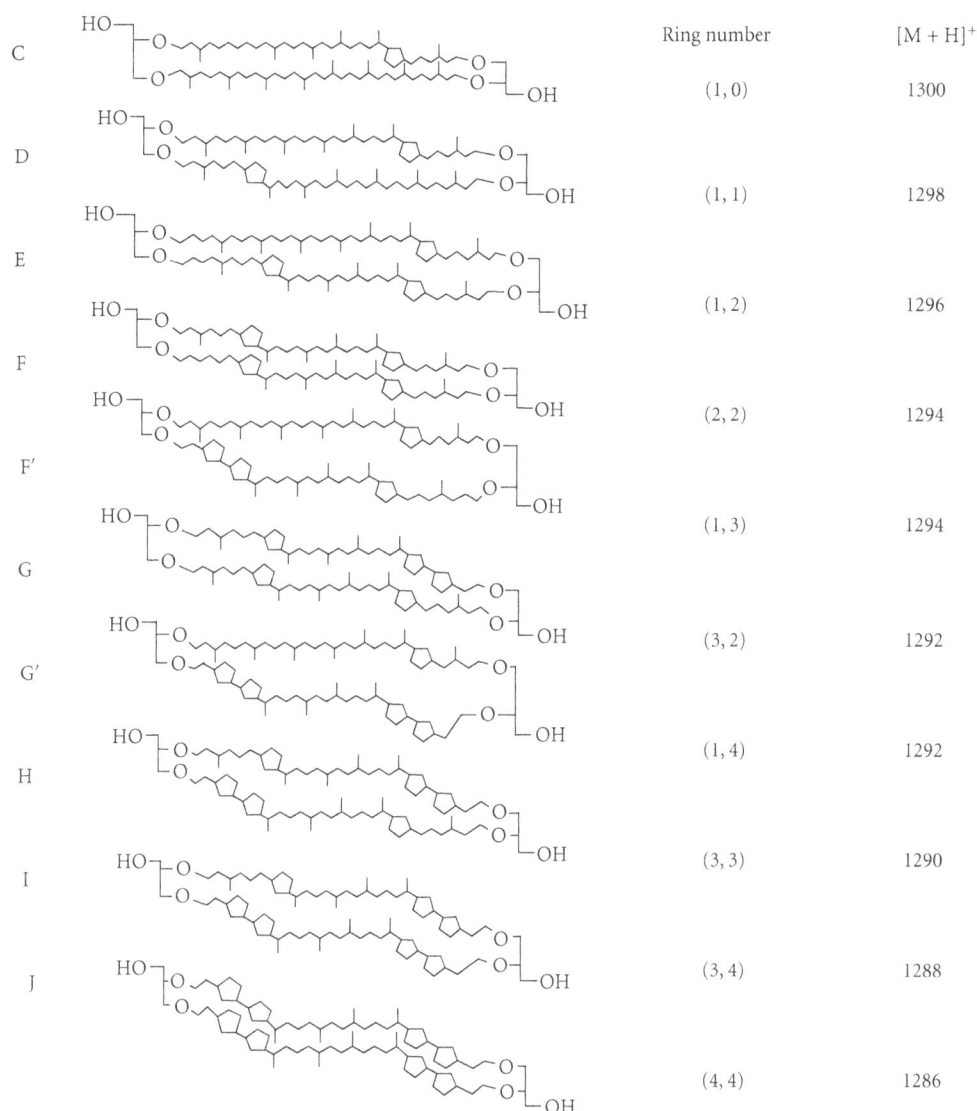

	Ring number	$[M + H]^+$
C	(1, 0)	1300
D	(1, 1)	1298
E	(1, 2)	1296
F	(2, 2)	1294
F'	(1, 3)	1294
G	(3, 2)	1292
G'	(1, 4)	1292
H	(3, 3)	1290
I	(3, 4)	1288
J	(4, 4)	1286

FIGURE 3: Structures of cyclopentane ring containing GDGTs previously reported to exist in archaea [7]. The number of cyclopentane rings in the first and second hydrocarbon chains is indicated in the parentheses. The mass-to-charge ratio (m/z) of the protonated form $[M+H]^+$ for each structure is also listed.

moieties) are not expected to be the same as those obtained from the liposomes made of intact BTLs [47].

2.2.6. Disruption of PLFE Liposome Stability. While BTL liposomes (such as PLFE liposomes) exhibit remarkable stability against a number of chemical and physical stressors as mentioned above, their stability can be attenuated or abolished under certain conditions. The most striking finding in this regard is that PLFE liposomes become excessively disrupted by the presence of two archaeal proteins, namely, CdvA and ESCRT-III (ESCRT: endosomal sorting complex required for transport) [57]. CdvA is a membrane interacting protein that forms structures at mid-cell prior to nucleoid segregation. CdvA recruits ESCRT-III to membranes in order to aid in the final steps of cell division in some species

of archaea. Negative stain electron microscopy revealed extensive deformation of PLFE liposomes in the presence of both CdvA and ESCRT-III together, but not individually [57]. The molecular mechanism underlying this disruption is not clear.

PLFE liposomes are "autoclavable." However, low pH (<4) and low salt concentrations (<50 mM) are unfavorable for autoclaving PLFE-based liposomes [47]. PLFE liposomes and PLFE-based stealth liposomes (e.g., 95 mol% PLFE, 3 mol% 1,2-distearoyl-*sn*-glycerol-3-phosphoethanolamine-polyethylene glycols (2000) (DSPE-PEG(2000)) and 2 mol% DSPE-PEG(2000)-maleimide) are extraordinarily stable against autoclaving between pH 4–10 [47]. These liposomes retained their particle size and morphology against multiple autoclaving cycles. One autoclaving cycle refers to the incubation of a sample for 20 min at 121°C under a steam pressure of ~18 psi. However, at pH 2-3, one or two autoclaving cycles appeared to disrupt these liposomal membranes, causing a significant increase in particle size [47]. PLFE liposomes were more resistant to dye leakage than the gel state of conventional diester liposomes under high salt and autoclaving conditions. As the salt concentration was decreased from 160 to 40 mM, the percent of dye molecules that leaked out from PLFE-based stealth liposomes after one autoclaving cycle increased from 10.8% to 56.3% [47].

As expected, PLFE-based liposomes can also be disrupted by surfactants. The effect of the surfactant n-tetradecyl-β-D-maltoside (TDM) on unilamellar vesicles composed of PLFE and POPC (1-palmitoyl-2-oleoyl-sn-glycero-3-phosphocholine, a monopolar diester lipid) has been examined [58]. TDM disrupts the POPC/cholesterol vesicles effectively; however, higher concentrations (~10 times) of TDM were required to disrupt PLFE/POPC vesicles.

2.2.7. Structural and Packing Properties of PLFE Monolayer Films Spread at the Interface between Air and Water. Effects of cell growth temperature, subphase temperature and pH, and lateral film pressure on PLFE lipid monolayers at the air-water interface have been examined using X-ray reflectivity (XRR) and grazing incidence X-ray diffraction (GIXD) [5]. XRR and GIXD determine the vertical and horizontal structure of the monolayers, respectively.

For PLFE derived from cells grown at 76°C, a total monolayer thickness of ~30 Å was found in the XRR measurements for all monolayers studied. This finding suggests that both head groups of a U-shaped conformation of the molecules are in contact with the subphase and that a single hydrocarbon chain region is protruded into the air. Similar U-shaped monolayer structures have been reported in other tetraether lipid membranes [59]. However, some other studies [60, 61] suggest that the U-shaped and the upright conformations may coexist in the monolayer at the same time or occur sequentially after spreading the TL lipids at the water-air interface.

At the subphase temperatures 10°C and 20°C, large, highly crystalline domains were observed by GIXD; and the thickness of the crystalline part of the monolayer is slightly larger than 30 Å, which indicates a tight packing of the whole lipid monolayer, including both the hydrocarbon chain and the head group regions. The area per hydrocarbon chain of PLFE (~19.3 Å^2) found by GIXD is significantly smaller than that of DPPC (~23.2 Å^2) or 1,2-dipalmitoyl-sn-glycero-3-phosphoglycerol (DPPG) (~22.6 Å^2). In fact, both the two hydrocarbon chains of a single PLFE lipid and the chains of neighbouring lipid molecules adopted an extremely tight packing.

For PLFE lipids derived from cells grown at higher temperatures, a slightly more rigid structure in the lipid dibiphytanyl chains was observed. However, the growth temperature, inferentially the number of cyclopentane rings, does not affect the parameters of the unit cell in GIXD measurements. This suggests that there exists a nearly identical crystalline packing of all the PLFE lipids examined and that, at high film pressures, membrane packing is primarily governed by the lipid headgroup region [5]. It is interesting to mention that the lack of cyclopentane rings in the bipolar tetraether lipids from *M. hungatei* has been suggested to be the cause of the U-shaped configuration adopted by these lipids in the monolayer film at the air-water interface [62]. Apparently, the presence of cyclopentane rings would hinder the dibiphytanyl hydrocarbon chains from bending to form the U-shaped configuration.

3. Physical Properties of Membranes Made of Synthetic Tetraether Lipids

The process of isolating well-defined archaeal tetraether lipids can be difficult and time consuming. In addition, archaeal tetraether lipids have several structural features distinctly different from conventional diester lipids. Therefore, it is rather difficult to elucidate the structure-activity relationship for each of the individual structural features when using native archaeal lipids. To resolve these problems to some extent, synthetic tetraether lipid analogues have been used [63–67].

3.1. Importance of the Stereochemistry of the Cyclopentane Ring. Jacquemet et al. were able to study the effect of the stereochemistry of the cyclopentane ring on BTL membrane properties by using two synthetic tetraether lipids [8, 9] (Compounds 1 and 2 in Figure 4). Both lipids have a bridging hydrocarbon chain with a single 1,3-disubstituted cyclopentane ring at the center. The substitutes on the ring are ether-linked to C3 of the two opposite glycerol moieties, while C2 of the glycerols is ether-linked to a phytanyl chain and C1 is linked to a lactosyl polar headgroup (Figure 4). The only difference between these two isomers is the configuration (*cis* or *trans*) of the 1,3-disubstituted cyclopentane ring [8, 9].

The *trans*-isoform showed multilamellar vesicles whereas the *cis*-counterpart led to nonspherical nanoparticles, as revealed by cryo-transmission electron microscopy [8]. Small angle X-ray scattering (SAXS) studies further showed that the *cis*-isomer exhibited L_c-L_α-Q_{II} (cystal, lamellar, and bicontinuous cubic phase (Pn3 m), resp.) phase transitions whereas the *trans*-isomer remained in L_α phase from 20 to 100°C. The electron density profiles calculated from the SAXS data were consistent with a stretched conformation of these synthetic BTLs within the L_α phase [9]. The difference in the phase behaviors was attributed to the conformation equilibrium of 1,3-disubstituted cyclopenatne rings. The dominant conformational motion in cyclopentane is pseudorotational [68]. Pseudorotation is more restricted for the *trans*-isomer whereas several more orientations of the two substituents on the ring can be created for the *cis*-1,3-dialkyl cyclopentane ring [9, 68, 69]. Even though this study shows that the stereochemistry at the cyclopentane ring has a dramatic influence on membrane properties, more work is still required in order to explain why liposomes made of PLFE, which naturally occurs and contains *trans*-1,3-disubstituted cyclopentyl rings, can undergo the L_α-to-Q_{II} phase transition [48, 49], while the synthetic *trans*

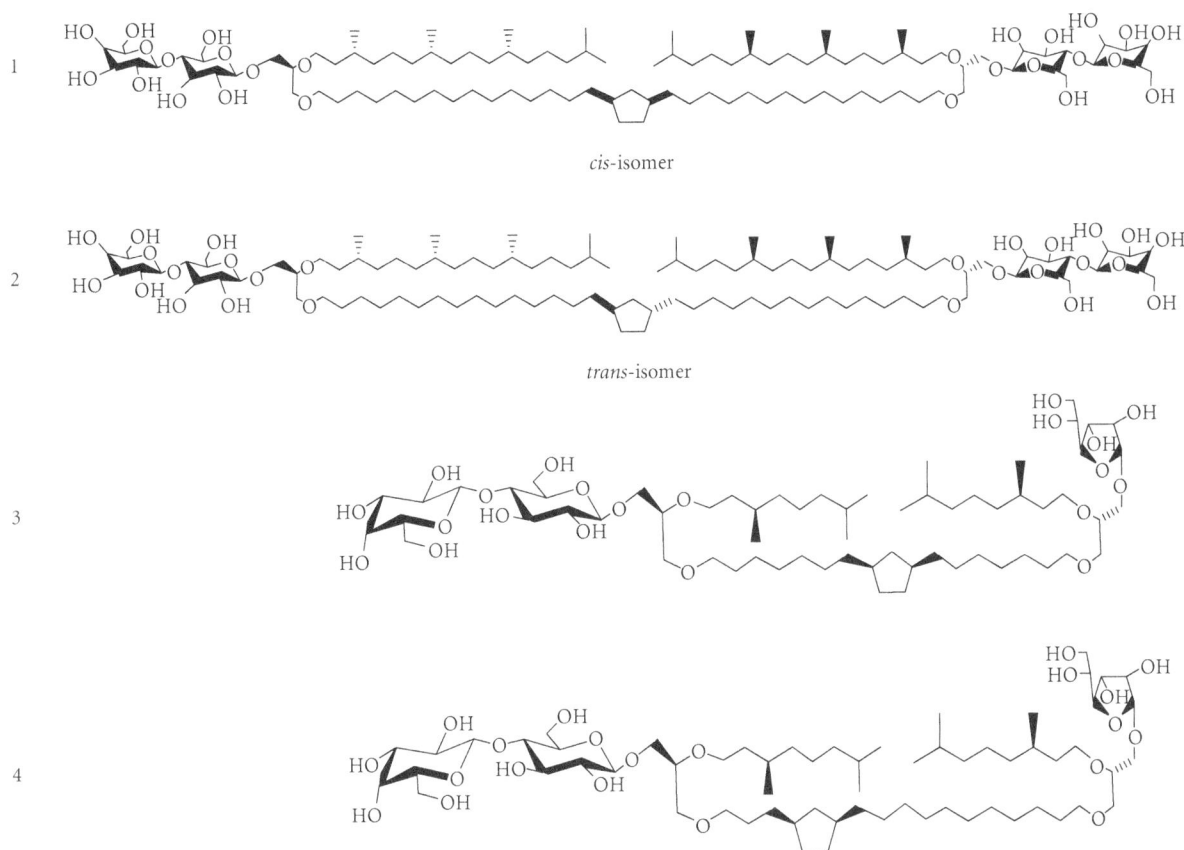

FIGURE 4: Synthetic tetraether lipids that have been used to study the effect of configuration (Compounds 1 and 2 [8, 9]) and position (Compounds 3 and 4 [10]) of the cyclopentane rings on membrane properties.

BTL (Figure 4) cannot [9]. Note that the placement and the number of cyclopentane rings are different between PLFE lipids (Figure 1) and the synthetic BTLs mentioned above (Figure 4). Apparently, BTLs with subtle differences in chemical structures can display distinctly different phase behaviors.

The difference in the polar headgroups between PLFE and the above-mentioned synthetic BTLs also leads to other subtle structural differences. The d-spacing of PLFE liposomes increases with increasing temperature [48], which is contrary to that obtained from the synthetic *trans*-isomer mentioned above (Compound 2 in Figure 4) [9]. The increased d-spacing with temperature is probably due to an increase in hydration at the polar headgroup of PLFE [48]. For unknown reasons, there is no change in hydration at the polar (lactosyl) headgroups in those two synthetic stereo-isomers [9].

3.2. Influence of the Position of the Cyclopentane Ring. Brard et al. studied the effect of the position of the cyclopentane ring on physical properties of tetraether lipid membranes [66]. They synthesized two tetraether glycolipids, each of which contains a single *cis*-1,3-disubstituted cyclopentane ring in the bridging chain. One glycolipid contained a cyclopentane ring in the middle of the bridging chain while the other had one at three methylene units from the glycerol

backbone (Compound 3 and 4 in Figure 4). This helped them determine the influences of the different positions of the cyclopentane ring.

The cyclopentane ring position appears to have a profound impact on hydration properties, lyotropic liquid crystalline behavior, and membrane organization [66]. Moreover, the synthetic BTL with the cyclopentane ring positioned at the center (Compound 3 in Figure 4) can be completely dispersed in water, and it can form sponge-like structures as revealed by electron microscopy. In contrast, the compound with the cyclopentane ring away from the center (Compound 4 in Figure 4) can only be partitially dispersed in water and it forms multilamellar vesicles. It has been suggested that the position of the cyclopentane ring in the bridging chain influences the orientation of the glycosidic polar headgroups attached to the glycerol backbone, which leads to different membrane organizations [66].

4. Membrane Properties Revealed by Computer Simulations

4.1. Effect of Cyclopentane Rings on Membrane Packing and Headgroup Orientation. An increase in growth temperature is known to increase the number of cyclopentane rings in the dibiphytanyl chains of archaeal lipids [23]. The number of cyclopentane rings may vary from 0 to 4 in each biphytanyl chain (i.e., 0 to 8 per dibiphytanyl unit). To evaluate how the number of cyclopentane rings might affect membrane packing, Gabriel and Chong have conducted molecular modeling studies on a membrane containing 4×4 GDNT molecules (with sugar moieties, Figure 1) [27]. It was found that when 8 cyclopentane rings are contained, the headgroup of GDNT runs almost parallel to the membrane surface. However, without containing any rings, the headgroup is oriented perpendicular to the membrane surface. The molecular modeling further showed that an increase in the number of cyclopentane rings in the dibiphytanyl chains of GDNT from 0 to 8 made GDNT membrane packing tighter, more rigid, and more negative in interaction energy (−156.5 kcal/mol for 0 cyclopentane ring to −191.6 kcal/mol with 8 rings [27]). The resulting energy lowering effect is neither due to the decrease in polar headgroup separation, nor the change in the van der Waals interactions. Instead, it is due to the more favorable hydrogen bonding, and bonded interactions including harmonic bending, theta expansion bond angle bending, dihedral angle torsion, and inversion [27].

4.2. Effect of Macrocyclic Linkage on Membrane Properties. Most archaeal BTLs are macrocyclic molecules with two biphytanyl hydrocarbon chains linked to two opposite glycerol or calditol backbones (illustrated in Figure 1 for the case of PLFE). The effect of the macrocyclic linkage on membrane properties has been studied by molecular dynamics simulations [10, 70, 71]. For simplicity, coarse graining approaches were employed and BTL molecules were modeled as di-monopolar lipids such as di-DPPC [10] and diphytanyl phosphatidylcholine (DPhPC) [70]. In essence, two monopolar molecules were tethered together either at one pair of the hydrocarbon chains (acyclic di-DPPC or di-DPhPC) or at both pairs (cyclic di-DPPC or di-DPhPC). The simulations showed that in the membranes composed of macrocyclic BTL-like molecules, the upright configuration gains favor over the U-shaped configuration [70]. The macrocyclic linkage also leads to a condensing effect on the membrane surface, increases the order of the lipid hydrocarbon chains, slows lateral mobility in the membrane, and increases membrane thickness [10, 70, 71]. Furthermore, the molecular dynamics simulations made by the dissipative particle dynamics method [71] revealed the formation of two types of membrane pores. Hydrophobic pores are unstable and transient and exist at the low temperature. Hydrophilic pores are more stable with much longer lifetimes and are observed at high temperatures. The simulation data [71] suggested that hydrophilic pores can lead to the rupture of membrane vesicles. More intriguingly, it was proposed that hydrophobic pores, which occur at low temperatures, may result in the permeation of encapsulated small molecules [71]. This implies that although BTL membranes are extremely stable and tightly packed, some small leakage of entrapped molecules can still occur due to either the formation of hydrophobic pores [71] or membrane volume fluctuations [6, 50] (discussed earlier).

5. Applications of Tetraether Lipid Membranes

The extraordinary stability of tetraether lipid membranes against a variety of physical and biochemical stressors has provided the basis for using these lipids to develop technological applications. BTLs can be used as a stable lipid matrix for biosensors [72], a light harvesting device [73], and nanoparticles for targeted imaging and therapy (reviewed in [12, 74]).

It has been proposed that liposomes made of archaeal lipids (also called archaeosomes) are taken up via a phagocytosis receptor in the plasma membrane of the target cells [75]. This uptake occurs in a liposomal composition-dependent manner [75]. Total polar lipids from the archaeon, *Halobacterium salinarum CECT 396*, have been used to make archaeosomes and archaeosomal hydrogels as a possible topical delivery system for antioxidants [76]. Compared to conventional liposomes, those archaeosomes and archaeosomal hydrogels showed better stability and more sustained drug release [76]. It is of interest to extend their study from diethers (abundant in *Halobacterium salinarum CECT 396*) to tetraether lipids (e.g., PLFE lipids isolated from thermoacidophiles). BTL-based liposomes are suitable for oral delivery of therapeutic agents because BTL liposomes are stable against the harsh conditions (such as bile salts, pancreatic enzymes, and low pH) in the gastrointestinal tract [77]. Tetraether lipid membranes have also been tailored and evaluated as an intranasal peptide delivery vehicle [78]. PEGylated tetraether lipids have been synthesized and tested for their stability in test tubes and for liposomal encapsulation potential [79]. Knowledge gained from the physical studies of cyclopentane rings, sugar moieties, and macrocyclic structures should help to optimize the numerous potential applications.

Abbreviations

AIP: archaetidylinositol phosphate
BTL: bipolar tetraether lipids
DMPC: 1,2-dimyristoyl-*sn*-glycero-3-phosphocholine
DPhPC: 1,2-di-O-phytanyl-*sn*-glycero-3-phosphocholine
DPPC: 1,2-dipalmitoyl-*sn*-glycero-3-phosphocholine
DPPG: 1,2-dipalmitoyl-*sn*-glycero-3-phosphoglycerol
DSC: differential scanning calorimetry
DSPE-PEG: 1,2-distearoyl-*sn*-glycerol-3-phosphoethanolamine-polyethylene glycols
ESCRT: endosomal sorting complex required for transport
GDGT: caldarchaeol
GDNT: calditoglycerocaldarchaeol
GIXD: grazing incidence X-ray diffraction
HPLC: high-performance liquid chromatography
PLFE: polar lipid fraction E

POPC: 1-palmitoyl-2-oleoyl-*sn*-glycero-3-phosphocholine
PPC: pressure perturbation calorimetry
TDM: n-tetradecyl-β-D-maltoside
TL: tetraether lipids
TPL: total polar lipids
XRR: X-ray reflectivity.

Acknowledgment

Financial support from the NSF (DMR-1105277) is gratefully acknowledged.

References

[1] C. R. Woese, O. Kandler, and M. L. Wheelis, "Towards a natural system of organisms: proposal for the domains Archaea, Bacteria, and Eucarya," *Proceedings of the National Academy of Sciences of the United States of America*, vol. 87, no. 12, pp. 4576–4579, 1990.

[2] A. S. Andrei, H. L. Banciu, and A. Oren, "Living with salt: metabolic and phylogenetic diversity of archaea inhabiting saline ecosystems," *FEMS Microbiology Letters*, vol. 330, no. 1, pp. 1–9, 2012.

[3] L. A. Powers, J. P. Werne, T. C. Johnson, E. C. Hopmans, J. S. S. Damsté, and S. Schouten, "Crenarchaeotal membrane lipids in lake sediments: a new paleotemperature proxy continental paleoclimate reconstruction?" *Geology*, vol. 32, no. 7, pp. 613–616, 2004.

[4] M. B. Karner, E. F. Delong, and D. M. Karl, "Archaeal dominance in the mesopelagic zone of the Pacific Ocean," *Nature*, vol. 409, no. 6819, pp. 507–510, 2001.

[5] C. Jeworrek, F. Evers, M. Erlkamp et al., "Structure and phase behavior of archaeal lipid monolayers," *Langmuir*, vol. 27, no. 21, pp. 13113–13121, 2011.

[6] Y. Zhai, P. L. G. Chong, L. J. Taylor et al., "Physical properties of archaeal tetraether lipid membranes as revealed by differential scanning and pressure perturbation calorimetry, molecular acoustics, and neutron reflectometry: effects of pressure and cell growth temperature," *Langmuir*, vol. 28, no. 11, pp. 5211–5217, 2012.

[7] E. C. Hopmans, S. Schouten, R. D. Pancost, M. T. van der Meer, and J. S. Sinninghe Damste, "Analysis of intact tetraether lipids in archaeal cell material and sediments by high performance liquid chromatography/atmospheric pressure chemical ionization mass spectrometry," *Rapid Communications in Mass Spectrometry*, vol. 14, no. 7, pp. 585–589, 2000.

[8] A. Jacquemet, L. Lemiegre, O. Lambert, and T. Benvegnu, "How the stereochemistry of a central cyclopentyl ring influences the self-assembling properties of archaeal lipid analogues: synthesis and cryoTEM observations," *Journal of Organic Chemistry*, vol. 76, no. 23, pp. 9738–9747, 2011.

[9] A. Jacquemet, C. Meriadec, L. Lemiegre, F. Artzner, and T. Benvegnu, "Stereochemical effect revealed in self-assemblies based on archaeal lipid analogues bearing a central five-membered carbocycle: a SAXS study," *Langmuir*, vol. 28, no. 20, pp. 7591–7597, 2012.

[10] M. Bulacu, X. Periole, and S. J. Marrink, "In silico design of robust bolalipid membranes," *Biomacromolecules*, vol. 13, no. 1, pp. 196–205, 2012.

[11] P. L.-G. Chong, "Physical properties of membranes composed of tetraether archaeal lipids, ? In," in *Thermophiles*, F. Robb,

[12] P. L.-G. Chong, "Archaebacterial bipolar tetraether lipids: physico-chemical and membrane properties," *Chemistry and Physics of Lipids*, vol. 163, no. 3, pp. 253–265, 2010.

[13] Y. Koga and H. Morii, "Recent advances in structural research on ether lipids from archaea including comparative and physiological aspects," *Bioscience, Biotechnology and Biochemistry*, vol. 69, no. 11, pp. 2019–2034, 2005.

[14] A. Sugai, R. Sakuma, I. Fukuda et al., "The structure of the core polyol of the ether lipids from Sulfolobus acidocaldarius," *Lipids*, vol. 30, no. 4, pp. 339–344, 1995.

[15] E. Untersteller, B. Fritz, Y. Blériot, and P. Sinaÿ, "The structure of calditol isolated from the thermoacidophilic archaebacterium Sulfolobus acidocaldarius," *Comptes Rendus de l'Academie des Sciences*, vol. 2, no. 7-8, pp. 429–433, 1999.

[16] A. Gambacorta, G. Caracciolo, D. Trabasso, I. Izzo, A. Spinella, and G. Sodano, "Biosynthesis of calditol, the cyclopentanoid containing moiety of the membrane lipids of the archaeon Sulfolobus solfataricus," *Tetrahedron Letters*, vol. 43, no. 3, pp. 451–453, 2002.

[17] Y. Bleriot, E. Untersteller, B. Fritz, and P. Sinay, "Total synthesis of calditol: structural clarification of this typical component of Archaea order Sulfolobales," *Chemistry*, vol. 8, no. 1, pp. 240–246, 2002.

[18] Y. H. Itoh, N. Kurosawa, I. Uda et al., "Metallosphaera sedula TA-2, a calditoglycerocaldarchaeol deletion strain of a thermoacidophilic archaeon," *Extremophiles*, vol. 5, no. 4, pp. 241–245, 2001.

[19] T. A. Langworthy and J. L. Pond, "Membranes and lipids of thermophiles," in *Thermophiles: General, Molecular, and Applied Microbiology*, T. D. Brock, Ed., pp. 107–134, John Wiley & Sons, New York, NY, USA, 1986.

[20] J. S. SinningheDamsté, S. Schouten, E. C. Hopmans, A. C. T. Van Duin, and J. A. J. Geenevasen, "Crenarchaeol: the characteristic core glycerol dibiphytanyl glycerol tetraether membrane lipid of cosmopolitan pelagic crenarchaeota," *Journal of Lipid Research*, vol. 43, no. 10, pp. 1641–1651, 2002.

[21] J. W. H. Weijers, S. Schouten, O. C. Spaargaren, and J. S. Sinninghe Damsté, "Occurrence and distribution of tetraether membrane lipids in soils: implications for the use of the TEX86 proxy and the BIT index," *Organic Geochemistry*, vol. 37, no. 12, pp. 1680–1693, 2006.

[22] M. De Rosa, A. Gambacorta, B. Nicolaus, B. Chappe, and P. Albrecht, "Isoprenoid ethers; backbone of complex lipids of the archaebacterium Sulfolobus solfataricus," *Biochimica et Biophysica Acta (BBA)/Lipids and Lipid Metabolism*, vol. 753, no. 2, pp. 249–256, 1983.

[23] M. De Rosa, E. Esposito, A. Gambacorta, B. Nicolaus, and J. D. Bu'Lock, "Effects of temperature on ether lipid composition of Caldariella acidophila," *Phytochemistry*, vol. 19, no. 5, pp. 827–831, 1980.

[24] L. L. Yang and A. Haug, "Structure of membrane lipids and physico-biochemical properties of the plasma membrane from Thermoplasma acidophilum, adapted to growth at 37∘C," *Biochimica et Biophysica Acta (BBA)/Lipids and Lipid Metabolism*, vol. 573, no. 2, pp. 308–320, 1979.

[25] I. Uda, A. Sugai, Y. H. Itoh, and T. Itoh, "Variation in molecular species of polar lipids from Thermoplasma acidophilum depends on growth temperature," *Lipids*, vol. 36, no. 1, pp. 103–105, 2001.

[26] H. Shimada, N. Nemoto, Y. Shida, T. Oshima, and A. Yamagishi, "Effects of pH and temperature on the composition of

G. Antranikian, D. Grogan, and A. Driessen, Eds., pp. 73–95, CRC Press, Boca Raton, Fla, USA, 2008.

polar lipids in Thermoplasma acidophilum HO-62," *Journal of Bacteriology*, vol. 190, no. 15, pp. 5404–5411, 2008.

[27] J. L. Gabriel and P. Lee Gau Chong, "Molecular modeling of archaebacterial bipolar tetraether lipid membranes," *Chemistry and Physics of Lipids*, vol. 105, no. 2, pp. 193–200, 2000.

[28] M. G. L. Elferink, J. G. de Wit, A. J. M. Driessen, and W. N. Konings, "Stability and proton-permeability of liposomes composed of archaeal tetraether lipids," *Biochimica et Biophysica Acta*, vol. 1193, no. 2, pp. 247–254, 1994.

[29] M. De Rosa, A. Gambacorta, B. Nicolaus, and S. Sodano, "Incorporation of labelled glycerols into ether lipids in Caldariella acidophila," *Phytochemistry*, vol. 21, no. 3, pp. 595–599, 1982.

[30] B. Nicolaus, A. Trincone, E. Esposito, M. R. Vaccaro, A. Gambacorta, and M. De Rosa, "Calditol tetraether lipids of the archaebacterium Sulfolobus solfataricus. Biosynthetic studies," *Biochemical Journal*, vol. 266, no. 3, pp. 785–791, 1990.

[31] H. Morii, S. Kiyonari, Y. Ishino, and Y. Koga, "A novel biosynthetic pathway of archaetidyl-myo-inositol via archaetidyl-myo-inositol phosphate from CDP-archaeol and D-glucose 6-phosphate in methanoarchaeon Methanothermobacter thermautotrophicus cells," *Journal of Biological Chemistry*, vol. 284, no. 45, pp. 30766–30774, 2009.

[32] N. Yamauchi, N. Kamada, and H. Ueoka, "The possibility of involvement of "cyclase" enzyme of the calditol carbocycle with broad substrate specificity in Sulfolobus acidcaldarius, a typical thermophilic archaea," *Chemistry Letters*, vol. 35, no. 11, pp. 1230–1231, 2006.

[33] L. Krishnan and G. D. Sprott, "Archaeosome adjuvants: immunological capabilities and mechanism(s) of action," *Vaccine*, vol. 26, no. 17, pp. 2043–2055, 2008.

[34] A. Gliozzi, A. Relini, and P. L. G. Chong, "Structure and permeability properties of biomimetic membranes of bolaform archaeal tetraether lipids," *Journal of Membrane Science*, vol. 206, no. 1-2, pp. 131–147, 2002.

[35] G. B. Patel, B. J. Agnew, L. Deschatelets, L. P. Fleming, and G. D. Sprott, "In vitro assessment of archaeosome stability for developing oral delivery systems," *International Journal of Pharmaceutics*, vol. 194, no. 1, pp. 39–49, 2000.

[36] J. C. Mathai, G. D. Sprott, and M. L. Zeidel, "Molecular Mechanisms of Water and Solute Transport across Archaebacterial Lipid Membranes," *Journal of Biological Chemistry*, vol. 276, no. 29, pp. 27266–27271, 2001.

[37] G. D. Sprott, J. P. Cote, and H. C. Jarrell, "Glycosidase-induced fusion of isoprenoid gentiobiosyl lipid membranes at acidic pH," *Glycobiology*, vol. 19, no. 3, pp. 267–276, 2009.

[38] A. Relini, D. Cassinadri, Q. Fan et al., "Effect of physical constraints on the mechanisms of membrane fusion: bolaform lipid vesicles as model systems," *Biophysical Journal*, vol. 71, no. 4, pp. 1789–1795, 1996.

[39] A. Relini, D. Cassinadri, Z. Mirghani et al., "Calcium-induced interaction and fusion of archaeobacterial lipid vesicles: a fluorescence study," *Biochimica et Biophysica Acta*, vol. 1194, no. 1, pp. 17–24, 1994.

[40] R. Kanichay, L. T. Boni, P. H. Cooke, T. K. Khan, and P. L. G. Chong, "Calcium-induced aggregation of archaeal bipolar tetraether liposomes derived from the thermoacidophilic archaeon Sulfolobus acidocaldarius," *Archaea*, vol. 1, no. 3, pp. 175–183, 2003.

[41] S. L. Lo and E. L. Chang, "Purification and characterization of a liposomal-forming tetraether lipid fraction," *Biochemical and Biophysical Research Communications*, vol. 167, no. 1, pp. 238–243, 1990.

[42] E. L. Chang and S. L. Lo, "Extraction and purification of tetraether lipids from Sulfolobus acidocaldarius," in *Protocols for Archaebacterial Research*, E. M. Fleischmann, A. R. Place, R. T. Robb, and H. J. Schreier, Eds., pp. 2.3.1–2.3.14, Maryland Biotechnology Institute, Baltimore, Md, USA, 1991.

[43] L. Bagatolli, E. Gratton, T. K. Khan, and P. L. G. Chong, "Two-photon fluorescence microscopy studies of bipolar tetraether giant liposomes from thermoacidophilic archaebacteria Sulfolobus acidocaldarius," *Biophysical Journal*, vol. 79, no. 1, pp. 416–425, 2000.

[44] M. G. L. Elferink, J. G. De Wit, R. Demel, A. J. M. Driessen, and W. N. Konings, "Functional reconstitution of membrane proteins in monolayer liposomes from bipolar lipids of Sulfolobus acidocaldarius," *Journal of Biological Chemistry*, vol. 267, no. 2, pp. 1375–1381, 1992.

[45] H. Komatsu and P. L. G. Chong, "Low permeability of liposomal membranes composed of bipolar tetraether lipids from thermoacidophilic archaebacterium Sulfolobus acidocaldarius," *Biochemistry*, vol. 37, no. 1, pp. 107–115, 1998.

[46] E. L. Chang, "Unusual thermal stability of liposomes made from bipolar tetraether lipids," *Biochemical and Biophysical Research Communications*, vol. 202, no. 2, pp. 673–679, 1994.

[47] D. A. Brown, B. Venegas, P. H. Cooke, V. English, and P. L. G. Chong, "Bipolar tetraether archaeosomes exhibit unusual stability against autoclaving as studied by dynamic light scattering and electron microscopy," *Chemistry and Physics of Lipids*, vol. 159, no. 2, pp. 95–103, 2009.

[48] P. L.-G. Chong, M. Zein, T. K. Khan, and R. Winter, "Structure and conformation of bipolar tetraether lipid membranes derived from thermoacidophilic archaeon Sulfolobus acidocaldarius as revealed by small-angle X-ray scattering and high-pressure FT-IR spectroscopy," *Journal of Physical Chemistry B*, vol. 107, no. 33, pp. 8694–8700, 2003.

[49] P. L. G. Chong, R. Ravindra, M. Khurana, V. English, and R. Winter, "Pressure perturbation and differential scanning calorimetric studies of bipolar tetraether liposomes derived from the thermoacidophilic archaeon Sulfolobus acidocaldarius," *Biophysical Journal*, vol. 89, no. 3, pp. 1841–1849, 2005.

[50] P. L.-G. Chong, M. Sulc, and R. Winter, "Compressibilities and volume fluctuations of archaeal tetraether liposomes," *Biophysical Journal*, vol. 99, no. 10, pp. 3319–3326, 2010.

[51] E. Falck, M. Patra, M. Karttunen, M. T. Hyvönen, and I. Vattulainen, "Impact of cholesterol on voids in phospholipid membranes," *Journal of Chemical Physics*, vol. 121, no. 24, pp. 12676–12689, 2004.

[52] P. F. F. Almeida, W. L. C. Vaz, and T. E. Thompson, "Lateral diffusion and percolation in two-phase, two-component lipid bilayers. topology of the solid-phase domains in-plane and across the lipid bilayer," *Biochemistry*, vol. 31, no. 31, pp. 7198–7210, 1992.

[53] Y. L. Kao, E. L. Chang, and P. L. G. Chong, "Unusual pressure dependence of the lateral motion of pyrene-labeled phosphatidylcholine in bipolar lipid vesicles," *Biochemical and Biophysical Research Communications*, vol. 188, no. 3, pp. 1241–1246, 1992.

[54] T. D. Brock, K. M. Brock, R. T. Belly, and R. L. Weiss, "Sulfolobus: a new genus of sulfur-oxidizing bacteria living at low pH and high temperature," *Archiv für Mikrobiologie*, vol. 84, no. 1, pp. 54–68, 1972.

[55] T. A. Langworthy, W. R. Mayberry, and P. F. Smith, "Long chain glycerol diether and polyol dialkyl glycerol triether lipids of Sulfolobus acidocaldarius," *Journal of Bacteriology*, vol. 119, no. 1, pp. 106–116, 1974.

[56] H. Shimada, N. Nemoto, Y. Shida, T. Oshima, and A. Yamagishi, "Complete polar lipid composition of Thermoplasma acidophilum HO-62 determined by high-performance liquid chromatography with evaporative light-scattering detection," *Journal of Bacteriology*, vol. 184, no. 2, pp. 556–563, 2002.

[57] R. Y. Samson, T. Obita, B. Hodgson et al., "Molecular and Structural Basis of ESCRT-III Recruitment to Membranes during Archaeal Cell Division," *Molecular Cell*, vol. 41, no. 2, pp. 186–196, 2011.

[58] A. Fafaj, J. Lam, L. Taylor, and P. L. G. Chong, "Unusual stability of archaeal tetraether liposomes against surfactants," *Biophysical Journal*, vol. 100, no. 3, p. 329a, 2011.

[59] M. De Rosa, "Archaeal lipids: structural features and supramolecular organization," *Thin Solid Films*, vol. 284-285, pp. 13–17, 1996.

[60] U. Bakowsky, U. Rothe, E. Antonopoulos, T. Martini, L. Henkel, and H. J. Freisleben, "Monomolecular organization of the main tetraether lipid from Thermoplasma acidophilum at the water-air interface," *Chemistry and Physics of Lipids*, vol. 105, no. 1, pp. 31–42, 2000.

[61] S. Vidawati, J. Sitterberg, U. Bakowsky, and U. Rothe, "AFM and ellipsometric studies on LB films of natural asymmetric and symmetric bolaamphiphilic archaebacterial tetraether lipids on silicon wafers," *Colloids and Surfaces B*, vol. 78, no. 2, pp. 303–309, 2010.

[62] A. Gliozzi, A. Relini, R. Rolandi, S. Dante, and A. Gambacorta, "Organization of bipolar lipids in monolayers at the air-water interface," *Thin Solid Films*, vol. 242, no. 1-2, pp. 208–212, 1994.

[63] T. Benvegnu, M. Brard, and D. Plusquellec, "Archaeabacteria bipolar lipid analogues: structure, synthesis and lyotropic properties," *Current Opinion in Colloid and Interface Science*, vol. 8, no. 6, pp. 469–479, 2004.

[64] T. Benvegnu, G. Rethore, M. Brard, W. Richter, and D. Plusquellec, "Archaeosomes based on novel synthetic tetraether-type lipids for the development of oral delivery systems," *Chemical Communications*, no. 44, pp. 5536–5538, 2005.

[65] M. Brard, C. Lainé, G. Réthoré et al., "Synthesis of archaeal bipolar lipid analogues: a way to versatile drug/gene delivery systems," *Journal of Organic Chemistry*, vol. 72, no. 22, pp. 8267–8279, 2007.

[66] M. Brard, W. Richter, T. Benvegnu, and D. Plusquellec, "Synthesis and supramolecular assemblies of bipolar archaeal glycolipid analogues containing a cis-1,3-disubstituted cyclopentane ring," *Journal of the American Chemical Society*, vol. 126, no. 32, pp. 10003–10012, 2004.

[67] G. Lecollinet, R. Auzély-Velty, M. Danel et al., "Synthetic approaches to novel archaeal tetraether glycolipid analogues," *Journal of Organic Chemistry*, vol. 64, no. 9, pp. 3139–3150, 1999.

[68] W. Cui, F. Li, and N. L. Allinger, "Simulation of conformational dynamics with the MM3 force field: the pseudorotation of cyclopentane," *Journal of the American Chemical Society*, vol. 115, no. 7, pp. 2943–2951, 1993.

[69] O. R. de Ballesteros, L. Cavallo, F. Auriemma, and G. Guerra, "Conformational analysis of poly(methylene-1,3-cyclopentane) and chain conformation in the crystalline phase," *Macromolecules*, vol. 28, no. 22, pp. 7355–7362, 1995.

[70] W. Shinoda, K. Shinoda, T. Baba, and M. Mikami, "Molecular dynamics study of bipolar tetraether lipid membranes," *Biophysical Journal*, vol. 89, no. 5, pp. 3195–3202, 2005.

[71] S. Li, F. Zheng, X. Zhang, and W. Wang, "Stability and rupture of archaebacterial cell membrane: a model study," *Journal of Physical Chemistry B*, vol. 113, no. 4, pp. 1143–1152, 2009.

[72] B. A. Cornell, V. L. B. Braach-Maksvytis, L. G. King et al., "A biosensor that uses ion-channel switches," *Nature*, vol. 387, no. 6633, pp. 580–583, 1997.

[73] K. Iida, H. Kiriyama, A. Fukai, W. N. Konings, and M. Nango, "Two-dimensional self-organization of the light-harvesting polypeptides/BChl a complex into a thermostable liposomal membrane," *Langmuir*, vol. 17, no. 9, pp. 2821–2827, 2001.

[74] T. Benvegnu, L. Lemiègre, and S. Cammas-Marion, "Archaeal lipids: innovative materials for biotechnological applications," *European Journal of Organic Chemistry*, no. 28, pp. 4725–4744, 2008.

[75] G. D. Sprott, S. Sad, L. P. Fleming, C. J. Dicaire, G. B. Patel, and L. Krishnan, "Archaeosomes varying in lipid composition differ in receptor-mediated endocytosis and differentially adjuvant immune responses to entrapped antigen," *Archaea*, vol. 1, no. 3, pp. 151–164, 2003.

[76] A. Gonzalez-Paredes, B. Clares-Naveros, M. A. Ruiz-Martinez, J. J. Durban-Fornieles, A. Ramos-Cormenzana, and M. Monteoliva-Sanchez, "Delivery systems for natural antioxidant compounds: archaeosomes and archaeosomal hydrogels characterization and release study," *International Journal of Pharmaeutics*, vol. 421, no. 2, pp. 321–331, 2011.

[77] J. Parmentier, B. Thewes, F. Gropp, and G. Fricker, "Oral peptide delivery by tetraether lipid liposomes," *International Journal of Pharmaceutics*, vol. 415, no. 1-2, pp. 150–157, 2011.

[78] G. B. Pate, H. Zhou, A. Ponce, G. Harris, and W. Chen, "Intranasal immunization with an archaeal lipid mucosal vaccine adjuvant and delivery formulation protects against a respiratory pathogen challenge," *PLoS ONE*, vol. 5, no. 12, Article ID e15574, 2010.

[79] J. Barbeau, S. Cammas-Marion, P. Auvray, and T. Benvegnu, "Preparation and characterization of stealth archaeosomes based on a synthetic PEGylated archaeal tetraether lipid," *Journal of Drug Delivery*, vol. 2011, Article ID 396068, 11 pages, 2011.

6

Archaeal Viruses, Not Archaeal Phages: An Archaeological Dig

Stephen T. Abedon and Kelly L. Murray

Department of Microbiology, The Ohio State University, 1680 University Drive, Mansfield, OH 44906, USA

Correspondence should be addressed to Stephen T. Abedon; abedon.1@osu.edu

Academic Editor: Naglis Malys

Viruses infect members of domains *Bacteria*, *Eukarya*, and *Archaea*. While those infecting domain *Eukarya* are nearly universally described as "Viruses", those of domain *Bacteria*, to a substantial extent, instead are called "Bacteriophages," or "Phages." Should the viruses of domain *Archaea* therefore be dubbed "Archaeal phages," "Archaeal viruses," or some other construct? Here we provide documentation of published, general descriptors of the viruses of domain *Archaea*. Though at first the term "Phage" or equivalent was used almost exclusively in the archaeal virus literature, there has been a nearly 30-year trend away from this usage, with some persistence of "Phage" to describe "Head-and-tail" archaeal viruses, "Halophage" to describe viruses of halophilic *Archaea*, use of "Prophage" rather than "Provirus," and so forth. We speculate on the root of the early 1980's transition from "Phage" to "Virus" to describe these infectious agents, consider the timing of introduction of "Archaeal virus" (which can be viewed as analogous to "Bacterial virus"), identify numerous proposed alternatives to "Archaeal virus," and also provide discussion of the general merits of the term, "Phage." Altogether we identify in excess of one dozen variations on how the viruses of domain *Archaea* are described, and document the timing of both their introduction and use.

1. Introduction

...most viruses infecting archaea have nothing in common with those infecting bacteria, although they are still considered as "bacteriophages" by many virologists, just because archaea and bacteria are both prokaryotes (without nucleus). [1]

For historical reasons, bacteriophage is widely used to refer to viruses of bacteria (and sometimes even archaea). The problem with such nomenclature is that it artificially divides the virosphere into two camps, with viruses of bacteria and archaea on one hand and viruses of eukaryotes on the other. [2]

Viruses are infectious agents that alternate between autonomous, encapsidated states known as virions, which are "packages of genes" [3], and unencapsidated, intracellular states known as infections [1], infected cells [3] or, more holistically, as "Virocells" or "Ribovirocells" [4, 5]. Numerous differences exist among viruses in terms of virion morphology, genome architecture, and infection strategy [6], and viruses also may be differentiated as a function of host range [7]. While it is possible to describe a virus's host range in terms of what species or even subspecies or strains of cellular hosts it is capable of infecting, it is also possible to distinguish between susceptible hosts more broadly. For example, one can, though with some ambiguity, distinguish between those hosts that are macroscopic versus those that instead are microscopic, with the latter hosting what can be described as viruses of microorganisms, or VoMs [8].

Among the viruses of microorganisms are those that infect microscopic eukaryotes along with those that instead infect prokaryotes. While the viruses of eukaryotes nearly exclusively are described as just that, that is, as viruses, the viruses of prokaryotes have been burdened with a more complicated naming history. Here we look at the naming conventions that have been applied to the latter, with some emphasis on considering the relative merits of the term "Phage" as a descriptor particularly of the viruses of domain *Archaea*. We agree with what we observe to be a near consensus within the field that the use of the term "Virus" as well as the qualification "Archae-" or "Archaeal"—as in, for instance, "Archaeal virus"—is both logical and reasonable, echoing, for example, the fairly common usage of "Bacterial virus" as an alternative to "Bacteriophage" or "Phage." For approximately half of the 40 or so years that these viruses have been studied, however, the explicit phrase "Archaeal virus" did not exist in the published literature. The absence of this phrase prior to the early 1990s reflects the replacement only in 1990 of "Archaebacteria" [9] with "*Archaea*" [10] as a descriptor of this cellular group.

Here we consider the history of the general naming of archaeal viruses as found within the published literature. In addition to exploring the timing of the introduction of the term "Archaeal virus," we also consider the transition from "Phage" to "Virus" as seen approximately a decade earlier as well as the use of various related terms including "Archaebacteria," "Halophage," and "Prophage." A summary of some of the terms that have been used to describe archaeal viruses, along with what we have been able to ascertain are their dates of introduction into the literature, is presented in Figure 1. Overall use in the literature—particularly as seen in journal articles—of "Virus" and "Phage" (including variations on these terms) as well as "Prophage" and "Halophage", is recorded in Table 1 (see Supplementary Materials available online at http://dx.doi.org/10.1155/2013/251245) as well as Table 2 (supplementary materials).

2. Kingdoms, Urkingdoms, Empires, Superkingdoms, and Domains

While schemes of organism classification have existed for millennia, prokaryotes were explicitly introduced into such systems only in 1938, by Copeland [36]. This "new" kingdom was dubbed the familiar Monera, as also used by Whittaker, in 1969, as part of his well-known five-kingdom system of organism classification [37]. Beyond Copeland's four-kingdom system and Whittaker's five-kingdom system, a modification was suggested in which cellular organisms were distinguished at a higher level than that of a kingdom, into simply prokaryotes versus eukaryotes [38], which can be designated as empires or superkingdoms [10, 39], that is, Prokaryota(e) versus Eukaryota(e). At that time, viruses similarly were distinguished, at least semantically, into phages (the viruses of prokaryotes) and viruses for everything else. For discussions of the early history of phages along with viruses more generally, see [3, 8, 40, 41] along with references cited.

As is well known especially to microbiologists, the five-kingdom system has given way to the three-domain system [10]. There the prokaryotes of kingdom *Monera* were, rightly or wrongly [42, 43], further differentiated into the domains *Bacteria* and *Archaea* versus the eukaryotes, with the latter dubbed domain *Eucarya* (or *Eukarya*) [44]. Note that the domain that would be named *Archaea* in 1990 [10] was first designated, in 1977, as "urkingdom... *archaebacteria*" by Woese and Fox [9]. From page 5089 of that publication (emphasis is theirs):

> The apparent antiquity of the methanogenic phenotype plus the fact that it seems well suited

to the type of environment presumed to exist on earth 3-4 billion years ago lead us tentatively to name this urkingdom the *archaebacteria*.

In terms of the viruses associated with each of these domains, an obvious question then was whether those infecting members of domain *Archaea* should retain the "Phage" designation or instead assume the more general term of "Virus." Here we consider the history of this issue especially in terms of published usage, focusing particularly on the relatively long transition from "Bacteriophage" to "Archaeal virus." We additionally consider overall usage as determined by examination of individual publications.

3. Methods

Myriad approaches were used to identify archaeal virus-associated publications. Where possible, publications were obtained, in various forms, and if not already searchable as PDFs (Portable Document Format) then digitally scanned and/or subject to optical character recognition (OCR). Articles were then computer searched using the Adobe Acrobat search function. This was initially done in bulk but ultimately individually, for various terms, where the context therefore could be observed. Terms found within reference lists were ignored and terms such as "Phage" that did not appear to be associated with considerations of archaeal viruses were also not considered (thus resulting in a "No" designation in Table 1, supplementary materials). Note that generally we equate "Virus" with "Viruses" as well as "Viral" but not with "Virion" and also consider the use of "Phage" as well as "Virus" both alone and as suffixes. We have also disregarded trivial misspellings such as "Archeal virus" as used in the same publication as "Archaeal virus."

We avoid describing a publication as containing the term "Virus" if that term was limited to within the name of a virus, for example, "*Acidianus* bottle-shaped virus" [45]. We have not applied the same "rule" to virus names that contain "phage," however, such as "*Methanobacterium* phage ΨM1", since technically this could be described instead as, for example, "*Methanobacterium* archaeal virus ΨM1" or simply "*Methanobacterium* virus ΨM1". The result of this bias is what we consider to be a conservative underestimation of the use of "Virus" along with an overestimation of use of "Phage." In terms of consideration of use of "Phage" within the literature, we feel that this approach is reasonable particularly since a publication will still be described as containing "Virus" if it also uses "Virus" to describe archaeal viruses outside of the names of specific viruses. The example at the beginning of this paragraph, on the other hand, was designated as *not* containing "Virus" since the quoted text was the sole use of "Virus" to describe an archaeal virus that we were able to locate in the nonreferences portion of that publication. In any case, "Virus" as well as "Phage" must have been used to describe one or more archaeal viruses to be counted towards these tallies.

Our working assumption is that our reference list (supplementary materials) is less than fully complete. Explicitly missed are references that were not published in English as

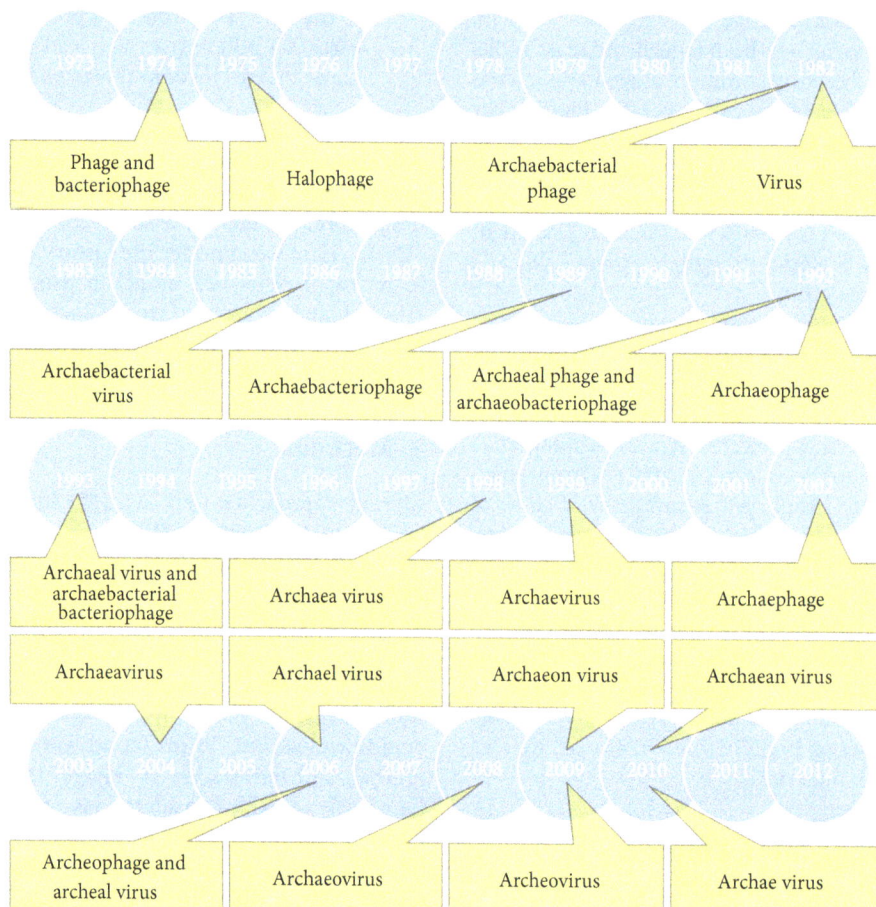

FIGURE 1: Various names that have been used to generally describe archaeal viruses as well as subsets of those viruses (e.g., "Halophage"). A sampling of those terms along with what to the best of our knowledge are their dates of introduction into the literature are presented in a timeline format. Note in particular the diversity as well as, even over the last decade, an apparently ongoing lack of consensus. References for first use, again to the best of our knowledge, are as follows: Bacteriophage (and Phage) [11], Halophage [12], Archaebacterial phage [13], Virus [13] (see, however, also [14]), Archaebacterial virus [15], Archaebacteriophage [16], Archaeobacteriophage [17], Archaeal phage [18], Archaeophage [19], Archaeal virus [20], Archaebacterial bacteriophage [21], Archaea virus [22], Archaevirus [23], Archaephage [24], Archaeavirus [25], Archael virus [26] (also found as a typo in a 1999 publication [27]), Archeal virus [28], Archeophage [29], Archaeovirus [30, 31], Archeovirus [32], Archaeon virus [33], Archae virus [34], and Archaean virus [35].

too patents. We also have likely missed various chapters along with whole books, especially those on subjects tangential to that of archaeal viruses. This is true also for materials that we were unable to obtain and therefore examine as well as any references that were not yet indexed nor easily identified prior to approximately January 1, 2013. Altogether, 694 references were identified, examined, and otherwise recorded as referring in some manner to archaeal viruses. All of these are listed in the reference section of the supplementary materials.

Just as human-generated indices can miss sought information, so too are digital searches fallible. It is possible also that relevant context was missed in the course of examining individual publications, though ultimately the decision as to whether a term such as "Phage" or "Virus" was used in a publication to refer to an archaeal virus, or not, was based on the experience of the lead author (S. T. Abedon). As with any "Experiment," the resulting output—given limitations in technology, retrieval, and subjective judgment as well as

simply "operator error"—should be viewed as representative, in this case of the archaeal virus literature, rather than either a complete or error-free sampling.

4. The "Phage" to "Virus" Transition

The earliest references to viruses of domain *Archaea* described them as phages or bacteriophages. This usage was appropriate, especially at the point of initial discovery, as the classification of their hosts as distinct from bacteria was not yet appreciated; that is, and as noted above, the term "Archaebacteria" dates to 1977 [9] but the first such archaeal virus publication was that of Torsvik and Dundas in 1974 [11]. That paper, published in *Nature*, was titled "Bacteriophage of *Halobacterium salinarum*."

The Torsvik and Dundas paper was followed by a publication during the next year, 1975, by Wais et al. [12]. This was also in *Nature* and was titled "Salt-dependent bacteriophage infecting *Halobacterium cutirubrum* and *H. halobium*." In each case neither "Virus," "Viral," nor the root "Arch..." appear in the article. With some irony, however, the article immediately following Torsvik and Dundas describes bacteriophage PM2—presumably, though not specified, of *Pseudoalteromonas* of domain *Bacteria*—as both a "Virus" and "Viral" [46]. Equivalent use of "Phage" as well as lack of use of "Virus", "Viral", and "Arch..." can be seen also with Torsvik and Dundas in 1978 [47] and 1980 [48] as well as Pauling in 1982 [49]. Pauling, however, does extensively employ "Virion", though we do not consider this term to be equivalent to "Virus" or "Viral" as phages too are routinely described as possessing virions, for example, bacteriophage virion(s) [50, 51].

Somewhat ambiguously, and in a report rather than an otherwise formal publication, Stube et al. [14] provide what to our knowledge is the first association of "Virus" with archaeal viruses. In their Table 2, which lists "Organisms of the Northern Arm of the Great Salt Lake," the term "Halophages" as a "Scientific Name" is associated with "Bacterial viruses" as a "Common Name." Later, on page 48, "Halophages" is then associated with *Halobacterium halobium* (see a footnote of Table 1, supplementary materials, for the full quotation of the latter). Of interest, for their Table 2, Stube et al. cite Post, 1975, in an article titled "Life in the Great Salt Lake" [52]. In that paper, reference is made only to "Bacterial viruses" (page 44) along with the statement (page 46) "Virus parasites of the bacteria also live in the lake." The bacteria or at least "Halophilic bacteria" are indicated, however, as "*Halobacterium—Halococcus*" (page 44).

In 1982, in an article by Schnabel et al. [13] titled "*Halobacterium halobium* phage ϕH," the word "Virus" to our knowledge makes it first unambiguous as well as mainstream appearance in what would become the archaeal virus literature. This is found in the first sentence of the abstract: "Phage ϕH, a novel virus of the archaebacterium *Halobacterium halobium*, resembles in size and morphology two other *Halobacterium* phages." In addition, under keywords is found "archaebacteria/virus," though otherwise "Phage" is used far more often than "Virus" in this publication. Furthermore, in two locations the term "Archaebacterial phage" can be found for the first time. This latter usage seems to reflect the citing of Fox et al. [53] as found in their Introduction (page 87).

> The genus *Halobacterium* belongs to a group of organisms now known as archaebacteria which differ in many respects from both eucaryotes and eubacteria (Fox et al., 1980)... This paper presents the first analysis of the DNA of an archaebacterial phage.

By contrast, and notably, this publication by Fox et al. is not cited in Pauling's article of the same year [49], along with a lack of use of "Virus" or "Viral" other than in the reference section in that other 1982 publication.

The Schnabel et al. [13] article represents something of a transition in usage towards the inclusion of "Virus" as a descriptor, though this could be viewed more as a matter of style rather than something that is particularly profound, as phages at this point in time (i.e., 1982) had been known to be

viruses for decades (see, however, the immediately following paragraph). Certainly more than a matter simply of style, though, is the first use of "Archae..." as a qualifier for the type of virus under study. A third article from 1982, also from Schnabel and Zillig [54], splits the difference between these two other 1982 studies by citing Fox et al. [53] and referring to "Archaebacteria" while using "Phage" but neither "Virus" nor "Viral."

The simultaneous timing of the introduction of the terms "Virus" and "Archaebacteria," and variants, into what up to that point had been nearly indistinguishable from the otherwise "Bacteriophage" literature was not necessarily coincidental. In addition to Fox et al. [53] having been published in the journal *Science* in 1980, and thus widely disseminated well prior to 1982, it can be argued (Anonymous, personal communication) that "This is because Zillig was firmly of the view that archaea have viruses not bacteriophages. Also, because of the influence of Carl Woese, as the first conference on Archaebacteria was held in Germany, around this time (June-July, 1981)....Munich (that Carl attended)...at the very institute that Zillig and Schnabel worked." As relayed to us by Stedman (personal communication), "the 1981 meeting was at the Max Planck Institute for Biochemistry, (actually just outside Munich in Martinsried) where both Wolfram Zillig (Departmental Director) and Heinke Schnabel (and her husband Ralf) worked as Group leaders." The concept of archaebacteria thus appears to have been very much a part of the conversation among those individuals who then introduced the concept of "Virus" to what would become the archaeal virus literature.

These trends in usage continued in the following year, 1983. In an article by Rohrmann et al. [55], titled "Bacteriophages of *Halobacterium halobium*: virion DNAs and proteins," the term "Viruses" is used ("The difference in sizes of the proteins between the two viruses, in addition to the restriction endonuclease fragment patterns, indicate [sic] that these two viruses are not closely related."). Also used is "Archaebacteria" ("These results indicate that these halophages, the host of which is included among the archaebacteria..."). This article does not cite Fox et al. [53] but does cite an earlier though less prominent article by that same group, one that is titled, simply, "Archaebacteria" [56].

Further though not yet complete trending away from use of "Phage" as a descriptor of these viruses is seen in an article published by Janekovic et al. [57], also in 1983 and on which Zillig serves as a middle author. This paper cites Fox et al. [53], along with the even earlier Woese and Fox [9], and is titled "TTV1, TTV2, and TTV3, a family of viruses of the extremely thermophilic, anaerobic, sulfur reducing archaebacterium, *Thermoproteus tenax*." The publication is interesting, etymologically, for at least two reasons.

(1) This is the first archaeal virus paper for which *Halobacterium* spp. did not serve as hosts. Related to that point, these are the first viruses to be described of kingdom Crenarchaeota, versus the kingdom Euryarchaeota [10], where genus *Halobacterium* is a member of the latter.

(2) There is an indication (page 45) of "particles also resembling viruses of eukaryotes rather than "normal" bacteriophages." In particular, "These viruses are unlike bacteriophages known to date, including halobacteriophage ϕH which resembles phages of eubacteria in many respects."

We thus have a new host genus and kingdom as well as a new paradigm for the nature of archaeal viruses which, in at least some cases, are somewhat divergent from what is seen among the viruses of bacteria. It is possible that this combined novelty provided some basis for a change in perspective, that is, from describing these infectious agents of domain *Archaea* as "Phages" to instead describing them primarily as "Viruses." In particular, the observation of virions that were not phage-like was suggestive of a kinship between the viruses of what would come to be known as domain *Archaea* and viruses that otherwise are described simply as "Viruses," that is, eukaryotic viruses.

Consistent with this perspective, though placing the date of the transition approximately ten years later than as indicated here, is this 2012 sentiment from Felisberto-Rodrigues et al. [58]:

> Although viruses infecting archaea are known since the early 1970s [11], they have only been studied in detail very recently. The notion that these viruses constitute a variety of bacteriophages with head and tail (Caudovirales), reinforced by the initial findings, was challenged by the analyses of samples isolated by Zillig and coworkers from extreme environments, rich in hyperthermophilic archaea, including the Icelandic solfatara [59]. These analyses revealed the presence of a large diversity of viral morphotypes, including viruses of linear, spindle-shaped, spherical and more exotic forms, such as drops and bottle-shapes.

5. Vestiges of "Phage" to Describe Archaeal Viruses

Use of "Phage" as a descriptor for viruses of domain *Archaea* would continue at a rate of at least one reference per year to the present. At this point, consideration of these publications is less relevant to the transition to "Archaeal viruses" except for the sake of documenting ongoing use. We thus present these publications primarily in graphical as well as tabular form (Figure 2, and also Table 1, supplementary materials, resp.). In the latter we provide both absolute numbers of usage, by calendar year, as well as relative numbers.

In Figure 2, note particularly the post-2000 rise in absolute numbers of publications that use "Virus" (Figure 2(a)) as well as the associated somewhat steady decline in the number of publications that use "Phage" as a *fraction* of the total number of publications considering archaeal viruses (Figure 2(c)). Indeed, even as "Phage" has persisted in this literature, most of the same publications have also used "Virus" and this has been the case for three quarters or more

of these publications that we have examined individually since 1997 (Figure 2(d)). We speculate that a driver of the *ongoing* transition from "Phage" to "Virus," beyond personal preference by core authors working in the field, is inclusion of the term "Virus" in many names of archaeal virus isolates, such as *Sulfolobus* spindle-shaped virus 1 (SSV1) or *Haloarcula hispanica* pleomorphic virus 1 (HHPV1), for example, as summarized by Krupovic et al. [60].

As there is no centralized authority governing of the naming of archaeal viruses nor central control over whether they are described in names as viruses versus phages, the use of "Virus" in these names not only may help to drive the increasing use of "Virus" in the archaeal virus literature but also can represent a consequence of that trend. Particularly, the "rules" for governing virus naming range from proposals for formal guidelines [61] to the whims of individual discoverers, for example, "Corndog" as the name of a bacterial virus [62], and both can be influenced by what usage otherwise is currently trending within a field. It is also conceivable that specific instances of retention or inclusion of the term "Phage" to describe archaeal viruses are a consequence of demands made by editors or reviewers, for example, as appears to be the case for the title of the 1999 Arnold et al. [63] encyclopedia article (K. Stedman, personal communication).

6. The "Archaebacterial Virus" to "Archaeal Virus" Transition

A number of descriptors exist that are synonymous with the term "Archaeal virus" as summarized in part in Figure 1, as well as numerous variations in spelling. Among these are terms that describe a subset of such viruses. The latter includes "Halophage" (as also presented in Figure 1), "Haloarchaeophage" (ditto), and "Haloarchaeal virus" as well as "Crenarchaeal virus" and "Euryarchaeal virus" plus additional variations. Notably, the latter three concepts are constructed of a combination of the term "Archaeal" and that of "Virus," that is, just as is "Archaeal virus" itself. Nonetheless, though use of the term "Archaeal virus" now dominates within publications today, that was not always the case (Table 2, supplementary materials).

As we have considered above, there first was a transition from use of "Phage" to use of "Virus" for what now are known, at least in part, as archaeal viruses. This transition occurred in the early 1980s and it appeared to have coincided—as discussed above—with the introduction of the term "Archaebacteria" into the literature considering their viruses. The transition to "Archaeal virus" by necessity, however, could not occur until the term "*Archaea*" and therefore "Archaeal" came into being, with the former not invented until 1990 [10]. When, then, did the transition to the now familiar "Archaeal virus" actually occur? The answer, also as indicated in Figure 1, and again to the best of our knowledge, is 1993, with a publication by Nölling et al. [20]. Highlighting the transitional aspects of that publication, note that "Archaeal phage" is also used by these authors. The first use of the term "Virus" and "Archaeal" in the same sentence also can be found

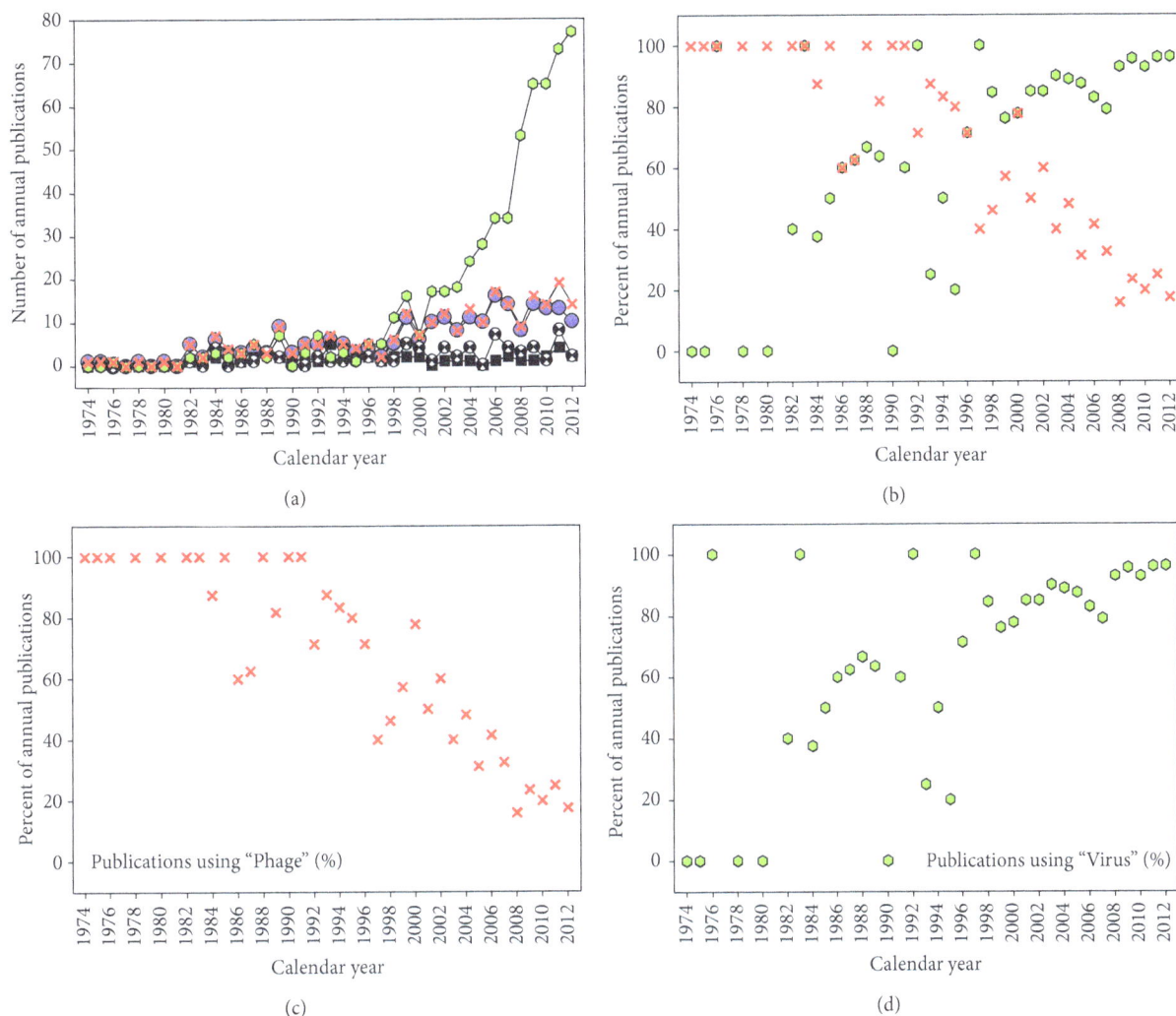

FIGURE 2: Counting publications using "Virus" (green hexagons), "Phage" (blue circles), "Halophage" (black squares), "Prophage" ("hourglass" circles), and an aggregate of all three phage terms, "All phage" (red Xs) to describe archaeal viruses. Presented are (a) the absolute number of publications, (b) the relative number of "All phage" versus "Virus", (c) just "All phage", and (d) just "Virus" (with (c) and (d) provided solely for clarity rather than to provide additional information). Shown are numbers of "Yes" statuses as indicated in Table 1, supplementary materials, but not "Yes/No" nor "No" entries.

in a Nölling et al. paper, from 1991 [64] (p. 1981): "The presence of viruses and virus-like particles has been described in various representatives of the archaeal domain…"

Given the forty-year span since the first archaeal virus publication in 1974 [11], the use of "Archaeal virus" thus is both 20 years old this year and has been in use for approximately half of the age of the archaeal virus literature. See Table 2 (supplementary materials) for consideration of various synonyms for "Archaeal virus," dates of use, and associated publications. All of the latter are also listed in Table 1 (supplementary materials).

7. Why Not Phage?

There clearly is a preference within the archaeoviral literature for describing the viruses of domain *Archaea* as "Viruses" rather than as "Phages" (Figure 2), and particularly as

"Archaeal viruses" (Table 2, supplementary materials). Even so, is this choice legitimate, particularly given the precedence of use of "Phage" as a description of viruses of prokaryotic organisms? That is, why not "Phage"? Rather than addressing the latter question directly, we instead consider its converse: Why "Phage"?

The term "Phage" appears to have first been applied to bacteriophages to describe a macroscopic phenomenon that is not necessarily always of viral origin, that is, the lysing of bacterial cultures such that, in particular, they are to clear from a turbid state. This transition can be legitimately described as an "eating" or "devouring" as is the Greek origin of "Phage," or alternatively something that "develops at the expense of something else" [65]. This "expense" is readily seen with phage-infected bacterial cultures, whether in broth or during the formation of plaques. Indeed, such macroscopically visible destruction of cells is not limited to phages as viruses. What *is* unique to phages, unlike other viruses, is that phages were discovered within the context of cell cultures, those of bacteria, rather than having only been subsequently studied within that context (e.g., tissue culture).

Contrasting bacterial phages, the first archaeal virus was discovered, at least initially, *not* specifically within the context of the destruction of a cell culture [11, pp. 680-681]:

> During an investigation of flagella from *H. salinarium*, phage particles were observed in some crudely purified flagellar preparations. Phage-containing preparations did not give rise to plaques when plated with exponentially growing *H. salinarium* cells. Attempts at ultraviolet induction of *H. salinarium* cultures resulted in normal death curves without any concomitant phage production. Maintenance of batch cultures, started from isolated *H. salinarium* colonies, in the logarithmic phase of growth, by means of serial transfer of cultures in late log phase to fresh media, resulted in eventual lysis of the cultures. Lysis may occur after the first transfer or be delayed for more than six transfers.

These first archaeal viruses even so were described as phages, as domain *Archaea* had yet to be defined, such that *H. salinarium* was considered to be a bacterium. Is it legitimate therefore for researchers who study these viruses nevertheless to *not* feel beholden to the concept of phages, but instead to explicitly tie these viruses to the larger phenomenon of viruses? Our answer to that question is that it is not that the use of the term "Archaeal virus" should be questioned by phage researches but instead that "Phage" as a descriptor should be tolerated and perhaps even appreciated by virologists due to its historical roots and ubiquity of use. There should be no requirement, in other words, for emulation of use of the term "Phage" by other virologists, whether those individuals choose to study the viruses of eukaryotes, archaeans, or even of bacteria. Indeed, it is quite clear that those who study the viruses of *Archaea* prefer "Virus" to "Phage" as a descriptor of those infectious agents (Figure 2).

These latter statements certainly should not be construed as any advocacy for avoidance, particularly within the bacterial virus literature, of the use of "Phage" to describe bacterial viruses. Indeed, its use such as in abstracts and titles can be helpful to the extent that it aids other phage biologists—that is, bacterial virus researchers—in identifying phage publications. It is important nevertheless to acknowledge both the legitimacy and, in many or most cases, the *greater* legitimacy of the term "Virus" as a general descriptor of acellular but encapsidated infectious agents. For example, from Prangishvili [65], page 551:

> It did not take d'Herelle long to realize that bacteriophages were the same type of biological entity as viruses of plants and animals. At this point, it would probably have been advantageous to play down the term "bacteriophage" and favor instead the more general term "bacterial virus". In any case, d'Herelle's concept of bacteriophages

as viruses of bacteria was widely accepted only much later, in the late 1930s.

Had viruses been discovered generally in a manner that was equivalent to that of phages, that is, within the context of the lysis and/or destruction of a cell culture, then it may have been legitimate to describe or at least to first have described all viruses as phages. Historically, however, this was not the case. The concept of viruses thus predates that of phages and so possesses greater legitimacy as a descriptor of what today universally are known as viruses, that is, than does the concept of phages.

8. Viruses, Tailed Viruses, Proviruses, and Haloviruses

An additional manner of considering the ideas covered in the previous section is that the concept of "Phage" is not synonymous with that of "Virus." Certainly there is overlap, and particularly so in terms of how we think of the viruses of domain *Bacteria*. It is important though to resist the temptation to use "Phage" simply as a means of avoiding being overly repetitive in the use of "Virus." If a virus of a prokaryotic host is not a bacterial virus then it is not a bacteriophage. Consequently, it would be suspect to call that virus a phage.

Similarly, it is important to keep in mind that the concept of "Tailed phage" is not identical to that of "Phage." There are in particular phages that possess tails along with phages that lack tails, for example, [22, 66, 67]. It therefore does not logically follow that if a virus is a phage then it possesses a tail, nor, given that there are a number of archaeal viruses which also possess tails, that if a virus possesses a tail then it is a phage, that is, as equivalent to bacteriophage or bacterial virus. One can also consider the lack of complete overlap between the concept of "Prophage" and that of "Provirus," with "Prophage" legitimately used to describe a provirus only to the extent that using "Phage" would also be a legitimate description of that virus. In each of these instances, to the extent that a trend exists among researchers to limit application of the term "Phage" in describing archaeal viruses (Figure 2), then it is reasonable to mostly avoid the use of "Phage" even as a qualified term in the archaeal virus literature. In particular, "Phage" as a synonym for "Tailed virus" is *not* a useful shorthand.

The term "Halophage" is by contrast less problematic and especially so when used as a general term, as in "Environmental halophages," though to our knowledge it is not sanctioned by any governing body, such as ICTV. Even here, though, it can be preferable to avoid the concept of "Phage" altogether by using "Halovirus" instead of "Halophage" since "Halovirus" implies neither that a virus is archaeal nor bacterial. Indeed, just as prophages are only a subset of proviruses, tailed phages only a subset of tailed viruses, and bacteriophages only a subset of all viruses, halophages could be considered to represent only a subset of haloviruses. Though complicated, it is also possible to use the phrase "Haloarchaeal virus" or its equivalent if there is a need to distinguish between haloviruses in terms of the domain of their hosts, versus, for example, "Halobacterial virus."

9. Conclusion

The viruses of domain *Archaea* were identified prior to appreciation of the existence of domain *Archaea* itself. Before introduction of the three-domain system of classification, it therefore was reasonable to describe these viruses as phages of bacteria, that is, as bacteriophages. Following the discovery of the concept of archaebacteria there appears to have been a shift away not just from "Bacteria" (as in bacteriophage) but also from use of "Phage" in describing these viruses, a shift that began in earnest especially during the early 1980s (Figure 2, and also Tables 1 and 2, supplementry materials). One then sees further movement during the early 1990s towards the use of "Archaeal viruses" as a descriptor (Table 2, supplementary materials).

Movement away from "Phage" has not been complete in considering archaeal viruses and we feel that there are various drivers towards retention of "Phage" or its derivatives to describe them. These drivers include inadvertent use (where an article otherwise uses "Virus" but in one or a few places "Phage" is substituted, in certain cases seemingly accidentally), failure to adequately distinguish between the concepts of "Phage" and "Virus" as applied to archaeal viruses, use of the term "Phage" to generally describe head-and-tail viruses, use of "Phage" within the context of "Halophage" (Table 1, supplementary materials, though "Halovirus" or "Haloarchaeal virus" are legitimate or even preferable substitutes), and the use of "Phage" within the context of "Prophage" (Table 1, supplementary materials, where "Provirus" would be preferable in the case of archaeal viruses). In addition, perhaps a special case is the use of "Phage" to describe specific archaeal virus isolates, such as *Methanobacterium* phage ΨM1". Also important towards impeding the transition from "Phage" to "Virus" in the archaeal virus literature has been a lumping together of bacteriophages and archaeal viruses in important databases [68].

Alternatives to "Archaeal virus" versus "Bacterial virus" versus "Eukaryotic virus" have been proposed by a number of authors. These include archaeovirus, bacteriovirus, and eukaryovirus as suggested by Raoult and Forterre [30] as well as Soler et al. [31], the slight variations of archeovirus, bacteriovirus, and eukaryavirus of Forterre and Prangishvili [32] or archeovirus, bacteriovirus, and eukaryovirus as presented as well by Pina et al. [69], and the bacterioviruses, archaealviruses, and eukaryalviruses of Comeau et al. [70]. See also archaebacteriophage versus eubacteriophage [71]. While we appreciate the unifying nature of these approaches, we do not necessarily advocate their general adoption if that would come at the expense of "Bacteriophage" (or "Phage"), "Archaeal virus," or simply "Virus" as descriptors of the viruses of members of domains *Bacteria*, *Archaea*, and *Eukarya*, respectively.

Within publications in which archaeal viruses are discussed, we nonetheless feel that the use of "Phage" should be limited in favor of "Archaeal virus" for viruses of domain *Archaea* and "Bacterial virus" for viruses of domain *Bacteria*, or alternatively the various "Virus-" based proposed alternatives as listed immediately above. We suggest this merely for the sake of limiting ambiguity within the archaeal virus literature, with a clear distinction therefore maintained between archaeal viruses on the one hand and bacterial viruses on the other. By contrast, within the bacterial virus literature, or when referring to specific phages (e.g., phage T4 or phage λ), it is neither likely nor necessarily desirable to abandon or reduce use of "Phage" or "Bacteriophage" as a general descriptor of the viruses of domain *Bacteria*.

Acknowledgments

Thanks are due to Ken Stedman as well as an anonymous reviewer for suggestions provided. For the sake of acknowledgement of potential conflict of interests, we note that S. T. Abedon "owns" the URLs http://www.phage.org/, http://www.bacteriophages.org/, http://www.thebacteriophages.org/, and http://www.archaealviruses.org/. The authors would like to dedicate this study to Inky, a.k.a., Inkblotch, the black silkie "travel chicken" (see http://www.biologyaspoetry.com/terms/birds.html), who lost her life to old injuries as well as old age in the course of completion of the paper. The authors also note sadly the passing of Carl Woese.

References

[1] P. Forterre, "Defining life: the virus viewpoint," *Origins of Life and Evolution of Biospheres*, vol. 40, no. 2, pp. 151–160, 2010.

[2] M. Krupovic and V. Cvirkaite-Krupovic, "Virophages or satellite viruses?" *Nature Reviews Microbiology*, vol. 9, no. 11, pp. 762–763, 2011.

[3] E. Norrby, "Nobel Prizes and the emerging virus concept," *Archives of Virology*, vol. 153, no. 6, pp. 1109–1123, 2008.

[4] P. Forterre, "Giant viruses: conflicts in revisiting the virus concept," *Intervirology*, vol. 53, no. 5, pp. 362–378, 2010.

[5] P. Forterre, "The virocell concept and environmental microbiology," *ISME Journal*, vol. 7, no. 2, pp. 233–236, 2012.

[6] A. M. Q. King, M. J. Adams, E. B. Carstens, and E. J. Lefkowitz, *Virus Taxonomy: Ninth Report of the International Committee on Taxonomy of Viruses*, Elsevier Academic Press, London, UK, 2012.

[7] P. Hyman and S. T. Abedon, "Bacteriophage host range and bacterial resistance," *Advances in Applied Microbiology*, vol. 70, pp. 217–248, 2010.

[8] P. Hyman and S. T. Abedon, "Smaller fleas: viruses of microorganisms," *Scientifica*, vol. 2012, Article ID 734023, 23 pages, 2012.

[9] C. R. Woese and G. E. Fox, "Phylogenetic structure of the prokaryotic domain: the primary kingdoms," *Proceedings of the National Academy of Sciences of the United States of America*, vol. 74, no. 11, pp. 5088–5090, 1977.

[10] C. R. Woese, O. Kandler, and M. L. Wheelis, "Towards a natural system of organisms: proposal for the domains *Archaea*, *Bacteria*, and *Eucarya*," *Proceedings of the National Academy of Sciences of the United States of America*, vol. 87, no. 12, pp. 4576–4579, 1990.

[11] T. Torsvik and I. D. Dundas, "Bacteriophage of *Halobacterium salinarium*," *Nature*, vol. 248, no. 5450, pp. 680–681, 1974.

[12] A. C. Wais, A. Kon, R. E. MacDonald, and B. D. Stollar, "Salt dependent bacteriophage infecting *Halobacterium cutirubrum* and *H. halobium*," *Nature*, vol. 256, no. 5515, pp. 314–315, 1975.

[13] H. Schnabel, W. Zillig, M. Pfaffle, R. Schnabel, H. Michel, and H. Delius, "Halobacterium halobium phage øH," *The EMBO Journal*, vol. 1, no. 1, pp. 87–92, 1982.

[14] J. C. Stube, F. J. Post, and D. B. Porcella, *Nitrogen Cycling in Microcosms and Application to the Biology of the Northern Arm of the Great Salt Lake*, Utah Water Research Laboratory, 1976.

[15] W. Zillig, F. Gropp, and A. Henschen, "Archaebacterial virus host systems," *Systematic and Applied Microbiology*, vol. 7, no. 1, pp. 58–66, 1986.

[16] L. Wünsche, "Importance of bacteriohphages in fermentation processes," *Acta Biotechnologica*, vol. 9, no. 5, pp. 395–419, 1989.

[17] F. Charbonnier, G. Erauso, T. Barbeyron, D. Prieur, and P. Forterre, "Evidence that a plasmid from a hyperthermophilic archaebacterium is relaxed at physiological temperatures," *Journal of Bacteriology*, vol. 174, no. 19, pp. 6103–6108, 1992.

[18] F. Gropp, B. Grampp, P. Stolt, P. Palm, and W. Zillig, "The immunity-conferring plasmid pφHL from the *Halobacterium salinarium* phage φH: nucleotide sequence and transcription," *Virology*, vol. 190, no. 1, pp. 45–54, 1992.

[19] J. N. Reeve, "Molecular biology of methanogens," *Annual Review of Microbiology*, vol. 46, pp. 165–191, 1992.

[20] J. Nölling, A. Groffen, and W. M. De Vos, "ΦF1 and ΦF3, two novel virulent, archaeal phages infecting different thermophilic strains of the genus Methanobacterium," *Journal of General Microbiology*, vol. 139, no. 10, pp. 2511–2516, 1993.

[21] S. D. Nuttall and M. L. Dyall-Smith, "HF1 and HF2: novel bacteriophages of halophilic archaea," *Virology*, vol. 197, no. 2, pp. 678–684, 1993.

[22] J. Maniloff and H. W. Ackermann, "Taxonomy of bacterial viruses: establishment of tailed virus genera and the order *Caudovirales*," *Archives of Virology*, vol. 143, no. 10, pp. 2051–2063, 1998.

[23] R. W. Hendrix, "Evolution: the long evolutionary reach of viruses," *Current Biology*, vol. 9, no. 24, pp. R914–R917, 1999.

[24] A. S. Lang, T. A. Taylor, and J. Thomas Beatty, "Evolutionary implications of phylogenetic analyses of the gene transfer agent (GTA) of *Rhodobacter capsulatus*," *Journal of Molecular Evolution*, vol. 55, no. 5, pp. 534–543, 2002.

[25] M. Ventura, C. Canchaya, R. D. Pridmore, and H. Brüssow, "The prophages of *Lactobacillus johnsonii* NCC 533: comparative genomics and transcription analysis," *Virology*, vol. 320, no. 2, pp. 229–242, 2004.

[26] U. Rass and S. C. West, "Synthetic junctions as tools to identify and characterize Holliday junction resolvases," *Methods in Enzymology*, vol. 408, pp. 485–501, 2006.

[27] J. Conrad, L. Niu, K. Rudd, B. G. Lane, and J. Ofengand, "16S ribosomal RNA pseudouridine synthase RsuA of *Escherichia coli*: deletion, mutation of the conserved Asp102 residue, and sequence comparison among all other pseudouridine synthases," *RNA*, vol. 5, no. 6, pp. 751–763, 1999.

[28] D. Ratel, J. L. Ravanat, F. Berger, and D. Wion, "N6-methyladenine: the other methylated base of DNA," *BioEssays*, vol. 28, no. 3, pp. 309–315, 2006.

[29] M. Skurnik and E. Strauch, "Phage therapy: facts and fiction," *International Journal of Medical Microbiology*, vol. 296, no. 1, pp. 5–14, 2006.

[30] D. Raoult and P. Forterre, "Redefining viruses: lessons from Mimivirus," *Nature Reviews Microbiology*, vol. 6, no. 4, pp. 315–319, 2008.

[31] N. Soler, E. Marguet, J. M. Verbavatz, and P. Forterre, "Virus-like vesicles and extracellular DNA produced by hyperthermophilic archaea of the order Thermococcales," *Research in Microbiology*, vol. 159, no. 5, pp. 390–399, 2008.

[32] P. Forterre and D. Prangishvili, "The great billion-year war between ribosome- and capsid-encoding organisms (cells and viruses) as the major source of evolutionary novelties," *Annals of the New York Academy of Sciences*, vol. 1178, pp. 65–77, 2009.

[33] B. Liu, F. Zhou, S. Wu, Y. Xu, and X. Zhang, "Genomic and proteomic characterization of a thermophilic *Geobacillus* bacteriophage GBSV1," *Research in Microbiology*, vol. 160, no. 2, pp. 166–171, 2009.

[34] D. Raoult, "Editorial: giant viruses from amoeba in a post-darwinist viral world," *Intervirology*, vol. 53, no. 5, pp. 251–253, 2010.

[35] V. I. Agol, "Which came first, the virus or the cell?" *Paleontological Journal*, vol. 44, no. 7, pp. 728–736, 2010.

[36] H. Copeland, "The kingdoms of organisms," *Quarterly Review of Biology*, vol. 13, pp. 383–420, 1938.

[37] R. H. Whittaker, "New concepts of kingdoms of organisms," *Science*, vol. 163, no. 3863, pp. 150–160, 1969.

[38] R. Y. Stanier and C. B. van Niel, "The concept of a bacterium," *Archiv für Mikrobiologie*, vol. 42, no. 1, pp. 17–35, 1962.

[39] J. Sapp, "The prokaryote-eukaryote dichotomy: meanings and mythology," *Microbiology and Molecular Biology Reviews*, vol. 69, no. 2, pp. 292–305, 2005.

[40] S. T. Abedon, C. Thomas-Abedon, A. Thomas, and H. Mazure, "Bacteriophage prehistory: is or is not Hankin, 1896, a phage reference?" *Bacteriophage*, vol. 1, no. 3, pp. 174–178.

[41] S. T. Abedon, "Salutary contributions of viruses to medicine and public health," in *Viruses: Essential Agents of Life*, G. Witzany, Ed., pp. 389–405, Springer, Heidelberg, Germany, 2012.

[42] H. Ochman, "Radical views of the tree of life: genomics update," *Environmental Microbiology*, vol. 11, no. 4, pp. 731–732, 2009.

[43] E. Mayr, "Two empires or three?" *Proceedings of the National Academy of Sciences of the United States of America*, vol. 95, no. 17, pp. 9720–9723, 1998.

[44] C. R. Woese, "There must be a prokaryote somewhere: microbiology's search for itself," *Microbiological Reviews*, vol. 58, no. 1, pp. 1–9, 1994.

[45] J. S. Koti, M. C. Morais, R. Rajagopal, B. A. L. Owen, C. T. McMurray, and D. L. Anderson, "DNA packaging motor assembly intermediate of bacteriophage φ29," *Journal of Molecular Biology*, vol. 381, no. 5, pp. 1114–1132, 2008.

[46] R. Schaefer, R. Hinnen, and R. M. Franklin, "Further observations on the structure of the lipid containing bacteriophage PM2," *Nature*, vol. 248, no. 5450, pp. 681–682, 1974.

[47] T. Torsvik and I. D. Dundas, "Halophilic phage specific for *Halobacterium salinarium* str. 1," in *Energetics and Structure of Halophilic Microorganisms*, S. R. Caplan and M. Ginzburg, Eds., pp. 609–614, Elsevier/North Holland, 1978.

[48] T. Torsvik and I. D. Dundas, "Persisting phage infection in *Halobacterium salinarium* str. 1," *Journal of General Virology*, vol. 47, no. 1, pp. 29–36, 1980.

[49] C. Pauling, "Bacteriophages of *Halobacterium halobium*: isolation from fermented fish sauce and primary characterization," *Canadian Journal of Microbiology*, vol. 28, no. 8, pp. 916–921, 1982.

[50] S. R. Casjens and E. B. Gilcrease, "Determining DNA packaging strategy by analysis of the termini of the chromosomes in tailed-bacteriophage virions," *Methods in Molecular Biology*, vol. 502, pp. 91–111, 2009.

[51] N. J. Bennett and J. Rakonjac, "Unlocking of the filamentous bacteriophage virion during infection is mediated by the C domain of pIII," *Journal of Molecular Biology*, vol. 356, no. 2, pp. 266–273, 2006.

[52] F. J. Post, "Life in the Great Salt Lake," *Utah Science*, vol. 36, no. 2, pp. 43–47, 1975.

[53] G. E. Fox, E. Stackebrandt, R. B. Hespell et al., "The phylogeny of prokaryotes," *Science*, vol. 209, no. 4455, pp. 457–463, 1980.

[54] H. Schnabel and W. Zillig, "Circular structure of the genome of phage φH in a lysogenic *Halobacterium halobium*," *Molecular and General Genetics*, vol. 193, no. 3, pp. 422–426, 1984.

[55] G. F. Rohrmann, R. Cheney, and C. Pauling, "Bacteriophages of *Halobacterium halobium*: virion DNAs and proteins," *Canadian Journal of Microbiology*, vol. 29, no. 5, pp. 627–629, 1983.

[56] C. R. Woese, L. J. Magrum, and G. E. Fox, "Archaebacteria," *Journal of Molecular Evolution*, vol. 11, no. 3, pp. 245–252, 1978.

[57] D. Janekovic, S. Wunderl, and I. Holz, "TTV1, TTV2 and TTV3, a family of viruses of the extremely thermophilic, anaerobic, sulfur reducing archaebacterium *Thermoproteus tenax*," *Molecular and General Genetics*, vol. 192, no. 1-2, pp. 39–45, 1983.

[58] C. Felisberto-Rodrigues, S. Blangy, A. Goulet et al., "Crystal structure of ATV^{ORF273}, a new fold for a thermo- and acidostable protein from the *Acidianus* two-tailed virus," *PLoS ONE*, vol. 7, no. 10, Article ID e45847, 2012.

[59] W. Zillig, A. Kletzin, C. Schleper et al., "Screening for *Sulfolobales*, their plasmids and their viruses in Icelandic solfataras," *Systematic and Applied Microbiology*, vol. 16, no. 4, pp. 609–628, 1994.

[60] M. Krupovic, M. F. White, P. Forterre, and D. Prangishvili, "Postcards from the edge: structural genomics of archaeal viruses," *Advances in Virus Research*, vol. 82, pp. 33–62, 2012.

[61] A. M. Kropinski, D. Prangishvili, and R. Lavigne, "Position paper: the creation of a rational scheme for the nomenclature of viruses of *Bacteria* and *Archaea*," *Environmental Microbiology*, vol. 11, no. 11, pp. 2775–2777, 2009.

[62] G. F. Hatfull, "Mycobacteriophages: genes and genomes," *Annual Review of Microbiology*, vol. 64, pp. 331–356, 2010.

[63] H. P. Arnold, K. M. Stedman, and W. Zillig, "Archaeal phages," in *Encyclopedia of Virology*, R. G. Webster and A. Granoff, Eds., pp. 76–89, Academic Press, London, UK, 1999.

[64] J. Nölling, M. Frijlink, and W. M. de Vos, "Isolation and characterization of plasmids from different strains of *Methanobacterium thermoformicicum*," *Journal of General Microbiology*, vol. 137, no. 8, pp. 1981–1986, 1991.

[65] D. Prangishvili, "Editorial: the 90th anniversary of "bacteriophage"," *Research in Microbiology*, vol. 158, no. 7, pp. 551–552, 2007.

[66] H. W. Ackermann, "Tailed bacteriophages: the order caudovirales," *Advances in Virus Research*, vol. 51, pp. 135–201, 1998.

[67] H. W. Ackermann, "Bacteriophage observations and evolution," *Research in Microbiology*, vol. 154, no. 4, pp. 245–251, 2003.

[68] M. R. Clokie, A. D. Millard, A. V. Letarov, and S. Heaphy, "Phages in nature," *Bacteriophage*, vol. 1, no. 1, pp. 31–45, 2011.

[69] M. Pina, A. Bize, P. Forterre, and D. Prangishvili, "The archeoviruses," *FEMS Microbiology Reviews*, vol. 35, no. 6, pp. 1035–1054, 2011.

[70] A. M. Comeau, G. F. Hatfull, H. M. Krisch, D. Lindell, N. H. Mann, and D. Prangishvili, "Exploring the prokaryotic virosphere," *Research in Microbiology*, vol. 159, no. 5, pp. 306–313, 2008.

[71] M. S. Mitchell and V. B. Rao, "Novel and deviant Walker A ATP-binding motifs in bacteriophage large terminase-DNA packaging proteins," *Virology*, vol. 321, no. 2, pp. 217–221, 2004.

tRNA-Derived Fragments Target the Ribosome and Function as Regulatory Non-Coding RNA in *Haloferax volcanii*

Jennifer Gebetsberger,[1,2] **Marek Zywicki,**[3,4] **Andrea Künzi,**[1] **and Norbert Polacek**[1,3]

[1] *Department of Chemistry and Biochemistry, University of Bern, Freiestraße 3, 3012 Bern, Switzerland*
[2] *Graduate School for Cellular and Biomedical Sciences, University of Bern, 3012 Bern, Switzerland*
[3] *Division of Genomics and RNomics, Innsbruck Biocenter, Innsbruck Medical University, Innrain 80/82, 6020 Innsbruck, Austria*
[4] *Laboratory of Computational Genomics, Institute of Molecular Biology and Biotechnology, Adam Mickiewicz University, 61-712 Poznan, Poland*

Correspondence should be addressed to Norbert Polacek, norbert.polacek@dcb.unibe.ch

Academic Editor: Anita Marchfelder

Nonprotein coding RNA (ncRNA) molecules have been recognized recently as major contributors to regulatory networks in controlling gene expression in a highly efficient manner. These RNAs either originate from their individual transcription units or are processing products from longer precursor RNAs. For example, tRNA-derived fragments (tRFs) have been identified in all domains of life and represent a growing, yet functionally poorly understood, class of ncRNA candidates. Here we present evidence that tRFs from the halophilic archaeon *Haloferax volcanii* directly bind to ribosomes. In the presented genomic screen of the ribosome-associated RNome, a 26-residue-long fragment originating from the 5′ part of valine tRNA was by far the most abundant tRF. The Val-tRF is processed in a stress-dependent manner and was found to primarily target the small ribosomal subunit *in vitro* and *in vivo*. As a consequence of ribosome binding, Val-tRF reduces protein synthesis by interfering with peptidyl transferase activity. Therefore this tRF functions as ribosome-bound small ncRNA capable of regulating gene expression in *H. volcanii* under environmental stress conditions probably by fine tuning the rate of protein production.

1. Introduction

Recent research revealed small nonprotein coding RNAs (ncRNAs) as pivotal players in regulatory networks shaping cellular life in all three phylogenetic domains (reviewed in [1]). Regulatory functions of ncRNAs are diverse ranging from chromosome biology, to epigenetics, transcription, and translation regulation [2]. This variety of ncRNA functions is mirrored by their complex genomics and biogenesis. Recent data suggest that a single ncRNA transcript can adopt different structures and thus perform distinct functional roles depending on different posttranscriptional processing events. Four years ago it could be shown that a functional small nucleolar RNA, which was initially processed from an mRNA intron, could function as a microRNA after further processing took place [3]. Many other reports followed expanding the list of ncRNA species that are target of further downstream processing into novel regulatory entities [4–7]

thus it appears that we have just started to disentangle the hidden layers of the transcriptome.

One recent example of ncRNA processing presents the emerging group of RNA fragments derived from mature tRNAs or precursor tRNAs [8]. These tRNA-derived fragments (tRFs) have long been regarded as random byproducts of tRNA biogenesis or degradation, but are now recognized as emerging players in tRNA biology. tRFs have been uncovered by deep sequencing projects in all three domains of life (reviewed in [8]). These tRFs have been shown to be produced under specific growth conditions and differ in size and sequence, thus indicating physiological relevance. Their biological functions, however, remained largely enigmatic. Several reports show a possible involvement of tRFs in cell proliferation [9], in the siRNA and microRNA pathway [8, 10–12] and in protein biosynthesis [13–16]. In those studies presenting functional experiments, it appears that tRFs are capable of globally downregulating protein synthesis. This tRF-induced inhibition of translation is not however due to a reduced pool of genuine tRNA molecules. A recent study by Ivanov et al. suggests that specific 5′ tRFs inhibit translation initiation in oxidatively stressed human cell lines by recruiting eIF4E/G/A from capped mRNAs or eIF4G/A from uncapped mRNAs [13]. If translation initiation is the sole target for tRFs in regulating protein synthesis or whether other phases of the ribosomal elongation cycle might be targeted by this ncRNA class remains to be seen.

In this study, we present evidence that in the archaeal model species *Haloferax volcanii* tRFs are processed in a stress-dependent manner and are capable of directly binding to the ribosome. Our deep-sequencing analysis of the ribosome-associated small RNome revealed four classes of tRFs. We present experimental evidence that one of these tRFs primarily target the small ribosomal subunit and demonstrate its inhibitory role in peptide bond formation and during *in vitro* translation. This is the first report of a tRF directly binding to the key enzyme of protein synthesis and therefore tRFs in *H. volcanii* represent ribosome-targeted regulatory ncRNA species.

2. Material and Methods

2.1. Strain and Growth Conditions. *Haloferax volcanii* strain H26 was grown aerobically at 42°C in complex medium (2.9 M NaCl, 150 mM $MgSO_4 \times 7 H_2O$, 60 mM KCl, 4 mM $CaCl_2$, 50 mM Tris-HCl (pH 7.2), 0.45% (w/v) tryptone, 0.275% (w/v) yeast extract). Cells were subjected to different temporary and permanent stress conditions as described [17]. In the case of temporary stress *H. volcanii* precultures were grown under standard conditions to the mid exponential phase ($OD_{600} = 0.5 \pm 0.1$) before environmental stress was applied for one respectively four hours. These stress conditions included cold shock at 30°C and heat shock at 60°C. For pH stress the cultures were either supplemented with 0.1 M Tris-HCl (pH 8.5-9.5) resulting in an elevated pH or with 12 mM Na(O)Ac for low pH conditions. Oxidative stress was induced by the addition of H_2O_2 to a final concentration of 0.78% (low oxidative stress) and 1.43% (high oxidative stress). For the ultraviolet (UV) stress the cells were irradiated with a UV dose of 120 J/m^2 for 30 seconds. For osmotic stress, the cells were collected by centrifugation and resuspended in the appropriate salt stress medium (0.9 M NaCl, 1.5 M NaCl, 300 mM $MgSO_4 \times 7 H_2O$). For setting a permanent stress the cells were challenged from the inoculation until an $OD_{600} = 1.0 \pm 0.2$ was reached. These stress conditions included UV irradiation, alkaline stress (pH 8.5), and growth under different salt concentrations (1.5 M NaCl, 300 mM $MgSO_4 \times 7 H_2O$) as described above. The growth was monitored at selected time points by measuring the absorbance at OD_{600}.

2.2. cDNA Library Generation. For the cDNA library preparation equal volumes of unstressed and stressed *H. volcanii* cells (for different stress conditions see above) were pooled and ribosomes were isolated basically as described [18] using buffer P [19] (3.4 M KCl, 100 mM Mg(OAc)₂, 6 mM

2-mercaptoethanol, 10 mM Tris-HCl pH 7.6). Ribosome-associated RNA was size-selected (ranging from 20–500 nucleotides), extracted, precipitated with ethanol, and used for cDNA library construction as described [7]. The cDNA library was deepsequenced (max. read lengths 76 bp) using the illumina platform (FASTERIS SA).

2.3. Data Analysis of the Deep Sequencing Results. The analysis of the sequence reads was performed using the APART pipeline (automated pipeline for annotation of RNA transcripts) [7]. In short, the sequences were cleaned by removal of the adaptor sequences and subsequently mapped to the *H. volcanii* genome where overlapping reads were assembled into contigs. From all 73.5 Mio raw reads originally obtained only reads with a minimal length of 18 nucleotides that contained both the 5′ and 3′ adaptors were further analyzed. At the genome mapping stage only a single mismatch was allowed.

2.4. Polysome Gradients. *H. volcanii* cells grown under normal conditions and permanent pH 8.5 stress condition (as described above) were collected by centrifugation, frozen in liquid nitrogen, and grounded three times in a precooled mortar. The resulting powder was resuspended in buffer A (3 M KCl, 150 mM $MgCl_2$, 6 mM 2-mercaptoethanol, 10 mM Tris-HCl pH 7.6, 0.5 mM DTT, 26 μM tetracycline hydrochloride) and the cell debris was removed by centrifugation (30,000 ×g, 15 min). The supernatant was treated with RNase-free DNase (10 min on ice) and subjected to a second centrifugation (30,000 ×g, 15 min). The supernatant, referred to as S30, was layered onto a linear sucrose gradient containing 10–40% (w/v) sucrose prepared in buffer A and centrifuged in a Beckman SW-41 rotor (4 h, 35,000 rpm, 4°C). Fractions containing polysomes, 50S, and 30S subunits were collected while monitoring the absorbance at 260 nm. For downstream northern blot analyses, the fractions were dialyzed two times against pure water (2 h, 4°C) and precipitated with ethanol before separation on 8% polyacrylamide gels.

2.5. Northern Blot Analysis. Total RNA from *H. volcanii* grown under selected conditions was isolated using TRI Reagent (Sigma Aldrich) according to the manufacturer's instruction. Size-selected ribosome-associated RNA was prepared by isolating RNA from crude ribosomes by phenol/chloroform extraction followed by size fractionation employing denaturing 8% PAGE. RNAs in the size range between 20 and 500 nt were excised from the gel, passively eluted into 0.3 M NaOAc at 4°C, and ethanol precipitated. Ribosomal particles (polysomes, 50S, or 30S subunits) were isolated using sucrose gradient centrifugation (as described above), followed by RNA extraction using phenol/chloroform and ethanol precipitation. For northern blot analysis the RNA (2.5–10 μg) was separated on 8% denaturing polyacrylamide gel (7 M Urea, 1 × TBE buffer), transferred onto nylon membranes (Amersham Hybond-N⁺, GE Healthcare) using a semidry blotting apparatus (V20-SDB, Scie-Plas). After immobilizing of RNA using a UV cross-linker (BLX-254, Vilber Lourmat) the nylon membranes were prehybridized for 30 minutes in hybridization buffer (1 M sodium phosphate buffer pH 6.2, 7% SDS). DNA oligonucleotides complementary to the RNA of interest were end-labeled with [γ-^{32}P]-ATP and T4 polynucleotide kinase. Hybridization was carried out in 1 M sodium phosphate buffer (pH 6.2, 7% SDS) over night at 52°C in a hybridization oven. Blots were washed once (room temperature) with washing buffer I (30 mM sodium citrate pH 7.0, 0.3 M NaCl, 0.1% SDS) and once (room temperature or 52°C) with washing buffer II (1.5 mM sodium citrate pH 7.0, 15 mM NaCl, 0.1% SDS) for 10 minutes each. Membranes were exposed to phosphor imaging screens. The signals were monitored with a phosphor imager (FLA-3000; Fuji Photo Film) and analyzed quantitatively with the densitometric program Aida Image Analyzer. Following DNA oligonucleotides were used for northern blotting targeting tRFs (depicted in 5′–3′ direction): Val(GAC) TCATAACCAGACTAGACCACCAACCC, Cys(GCA) CCG-AACTCTGCCACCTTGGC Ser(GCT) CCAGGCTTGGCT-ACCGCAAC, Leu(CAG) CCAGACTTGGCTATCCCTGC, Arg(GCG) ATAGTCCACTACCCTATCAGGAC, Ala(GGC) ATCTACCCCTGATCTACGAGCCC, Leu(GAG) CCTGGC-TTGGCTACCCACGC, Val(CAC) TCATAACCTGGCTAG-ACCACCAACCC, Gly(CCC) CTACCACTGGACCATCGG-CGC, Asp(GTC) TCGTATGATGGGCCACTACACCAC-CCGGGC, Leu(CAA) ATTTGAACCCACGGACCCCTA-CGGGAGCGGAT, Ser(GGA) CCTTACCGCTCGGCCATC-CTGGC, Val(GAC) TGGGTTGGGGCAGATTTGAA.

2.6. In Vitro Binding Studies. Binding studies of Val-tRF were performed using a dot blot-filtering device. For the experiments 10 pmol of *H. volcanii* ribosomal particles (70S, 50S, 30S) were incubated with 4 pmol 5′[^{32}P]-end-labeled synthetic RNA (1.750 cpm/pmol) (Val-tRF 5′-GGGUUGGUGGUCUAGUCUGGUUAUGA-3′, Ile-tRF 5′.GGGCCAAUAGCUCAGUCAGGUUGAGC-3′) in 25 µL binding buffer (f.c. 20 mM Hepes/KOH pH 7.6, 6 mM MgAc$_2$, 150 mM NH$_4$Cl, 4 mM 2-mercaptoethanol, 2 mM spermidine, 50 µM spermine). As positive control *E. coli* 70S were incubated with 50 µg polyU-mRNA and 5′[^{32}P]-end-labeled deacylated tRNA^{Phe} (7,000 cpm). After 30 min incubation on ice the reactions were filtered through a nitrocellulose membrane (0.45 µm diameter) using a vacuum device, followed by two washing steps with cold binding buffer. The membrane was quantified as described above.

2.7. In Vitro Translation. For *in vitro* translation an S30 extract from *H. volcanii* was prepared as described above and stored in aliquots at −80°C. For 16 translation reactions the following components were added to 150 µL S30 extract: 25 µL of 10 × translation cocktail (100 mM HEPES-KOH pH 7.4, 15 mM Mg(OAc)$_2$, 750 mM KOAc, 4 mM GTP, 10 mM ATP, 500 µM of all amino acids except methionine and cysteine), 2.5 µL creatine phosphokinase (10 mg/mL, Roche), 5 µL creatine phosphate (0.6 M, Roche), 2.5 µL Mg(OAc)$_2$ (100 mM), and 5 µL bulk brewer's yeast tRNA (10 mg/mL, Roche). Before the addition of 10 µL [^{35}S]

cysteine/methionine (10 µCi/µL), a preincubation step for 10 min at 23°C was performed. 12 µL of translation mixture were used per reaction to which 50 pmol tRFs or H$_2$O was added to a final volume of 15 µL. *In vitro* translation was performed at 23°C for 30 min and was stopped by the addition of 1 mL 20% TCA and incubation at 95°C for 15 minutes. The TCA-precipitated proteins were filtered through glass-fiber filter and quantified by liquid scintillation counting. The background counts observed in analogous samples but in the absence of S30 extract (in average 30-fold lower than in complete *in vitro* translation reactions containing S30) was always subtracted.

2.8. Peptidyl Transferase Assay. Peptide bond formation activity was tested using a modified peptidyl transferase assay under fragment reaction conditions [20] with N-acetyl-[^{3}H]Phe-tRNA (15,000 cpm/pmol) [21] as the donor substrate and puromycin as acceptor. The assay was performed in 25 µL of high salt buffer (2 M KCl, 30 mM Tris-HCl pH 7.6, 0.4 M NH$_4$Cl, 60 mM Mg(OAc)$_2$, 7 mM 2-mercaptoethanol, 2 µM spermidine, 0.05 mM spermine) containing 10 pmol of *H. volcanii* 70S ribosomes, respectively, 50S subunits, 0.8 pmol N-acetyl-[^{3}H]Phe-tRNA, puromycin (f.c. 1 mM) and, as indicated, 100 pmol of the RNA of interest (Val-tRF, Ile-tRF, or scr 5′-GUUGUUCGCUGUGAGGUGGUGAAUUG-3′). The transpeptidation reaction was initiated by the addition of 12.5 µL cold methanol (f.c. 33%) and incubated for 3.5 h on ice. In order to test for potential P-site competition between N-acetyl-[^{3}H]Phe-tRNA and Val-tRF, 0.8 pmol of the peptidyl-tRNA analog was preincubated in the presence of 100 pmol Val-tRF and 33% methanol for 20 minutes on ice before the reaction was initiated by the addition of puromycin. In a control experiment, the Val-tRF was added simultaneously with the A-site substrate puromycin to ribosomal complexes that carried already a prebound N-acetyl-[^{3}H]Phe-tRNA (0.8 pmol). In all cases the reaction was terminated by the addition of 5 µL 10 M KOH followed by an incubation for 15 min at 37°C. After the addition of 100 µL 1 M KH$_2$PO$_4$ the reaction product (N-acetyl-[^{3}H]Phe-puromycin) was extracted into 1 mL cold ethyl acetate by vortexing for 1 min. 800 µL of the upper organic phase were measured using a liquid scintillation counter.

3. Results

3.1. tRNA-Derived Fragments Are Abundant in the RNA Interactome of H. volcanii Ribosomes. In the course of studying the ncRNA interactome of the archaeal ribosome, we have constructed a specialized cDNA library from small RNAs (sized 20–500 nt) that copurifies with ribosomes of *H. volcanii* under different environmental stress conditions. In order to select for functional ncRNAs putatively involved in translation regulation we have set temporary stress (UV, cold shock, heat shock, low pH, high pH, oxidative stress, low NaCl, high MgSO$_4$) for one, respectively four hours, as well as permanent stress (UV, high pH, low NaCl, high MgSO$_4$). For the library construction small RNAs that copurified with

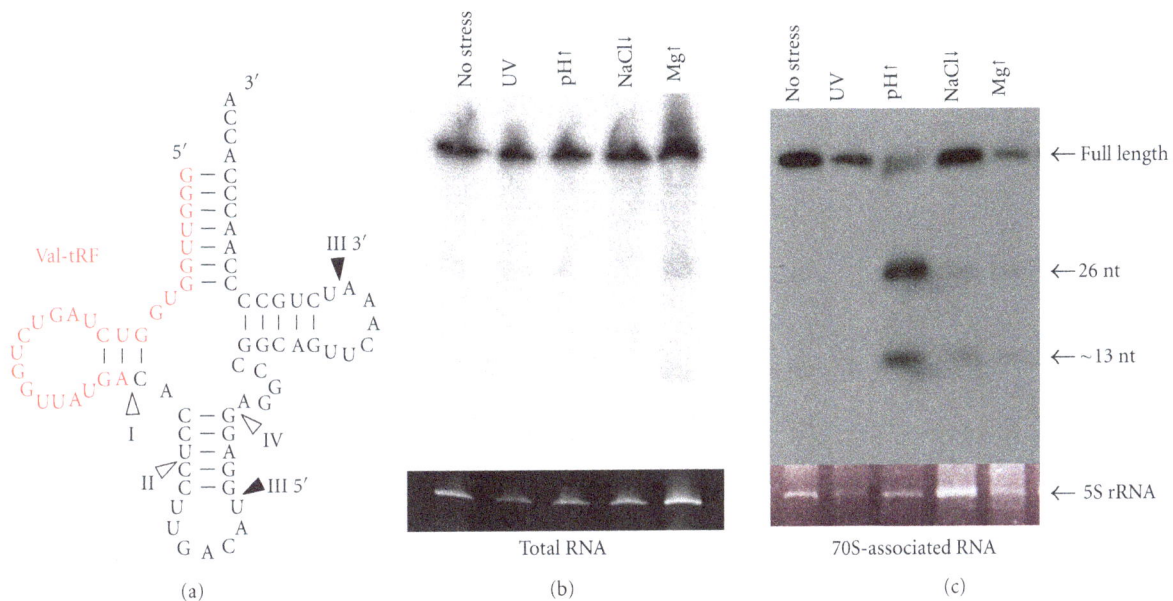

FIGURE 1: Processing and expression of Val-tRF. (a) Secondary structure of *H. volcanii* Val-tRNA with the Val-tRF depicted in red. Arrowheads indicate the processing positions for the four different observed tRF classes (I–IV). Open arrowheads on the Val-tRNA structure indicate the 3′ ends of the tRFs for class I, as well as the analogous positions for the other tRF classes II, and IV. tRFs from classes I, II, and IV are all processed from the 5′ end of mature tRNAs. Filled arrowheads mark the 5′ and 3′ ends of tRFs derived from class III. Northern blot analyses for Val-tRF were performed using (b) total RNA or (c) ribosome-associated RNA. RNA was isolated from unstressed *H. volcanii* cells (no stress), or cells grown under different permanent environmental stress conditions (ultraviolet stress (UV), high pH (pH↑), 0.9 M NaCl (NaCl↓), 300 mM MgSO$_4$ (Mg↑)). Arrows indicate the full-length tRNA and the detected processing products. In all panels 5S rRNA served as internal loading control.

H. volcanii ribosomes isolated from stressed and unstressed cultures were reverse transcribed into cDNA. The subsequent deep sequencing yielded 73.5 million raw reads which were analyzed using the APART pipeline [7], ending up with 19.2 million reads for downstream analysis which were grouped into 6.250 putative ribosome-associated ncRNA candidates (our unpublished data). Among other potential novel ncRNAs we significantly observed the emersed presence of tRNA-derived fragments (tRFs) originating from 14 different tRNAs (Table 1). Based on their sizes and processing ends we have categorized them into four different classes of which class I, with a size distribution of 20–26 nt and the 3′-end located in the D-stem, is the most prominent one (1.3 million reads). Interestingly, tRFs of all classes (with the exception of class IV) are processed from the 5′ end of the mature tRNA (for a schematic representation see Figure 1(a)). No reads encompassing 3′-derived tRFs were obtained in our library. The presence of tRFs in a library constructed from the ribosome-bound RNome suggests that 5′ tRFs (but not 3′ tRFs) are capable of interacting with the translation machinery. Bioinformatic analyses failed to detect any correlations between the processed tRNAs and the codon usage statistics in *H. volcanii*.

3.2. Valine tRF Is Highly Abundant and Associates with the Ribosome.
With more than 1.1 million reads (85% of all detected tRF reads) a 26-residue-long fragment deriving from valine tRNA showed unexpected high abundancy

(Table 1). Whereas in *H. volcanii* four genes give rise to Val-tRNAs, our observed tRF originates from two paralogous tRNA genes decoding for valine (GAC), positioned adjacent in the genome. To confirm the presence of this tRF and to investigate its potential association with ribosomes, northern blot analyses were performed. While only very faint bands for Val-tRNA processing products were seen when total RNA was used for northern blotting (Figure 1(b)), clear bands of the expected size of about 26 nucleotides were detected in blots using ribosome-associated RNAs (Figure 1(c)). Obviously the Val-tRF is enriched in the latter RNA preparation thus supporting our assumption of Val-tRF being a ribosome-bound RNA species. Even extended incubation times with a northern blot probe targeting the 3′ part of Val-tRNA failed to identify any processing product (data not shown). Thus only the 5′ tRF has the potential to bind to ribosomes. From Figure 1(c), it is also obvious that Val-tRF is processed differently in response to diverse stress conditions. At elevated pH this tRF is most abundant.

3.3. tRFs Are Differentially Expressed under Stress.
In order to clarify whether tRFs from all four classes found in our library are part of the ribosome-derived RNome, northern blot analyses were performed (Figure 2). It turned out that indeed signals for almost all tested tRFs were observed; however, the processing pattern as well as the abundance was distinct. While the class I tRF derived from Ala-tRNA showed a very similar processing pattern to Val-tRF at elevated pH as

TABLE 1: Overview on tRF sequence reads in the *H. volcanii* cDNA library.

	tRNA	tRNA #	tRNA-derived fragment		Sequence (5′–3′)	Length	Reads	Reads/million
			Begin	End				
	ValGAC	19	2,328,216	2,328,241	GGGUUGGGUGGUCUAGUCUGGUUAUGA	26 nt	1,149,000	59,548
	ValGAC	20	2,328,336	2,328,361				
	CysGCA	13	1,603,383	1,603,402	GCCAAGGUGGCAGAGUUCGG	20 nt	95,214	4,935
	SerGCT	29	2,620,371	2,620,352	GUUGCGGUAGCCAAGCCUGG	20 nt	44,220	2,292
	LeuCAG	30	2,617,916	2,617,897	GCAGGGAUAGCCAAGUCUGG	20 nt	42,006	2,177
Class I	ArgGCG	43	452,423	452,401	GUCCUGAUAGGGUAGUGGACUAU	23 nt	5,144	267
	AlaGGC	10	1,048,559	1,048,581	GGGCUCGUAGAUCAGGGGUAGAU	23 nt	4,332	225
	LeuGAG	31	2,564,858	2,564,839	GCGUGGGUAGCCAAGCCAGG	20 nt	3,606	187
	ValCAC	33	2,264,075	2,264,050	GGGUUGGUGGGUCUAGCCAGGUUAUGA	26 nt	3,576	185
	GlyCCC	35	1,733,386	1,733,366	GCGCCGAUGGUGGUCCAGUGGUAG	21 nt	91	5
Class II	AspGTC	48	311,827	311,755	GCCCGGGUGGUGUAGUGGCCCAUCAUACGA	30 nt	2,094	109
	AspGTC	49	311,725	311,696		30 nt	1,962	102
Class III	LeuCAA	26	2,693,522	2,693,553	AUCCGCUCCGUAGGGUCCGUGGGUUCAAAU	32 nt	1,425	74
Class IV	SerGGA	47	408,423	408,380	GCCAGGAUGGCCGAGCGGUAAGGCGCCUGGAAAGCGUGUU	44 nt	291	15

The tRNA genes for which tRFs have been detected are shown. Columns "begin" and "end" list the position of the respective tRF on the *H. volcanii* chromosome. Sequence, length, and read numbers for each tRF are depicted. The presence of all tRFs could be verified by northern blot analyses (Figure 2 and data not shown), with the exception of ValCAC, GlyCCC, and SerGGA.

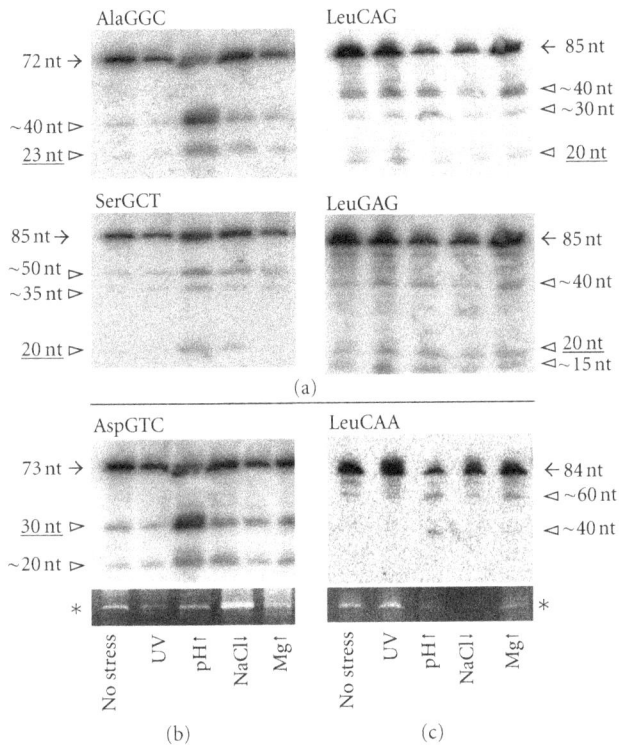

FIGURE 2: tRFs are present in the ribosome-associated RNome. Northern blot analyses of tRFs confirm the presence of class I (a), class II (b), and class III (c) tRFs in the ribosome-associated RNA fraction. The full-length tRNA signals are depicted by black arrows and tRFs by open arrow heads. Approximate lengths of the fragments, as deduced from RNA markers, are indicated. tRFs corresponding to the sequence reads of the cDNA library (Table 1) are underlined. RNA was isolated from unstressed *H. volcanii* cells (no stress), or cells grown under different permanent environmental stress conditions (ultraviolet stress (UV), high pH (pH↑), 0.9 M NaCl (NaCl↓), 300 mM MgSO₄ (Mg↑)). The 5S rRNA (asterisk) of ethidium bromide stained gels served as loading controls.

the condition of most significant tRF processing, the class I tRFs from Leu-tRNA were markedly different. Nevertheless, the fact that these tRFs could be readily detected via northern blot analyses on ribosome-derived RNA suggests that the tRFs identified in our genomic screen are capable of interacting with the ribosome in *H. volcanii*.

3.4. Valine tRF Primarily Binds to the Small Ribosomal Subunit. To more precisely investigate the tRF interaction with the ribosome, polysome gradient analyses and *in vitro* binding studies were performed. *H. volcanii* cell lysates were passed through a linear sucrose gradient in order to separate the polysomal fraction from nontranslating 70S ribosomes and ribosomal subunits. RNA isolated from these fractions was used for northern blot analysis employing a radiolabelled antisense probe targeting Val-tRF (Figures 3(a) and 3(b)). The data revealed that Val-tRF was primarily comigrating with the 30S subunit fraction (Figure 3(b); unstressed and pH 8.5 stress) in density gradients. Only after high pH stress, Val-tRF signals were also obtained in the polysomal fraction

(Figure 3(b)). This is evidence for a putative functional role of Val-tRF during stress response. To more directly investigate ribosome-association, radiolabelled synthetic Val-tRF was bound to purified *H. volcanii* ribosomes or subunits in a filter binding setup. The binding data show that Val-tRF can indeed associate with 70S ribosomes and 30S subunits, while 50S particles were less efficiently targeted (Figure 3(c)). Obviously, no additional cellular proteins are required for ribosome binding. As a negative control, an analogous tRF deriving from a comparable region of isoleucine tRNA, a fragment that was not represented in our deep-sequence library, was used. This isoleucine tRNA-derived fragment (Ile-tRF) was unable to bind to the ribosome in a comparable manner (Figure 3) thus highlighting binding specificity of the Val-tRF.

3.5. Valine tRF Inhibits In Vitro Translation by Interfering with Peptide Bond Formation. The observation that Val-tRF strongly binds to ribosomes led to the speculation of a potential effect as regulatory ncRNA during protein biosynthesis. To clarify this we set up an *in vitro* translation system for *H. volcanii* and quantified the amount of synthesized proteins in the presence or absence of Val-tRF. Repeatedly, Val-tRF decreased translation efficiency of the total *H. volcanii* mRNA population by about 45% (Figure 4(a)). Importantly a 26-mer composed of a scrambled sequence of the Val-tRF did not show any inhibition of protein synthesis, thus hinting at a sequence-specific effect. This effect of Val-tRF on *in vitro* translation also helps explaining the observed severe growth defect, when *H. volcanii* was grown in alkaline media (Figure 4(b)). Under these environmental conditions Val-tRF is most strongly associated with ribosomes *in vivo* (Figure 1(b)).

Protein synthesis is a multistep process involving the orchestrated action of several protein cofactors, including initiation, elongation, and termination factors. To reveal whether the ribosome itself or any of the translation factors needed for protein synthesis are inhibited by Val-tRF, we performed a peptidyl transfer assay utilizing purified *H. volcanii* ribosomes. The transpeptidation assay was performed using N-acetyl-[³H]Phe-tRNA as donor and the antibiotic puromycin as acceptor substrate. In this experimental setup, Val-tRF inhibited the extent of peptide bond formation by about 60%, while the scrambled sequence of the Val-tRF had hardly any inhibitory effect (Figure 5). As second specificity control served the Ile-tRF previously used in the binding studies (see Figure 3(c)). As shown before this RNA was unable to efficiently bind to *H. volcanii* ribosomes (Figure 3(c)) and had also no inhibitory effects on the peptidyl transferase reaction (Figure 5).

To functionally define the site of action of the Val-tRF, the puromycin reaction was repeated under identical reaction conditions, but in the presence of 50S subunits alone. In this experimental setup, Val-tRF did not inhibit transpeptidation, thus confirming that the prime binding target for Val-tRF is the small ribosomal subunit. A possible mechanism for inhibiting the puromycin reaction is that the Val-tRF competes with the P-site substrate N-acetyl-[³H]Phe-tRNA for ribosome binding. This could occur if

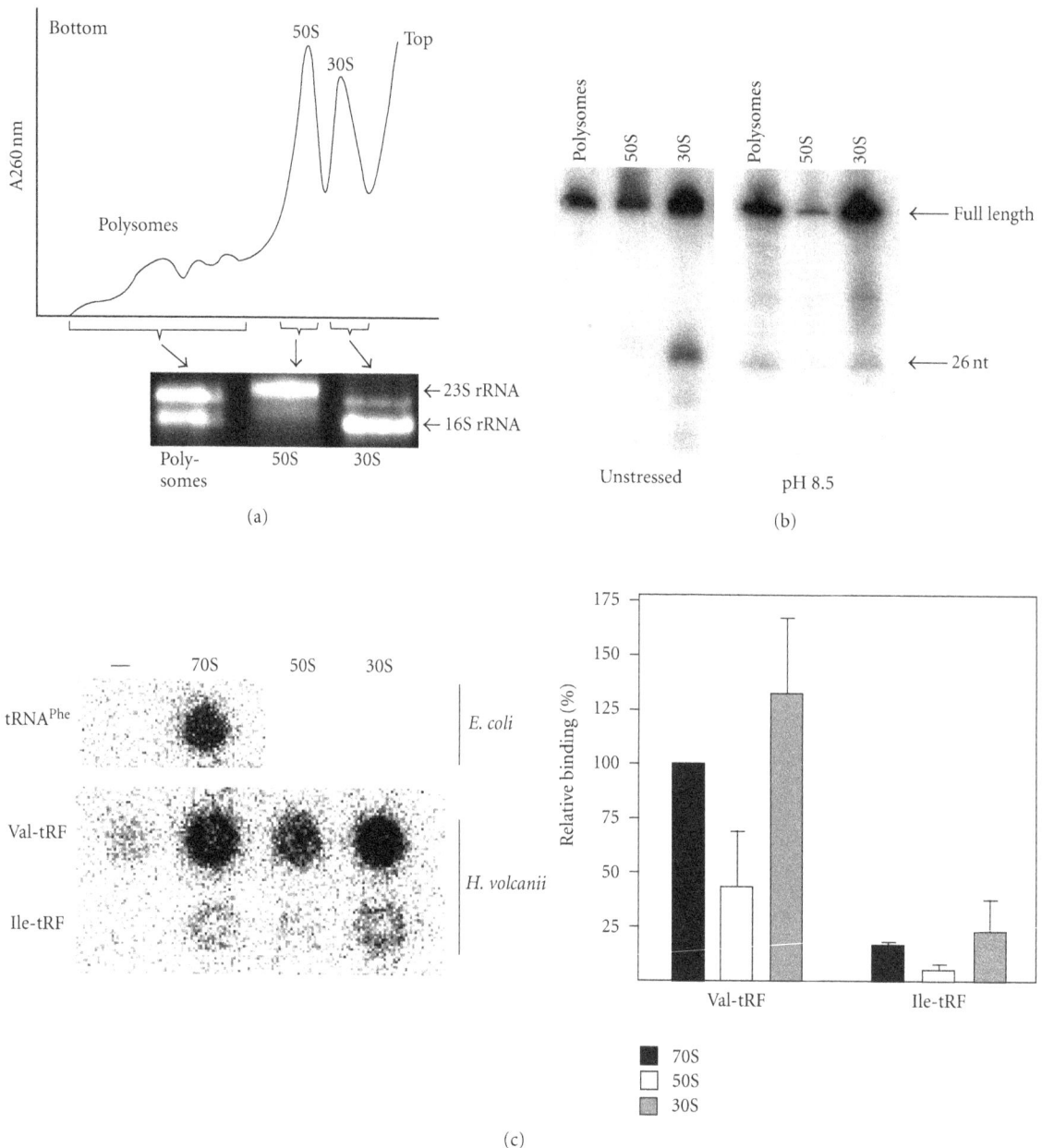

FIGURE 3: Val-tRF associates with ribosomes *in vitro* and *in vivo*. (a) A representative polysome gradient of *H. volcanii*. Fractions containing polysomes, 50S, or 30S subunits were collected and used for northern blot analyses. The identity of the individual fractions was confirmed by agarose gel electrophoresis. (b) The presence of the Val-tRF in the different gradient fractions was investigated by northern blot analysis using RNA from unstressed cells or from cultures incubated at high pH (pH 8.5). Arrows indicate the full-length Val-tRNA and the 26 nt long fragment detected in the cDNA library. (c) *In vitro* filter binding studies of radiolabelled synthetic Val-tRF on ribosomal particles (70S, 50S, 30S) from *H. volcanii* (left panel). As a positive control, binding of tRNA[Phe] to *E. coli* 70S was monitored. To confirm specific binding of Val-tRF an equally long fragment of isoleucine tRNA (Ile-tRF), an RNA sequence not found in our cDNA screen, served as negative control. (Right panel) Quantification of relative binding whereas association of Val-tRF to 70S was normalized to 100%. Signals measured in the absence of any ribosomal particles (-) were subtracted from all experimental points. Error bars show the mean and standard deviation of at least four independent experiments.

the Val-tRF would be able to refold into a hairpin structure resembling an anticodon stem-loop. Even though secondary structure analyses by M-fold did not indicate any stable stem-loop structure of Val-tRF (data not shown), we decided to experimentally test the P-site competition hypothesis. Therefore we slightly modified the peptidyl transferase assay

namely; we incubated N-acetyl-[³H]Phe-tRNA simultaneously with a 125-fold excess of Val-tRF for 20 minutes before the reaction was initiated by the addition of puromycin. Under these conditions significant inhibition of transpeptidation was observed (Figure 5(b)), thus potentially P-site competition occurred. If this Val-tRF triggered inhibition

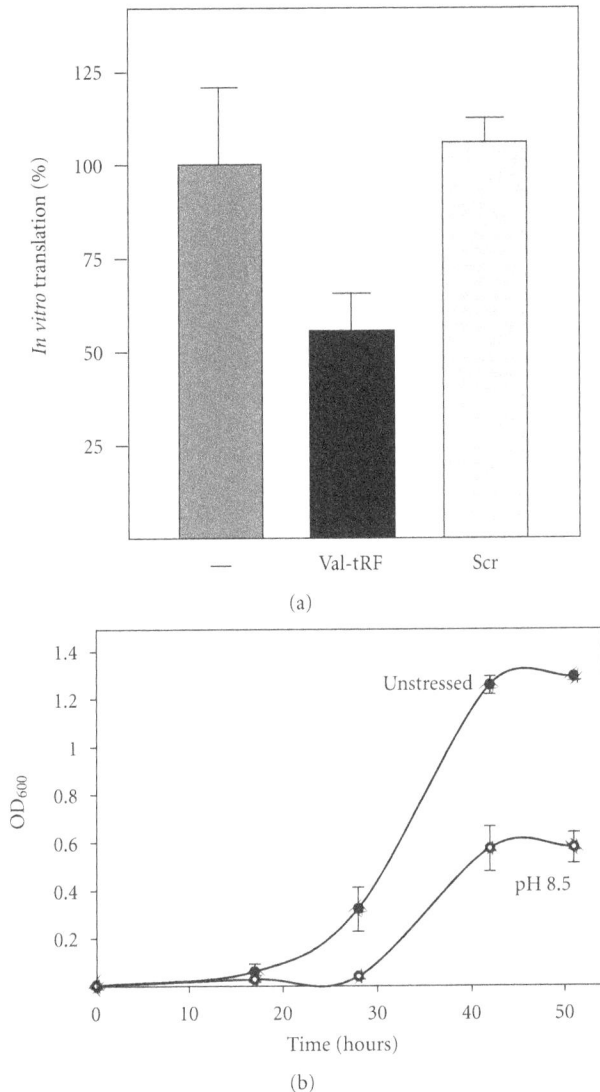

FIGURE 4: Effects of Val-tRF or elevated pH on protein synthesis and cell growth. (a) Val-tRF inhibits *H. volcanii in vitro* translation. The relative amount of radiolabeled proteins in the absence (-) or in the presence of 3.3 μM Val-tRF is shown. The scrambled version of the Val-tRF (scr) served as specificity control. In all cases, the background values measured in reactions without S30 extracts were subtracted from all experimental points. Error bars represent the mean and standard deviation of at least three independent experiments. (b) Cell growth of unstressed *H. volcanii* cultures and of cells grown under alkaline conditions (pH 8.5) is shown. Cell density was measured at 600 nm and the average values of two independent growth curves each and their standard deviations are given.

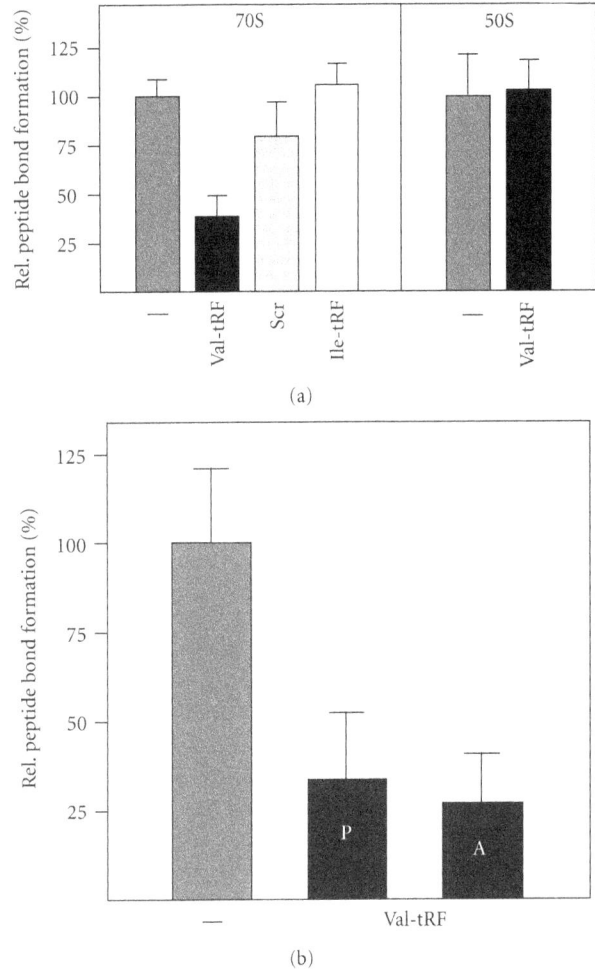

FIGURE 5: Val-tRF inhibits peptide bond formation. (a) Peptidyl transferase reactions catalyzed by *H. volcanii* 70S ribosomes (left panel) or 50S subunits (right panel) in the absence (-) or presence of 2.7 μM Val-tRF (a 10-fold molar excess) were performed as described in Material and Methods. The scrambled version of the Val-tRF (scr) as well as Ile-tRF served as specificity controls. The relative amount of the reaction product N-acetyl-[^{3}H]Phe-puromycin (in %) is shown. The product formed in the absence of any tRF (-) was taken as 100%. (b) Val-tRF inhibits peptide bond formation to the same extent regardless if it was added simultaneously with the P-site donor substrate (P) or with the A-site acceptor substrate (A). In all cases, the background values measured in reactions without any ribosomal particles were subtracted from all experimental points. Error bars represent the mean and standard deviation of three to five independent experiments.

was indeed the cause of a P-site competition, we expected no reduction in the peptidyl transferase activity when Val-tRF was added together with the A-site substrate puromycin to a ribosomal complex already prebound with N-acetyl-[^{3}H]Phe-tRNA at the P-site. However, also under these experimental conditions a significant drop of peptide bond formation was observed (Figure 5(b)). Since the puromycin assay under these conditions has been shown before to be

a single-turnover reaction [22], we conclude that Val-tRF does not interfere with peptide bond formation due to a competition with N-acetyl-[^{3}H]Phe-tRNA for the ribosomal P-site.

4. Discussion

Translation is essentially the last step where regulatory input can be given in the course of gene expression. Typically translation initiation, the rate limiting step of protein biosynthesis, is targeted by regulatory molecules which lead to either global or mRNA-specific effects on protein production [23]. Many regulatory proteins have been identified whose actions result in fine tuning protein synthesis rates thus enabling organisms and cells to adapt to changing environments or to fulfill tissue specific tasks. More recently also ncRNAs have been added to the list of regulatory entities that modulate translation with microRNAs and antisense RNAs being the most prominent classes in eukaryal and prokaryotic species, respectively [24]. Interestingly essentially all of these ncRNAs that control protein biosynthesis target the mRNAs rather than the ribosome itself, with 7SL RNA and tmRNA as notable exceptions (reviewed in [25, 26]). This is unexpected given the central position the ribosome, a highly conserved multifunctional ribonucleoprotein (RNP) complex plays during gene expression. The fact that more than 50% of all natural antibiotic target the ribosome clearly shows that this highly conserved RNA enzyme is in principal an ideal target for manipulating the rate of gene expression [27]. Furthermore, it is assumed that the ribosome is an evolutionary relic from the RNA world where regulatory input likely was given by nonproteinous molecules, such as small ncRNAs.

In order to investigate whether or not small ncRNAs are still used by contemporary cells and organisms to regulate protein production, we have analyzed the RNA interactome of translating ribosomes. Our previous genomic screen in *S. cerevisiae* already indicated that such RNAs do exist [7]; thus we set out to look for this potentially novel class of ncRNAs in the halophilic archaeon *Haloferax volcanii*. Ribosomes were isolated from environmentally stressed cells and the small RNAs that copurified were analyzed by deep-sequencing analysis. Besides numerous other potential ncRNA regulators found to be associated with the ribosomes (our unpublished data), 26% of all obtained reads originated from tRNA loci. This by itself was not unexpected since tRNAs are abundant ncRNAs and well-known substrates of the ribosomes. More unforeseen, however, was the detection of 1.3 million reads originating from tRNA fragments (tRFs) (Table 1). 14 different tRNA species were found to be processed into tRFs almost exclusively from their 5′ ends (Figure 1). Stress-depandant processing of tRFs has been reported before in other model systems including yeast [7, 28], human cell lines [13, 15, 29], pumpkin [16], *Aspergillus* [14], *H. volcanii* [30], and others (see [8] and references therein). While the enzymes responsible for such tRNA processing in yeast (Rny1) [31] and mammalian cells (angiogenin) [29] have just very recently been identified, essentially nothing is known about tRF processing in archaea. In most of the cases the cellular target and the molecular function of tRFs remained unclear. In this paper, we show that tRFs from *H. volcanii* bind to the ribosome *in vivo* and *in vitro* (Figures 1–3). Importantly only tRFs derived from the 5′-part of genuine tRNA were capable for this interaction. With 1.1 million essentially identical sequence reads, the 26 nucleotide long tRF from the Val-tRNA (GAC) locus was by far the most abundant tRF found in our screen (Table 1). The northern blot signal for Val-tRF was faint

when total cellular RNA was used (Figure 1(b)) but increased significantly when ribosome-associated RNA was employed, especially upon alkaline stress (Figure 1(c)). These findings suggest that tRNA processing is stress-independent, a finding that is in good agreement with a very recent study by Saikia et al. [32] and previous reports [5, 8, 9, 14, 16] and shows that Val-tRF is preferentially bound to ribosomes upon shifting the cells to elevated pH. Additional enrichment for functional ribosomal particles by polysome gradient analysis led to a further increase of the northern blot signal for Val-tRF (Figure 3(b)). This polysome profiling approach and direct binding studies demonstrated the association of this tRF primarily with the small ribosomal subunit (Figure 3). Only under elevated pH conditions, Val-tRF signals were also found in the polysome fraction. Val-tRF was able to inhibit the translation of *H. volcanii* proteins *in vitro* (Figure 4(a)), most likely by interfering with peptide bond formation (Figure 5). Inhibition of amide bond synthesis was only seen in the context of 70S ribosomes, but not in 50S subunits (Figure 5). These functional data support our findings that Val-tRF mainly binds to and acts on the small ribosomal subunit of *H. volcanii*. Since Val-tRF shifts into the polysomal fraction during pH stress (Figure 3(a)), it is possible that this interaction downregulates translational activity by interfering with efficient transpeptidation. This possible functional scenario can explain the slow growth phenotype of *H. volcanii* at elevated pH (Figure 4(b)), conditions where Val-tRF most strongly associates with ribosomes (Figure 1(c)).

The ever growing class of tRFs is not by far a homogenous group of RNA molecules. Markedly different processing patterns, sequence motifs, stress response behaviors, and (putative) mode of actions have been reported. In human cells, it was shown that a run of consecutive G residues on the 5′-end of tRFs was crucial for inhibiting the translation initiation step [13]. However, most other tRF identified so far lack any 5′ homopolymeric stretches of nucleotides. Also the fate of the two potential tRNA halves (5′ or 3′) after processing seems to be diverse. In some cases both tRNA halves are stable in the cell and thus can be considered ncRNA candidates [7, 14, 33, 34] while in other cases ([7, 13] and this study), primarily 5′ tRNA fragments were observed. Even within one species, the pattern of tRF biogenesis appears to vary evidently. A recent study by the Marchfelder group reported on the detection of 11 different tRFs in *H. volcanii* [30]. In that study total RNA of differently grown *H. volcanii* cells was used to generate the cDNA library for deep-sequencing analysis. Significantly, not a single tRF was identical when compared to our data despite the fact that the same archaeal strain was used and similar stress conditions were applied. Furthermore the tRF processing pattern in Heyer et al. was strikingly different [30]. All but one tRF derived from the 3′-trailer sequence of pre-tRNAs, and only one from the 5′-part of the tRNA. In our screen, however, nearly all tRFs were processed out from the 5′-ends of full length tRNA. One likely explanation for these apparent differences is that the ribosome selects specifically only tRFs originating from the 5′-ends of tRNAs. This strongly indicates that markedly different RNomes have been analyzed in these two studies, possibly reflecting the highly diverse ncRNA biology of tRNA-derived fragments. Additionally, recent evidence demonstrated [35] that the fraction of biologically functional ncRNA molecules can be increased when cDNA libraries were prepared from RNP particles (our approach) compared to naked total RNA [30].

In summary, we describe herein one of the first direct cellular target for tRFs, namely the ribosome. Furthermore we present evidence that the 26-residue-long Val-tRF is processed in a stress-dependent manner, primarily at elevated pH values and show that ribosome association of this RNA species downregulates protein biosynthesis. Thus we think that this RNA species fulfills all criteria to be referred to as regulatory ncRNA in *H. volcanii*.

Acknowledgments

Jörg Soppa and Anita Marchfelder are acknowledged for providing strains and reagents and for insightful discussions. The authors thank Kamilla Bakowska-Zywicka for initial help with the cDNA library preparation and Miriam Koch, Andreas Pircher, Nina Clementi, Matthias Erlacher, and Ronald Micura for critical comments on the paper. Our thanks are extended to Lavinia Furrer for her valuable help in optimizing polysome gradient conditions during her bachelor thesis. This work was supported by Grants from the Austrian Science Foundation FWF (Y315) and the Austrian Ministry of Science and Research (GenAU project consortium "non-coding RNAs" D-110420-012-012) to N.Polacek.

References

[1] A. Hüttenhofer, P. Schattner, and N. Polacek, "Non-coding RNAs: hope or hype?" *Trends in Genetics*, vol. 21, no. 5, pp. 289–297, 2005.

[2] P. P. Amaral, M. E. Dinger, T. R. Mercer, and J. S. Mattick, "The eukaryotic genome as an RNA machine," *Science*, vol. 319, no. 5871, pp. 1787–1789, 2008.

[3] C. Ender, A. Krek, M. R. Friedländer et al., "A human snoRNA with microRNA-like functions," *Molecular Cell*, vol. 32, no. 4, pp. 519–528, 2008.

[4] M. Brameier, A. Herwig, R. Reinhardt, L. Walter, and J. Gruber, "Human box C/D snoRNAs with miRNA like functions: expanding the range of regulatory RNAs," *Nucleic Acids Research*, vol. 39, no. 2, pp. 675–686, 2011.

[5] H. Kawaji, M. Nakamura, Y. Takahashi et al., "Hidden layers of human small RNAs," *BMC Genomics*, vol. 9, article 157, 2008.

[6] Z. Li, C. Ender, G. Meister, P. S. Moore, Y. Chang, and B. John, "Extensive terminal and asymmetric processing of small RNAs from rRNAs, snoRNAs, snRNAs, and tRNAs," *Nucleic Acids Research*, vol. 40, no. 14, pp. 6787–6799, 2012.

[7] M. Zywicki, K. Bakowska-Zywicka, and N. Polacek, "Revealing stable processing products from ribosome-associated small RNAs by deep-sequencing data analysis," *Nucleic Acids Research*, vol. 40, no. 9, pp. 4013–4024, 2012.

[8] D. M. Thompson and R. Parker, "Stressing out over tRNA cleavage," *Cell*, vol. 138, no. 2, pp. 215–219, 2009.

[9] Y. S. Lee, Y. Shibata, A. Malhotra, and A. Dutta, "A novel class of small RNAs: tRNA-derived RNA fragments (tRFs)," *Genes and Development*, vol. 23, no. 22, pp. 2639–2649, 2009.

[10] M. Bühler, N. Spies, D. P. Bartel, and D. Moazed, "TRAMP-mediated RNA surveillance prevents spurious entry of RNAs into the *Schizosaccharomyces pombe* siRNA pathway," *Nature Structural and Molecular Biology*, vol. 15, no. 10, pp. 1015–1023, 2008.

[11] C. Cole, A. Sobala, C. Lu et al., "Filtering of deep sequencing data reveals the existence of abundant Dicer-dependent small RNAs derived from tRNAs," *RNA*, vol. 15, no. 12, pp. 2147–2160, 2009.

[12] D. Haussecker, Y. Huang, A. Lau, P. Parameswaran, A. Z. Fire, and M. A. Kay, "Human tRNA-derived small RNAs in the global regulation of RNA silencing," *RNA*, vol. 16, no. 4, pp. 673–695, 2010.

[13] P. Ivanov, M. M. Emara, J. Villen, S. P. Gygi, and P. Anderson, "Angiogenin-induced tRNA fragments inhibit translation initiation," *Molecular Cell*, vol. 43, no. 4, pp. 613–623, 2011.

[14] C. Jöchl, M. Rederstorff, J. Hertel et al., "Small ncRNA transcriptome analysis from *Aspergillus fumigatus* suggests a novel mechanism for regulation of protein synthesis," *Nucleic Acids Research*, vol. 36, no. 8, pp. 2677–2689, 2008.

[15] S. Yamasaki, P. Ivanov, G. F. Hu, and P. Anderson, "Angiogenin cleaves tRNA and promotes stress-induced translational repression," *Journal of Cell Biology*, vol. 185, no. 1, pp. 35–42, 2009.

[16] S. Zhang, L. Sun, and F. Kragler, "The phloem-delivered RNA pool contains small noncoding RNAs and interferes with translation," *Plant Physiology*, vol. 150, no. 1, pp. 378–387, 2009.

[17] J. Straub, M. Brenneis, A. Jellen-Ritter, R. Heyer, J. Soppa, and A. Marchfelder, "Small RNAs in haloarchaea: identification, differential expression and biological function," *RNA Biology*, vol. 6, no. 3, pp. 281–292, 2009.

[18] P. Khaitovich, T. Tenson, P. Kloss, and A. S. Mankin, "Reconstitution of functionally active *Thermus aquaticus* large ribosomal subunits with *in vitro*-transcribed rRNA," *Biochemistry*, vol. 38, no. 6, pp. 1780–1788, 1999.

[19] G. Ring, P. Londei, and J. Eichler, "Protein biogenesis in Archaea: addressing translation initiation using an *in vitro* protein synthesis system for *Haloferax volcanii*," *FEMS Microbiology Letters*, vol. 270, no. 1, pp. 34–41, 2007.

[20] R. E. Monro and K. A. Marcker, "Ribosome-catalysed reaction of puromycin with a formylmethionine-containing oligonucleotide," *Journal of Molecular Biology*, vol. 25, no. 2, pp. 347–350, 1967.

[21] N. Polacek, S. Swaney, D. Shinabarger, and A. S. Mankin, "SPARK—a novel method to monitor ribosomal peptidyl transferase activity," *Biochemistry*, vol. 41, no. 39, pp. 11602–11610, 2002.

[22] M. D. Erlacher, K. Lang, N. Shankaran et al., "Chemical engineering of the peptidyl transferase center reveals an important role of the 2′-hydroxyl group of A2451," *Nucleic Acids Research*, vol. 33, no. 5, pp. 1618–1627, 2005.

[23] F. Gebauer and M. W. Hentze, "Molecular mechanisms of translational control," *Nature Reviews Molecular Cell Biology*, vol. 5, no. 10, pp. 827–835, 2004.

[24] J. S. Mattick and I. V. Makunin, "Non-coding RNA," *Human molecular genetics*, vol. 15, supplement 1, pp. R17–R29, 2006.

[25] C. Zwieb and S. Bhuiyan, "Archaea signal recognition particle shows the way," *Archaea*, vol. 2010, Article ID 485051, 11 pages, 2010.

[26] B. D. Janssen and C. S. Hayes, "The tmRNA ribosome-rescue system," *Advances in Protein Chemistry and Structural Biology*, vol. 86, pp. 151–191, 2012.

[27] D. N. Wilson, "The A-Z of bacterial translation inhibitors," *Critical Reviews in Biochemistry and Molecular Biology*, vol. 44, no. 6, pp. 393–433, 2009.

[28] D. M. Thompson, C. Lu, P. J. Green, and R. Parker, "tRNA cleavage is a conserved response to oxidative stress in eukaryotes," *RNA*, vol. 14, no. 10, pp. 2095–2103, 2008.

[29] H. Fu, J. Feng, Q. Liu et al., "Stress induces tRNA cleavage by angiogenin in mammalian cells," *FEBS Letters*, vol. 583, no. 2, pp. 437–442, 2009.

[30] R. Heyer, M. Dorr, A. Jellen-Ritter et al., "High throughput sequencing reveals a plethora of small RNAs including tRNA derived fragments in *Haloferax volcanii*," *RNA Biology*, vol. 9, no. 7, pp. 1011–1018, 2012.

[31] D. M. Thompson and R. Parker, "The RNase Rny1p cleaves tRNAs and promotes cell death during oxidative stress in *Saccharomyces cerevisiae*," *Journal of Cell Biology*, vol. 185, no. 1, pp. 43–50, 2009.

[32] M. Saikia, D. Krokowski, B. J. Guan et al., "Genome-wide identification and quantitative analysis of cleaved tRNA fragments induced by cellular stress," *The Journal of Biological Chemistry*, vol. 287, no. 51, pp. 42708–42725, 2012.

[33] M. R. Garcia-Silva, M. Frugier, J. P. Tosar et al., "A population of tRNA-derived small RNAs is actively produced in *Trypanosoma cruzi* and recruited to specific cytoplasmic granules," *Molecular and Biochemical Parasitology*, vol. 171, no. 2, pp. 64–73, 2010.

[34] H. J. Haiser, F. V. Karginov, G. J. Hannon, and M. A. Elliot, "Developmentally regulated cleavage of tRNAs in the bacterium *Streptomyces coelicolor*," *Nucleic Acids Research*, vol. 36, no. 3, pp. 732–741, 2008.

[35] M. Rederstorff and A. Hüttenhofer, "CDNA library generation from ribonucleoprotein particles," *Nature Protocols*, vol. 6, no. 2, pp. 166–174, 2011.

8

Effect of Growth Medium pH of *Aeropyrum pernix* on Structural Properties and Fluidity of Archaeosomes

Ajda Ota,[1] Dejan Gmajner,[1] Marjeta Šentjurc,[2] and Nataša Poklar Ulrih[1, 3]

[1] *Department of Food Science and Technology, Biotechnical Faculty, University of Ljubljana, Jamnikarjeva 101,*
  1000 Ljubljana, Slovenia
[2] *EPR Center, Institute Jožef Stefan, Jamova 39, 1000 Ljubljana, Slovenia*
[3] *Centre of Excellence for Integrated Approaches in Chemistry and Biology of Proteins (CipKeBiP), Jamova 39, 1000 Ljubljana, Slovenia*

Correspondence should be addressed to Nataša Poklar Ulrih, natasa.poklar@bf.uni-lj.si

Academic Editor: Parkson Chong

The influence of pH (6.0; 7.0; 8.0) of the growth medium of *Aeropyrum pernix* K1 on the structural organization and fluidity of archaeosomes prepared from a polar-lipid methanol fraction (PLMF) was investigated using fluorescence anisotropy and electron paramagnetic resonance (EPR) spectroscopy. Fluorescence anisotropy of the lipophilic fluorofore 1,6-diphenyl-1,3,5-hexatriene and empirical correlation time of the spin probe methylester of 5-doxylpalmitate revealed gradual changes with increasing temperature for the pH. A similar effect has been observed by using the trimethylammonium-6-diphenyl-1,3,5-hexatriene, although the temperature changes were much smaller. As the fluorescence steady-state anisotropy and the empirical correlation time obtained directly from the EPR spectra alone did not provide detailed structural information, the EPR spectra were analysed by computer simulation. This analysis showed that the archaeosome membranes are heterogeneous and composed of several regions with different modes of spin-probe motion at temperatures below $70°$C. At higher temperatures, these membranes become more homogeneous and can be described by only one spectral component. Both methods indicate that the pH of the growth medium of *A. pernix* does not significantly influence its average membrane fluidity. These results are in accordance with TLC analysis of isolated lipids, which show no significant differences between PLMF isolated from *A. pernix* grown in medium with different pH.

1. Introduction

Archaea are the third domain of living organisms, and they have cell structures and components that are markedly different from those of bacteria and eukaryotes. The glycerol ether lipids are the main feature that distinguishes the members of archaea from bacteria and eukarya [1]. In contrast to bacteria and eukarya, where the acyl chains of the membrane phospholipids are ester-linked to the *sn*-glycerol-3-phosphate scaffold, the backbone of archaeal lipids is composed of *sn*-glycerol-1-phosphate, with isoprenoid groups connected *via* ether linkages [2–7].

Aeropyrum pernix K1 was the first absolutely aerobic, hyperthermophilic archaeon that was isolated from a costal solfataric thermal vent in Japan [8]. The polar lipids of *A. pernix* K1 consist solely of $C_{25,25}$-archaeol (2,3-di-sesterpanyl-*sn*-glycerol), with $C_{25,25}$-archetidyl(glucosyl)inositol (AGI) accounting for 91 mol%, and $C_{25,25}$-archetidylinositol (AI) accounting for the remaining 9 mol% (Figure 1). Membranes composed of such $C_{25,25}$ diether lipids have 20% greater thickness than those composed of tetraether $C_{20,20}$ archaeal-based lipids [9].

Over the last five years, we have investigated the influence of some environmental factors on the structural properties of the membrane of *A. pernix in vivo* using electron paramagnetic resonance (EPR) and fluorescence emission spectrometry [10]. These studies included the influence of pH and temperature on the physiochemical properties of bilayer archaeosomes prepared from a polar-lipid methanol fraction (PLMF) isolated from *A. pernix* cells grown at $92°$C at pH 7.0, of mixed liposomes prepared from mixtures of this PLMF and 1,2-dipalmitoyl-*sn*-glycero-3-phosphoholine

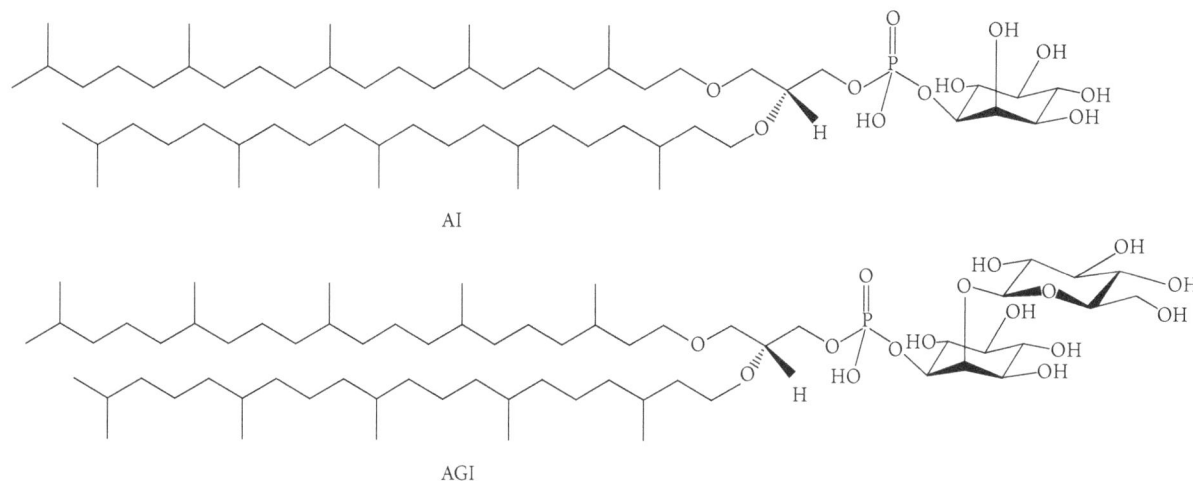

FIGURE 1: Structural formulas of 2,3-di-*O*-sesterpanyl-*sn*-glycerol-1-phospho-myo-inositol ($C_{25,25}$-archetidylinositol) (top: AI) and 2,3-di-*O*-sesterpanyl-*sn*-glycerol-1-phospho-1′-(2′*O*-α-D-glucosyl)-myo-inositol ($C_{25,25}$-archetidyl(glucosyl)inositol) (bottom: AGI).

(DPPC) at different ratios [11–13]. The major conclusion based on our differential scanning calorimetry (DSC) was that the archaeosomes do not show gel to liquid crystalline phase transition in the temperature range from 0 to 100°C [11].

Through these investigations of *A. pernix in vivo*, we have shown that the growth medium pH influences the initial growth rate and cell density [14]. A pH below 7.0 was less favourable than pH 8.0, and there was no growth of *A. pernix* at pH 5.0. Using the EPR and fluorescence emission measurements, changes in the distribution of the spin probes and their motional characteristic were monitored. These changes reflect the changes in the membrane domain structure with temperature, and they were different for *A. pernix* grown at pH 6.0 than at pH 7.0 and 8.0 [10]. Macalady and coworkers (2004) [15] suggested that there is a strong correlation between core-lipid composition and optimal pH of the growth medium.

In the present study, we have extended our EPR and fluorescence emission spectrometry to investigate the influence of growth medium pH (6.0; 7.0; 8.0) on the physiochemical properties of bilayer archaeosomes prepared from this PLMF from *A. pernix*.

2. Materials and Methods

2.1. Growth of A. pernix K1.
A. pernix K1 was purchased from Japan Collection of Microorganisms (number 9820; Wako-shi, Japan). The culture medium comprised (per litre): 34.0 g marine broth 2216 (Difco Becton, Dickinson & Co., Franklin Lakes, NJ, USA), 5.0 g Trypticase Peptone (Becton, Dickinson and Company, Sparks, USA), 1.0 g yeast extract (Becton, Dickinson and Company, Sparks, USA) and 1.0 g $Na_2S_2O_3 \bullet 5H_2O$ (Sigma-Aldrich, St. Louis, USA). The buffer systems used were 20 mM MES [2-(N-morpholino)ethanesulfonic acid; Acros Organics, Geel, Belgium] for growth at pH 6.0, and 20 mM HEPES [4-(2-hydroxyethyl)-1-piperazineethanesulfonic acid; Sigma-Aldrich Chemie GmbH, Steinheim, Germany] for growth at

pH 7.0 and pH 8.0. The *A. pernix* cells were grown in 800 mL growth medium in 1000 mL heavy-walled flasks, with a magnetic stirring hot plate and forced aeration ($0.5 \, L \cdot min^{-1}$) at 92°C, as described previously [14].

2.2. Isolation and Purification of Lipids.
The PLMF that is composed of approximately 91% AGI and 9% AI (average molecular weight of $1181.42 \, g \cdot mol^{-1}$) was prepared from the lyophilised *A. pernix* cells as described previously [11]. The lipids were fractionated using adsorption chromatography and analysed by TLC with the chloroform/methanol/acetic acid/water (85/30/15/5) solvent. Analysis was performed by 0.04 mg of PLMF isolated from *A. pernix* grown at different pH. TLC plate was developed and sprayed with 20% H_2SO_4. Lipid spots were visualized by heating at 180°C for 20 minutes [9]. TLC plates were analysed using JustTLC software (Version 3.5.3. http://www.sweday.com/), where intensity ratio of the two lipid components was compared. No differences between PLMF isolated from *A. pernix* grown in medium with different pH were observed (Figure 2).

The methanol fraction containing the polar lipids (PMLF) was used for further analysis. This lipid solution was dried by slow evaporation under a constant flow of dry nitrogen, followed by vacuum evaporation of solvent residues.

2.3. Preparation of Archaeosomes.
The appropriate weights of the dried PLMF were dissolved in chloroform and transferred into glass round-bottomed flasks, where the solvent was evaporated under reduced pressure (17 mbar). The dried lipid films were then hydrated with the aqueous buffer solutions. As indicated above, the following 20 mM buffer solutions were used: MES for pH 6.0 and HEPES for pH 7.0 and 8.0. The final concentration of the lipids was $10 \, mg \cdot mL^{-1}$. Multilamellar vesicles (MLVs) were prepared by vortexing the lipid suspensions for 10 min. The MLVs were further transformed into small unilamellar vesicles

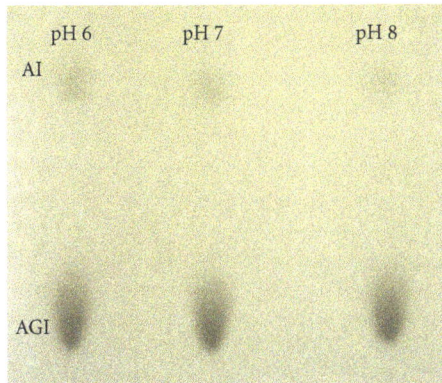

FIGURE 2: TLC results of PLMF from *A. pernix* grown at different pH as marked. AI and AGI stands for $C_{25,25}$-archetidylinositol and $C_{25,25}$-archetidyl(glucosyl)inositol, respectively.

(SUVs) by 30 min sonication with 10 s on-off cycles at 50% amplitude with a Vibracell Ultrasonic Disintegrator VCX 750 (Sonics and Materials, Newtown, USA). To separate the debris from SUVs after sonification, the sample was centrifuged for 10 min at 14.000 rpm (Eppendorf Centrifuge 5415C).

2.4. Fluorescence Anisotropy Measurements. Fluorescence anisotropy measurements of 1,6-diphenyl-1,3,5-hexatriene (DPH) and trimethylammonium-6-phenyl-1,3,5-hexatriene (TMA-DPH) (Figure 3) in PLMF archaeosomes were performed in a 10 mm-path-length cuvette using a Cary Eclipse fluorescence spectrophotometer (Varian, Mulgrave, Australia), in the temperature range from 20°C to 90°C, and the pH range from 6.0 to 8.0 in the relevant buffer solutions. Varian autopolarizers were used, with slit widths with a nominal band-pass of 5 nm for both excitation and emission. Here, 10 μL DPH or TMA-DPH (Sigma-Aldrich Chemie GmbH, Steinheim, Germany) in dimethyl sulphoxide (Merck KGaA, Darmstadt, Germany) was added to 2.5 mL 100 μM solutions of SUVs prepared from the PLMF from *A. pernix* in the relevant buffer, to reach a final concentration of 0.5 μM DPH and 1.0 μM TMA-DPH. DPH and TMA-DPH fluorescence anisotropy was measured at the excitation wavelength of 358 nm, with the excitation polarizer oriented in the vertical position, while the vertical and horizontal components of the polarized emission light were recorded through a monochromator at 410 nm for both probes. The emission fluorescence of DPH and TMA-DPH in aqueous solution is negligible. The anisotropy (*r*) was calculated using the built-in software of the instrument (1):

$$r = \frac{I_{||} - I_{\perp}}{I_{||} + 2I_{\perp}}, \qquad (1)$$

where, $I_{||}$ and $I_{\perp}$ are the parallel and perpendicular emission intensities, respectively. The values of the G-factor [the ratio of the sensitivities of the detection system for vertically (I_{HV}) and horizontally polarized light (I_{HH})] were determined for each sample separately.

The lipid-order parameter *S* was calculated from the anisotropy using the analytical expression given in (2) [16]:

$$S = \frac{\left[1 - 2(r/r_0) + 5(r/r_0)^2\right]^{1/2} - 1 + r/r_0}{2(r/r_0)}, \qquad (2)$$

where r_0 is the fluorescence anisotropy of DPH in the absence of any rotational motion of the probe. The theoretical value of r_0 of DPH is 0.4, while the experimental values of r_0 lie between 0.362 and 0.394 [16]. In our calculation, the experimental value of $r_0 = 0.370$ and $r_0 = 0.369$ for DPH and TMA-DPH in DPPC at 5°C was used, respectively.

2.5. Electron Paramagnetic Resonance Measurements. For the EPR measurements, the PLMF SUVs were spin-labelled with a methylester of 5-doxyl palmitic acid [MeFASL(10,3)] (Figure 3), and the EPR spectra recorded with a Bruker ESP 300 X-band spectrometer (Bruker Analytische Messtechnik, Rheinstein, Germany). The MeFASL(10,3) lipophilic probe was selected due to its moderate stability in the membrane and its relatively high-resolution capability for local membrane ordering and dynamics. It is dissolved in the phospholipid bilayer with nitroxide group located in the upper part of the layers.

With the MeFASL(10,3) film dried on the wall of a glass tube, 50 μL 10 mg·mL^{-1} PLMF SUVs in the relevant buffer was added, and the sample was vortexed for 15 min. This was designed for a final molar ratio of MeFASL(10,3): lipids of 1 : 250. The sample was transferred to a capillary (75 mm; Euroglas, Slovenia), and the EPR spectra were recorded using the following parameters: centre field, 332 mT; scan range, 10 mT; microwave power, 20.05 mW; microwave frequency, 9.32 GHz; modulation frequency, 100 kHz; modulation amplitude, 0.2 mT; temperature range; 5°C to 95°C. Each spectrum was the average of 10 scans, to improve the signal-to-noise ratio. From the EPR spectra, the mean empirical correlation time (τ_c) was calculated using (3) [17]:

$$\tau_c = k\Delta H_0 \left[(h_0/h_{-1})^{1/2} - 1 \right]. \qquad (3)$$

The line width (ΔH_0; in mT) and the heights of the mid-field (h_0) and high-field (h_{-1}) lines were obtained from the EPR spectrum (Figure 6); *k* is a constant typical for the spin probe, which is 5.9387×10^{-11} mT^{-1} for MeFASL (10,3) [17].

2.6. Computer Simulation of the EPR Spectra. For more precise descriptions of the membrane characteristics, computer simulations of the EPR spectra line shapes were performed using the EPRSIM programme (Janez Štrancar, 1996-2003, http://www2.ijs.si/~jstrancar/software.htm). Generally, to describe the EPR spectra of spin labels, the stochastic Liouville equation is used [18–20]. However, in a membrane system labelled with fatty acid spin probes, local rotational motions are fast with respect to the EPR time scale. Modeling of the spectra taken at physiological temperature is therefore simplified by restricting the motions to the fast motional regime. Since the basic approach was already discussed elsewhere [21, 22], it is only summarized here. The model

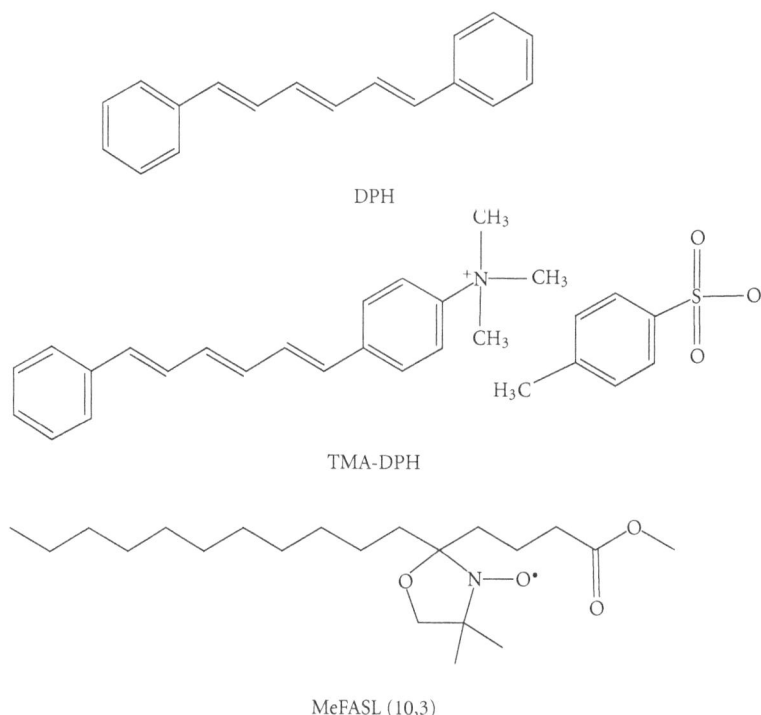

FIGURE 3: Structural formulas of 1,6-diphenyl-1,3,5-hexatriene (DPH), trimethylammonium-6-phenyl-1,3,5-hexatriene (TMA-DPH), and methylester of 5-doxyl palmitic acid [MeFASL (10,3)].

takes into account that the membrane is heterogeneous, and is composed of several regions that have different fluidity characteristics. Therefore, the EPR spectra are composed of several spectral components that reflect the different modes of restricted rotational motion of the spin probe molecules in the different membrane environments. Each spectral component is described by a set of spectral parameters that define the line shape. These are the order parameter (S), the rotational correlation time (τ_c), the line width correction (W), and the polarity correction factors of the magnetic tensors g and A (p_g and p_A, resp.). The S describes the orientational order of the phospholipid alkyl chains in the membrane domains, with S = 1 for perfectly ordered chains and S = 0 for isotropic alignment of the chains. Membrane domains that are more fluid are characterized by a smaller S. The τ_c describes the dynamics of the alky chain motion, with the W due to the unresolved hydrogen superhyperfine interactions, and contributions from other paramagnetic impurities (e.g., oxygen, external magnetic field inhomogeneities). The p_g and p_A polarity correction factors arise from the polarity of the environment of the spin probe nitroxide group (p_g and p_A are large in more polar environment and are below 1 in hydrophobic region). Beside these parameters, the line shape of the EPR spectra is defined by the relative proportions of each of the spectral components (d), which describes the relative amount of the spin probe with a particular motional mode, and which depends on the distribution of the spin probe between the coexisting domains with different fluidity characteristics. As the partition of the MeFASL (10,3) was found to be approximately equal between the different types of domains

of phospholipid/cholesterol vesicles [23], we assumed that the same is valid also for these PLMF liposomes.

It should be stressed that the lateral motion of the spin probe is slow on the time scale of the EPR spectra [24]. Therefore, an EPR spectrum describes only the properties of the nearest surroundings of a spin probe on the nm scale. All of the regions in the membrane with similar modes of spin probe motion contribute to one and the same spectral component. Thus, each spectral component reflects the fluidity characteristics of a certain type of membrane nanodomain (with dimensions of several nm) [25].

To obtain best fit of calculated-to-experimental spectra, stochastic and population-based genetic algorithm is combined with Simplex Downhill optimization method into the evolutionary optimization method (HEO), which requires no special starting points and no user intervention [26]. In order to get a reasonable characterization one still has to define the number of spectral components before applying the optimization. To resolve this problem multirun HEO optimization is used together with a newly developed GHOST condensation procedure. According to this method, 200 independent HEO simulation runs for each EPR spectrum were applied, taking into account 4 different motional modes of spin probe (23 spectral parameters), which is around the resolution limit of EPR nitroxide experiments. From these runs only the set of parameters, which correspond to the best fits were used. All the best-fit sets of parameters obtained by 200 optimizations were evaluated according to the goodness of the fit (χ^2 filter) and according to the similarity of the parameter values of best fits (density filter). The parameters of the best fits were presented by three two-dimensional cross-section plots using four spectral parameters: order parameter S, rotational correlation time τ_c, line broadening W, and polarity correction factor p_A (S-τ_c, S-W, and S-p_A). Groups of solutions, which represent the motional modes of spin probes in particular surrounding and which could correspond to different membrane regions, can be resolved either graphically on GHOST diagrams or numerically within GHOST condensation. Starting values of parameters of spectral components were defined using the average parameters taken from the GHOST diagrams [27]. From these plots information about the motional patterns, defined with S, τ_c, W, and p_A in different membrane regions can be obtained. In this way, the changes in the spin probe motional patterns in different membrane regions, due to the interaction of membrane with biologically active compound, due to temperature, pH, and changes in membrane composition, can be studied.

3. Results and Discussion

3.1. Fluorescence Anisotropy Measurements. Fluorescence probes have been widely used in the study of the structure and dynamic of biological membranes [28]. Their photophysical properties are affected by the physicochemical changes of the microenvironment where the probes are located. Two common probes for the study of membrane properties are DPH and its cationic derivative TMA-DPH. Since DPH is a hydrophobic probe, it is incorporated in the inner apolar core at different positions along the membrane, while the polar group region of TMA-DPH remains anchored at the lipid-water interface of the membrane with the hydrocarbon chain entering the lipid part of the membrane. Xu and London [29] showed that anisotropy values are highest in gel states, lowest in liquid-disordered states, and intermediate in liquid-ordered states. DPH and TMA-DPH r depend on the degree of molecular packing of membrane chains and can be related to the order parameter S. The fluidity may be defined as the reciprocal of the lipid order parameter S [30].

The levels of order in the SUVs composed of PLMF isolated from *A. pernix* grown at pH 6.0, pH 7.0, and pH 8.0 and measured at the same pHs or at pH 7.0 were calculated from anisotropy measurements of DPH (Figures 4(a) and 4(b)) and TMA-DPH (Figures 5(a) and 5(b)), respectively. No significant differences in the order parameters of archaeosomes were observed regardless the growth medium of the *A. pernix* cells or measured pH in the tested temperature range. The order parameter determined by applying DPH of these archaeosomes steadily decreased with increasing temperature, which indicates a gradual increase in membrane fluidity (Figures 4(a) and 4(b)). Previously, we have shown also by applying DSC that in the range from 0°C to 100°C, the archaeosomes do not undergo gel-to-liquid crystalline phase transition [11]. The initial values of the order parameter of DPH at 20°C were: pH 6.0, 0.72 ± 0.1; pH 7.0, 0.72 ± 0.1; pH 8.0, 0.73 ± 0.1. Similarly, we have not detected the significant differences in the order parameter determined by applying TMA-DPH in archaeosomes regardless

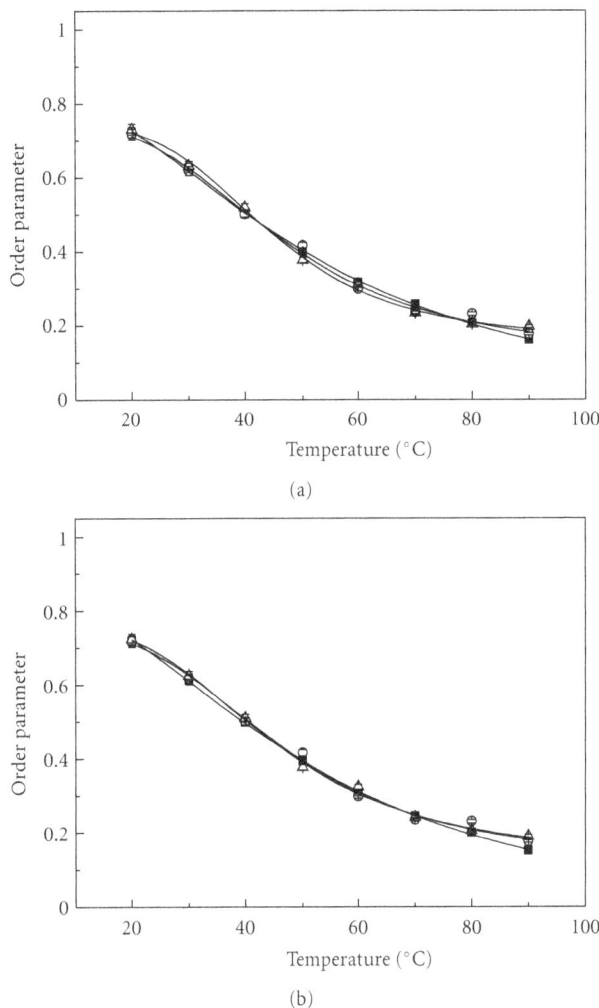

FIGURE 4: Temperature dependence of the lipid-order parameter of the PLMF from *A. pernix* grown in medium with different pH (■ pH 6.0; ○ pH 7.0; △ pH 8.0) determined by measuring the anisotropy of DPH. The lines represent nonlinear curve fitting to the data points. (a) pH of measured samples was the same as the pH of growth medium; (b) experiments were performed at pH 7.0.

the growth or measured pH values. The initial value of order parameter of TMA-DPH in comparison to DPH in archaeal lipids at the same temperature and pH was higher: pH 6.0, 0.93 ± 0.1; pH 7.0, 0.91 ± 0.1; pH 8.0, 0.91 ± 0.1. Another observation, which should be stressed is that the changing in the order parameter determined by TMA-DPH is less temperature sensitive (Figures 5(a) and 5(b)). This might not be surprising since TMA-DPH is cationic probe located at the lipid-water interface of the membrane and the archaeosomes (SUV) have zeta potential of −50 mV [11]. The zeta potential of archaeosomes (LUV) was not changed with pH in the pH range from 5.0 to 10.0 [11]. It is likely that in the studied pH range from 6.0 to 8.0 the zeta potential of SUV archaeosomes is also not changing, which correlate with no observed changes in TMA-DPH anisotropy with pH.

The fact that we have not determined any significant differences in the behaviour of two fluorescence probes regardless the pH of growth medium of *A. pernix*, suggest

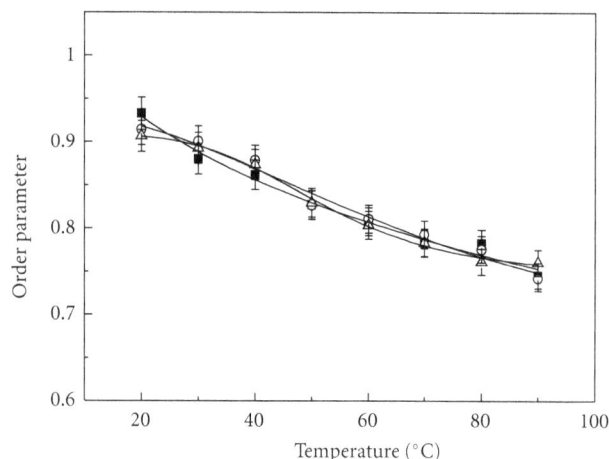

(a)

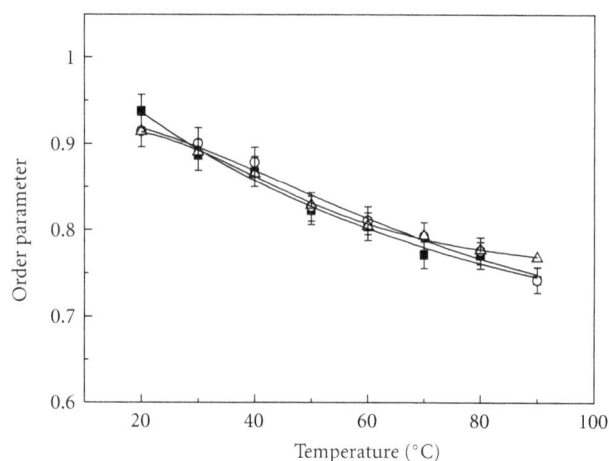

(b)

FIGURE 5: Temperature dependence of the lipid order parameter of the PLMF from *A. pernix* grown in medium with different pH (■ pH 6.0; ○ pH 7.0; △ pH 8.0) determined by measuring the anisotropy of TMA-DPH. The lines represent nonlinear curve fitting to the data points. (a) pH of measured samples was the same as the pH of growth medium; (b) experiments were performed at pH 7.0.

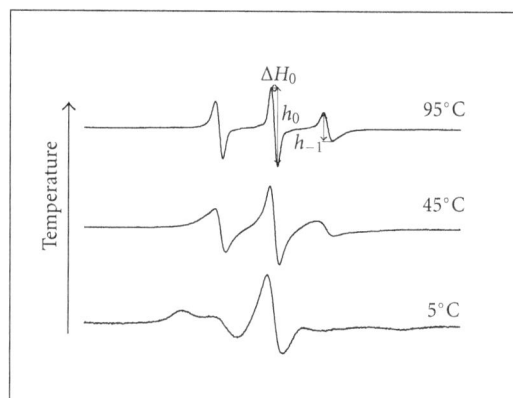

FIGURE 6: Representative EPR spectra of MeFASL(10,3) in the membrane of the SUV archaeosomes at pH 7.0 prepared from the PLMF isolated from *A. pernix* grown at pH 7.0.

medium (Figure 7). The empirical correlation time reflect an average ordering and dynamics of the phospholipid alkyl chains in the spin-probe nitroxide group surrounding and is in inverse relation to membrane fluidity. The data correlate well with fluorescence anisotropy measurements of DPH incorporated into archaeosomes, which shows that the membrane fluidity increases with temperature, but on average it does not depend on the pH of the growth medium. Similar results have been reported for archaeosomes composed of bipolar tetraether lipids [31].

To obtain more detailed information about the possible influences of different growth medium pH on the membrane structural characteristics and on their changes with temperature, computer simulations of the EPR spectra were performed. At temperatures below 70°C, good fits with the experimental spectra were obtained taking into account that the spectra are composed of at least three spectral components. This indicates that the archaeosome membranes are heterogeneous and composed of several regions with different modes of spin-probe motions. All of the regions in the membranes with the same fluidity characteristics are described by a single spectral component. The corresponding EPR parameters determine motional pattern of the spin probe, irrespective to its location in the membrane. Smaller regions with the same physical characteristics could not be distinguished from few large regions. This also means that EPR does not necessarily reflect directly the macroscopic properties of the membrane or large membrane domains, but reflects also the membrane superstructure on nm scale. The three motional patterns of spin probe observed could be due to the two-component lipid composition (AI and AGI) of the membrane. Additionally, some dynamic fluctuations of phospholipids or vertical motion of spin probe within the membrane can be detected as a specific motional pattern of spin probe. These motional patterns could be altered if the membrane is influenced by some external perturbations or if the membrane composition is changed. At higher temperatures, the membranes become more homogeneous and can be described by only one spectral component. The changes in the order parameters of the different membrane

that the lipid composition is not changing in the studied pH range of growing (from pH 6.0 to 8.0) or in the measured pH range from 6.0 to 8.0. This statement was supported by the TLC results of PLMF of *A. pernix* growth at different pHs (Figure 2). The ratio between two major lipids component in *A. pernix* membrane $C_{25,25}$-archetidylinositol (AI) and $C_{25,25}$-archetidyl(glucosyl)inositol (AGI) is at growth pH 6.0 and 7.0: 9 ± 1% of AI and 91 ± 1% of AGI and at growth pH 8.0: 8 ± 1% of AI and 92 ± 1% of AGI.

3.2. Electron Paramagnetic Resonance Measurements. The empirical correlation times of MeFASL(10,3) in these liposomes prepared from the PLMF isolated from *A. pernix* were measured directly from the EPR spectra (Figure 6). These decreased gradually with temperature and did not show significant differences with respect to the pH of the growth

(a)

(b)

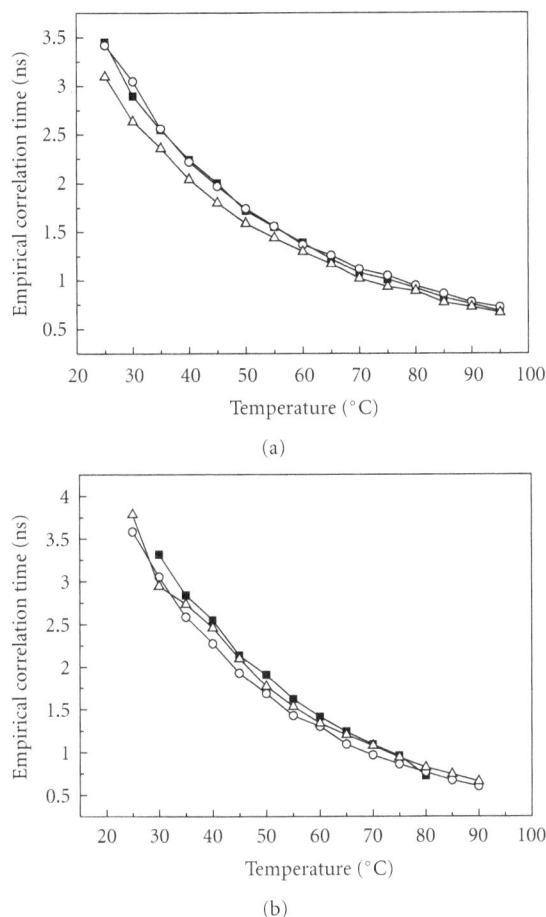

FIGURE 7: Temperature dependence of empirical correlation time (τ_c) of MeFASL(10,3) in SUV archaeosomes prepared from PLMF lipids isolated from *A. pernix* grown in medium with different pH (■ pH 6.0; ○ pH 7.0; △ pH 8.0) and measured at the same pHs (a) and at pH 7.0 (b). Empirical correlation time was calculated directly from the EPR spectra according to Equation (3).

(a)

(b)

FIGURE 8: Temperature dependence of the order parameters (S) and proportions of MeFASL(10,3) in the membrane regions of the SUV archaeosomes prepared from the PLMF from *A. pernix* grown at pH 6.0 (black circles) and 7.0 (grey circles) (a), and at pH 7.0 (grey circles) and 8.0 (black circles) (b). The diameters of the symbols indicate the proportions of each region. D1, D2, and D3 indicate the regions with the highest, intermediate-and lowest-order parameters, respectively. Experiments were performed at pH 7.0.

regions and their proportions with temperature are shown in the form of bubble diagrams in Figure 8, where the dimensions of each symbol represent the proportions of the spin probes in the corresponding membrane regions. With increasing temperature, the order parameter of the most ordered region decreases, its proportion decreases and disappears in the temperature region between 55°C and 65°C. The proportions of the less ordered regions increase with increasing temperature, and above 70°C these remain unchanged. The calculated order parameters for the samples grown at different pH and measured at pH 7.0 are in the range uncertainty of the calculation.

Order parameters obtained by fluorescence measurement (Figures 4 and 5) and those obtained by computer simulation of EPR spectra (Figure 8) cannot be directly compared since the three probes (Figure 3), which differ appreciably in their shape and dimensions cause different perturbations in their surrounding and monitors the properties at different depth of the membrane. DPH is highly hydrophobic and reflects the properties in the inner apolar core at different positions

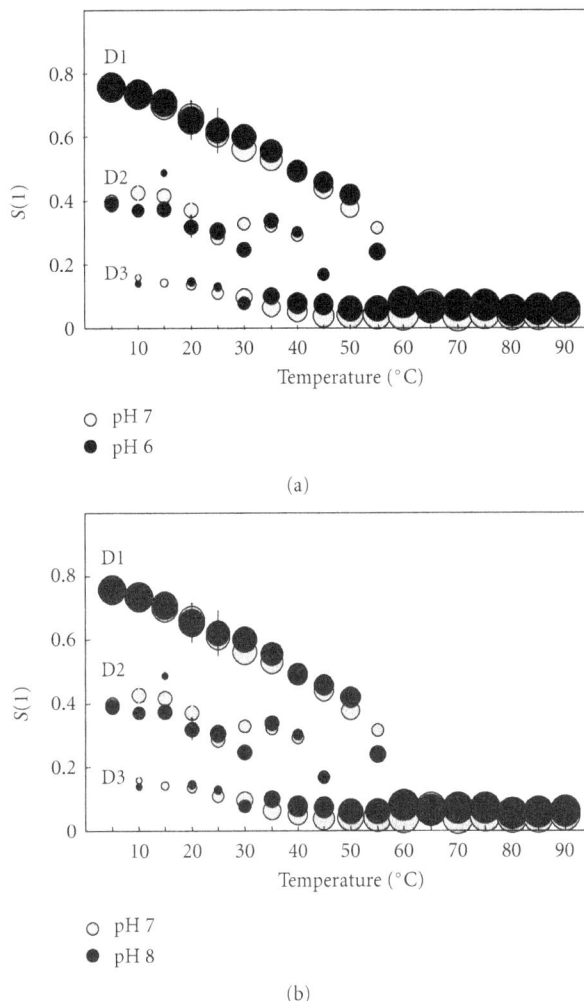

along the membrane, TMA-DPH is anchored at water-lipid interface, while MeFASL(10,3) with nitroxide group on the 5th C atom (counting from the methyl-ester group) monitors the properties in the upper part of phospholipid layers but exhibit also some translational motion within the membrane. Besides by fluorescence polarization measurements an average order parameter in the membrane is obtained, while by computer simulation of EPR spectra order parameter is distinguished from the rotational rate and reflects differe surroundings of the spin probe at lower temperatures, which could be due to membrane heterogeneity produced by distribution between AI and AGI of the membrane but can as well be the consequence of some fluctuations or vertical motion of the spin probe within the bilayer, which seems to be influenced by temperature.

4. Conclusions

Fluorescence anisotropy measurements of DPH and TMA-DPH in addition to EPR spectrometry here showed steady decreases in the order parameter of archaeal lipids with increasing temperature, regardless the pH of growth of archaea or measuring pH. This indicates a gradual increase in the membrane fluidity in all of these samples, although no significant differences were seen for the influence of the *A. pernix* growth medium pH. TMA-DPH located close to water-lipid interface shows less temperature dependence in comparison of DPH or MeFASL(10,3). The more detailed analysis using computer simulation of the EPR spectra revealed membrane heterogeneity at temperatures below $55°C$, which disappears at higher temperatures. But the EPR parameters calculated from the spectra of archaeosomes obtained from the PLMF from *A. pernix* grown at different pH and measured at pH 7.0 remains in the range of the calculation uncertainty. The results are supported by TLC analysis of isolated lipids, which show no significant differences between PLMF isolated from *A. pernix* grown in medium with different pHs.

To summarize, the present data showed that cell growth pH has no effect on membrane properties being examined. The previous *in vivo* study [14] showed that the cell growth varies with medium pH. This discrepancy is interesting since the polar lipids of *A. pernix* K1 consist solely of $C_{25,25}$-archaeol, which has not been changed by growth pH according to our data presented here. Previously, we have reported that the maximum cell density of *A. pernix* growth at pH 7.0 and 8.0 conditions were similar, while a significantly lower maximum cell density was obtained at pH 6.0 and no growth at pH 5.0 [14]. It is likely that at pHs lower than 6.0 the membranes of the neutrophilic *A. pernix* composed of $C_{25,25}$-archaeol becomes proton permeable and that the permeability is not regulated by lipid composition.

Acknowledgments

The authors would like to express their gratitude for financial support from the Slovenian Research Agency through the P4-0121 research programme and the J2-3639 project. A. Ota was partly financed by the European Social Fund of the European Union.

References

[1] C. R. Woese, O. Kandler, and M. L. Wheelis, "Towards a natural system of organisms: proposal for the domains Archaea, Bacteria, and Eucarya," *Proceedings of the National Academy of Sciences of the United States of America*, vol. 87, no. 12, pp. 4576–4579, 1990.

[2] A. A. H. Pakchung, P. J. L. Simpson, and R. Codd, "Life on earth. Extremophiles continue to move the goal posts," *Environmental Chemistry*, vol. 3, no. 2, pp. 77–93, 2006.

[3] J. Peretó, P. López-García, and D. Moreira, "Ancestral lipid biosynthesis and early membrane evolution," *Trends in Biochemical Sciences*, vol. 29, no. 9, pp. 469–477, 2004.

[4] N. P. Ulrih, D. Gmajner, and P. Raspor, "Structural and physicochemical properties of polar lipids from thermophilic archaea," *Applied Microbiology and Biotechnology*, vol. 84, no. 2, pp. 249–260, 2009.

[5] Y. Koga and H. Morii, "Recent advances in structural research on ether lipids from archaea including comparative and physiological aspects," *Bioscience, Biotechnology and Biochemistry*, vol. 69, no. 11, pp. 2019–2034, 2005.

[6] Y. Koga and H. Morii, "Biosynthesis of ether-type polar lipids in archaea and evolutionary considerations," *Microbiology and Molecular Biology Reviews*, vol. 71, no. 1, pp. 97–120, 2007.

[7] A. Cambacorta, A. Trincone, B. Nicolaus, L. Lama, and M. De Rosa, "Unique features of lipids of Archaea," *Systematic and Applied Microbiology*, vol. 16, no. 4, pp. 518–527, 1994.

[8] Y. Sako, N. Nomura, A. Uchida et al., "Aeropyrum pernix gen. nov., sp. nov., a novel aerobic hyperthermophilic archaeon growing at temperatures up to $100°C$," *International Journal of Systematic Bacteriology*, vol. 46, no. 4, pp. 1070–1077, 1996.

[9] H. Morii, H. Yagi, H. Akutsu, N. Nomura, Y. Sako, and Y. Koga, "A novel phosphoglycolipid archaetidyl(glucosyl)inositol with two sesterterpanyl chains from the aerobic hyperthermophilic archaeon Aeropyrum pernix K1," *Biochimica et Biophysica Acta*, vol. 1436, no. 3, pp. 426–436, 1999.

[10] N. P. Ulrih, U. Adamlje, M. Nemec, and M. Šentjurc, "Temperature- and pH-induced structural changes in the membrane of the hyperthermophilic archaeon Aeropyrum pernix K1," *Journal of Membrane Biology*, vol. 219, no. 1–3, pp. 1–8, 2007.

[11] D. Gmajner, A. Ota, M. Šentjurc, and N. P. Ulrih, "Stability of diether C25,25 liposomes from the hyperthermophilic archaeon Aeropyrum pernix K1," *Chemistry and Physics of Lipids*, vol. 164, no. 3, pp. 236–245, 2011.

[12] D. Gmajner and N. P. Ulrih, "Thermotropic phase behaviour of mixed liposomes of archaeal diether and conventional diester lipids," *Journal of Thermal Analysis and Calorimetry*, vol. 106, no. 1, pp. 255–260, 2011.

[13] D. Gmajner, P. A. Grabnar, M. T. Žnidarič, J. Štrus, M. Šentjurc, and N. P. Ulrih, "Structural characterization of liposomes made of diether archaeal lipids and dipalmitoyl-L-α-phosphatidylcholine," *Biophysical Chemistry*, vol. 158, no. 2-3, pp. 150–156, 2011.

[14] I. Milek, B. Cigić, M. Skrt, G. Kaletunç, and N. P. Ulrih, "Optimization of growth for the hyperthermophilic archaeon Aeropyrum pernix on a small-batch scale," *Canadian Journal of Microbiology*, vol. 51, no. 9, pp. 805–809, 2005.

[15] J. L. Macalady, M. M. Vestling, D. Baumler, N. Boekelheide, C. W. Kaspar, and J. F. Banfield, "Tetraether-linked membrane monolayers in Ferroplasma spp: a key to survival in acid," *Extremophiles*, vol. 8, no. 5, pp. 411–419, 2004.

[16] H. Pottel, W. van der Meer, and W. Herreman, "Correlation between the order parameter and the steady-state fluorescence anisotropy of 1,6-diphenyl-1,3,5-hexatriene and an evaluation of membrane fluidity," *Biochimica et Biophysica Acta*, vol. 730, no. 2, pp. 181–186, 1983.

[17] L. Coderch, J. Fonollosa, M. De Pera, J. Estelrich, A. De La Maza, and J. L. Parra, "Influence of cholesterol on liposome fluidity by EPR. Relationship with percutaneous absorption," *Journal of Controlled Release*, vol. 68, no. 1, pp. 85–95, 2000.

[18] D. E. Budil, L. Sanghyuk, S. Saxena, and J. H. Freed, "Nonlinear-least-squares analysis of slow-motion EPR spectra in one and two dimensions using a modified levenberg-marquardt algorithm," *Journal of Magnetic Resonance—Series A*, vol. 120, no. 2, pp. 155–189, 1996.

[19] B. Robinson, H. Thomann, A. Beth, and L. R. Dalton, "The phenomenon of magnetic resonance: theoretical considerations," in *EPR and Advanced EPR Studies of Biological Systems*,

L. R. Dalton, Ed., pp. 11–110, CRC Press, Boca Raton, Fla, USA, 1985.

[20] D. J. Schneider and J. H. Freed, "Calculating slow motional magnetic resonance spectra: a user's guide," in *Biological Magnetic Resonance: Spin Labeling, Theory and Applications*, L. J. Berliner and J. Reuben, Eds., pp. 1–76, Plenum Press, New York, NY, USA, 1989.

[21] J. Štrancar, M. Šentjurc, and M. Schara, "Fast and accurate characterization of biological membranes by EPR spectral simulations of nitroxides," *Journal of Magnetic Resonance*, vol. 142, no. 2, pp. 254–265, 2000.

[22] H. Schindler and J. Seelig, "EPR spectra of spin labels in lipid bilayers," *Journal of Chemical Physics*, vol. 59, no. 4, pp. 1841–1850, 1973.

[23] Z. Arsov and J. Štrancar, "Determination of partition coefficient of spin probe Between different lipid membrane phases," *Journal of Chemical Information and Modeling*, vol. 45, no. 6, pp. 1662–1671, 2005.

[24] M. E. Johnson, D. A. Berk, D. Blankschtein, D. E. Golan, R. K. Jain, and R. S. Langer, "Lateral diffusion of small compounds in human stratum corneum and model lipid bilayer systems," *Biophysical Journal*, vol. 71, no. 5, pp. 2656–2668, 1996.

[25] J. Štrancar, T. Koklič, and Z. Arsov, "Soft picture of lateral heterogeneity in biomembranes," *Journal of Membrane Biology*, vol. 196, no. 2, pp. 135–146, 2003.

[26] B. Filipič and J. Štrancar, "Tuning EPR spectral parameters with a genetic algorithm," *Applied Soft Computing*, vol. 1, no. 1, pp. 83–90, 2001.

[27] J. Štrancar, T. Koklič, Z. Arsov, B. Filipič, D. Stopar, and M. A. Hemminga, "Spin label EPR-based characterization of biosystem complexity," *Journal of Chemical Information and Modeling*, vol. 45, no. 2, pp. 394–406, 2005.

[28] J. R. Lakowitz, *Principles of Fluorescence Spectroscopy*, Kluwer Academic/Plenum, New York, NY, USA, 2nd edition, 1999.

[29] X. Xu and E. London, "The effect of sterol structure on membrane lipid domains reveals how cholesterol can induce lipid domain formation," *Biochemistry*, vol. 39, no. 5, pp. 843–849, 2000.

[30] J. G. Kuhry, P. Fonteneau, G. Duportail, C. Maechling, and G. Laustriat, "TMA-DPH: a suitable fluorescence polarization probe for specific plasma membrane fluidity studies in intact living cells," *Cell Biophysics*, vol. 5, no. 2, pp. 129–140, 1983.

[31] R. Bartucci, A. Gambacorta, A. Gliozzi, D. Marsh, and L. Sportelli, "Bipolar tetraether lipids: chain flexibility and membrane polarity gradients from spin-label electron spin resonance," *Biochemistry*, vol. 44, no. 45, pp. 15017–15023, 2005.

Molecular Characterization of Copper and Cadmium Resistance Determinants in the Biomining Thermoacidophilic Archaeon *Sulfolobus metallicus*

Alvaro Orell,[1] **Francisco Remonsellez,**[1,2] **Rafaela Arancibia,**[1] and **Carlos A. Jerez**[1]

[1] *Laboratory of Molecular Microbiology and Biotechnology, Department of Biology and Millennium Institute for Cell Dynamics and Biotechnology, Faculty of Sciences, University of Chile, Santiago, Chile*
[2] *Department of Chemical Engineering, North Catholic University, Antofagasta, Chile*

Correspondence should be addressed to Carlos A. Jerez; cjerez@uchile.cl

Academic Editor: Elisabetta Bini

Sulfolobus metallicus is a thermoacidophilic crenarchaeon used in high-temperature bioleaching processes that is able to grow under stressing conditions such as high concentrations of heavy metals. Nevertheless, the genetic and biochemical mechanisms responsible for heavy metal resistance in *S. metallicus* remain uncharacterized. Proteomic analysis of *S. metallicus* cells exposed to 100 mM Cu revealed that 18 out of 30 upregulated proteins are related to the production and conversion of energy, amino acids biosynthesis, and stress responses. Ten of these last proteins were also up-regulated in *S. metallicus* treated in the presence of 1 mM Cd suggesting that at least in part, a common general response to these two heavy metals. The *S. metallicus* genome contained two complete *cop* gene clusters, each encoding a metallochaperone (CopM), a Cu-exporting ATPase (CopA), and a transcriptional regulator (CopT). Transcriptional expression analysis revealed that *copM* and *copA* from each *cop* gene cluster were cotranscribed and their transcript levels increased when *S. metallicus* was grown either in the presence of Cu or using chalcopyrite (CuFeS$_2$) as oxidizable substrate. This study shows for the first time the presence of a duplicated version of the *cop* gene cluster in *Archaea* and characterizes some of the Cu and Cd resistance determinants in a thermophilic archaeon employed for industrial biomining.

1. Introduction

Bioleaching is the biological conversion of an insoluble metal compound into a water soluble form [1, 2]. Microbe-based processes have clear economic advantages in the extraction of metals from many low-grade deposits [3], and these metal-extraction processes are usually more environmentally friendly than physical-chemical processes [3–5]. Some ores are refractory to mesophilic leaching and temperatures preferably as high as 75–85°C are required [6, 7]. At high temperatures, biomining consortia are dominated by thermoacidophilic *Archaea* from the genus *Sulfolobus, Acidianus,* and *Metallosphaera* [8].

Metals play an integral role in the life process of microorganisms, but at high levels both essential and nonessential metals can damage cell membranes, alter enzyme specificity, disrupt cellular functions, and damage the structure of DNA

[9, 10]. Acid-leaching solutions are characterized by high metal concentrations that are toxic to most life, and as might be expected, microorganisms that grow in mineral-rich environments are, in most cases, remarkably tolerant to a wide range of metal ions [3, 11] and should possess robust metal resistance mechanisms [11–15].

Despite this, only some metal tolerance values have been reported [6] and the genetic and biochemical mechanisms responsible for metal resistance in biomining acidophilic *Archaea* are just beginning to be characterized [16]. It is therefore important to further understand the mechanisms used by these microorganisms to adapt to and to resist high concentrations of heavy metals.

Related to archaeal copper (Cu) resistance mechanisms, a few metal efflux pumps have been identified from sequenced genomes of some members of this domain [17]. A Cu-resistance (*cop*) locus has been described in *Archaea*,

Molecular Characterization of Copper and Cadmium Resistance Determinants in the Biomining
Thermoacidophilic Archaeon Sulfolobus metallicus

83

which includes genes encoding a new type of archaeal transcriptional regulator (CopT), a putative metal-binding chaperone (CopM), and a putative Cu-transporting P-type ATPase (CopA) [10]. The same Cu-resistance mechanism was described in *Sulfolobus solfataricus* P2 and *Ferroplasma acidarmanus*. In both microorganisms, the putative metal chaperones and the ATPase are cotranscribed and their transcriptional levels increase significantly in response to Cu exposure, suggesting that the transport system is operating for Cu efflux [18, 19]. Recently, it was described that the *copRTA* operon from *S. solfataricus* strain 98/2 (*copTMA* in *S. solfataricus* P2) is cotranscribed at low levels from the *copR* promoter under all conditions, whereas increased transcription from the *copTA* promoter took place in the presence of Cu excess. These authors proposed a model for Cu homeostasis in *Sulfolobus* which relies on Cu efflux and sequestration [20].

In *silico* studies have further identified a CPx-ATPase which most likely mediates the efflux of heavy metal cations in the biomining archaeon *Metallosphaera sedula* [21]. This putative protein has significant identity to a P-type ATPase from *S. solfataricus* (CopA) [19]. Moreover, *M. sedula* contains ORFs with significant similarity to both CopM (Msed0491) and CopT (Msed0492) from *S. solfataricus* [21]. Very recently, Maezato et al. [22] have reported a genetic approach to investigate the specific relationship between metal resistance and lithoautotrophy during biotransformation of chalcopyrite by *M. sedula*. The functional role of its *copRTA* operon was demonstrated by cross-species complementation of a Cu-sensitive *S. solfataricus copR* mutant [22].

Cadmium (Cd) is very toxic and probably carcinogenic at low concentrations. However, the biological effects of this metal and the mechanism of its toxicity are not yet clearly understood [23–26]. In some neutrophilic microorganisms, Cd is taken up via the magnesium or manganese uptake systems [23]. Although the mechanisms in acidophiles have not been elucidated, putative Cd resistance operons in some sequenced genomes from acidophilic microorganisms have been identified. The species with the highest homology to the *cadA* motif were *Acidithiobacillus ferrooxidans* and *Thermoplasma* spp. These high similarities suggest that Cd export may be a common resistance mechanism among acidophiles [11]. Recently, a time-dependent transcriptomic analysis using microarrays in the radioresistant archaeon *Thermococcus gammatolerans* cells exposed to Cd showed the induction of genes related to metal homeostasis, drug detoxification, reoxidation of cofactors, ATP production, and DNA repair [27].

One alternative mechanism proposed for metal tolerance in microorganisms is the sequestration of metal cations by inorganic polyphosphates (polyP) [28], and at the same time the intracellular cations concentration would regulate the hydrolysis of this polymer [29]. *S. metallicus* can tolerate very high concentrations of copper and accumulates high amounts of polyP granules [14, 30]. Furthermore, the levels of intracellular polyP are greatly decreased when this archaeon is either grown in 200 mM Cu or shifted to 100 mM Cu [14, 30]. An increase in exopolyphosphatase (PPX) activity and Pi efflux due to the presence of Cu suggests a

metal tolerance mechanism mediated through polyP [14, 30]. Actual evidence suggests that polyP may provide mechanistic alternatives in tuning microbial fitness for the adaptation under stressful environmental situations and may be of crucial relevance amongst extremophiles. The genes involved in polyP metabolism in Crenarchaeota have been only partially elucidated, as long as a polyP synthase activity is still to be reported and characterized in this kingdom [15].

Thus far, there are no studies on the prospective genetic and biochemical mechanisms that enable *S. metallicus* to thrive in such high concentrations of Cu and other metals. Understanding these mechanisms could be particularly useful in potential improvement of the bioleaching microorganisms, which could likely increase the efficiency of bioleaching processes in due course. Since the genome of *S. metallicus* is not currently available, possible genes involved in Cu resistance were searched by using a CODEHOP and "genome walking" approaches and the transcriptional expression of these isolated genes was assessed by real-time RT-PCR. Furthermore, a proteomic approach was used to identify possible proteins involved in resistance to Cu and Cd in *S. metallicus*. To our knowledge, this is the first paper that shows the occurrence of a duplicated version of the *cop* gene cluster in *Archaea* and gives insights into the molecular Cu and Cd resistance determinants in a thermophilic archaeon employed for industrial biomining.

2. Materials and Methods

2.1. Strains and Growth Conditions. *S. metallicus* DSM 6482 was grown at 65°C in medium 88 (Deutsche Sammlung von Mikroorganismen und Zellkulturen) containing 0.05% (w/v) elemental sulfur and 0.02% (w/v) yeast extract. *S. solfataricus* DSM 1617 was grown at 70°C in medium 182 (Deutsche Sammlung von Mikroorganismen und Zellkulturen) with 0.1% (w/v) yeast extract and with 0.1% (w/v) Casamino acids. To study Cd tolerance of *S. metallicus* and *S. solfataricus*, the microoganisms were grown in their respective media, except that different concentrations of Cd (0.005–5 mM) were present initially, as indicated in the corresponding experiment.

For differential expression assays, *S. metallicus* cells were grown in the absence of Cu or Cd to the early stationary phase, and after removing the medium from the cells by centrifugation, they were then shifted to a new medium containing 100–200 mM $CuSO_4$ (Cu from now on) or 1 mM $CdSO_4$ (Cd from now on) during 24 h. After this time, cells were treated to obtain protein extracts. Growth was monitored by measuring unstained cells numbers by means of a Petroff-Hausser chamber under a phase contrast microscope.

2.2. Preparation of Protein Extracts from S. metallicus. Cells from 800 mL of a control culture grown to 10^8 cells/mL (early stationary phase), or cultures shifted to 100–200 mM Cu or 1 mM, were harvested by centrifugation at 7,700 g for 15 min. The pellets were washed with medium 88 to remove the sulfur. Cells were then resuspended in 50 mM Tris-HCl pH 8.15, 10 mM EDTA, 100 μg/mL PMSF buffer (20 μL per mg wet weight), frozen, and sonicated six times for 30 s each time. The lysates were centrifuged at 4,300 g for 5 min to eliminate cellular debris. The protein concentration of supernatants was determined by the method of Bradford (Coomasie Plus protein assay reagent, Pierce). Between 120 and 500 μg of proteins from the protein extracts were mixed with rehydration IEF buffer, as described by Hatzimanikatis et al. [31] and Choe and Lee [32] with some modifications, including urea 8 M, thiourea 2 M, CHAPS 2% (w/v), Bio-Lyte 3–10 0.27% (v/v), Bio-Lyte 5–8 0.13% (v/v), and bromophenol blue 0.001% (w/v), followed by incubation at 25°C for 30 min. DTT (0.03 g) and sterile nanopure water were then added to complete a final volume of 300 μL. The samples were then incubated at room temperature for 1 h.

2.3. Isoelectric Focussing (IEF). Each sample (300 μL) was loaded onto the pH 3–10 (nonlinear) 17 cm IPG strips (Biorad) in the first dimension chamber and was incubated at room temperature for 1 h. Mineral oil (2.5 mL) was then added to prevent evaporation of the sample and precipitation of urea, and strips were passively rehydrated for 18 h. The isoelectric focusing was performed with the PROTEAN IEF (Biorad) using the following conditions: 250 V for 15 min, 2,000 V for 2 h, 8,000 V for 4 h, 10,000 V for 11 h, and 50 V for 4 h, reaching a total of 120,000 V/h.

2.4. SDS-PAGE. Prior to the second-dimensional electrophoresis, strips were equilibrated as described by Hatzimanikatis et al. [31] with some modifications. Strips were incubated for 15 min with a solution containing 6 M urea, 156 mM DTT, 30% (v/v) glycerol, 2% (w/v) SDS and 24 mM Tris-HCl pH 6.8 and subsequently for 15 more min in a solution containing 6 M urea, 135 mM iodoacetamide, 30% (v/v) glycerol, 2% (p/v) SDS, and 24 mM Tris-HCl pH 6.8. Finally, strips were incubated with electrophoresis buffer containing 192 mM glycine, 1% (w/v) SDS, and 250 mM Tris-HCl pH 8.3 until the second-dimensional run. SDS-PAGE was performed using the PROTEAN II xi cell as described by Laemmli [33], and gels consisted of 11.5 or 15% (w/v) polyacrylamide. Strips were then overlaid onto the second-dimensional gels sealed with 1% (w/v) agarose in electrophoresis buffer containing a trace amount of bromophenol blue. Electrophoresis was carried out at constant 70 V for 15 h. All experiments were performed in triplicate. Gels were stained with silver or Coomasie Blue G-250 as described by Shevchenko et al. [34] and Giavalisco et al. [35], respectively.

2.5. Gels Analysis and Mass Spectrometry. Gel images were digitized by scanning (Epson) and analyzed with the Delta 2D software (Decodon) to identify the spots differentially expressed due to the presence of toxic metals. An estimate of relative quantitative changes was made on the basis of the change in percent volume among silver stained gels. Spots of interest were recovered from Coomasie Blue G-250 stained gels manually and were sent to electron spray tandem ionization mass spectrometric analysis (tandem MS-MS: ESI-QUAD-TOF). The results obtained were analyzed with Mascot algorithm (http://www.matrixscience.com/index.html)

by using the MS/MS Ion search, and all genomes available at databases were used as queries. The entire MS analysis was performed at the Cambridge Center for Proteomics, University of Cambridge, UK.

2.6. CODEHOP-PCR. CopA and CopM amino acid sequences from S. *acidocaldarius*, S. *tokodaii* and S. *solfataricus* were obtained from NCBI (http://www.ncbi.nih.gov/). After aligning the sequences by using the CLUSTAL X program to identify areas of homology, consensus-degenerate PCR primers were designed according to the CODEHOP strategy [36, 37], using the WWW access at http://blocks.fhcrc.org/blocks/codehop.html. Two regions of high sequence similarity were identified for both CopA and CopM sequences, respectively (Figure S1 see Supplementry Material available online at http://dx.doi.org/10.1155/2013/289236) and used to design the consensus-degenerate hybrid oligonucleotide primers (Table 1). Consensus-degenerate hybrid oligonucleotide primers were designed for the N- and C-terminus of CopA, while a pair of degenerate hybrid oligonucleotide primers was designed for CopM (Figure S1).

Amplification reactions contained 1x thermophilic DNA polymerase buffer with 2 mM MgCl$_2$, 0.2 mM dNTPs, 0.5 μM of each primer, 5 U Taq DNA polymerase, 40 ng of S. *metallicus* genomic DNA as a template and water to a final volume of 50 μL. The thermal cycling conditions were 3 min at 95°C; following 30 cycles of 95°C for 30 s, 50°C for 30 s, and 72°C for 1 min; 1 final additional cycle at 72°C for 10 min. Products of amplification were applied onto 1.0% (w/v) agarose gels, and main amplification bands were excised, purified, and TA-cloned into pCR2.1-TOPO (Invitrogen) vector and finally sequenced.

2.7. Genome Walking Experiments. Genome walking strategy was performed as described by Acevedo et al. [38]. Thus, a double-stranded oligo-cassette AdaptT adapter was constructed by annealing of the two unphosphorylated primers AdaptF: (5′-CTAGGCCACGCGTCGACTAGTACTA-G-CTT-3′) and AdaptR: (5′-AGCTAGTACTAGTCGACGCG-TGGCCTAG-3′). Annealing was performed by heating the primers (10 μM) in a boiling water bath for 5 min, and then slowly cooling to room temperature.

Six different DNA libraries were constructed by means of digesting 1 μg of S. *metallicus* genomic DNA with the following restriction enzymes: *Hind*III, *Bam*HI, *Eco*RI, *Eco*RV, *Nco*I, and *Pst*I, respectively. DNA digestion reactions were carried out using 10 U of restriction enzyme, 2 μL of the corresponding enzyme reaction buffer in 20 μL of total reaction volume. The reaction mix was incubated at 16°C during 16 h. To complete the 3′ recessive end of the DNA fragments and to add a 3′ overhanging adenine, 500 ng of the digested and purified DNA were incubated with 5 U of Taq DNA polymerase, 1 μL of 10 mM dNTPs mix, and 5 μL of 10x thermophilic DNA polymerase buffer in 50 μL total volume, at 70°C for 45 min. Seven μL of this mixture was then incubated with 15 pmol of AdaptT oligo-cassette, 1 U T4 DNA ligase (Promega) and 2 μL of 5x ligase buffer, in a total volume

Molecular Characterization of Copper and Cadmium Resistance Determinants in the Biomining
Thermoacidophilic Archaeon Sulfolobus metallicus

85

TABLE 1: Oligonucleotides.

Name	Sequence	Description
copA_cdegF	5′-gatgtagtaatagtaaaaactggagaaataataccngcngaygg	CODEHOP-PCR
copA_cdegR	5′-tcatcagcaaaattagaagaatctccngtngcdat	CODEHOP-PCR
copT_cdegF1	5′-ctcaaatagaatataaagtattacaaatgttaaaagargaywsnmg	CODEHOP-PCR
copT_cdegR1	5′-ggattaccatttatttcatttccacartartcrca	CODEHOP-PCR
copT_cdegR2	5′-cttatcatattcataaatcttccatctattaatttrtarcaytc	CODEHOP-PCR
copT_degF1	5′-gartgytayaarctnat	DOP-PCR
copT_degR1	5′-atnagyttrtarcayct	DOP-PCR
copM_degF	5′-gayccngtntgyggnatgga	DOP-PCR
copM_degR	5′-ccnggnttyccntacgg	DOP-PCR
AdaptF2	5′-cacgcgtcgactagtactagctt	Genome Walking
SP1copA1_3′	5′-aaggatgaggggggaccttatgg	Genome Walking
SP2copA1_3′	5′-ggagataagaaatggggtaaaagag	Genome Walking
SP1copA1_5′	5′-tgataccatcatggaacctgtcag	Genome Walking
SP2copA1_5′	5′-tcctccacaatcccatccgctg	Genome Walking
SP1copA1_5′_2	5′-gattgtagctaagttaacctcggcctcg	Genome Walking
SP2copA1_5′_2	5′-cttctcaccctcagtctggttgg	Genome Walking
SP1copA2_3′	5′-gaaagaggaatatatgcaagggtaaacgg	Genome Walking
SP2copA2_3′	5′-gtgttaatgggagagctggaggg	Genome Walking
SP1copA2_5′	5′-cttctctgtggcaacatcataaccagcc	Genome Walking
SP2copA2_5′	5′-acgcatgtggcgcaatgcattcc	Genome Walking
SP1copT1_5′	5′-cattcctcgcaccagcttgcacactctc	Genome Walking
SP1copT2_5′	5′-cctatgaatactagatcttttccctgaac	Genome Walking
SP2copT2_5′	5′-aacagcttataacactcgtcactttggc	Genome Walking
copM2_RT_F	5′-gatgaaaaagccaatataagac	RT-PCR
copA2_Rv	5′-gaacactaactaacatcgcc	RT-PCR
copM1_RT_Fw	5′-ctatcgtttttgttccgaagcttg	RT-PCR
copA1_Rv	5′-cagcagcaagaacagagacgcc	RT-PCR
*SM16Sf	5′-acgctctaaaaaggcgtgggaata	RT-qPCR
*SM16Sr	5′-ttgagctcggggtctttaagcagtg	RT-qPCR
copA1Sm_qRT_F1	5′-gctaaggtaatagagagcgg	RT-qPCR
copA1Sm_qRT_R1	5′-tgaacaggaatggacagg	RT-qPCR
copA2Sm_qRT_F	5′-tgtgcttgtctccttagcgt	RT-qPCR
copA2Sm_qRT_R	5′-actcttccgtctttcggagt	RT-qPCR
copT1Sm_qRT_F	5′-tgtaggagagtgtgcaagct	RT-qPCRl
copT1Sm_qRT_R	5′-tcgcaagtgagggttatggt	RT-qPCR
copT2Sm_qRT_F	5′-gtgttacggagcttgca	RT-qPCR
copT2Sm_qRT_R	5′-acactcgtcactttggc	RT-qPCR

All oligonucleotides were synthesized by Invitrogen. *Oligonucleotides designed by Bathe and Norris [39].

of 10 μL. The ligation reaction was incubated at 16°C during 16 h.

Two consecutive amplification reactions were then performed. The first PCR reaction was done with 1x Elongase mix buffer, 1.9 mM MgCl$_2$, 0.2 mM dNTPs, 0.5 μM first specific primer (SP1) (designed from the known sequence of the target gene; a forward primer was used to amplify the 3′ end a reverse primer for 5′ end amplification), 5 μL of the ligated DNA diluted 10-fold and 1 μL of Elongase (Invitrogen) and water to a final volume of 50 μL. The thermal cycling conditions were as follows: 1 cycle at 94°C for 1 min, 20 cycles of 94°C for 32 s and 68°C for 5 min, and one final additional cycle at 70°C for 7 min. The PCR product was diluted 10 fold and 3 μL were used as a DNA template for a second PCR, which was performed using the same conditions as the first PCR with 0.5 μM second specific primer (SP2) and 0.2 μM oligo-cassette-specific primer AdaptF2 (5′-CACGCGTCGACTAGTACTAGCTT-3′) and 1 μL of Elongase. The thermal cycling conditions were 1 cycle at 94°C for 1 min, 35 cycles of 94°C for 32 s, and 68°C for 5 min, and 1 final additional cycle at 70°C for 7 min. PCR products were excised, purified, TA-cloned into pCR2.1-TOPO (Invitrogen) vector, and finally sequenced.

2.8. Isolation of Total RNA. Cell pellets (10 mg wet weight) were collected and diluted in $60\,\mu L$ TEN buffer (20 mM Tris-HCl pH 8.0, 1 mM EDTA, 100 mM NaCl) followed by addition of $60\,\mu L$ TENST buffer (20 mM Tris-HCl pH 8.0, 1 mM EDTA, 100 mM NaCl, 1.6% Na-lauroyl sarcosine, and 0.12% Triton X-100). This suspension was incubated at room temperature for 15 min to allow cell lysis. Total RNA was then extracted using the TRIzol reagent (Invitrogen) as recommended by the manufacturer. DNA contamination in RNA preparations was removed by DNase I treatment (Roche), following the manufacturer's instructions. RNA was then purified, precipitated, and finally resuspended in diethylpyrocarbonate (DEPC)-treated water. Total RNA concentrations were estimated by spectrophotometric measurements (OD_{260}) and its quality was evaluated by determining the ratio of absorption at 260 nm and 280 nm, which was within the preferred range of 1.8–2.1.

2.9. Northern Blotting. For differential gene expression by Northern blotting experiments, the media were supplemented with either different concentrations of Cu (5–50 mM), Cd (5 mM), $NiSO_4$ (15 mM), $ZnSO_4$ (50 mM), or Ag_2SO_4 (0.08 mM), respectively, as indicated. For some experiments, elemental sulfur was replaced by a chalcopyrite ($CuFeS_2$) concentrate, which was used as the only energy source at 1% (w/v).

Total RNA ($5\,\mu g$) was separated by electrophoresis on a 1% formaldehyde agarose gel followed by blotting onto Hybond-N nylon membranes (Amersham Pharmacia Biotech, Bucking-hamshire, UK). Hybridization was conducted as described by Sambrook et al. [40]. DNA probes were labeled by random primer DNA labeling kit (Fermentas) using [α-32 p] dCTP. The hybridization signal was detected and analyzed by using the molecular Imager FX system and Quantity One software. The primer sequences for amplification of these gene probes are listed in Table 1.

2.10. Cotranscriptional Analysis by RT-PCR. To study the expression of adjacent genes *copM* and *copA*, cDNA was synthesized by using $0.8\,\mu g$ of total RNA from a *S. metallicus* culture grown in the presence of 20 mM Cu and a reverse primer hybridizing to *copA* sequence. A forward primer annealing on *copM* sequence was used for the PCR reactions (Table 1). PCR amplifications were performed with $1\,\mu L$ of a 1/10 dilution of the cDNA and 25 pmol of each primer. Amplification conditions included an initial 3 min of denaturation at 95°C, followed by 35 cycles of 30 s at 95°C, 30 s at 55°C, and 1.5 min at 72°C and finished by 10 min at 72°C. A reverse transcriptase reaction without enzyme was carried out in order to exclude amplification due to genomic DNA contamination.

2.11. Quantitative RT-PCR. Single stranded cDNA was synthesized from 0.8 ng of DNA-free RNA samples using random hexamers (Fermentas) and ImProm-II reverse transcription system (Promega) following manufacturer's instructions. The software tool IDT Scitools (Integrated DNA Technologies) was used to design primers producing

amplicons of 150–200 bp (Table 1). qPCR was performed with $2\,\mu L$ of 1 : 10 diluted cDNA samples, $12.5\,\mu L$ 2x QuantiFast SYBRGreen PCR Master Mix (Quiagen), $1\,\mu M$ of each primer, and water to a final volume of $25\,\mu L$. The efficiency of each primer pair was calculated from the average slope of a linear regression curve, which resulted from qPCRs using a 10-fold dilution series (10 pg–10 ng) of *S. metallicus* chromosomal DNA as template. Cq values (quantification cycle) were automatically determined by Real-Time Rotor-gene 6000 PCR software (Corbett Life Sciences) after 40 cycles. Cq values of each transcript of interest was standardized to the Cq value of the 16S *rRNA* gene [39]. At least 3 biological replicates of each assessed condition and 2 technical replicates per qPCR reaction were performed.

3. Results and Discussion

3.1. Tolerance of S. metallicus and S. solfataricus to Cu and Cd. In a previous work, it was determined that *S. metallicus* was able to tolerate high Cu concentrations. While the presence of 100 mM Cu did not affect *S. metallicus* growth kinetics, a decreased in cell biomass of only 30% was observed when exposed to 200 mM Cu [30]. The high tolerance to Cu has been described in other acidophilic Bacteria and Archaea compared mostly with neutrophilic microorganisms [11, 14]. Here, this analysis was further extended to characterize the response of *S. metallicus* to Cd. Thus, it was determined that *S. metallicus* growth was not affected in the presence of either 0.5 or 1 mM Cd when compared with the control condition in the absence of the metal (Figure 1(a)). In addition, when *S. metallicus* was challenged with 2 and 3 mM Cd, it was observed that at the late exponential growth phase, the cell numbers decreased by 30 and 50%, respectively (Figure 1(a)). Moreover, growth of *S. metallicus* was completely inhibited in the presence of 5 mM Cd (Figure 1(a)). On the other hand, *S. solfataricus* was not able to grow at Cd concentrations greater than 0.05 mM (Figure 1(b)). At 0.01 mM Cd, *S. solfataricus* cell numbers decreased around 35% (Figure 1(b)). These results are in agreement with previous reports in which *S. solfataricus* was shown to be able to grow in up to 0.01 mM Cd [41]. Other acidophilic archaeons involved in bioleaching processes, such as *Metallosphaera sedula* and *Ferroplasma acidarmanus*, were found to be able to tolerate up to 0.9 and 9 mM Cd, respectively [18, 42]. The minimal inhibitory concentration (MIC) to Cd has been described to be not higher than 1 mM for several thermophilic and neutrophilic *Thermococcus* species [43]. Recently, it was reported that the most radioresistant archaeon, *Thermococcus gammatolerans*, stands Cd concentrations with an MIC of 2 mM [27]. To date, despite the heavy metals tolerance showed by some of these microorganisms, strategies to withstand stress from transition metals have been most widely studied only in haloarchaea [44].

3.2. Effect of Cu and Cd on the Global Proteome of S. metallicus. The proteomic response to Cu and Cd was analyzed to identify proteins which could be involved in heavy metals resistance in *S. metallicus*. Early stationary phase growing

Molecular Characterization of Copper and Cadmium Resistance Determinants in the Biomining
Thermoacidophilic Archaeon Sulfolobus metallicus

87

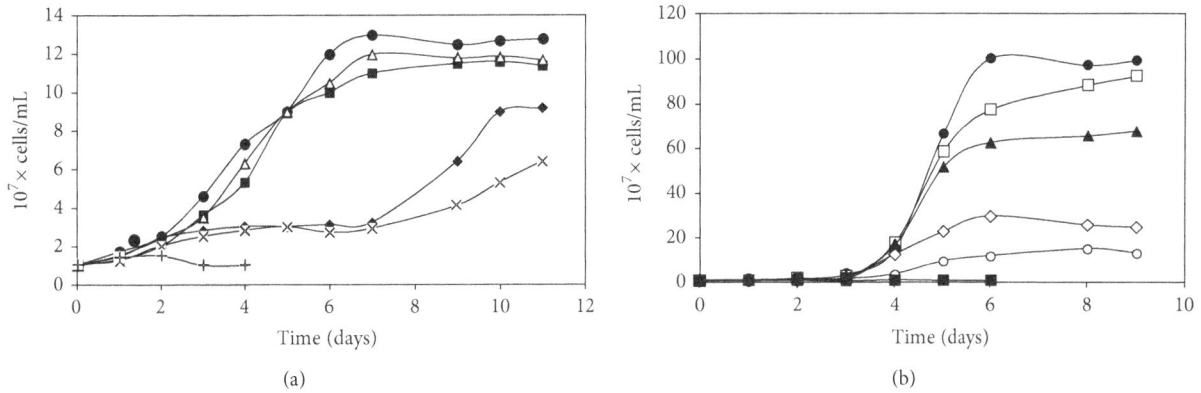

FIGURE 1: Growth of *S. metallicus* (a) and *S. solfataricus* (b) in the presence of Cd. Cells were inoculated in their respective growth media in absence of added Cd (•), or were supplemented with 0.005 mM (□), 0.01 mM (▲), 0.05 mM (◇), 0.1 mM (○), 0.5 mM (■), 1 mM (△), 2 mM (♦), 3 mM (×), or 5 mM (+) Cd, and cells were counted daily.

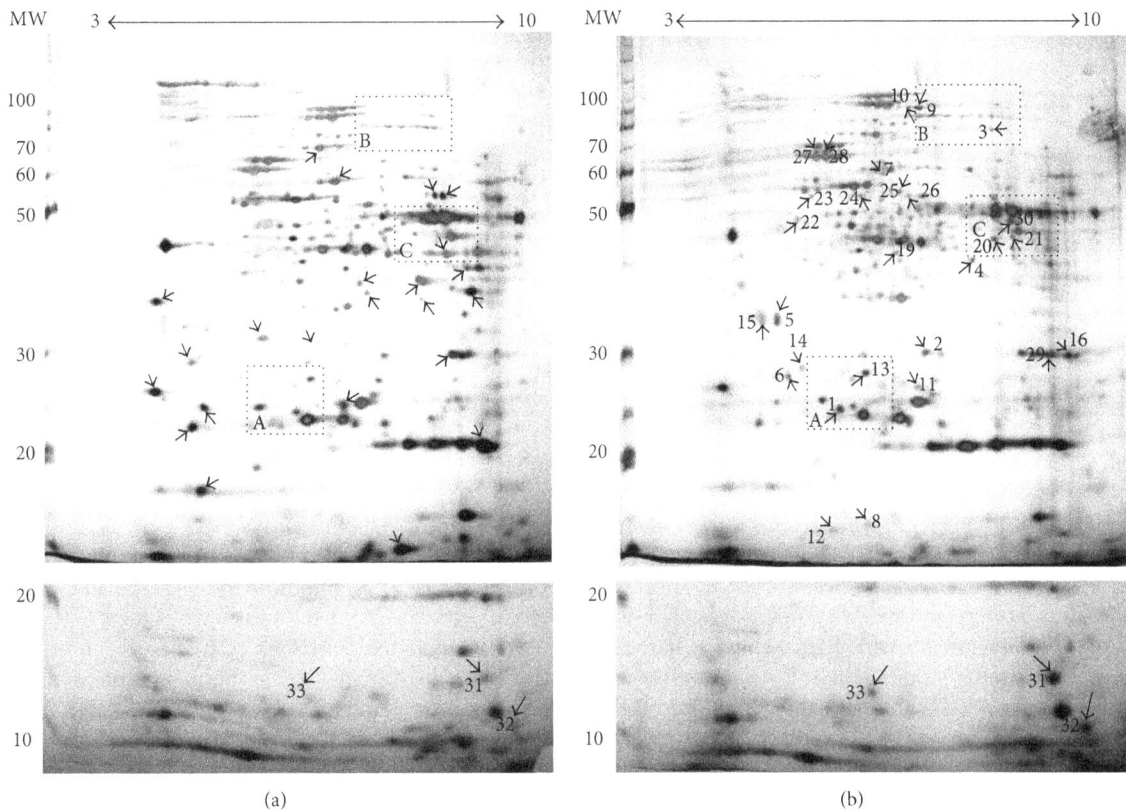

FIGURE 2: Changes in the proteome of *S. metallicus* grown in the presence of Cu. Cells were incubated in the absence of any added metal (a) or in the presence of 100 mM Cu (b). Arrows indicate the spots that were downregulated (a) or upregulated (b) in the presence of Cu. Numbers indicate the spots with increased intensity in cells treated with copper. The dashed boxes are shown as enlarged areas that include some proteins upregulated in the presence of Cu and Cd (Figure 3). The bottom panels show low molecular weight proteins upregulated in the presence of Cu separated by using a 15% polyacrylamide gel in the second dimension.

cells were untreated or treated with either 100 or 200 mM Cu or 1 mM Cd during 24 h, and analyzed by comparative two-dimensional gel electrophoresis as indicated in experimental procedures. Twenty-three proteins were found to be down-regulated after 100 mM Cu treatment (Figure 2(a)). A similar pattern was obtained at 200 mM Cu (not shown). Further-more, by using the Delta 2D software (Decodon), eleven of these proteins were found to be completely absent in the gels and more than 8 proteins decreased their intensity in the range from 1.5-to 5-fold compared with control cells grown

Control

Copper

Cadmium

(a)

(b)

(c)

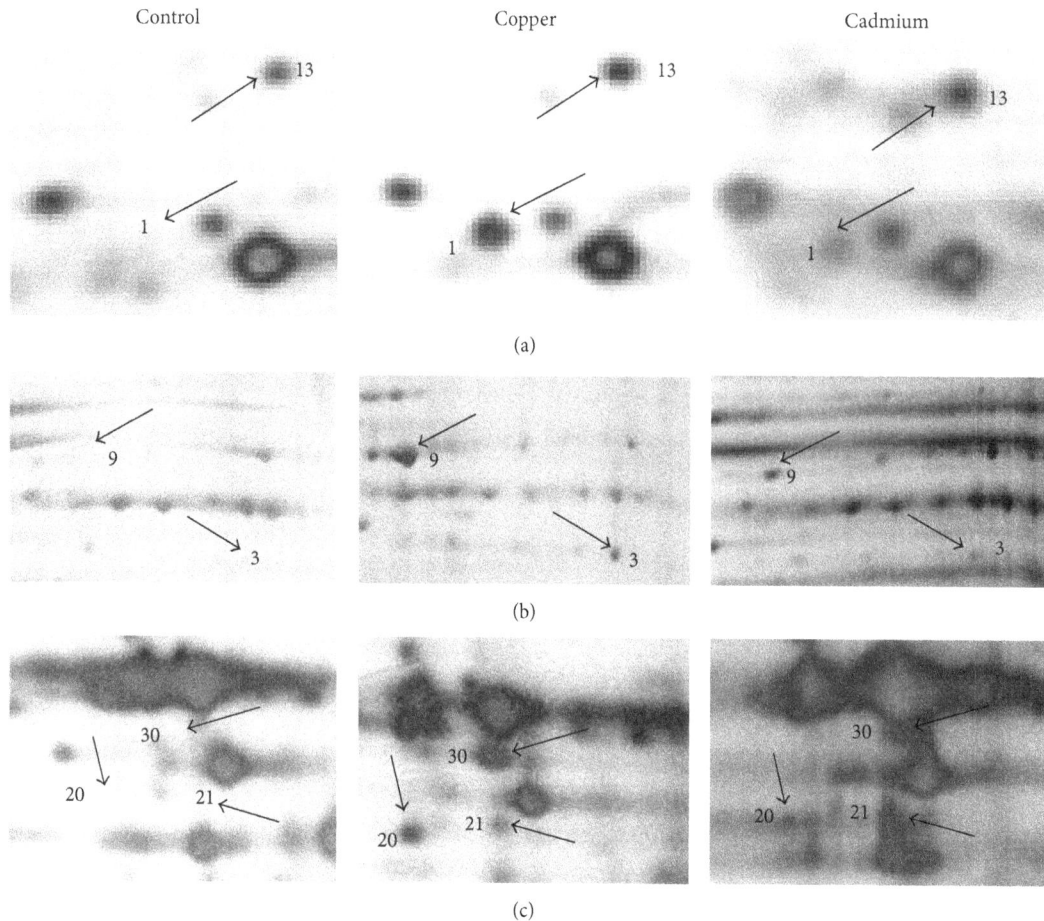

FIGURE 3: Comparison of selected proteins of *S. metallicus* cells treated with Cu or Cd. Protein extracts were obtained from cells treated without metals and with 100 mM Cu or 1 mM Cd for 24 h and the proteins were separated by 2D-PAGE. The segments (a), (b), and (c) are the enlarged dashed boxes in Figure 2 under the three conditions indicated. Numbers show some of the spots with increased intensity in cells treated with Cu or Cd.

in the absence of Cu (Figure 2(a)). Therefore, these results show that a number of proteins became non-detectable or decreased their levels when *S. metallicus* faced Cu. This kind of response was also seen when the microorganism was exposed to Cd (data not shown). This behavior has also been observed in similar studies where *F. acidarmanus* was challenged by either As(III) or Cu [16, 18]. In addition, the expression of 30 other proteins was found to be upregulated after Cu treatment (100 or 200 mM) (Figure 2(b)). Most of these proteins could be identified only in the condition with Cu, and spots 5, 13, and 15 were induced 2.6, 3.4, and 5.5-fold, respectively, compared to cells without Cu treatment. Moreover, 3 proteins of low molecular weight (spots 31, 3, and 33) that increased their expression when cells were exposed to 100 mM Cu were also identified (Figure 2, bottom panels). These proteins were only detected when 15% polyacrylamide was used during the second-dimensional separation (SDS-PAGE). On the other hand, a large decrease in overall expression of proteins after 1 mM Cd treatment was observed (not shown). Interestingly, 10 out of the 13 proteins identified as up-regulated when cells were exposed to Cd and also showed increased levels after Cu treatment. Some of them are shown in Figure 3.

3.3. Identification of Proteins Upregulated in Cells Treated with Cu and Cd in S. metallicus. A total of 18 *S. metallicus* proteins whose levels were found to be up-regulated in response to Cu were analyzed by mass spectrometry. The identified proteins included functions related to the production and transport of energy, biosynthesis of amino acids, stress responses, and transcription regulation (Table 2). Amongst the proteins related to production and transport of energy, a putative ATP synthase subunit B (spot 23) that has been described as playing a fundamental role in ATP synthesis was identified (Table 2). When cells are subjected to some stressing conditions such as the presence of heavy metals, a greater cellular demand for energy has been reported to occur [18]. Most likely, this phenomenon might be due to ATP-driven Cu transport via ATPases that have a substantial interplay during metal detoxification [45]. Furthermore, the ATP synthase subunit B was also identified as up-regulated when *S. metallicus* was treated with Cd (Table 2).

Two other proteins corresponded to putative oxidoreductases such as ferredoxin oxidoreductase (spot 3) and alcohol dehydrogenase (spot 4), that have been generally involved in electron transporter chains and use NAD^+ as an electron acceptor. The levels of ferredoxin oxidoreductase were also

TABLE 2: Proteins upregulated in *S. metallicus* cells exposed to 100 mM Cu or 1 mM Cd.

Spot	Molecular weight (kDa)	Cu induction levels	Putative function	Microorganism related	Accesion number
1[†]	26.6	∞	Proteasome subunit	*S. solfataricus*	AAK41034
3[†]	70.1	∞	Ferredoxin oxidoreductase	*S. solfataricus*	AAK42926
4[†]	42.9	∞	Alcohol dehydrogenase	*S. solfataricus*	AAK43154
5	35	2.6	Hypothetical protein	*S. acidocaldarius*	YP_254976
6	27	∞	Hypothetical protein	*P. aerophilum*	NP_559889
9[†]	57	∞	Phosphoglycerate dehydrogenase	*M. thermautotrophicus*	AAB85466
13[†]	27	3.4	Unknown	—	—
14	28	∞	Unknown	—	—
15	35	5.5	Unknown	—	—
20[†]	46.1	∞	Glutamate dehydrogenase	*S. solfataricus*	AAK42230
21[†]	45	∞	Hypothetical protein	*M. acetivorans*	NP_618380.1
22	49	∞	Hypothetical protein	*P. torridus*	YP_022807
23[†]	51	∞	ATP synthase subunit B	*S. solfataricus*	AAK40880
24	58.5	∞	HSP60 subunit	*S. shibatae*	AAG37273.1
30[†]	48	∞	Unknown	—	—
31	16	ND	Unknown	—	—
32	14.4	ND	Putative transcription regulator	*S. solfataricus*	AAK40413
33	13.8	ND	Transcription factor nusA	*S. solfataricus*	AAK40563

Spots refer to those numbered in Figure 2. Proteins in Figure 2 which are not included in this table were not subjected to sequencing due to their low intensities in the gels.
[†]Spot up-regulated in cells exposed to Cu and Cd.
ND: not determined.
Infinity symbol indicates that proteins were expressed only in presence of copper or cadmium.
M. thermoautotrophicus: *Methanobacterium thermoautotrophicus*; *M. acetivorans*: *Methanosarcina acetivorans*; *S. shibatae*: *Sulfolobus shibatae*; *P. torridus*: *Picrophilus torridus*; *P. aerophilum*: *Pyrobaculum aerophilum*; *S. acidocaldarius*: *Sulfolobus acidocaldarius*.

up-regulated in response to Cd (Table 2). Several previous reports have suggested that oxidoreductases contribute to an oxidative protection in both Bacteria and Archaea in response to heavy metals [46–48]. In the neutrophilic microorganisms *Escherichia coli* and *Staphylococcus aureus*, it has been described that oxidases and dehydrogenases contribute to oxidative protection, Cu homeostasis, and stress responses [46, 47]. Furthermore, some proteins involved in oxidative damage repair, such as NADH-dependent oxidases and thioredoxin reductases, were expressed in cells of *F. acidarmanus* exposed to As(III) [16]. The expression of this group of proteins has also been observed when the same microorganism was exposed to Cu [18]. Therefore, oxidoreductases play an important role against oxidative stress and may eliminate reactive oxygen species, which constitute the major component of the stress caused by transition metals [44, 49, 50].

The enzymes related to biosynthesis of amino acids were found as commonly up-regulated either in response to Cu or Cd. They corresponded to a phosphoglycerate dehydrogenase (spot 9) and a glutamate dehydrogenase (spot 20). The former one catalyzes the NAD$^+$-dependent oxidation of 3-phosphoglycerate into 3-phosphohydroxypyruvate, a branch point from the glycolytic pathway and the initial reaction in L-serine biosynthesis [51]. Glutamate dehydrogenase catalyzes the oxidative deamination of glutamate to produce 2-oxoglutarate and ammonia with reduction of NAD$^+$ [52]. Proteins involved in amino acids synthesis were expressed

in cells of *F. acidarmanus* exposed to As(III) [16]. Moreover, increased levels of proteins involved in the biosynthesis of sulfur-containing amino acids have been observed in *Saccharomyces cerevisiae* cells exposed to Cd [24], suggesting that these proteins might also have a role in the cellular response to adverse conditions.

Two proteins showed identity to components of the stress response mechanism such as one subunit of the HSP60 chaperonin (spot 24) and one subunit of the proteasome (spot 1). In Archaea, the stress protein HSP60 is directly involved in protein folding processes [53]. Proteomic analysis of *F. acidarmanus* cells exposed to As(III) and Cd revealed the expression of proteins involved in protein folding and DNA repair, including HSP60 chaperonin and DnaK heat-shock protein (HSP70); thereby the authors suggested that this microorganism uses multiple mechanisms to resist high levels of Cu [16, 18]. On the other hand, proteasomes are described as nonspecific proteolytic nanomachines associated with protein catabolism. Proteasomes are known to be closely involved in maintaining protein quality control by degrading miss folded and denatured proteins in response to cell stress in all three domains of life [54]. A proteomic analysis in the thermoacidophilic archaeon *Thermoplasma acidophilum* showed that proteasomes and chaperone-related proteins were highly induced against stress conditions, indicating a high turnover rate of proteins [55]. In *P. furiosus*, the expression of the β1 subunit of the proteasome was induced in cells subjected to heat shock stress [56]. Additionally, the

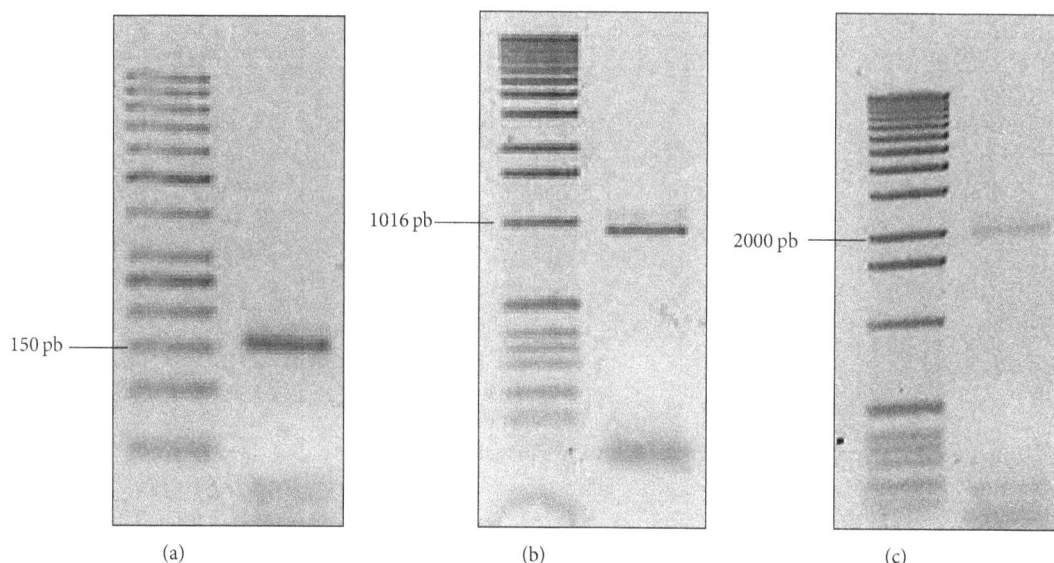

(a)   (b)   (c)

FIGURE 4: CODEHOP-based PCR for the amplification of the putative genes *copA* and *copM* from *S. metallicus*. Amplification products, (a) using degenerate PCR primers (DOP) copM_degF and copM_degR designed from amino acid sequence conserved blocks of *Sulfolobales* CopM proteins, (b) using primers copA_cdegF and copA_cdegR designed by CODEHOP strategy for amplification of a *copA*-like gene. (c) Primers copM_degF and copA_cdegR were assesed for the amplification of a *copMA*-like DNA region. Those primers were tested with genomic *S. metallicus* DNA samples. Amplicons expected sizes were as follows: ca. 150 bp for *copM*, ca. 950 bp for *copA*, and ca. 2,000 bp for *copMA*. PCR products were excised and cloned for later sequencing.

proteasome subunit was also found to be up-regulated when *S. metallicus* was subjected to Cd stress (Table 2).

The transcriptional regulation-related proteins corresponded to a putative transcription regulator (spot 32) and the transcription factor NusA (spot 33) involved in the intrinsic termination of transcription [57]. NusA has been also reported to display important functions in the cellular response to adverse factors [58]. The expression of two transcriptional regulators related to amino acids biosynthesis and metal-dependent genetic repression were found to be induced in *P. furiosus* cells subjected to oxidative stress [48]. Moreover, the expression of several genes encoding predicted transcriptional regulators were induced following Cd exposure in *T. gammatolerans* cells [27].

Spots 5, 6, 13, 14, 15, 21, 22, 30, and 31 (Table 2) did not show significant scores with any protein sequences currently available in the data bases. Interestingly, spots 2, 4, 9, 13, 21, and 30 were found to be commonly upregulated when *S. metallicus* cells were treated with either Cu or Cd (Figure 3, Table 2).

The results presented here are consistent with previous work describing global protein expression profiles in response to heavy metals as reported for *F. acidarmanus* [16, 18], microbial-biofilm communities in an acid mine drainage site [59], and other prokaryotic and eukaryotic microorganisms [24, 60]. Therefore, a coordinated expression of different groups of genes suggests the existence of regulatory networks such as stress response mechanisms and respiratory chain adjustments to cope with the presence of heavy metal ions [61].

3.4. Cu-Resistance (Cop) Genes Cluster Is Duplicated in S. metallicus Genome.
A Cu-resistance (*cop*) locus has been described to be highly conserved in archaeal genomes, which consists of genes encoding a new type of archaeal transcriptional regulator (CopT), a metal-binding chaperone (CopM), and a Cu-transporting P-type ATPase (CopA) [19]. Since the genome sequence of *S. metallicus* is not yet available, we searched for the presence of *cop* genes by using consensus degenerate hybrid oligonucleotide primers-based PCR (CODEHOP-PCR). CopA and CopM amino acid sequence alignments from *S. acidocaldarius*, *S. tokodaii*, and *S. solfataricus* were used to identify conserved sequence blocks and therefore to design either CODEHOP or DOP (degenerate hybrid oligonucleotide) primers (Figure S1, Table 1). Thereby PCR assays using *S. metallicus* genomic DNA as template yielded amplicons of ca. 150 bp and ca. 950 bp for *copM* and *copA* CODEHOP pair primers, respectively (Figures 4(a), 4(b)). As estimated from primer pair's position on amino acid sequence alignments, the obtained PCR products corresponded to the expected sizes, suggesting the presence of *copM*- and *copA*-like genes in *S. metallicus* genome. In order to determine the identity of the amplified DNA fragments, they were TA-cloned and sequenced. Blastx sequence analysis showed that the 150 bp DNA fragment coded for an incomplete ORF sharing 63% homology with *S. solfataricus* CopM (SSO10823), whereas the analysis of the 950 pb DNA fragment yielded a partial ORF sharing 54% homology with *S. solfataricus* CopA protein (SSO2651).

The genomic organization of the putative *cop* genes in *S. metallicus* was analyzed to find out its similarity with those

Molecular Characterization of Copper and Cadmium Resistance Determinants in the Biomining
Thermoacidophilic Archaeon Sulfolobus metallicus

91

TABLE 3: Sequence identity comparison of S. metallicus Cop proteins to those in other Sulfolobales.

S. metallicus protein	% Identity to homologue in other Sulfolobales			
	S. solfataricus P2	S. acidocaldarius	S. tokodaii	M. sedula
CopA1	**51**	43	44	45
CopA2	**88**	46	49	50
CopM1	63	69	**71**	**71**
CopM2	**89**	72	70	77
CopT1	**36**	27	30	32
CopT2*	**90**	52	52	53

Boldface type indicates to which organism the S. metallicus homologue shows the highest sequence identity.
*Refers to the analysis obtained using the uncompleted amino acid sequence of CopT2, which represents ~90% in length when compared with S. solfataricus CopT sequence.

previously described in S. solfataricus P2 and F. acidarmanus, where the cop cluster consists of tandem-orientated genes as copTMA [18, 19]. Thus, using copM_degF (forward) and copA_cdegR (reverse) primers, we attempted to amplify a putative copMA DNA region. This PCR yielded a product of ca. 2,000 pb, corresponding to the expected size, and resulted in 2 ORFs showing high sequence homology with CopM and CopA from S. solfataricus (Figure 4(c)). Interestingly, when aligning this copMA sequence with the copA and copM sequences obtained previously, they were not identical, which strongly indicated the presence of duplicated putative cop genes encoding for both Cu-ATPases (copA1 and copA2) and metallochaperones (copM1 and copM2) in the S. metallicus genome.

To isolate the entire cop gene sequences, the genome walking method described by Acevedo et al. [38] was used. Six different DNA libraries were constructed from S. metallicus genomic DNA digested with several restriction enzymes, including 5′ overhang and blunt end restriction enzymes. The DNA libraries were used as templates in 2 successive PCR reactions. From the previously known sequences, two specific sense primers for the 3′ end amplification and two specific antisense primers for the 5′ end amplification were designed (abbreviated SP1 and SP2). Therefore, the 5′ and 3′ ends of the putative genes copA1 and copA2 that had been partially identified through the genome walking technique were amplified.

By overlapping the sequence isolated by degenerate PCR and the lateral sequences obtained by genome walking (Figure S2), it was possible to complete the whole nucleotide sequence of the putative gene copA1. Additionally, it was possible to confirm the presence of the upstream copM1 gene and determine the occurrence of a partial copT1 gene tandem-orientated upstream of copM1 (Figure 5). Through additional genome walking experiments, the partial copT1 sequence could be further completed. The same was true when isolating the whole nucleotide sequence of the putative gene copA2 as the presence of copM2 and copT2 genes were confirmed upstream of copA2. However, it was not possible to complete the whole 5′ end of the putative gene copT2 as its determined length corresponded to only 90% of the full length when compared with S. solfataricus copT gene (Figure 5).

In conclusion, degenerate PCR together with genome walking experiments allowed to describe the occurrence in S. metallicus genome of 2 cop genes loci (named as locus cop1 and locus cop2) coding for paralogous genes whose products may be involved in Cu-resistance (Figure 5). Although some archaeal genomes exhibit two copies of putative Cu-P-type ATPases as described for S. solfataricus [62], the genomic arrangement displayed as a cop gene cluster (copTMA) has been reported to be represented in only one copy for many archaeal genomes [10, 19, 21]. Moreover, we further updated this analysis by searching for duplications of the cop genes cluster in all available archaeal genome sequences up to date (October, 2012), retrieving only one cop locus in each case. Thereby we propose that the discovery of this cop gene cluster duplication in the genome of S. metallicus constitutes so far an unprecedented feature for a representative of the Archaea domain and might contribute to its high metal resistance.

In this context, it has been widely reported that increased gene copy number can increase gene expression allowing prokaryotic to thrive under growth limiting conditions [13, 63–68]. However, alongside the augmented gene copy, one cannot exclude that the cop operon duplication in S. metallicus may offer among others: a wider repertoire of Cu-ATPases, which might differ in terms of metal affinities and/or specificity, efflux rates, and/or differences in their abundance regulation.

Throughout amino acid sequence analysis, it was further determined that each polypeptide encoded by S. metallicus, with the exception of CopT1, showed to be highly homologous to their orthologous counterparts encoded by others Sulfolobales (Table 3). Moreover, the 3 ORFs encoded by the locus cop2 shared about 90% identity with the corresponding S. solfataricus orthologous peptides (Table 3). S. metallicus cop genes products showed characteristic domains, referred to as critical for their respective proposed biological activities. S. metallicus paralogous gene products CopA1 (749 aa′) and CopA2 (747 aa′) shared 51.3% of identity. These two putative Cu-ATPases (CopA1 and CopA2) contain the amino acid sequence motif CPCALGLA which has been proposed to confer Cu-transporting specificity in CPx-ATPases [48] and has been also found in other CopA-like proteins from biomining Bacteria and Archaea [14].

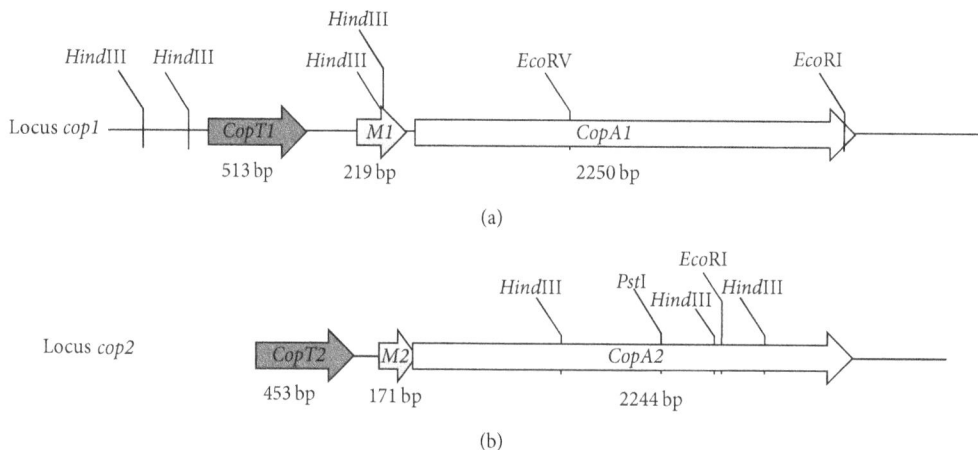

(a)

(b)

FIGURE 5: Schematic representation of the two *cop* gene clusters isolated from *S. metallicus* genome. Each cluster codes for an *Archaea*-specific transcriptional regulator (*copT*), a metallochaperone (*copM*), and a P-type Cu-exporting ATPase (*copA*). Lengths of each gene are indicated. Locus *cop1* corresponds to 4,346 sequenced base pairs and shows an intergenic region *copT1-M1* of 232 base pairs. Locus *cop2* corresponds to 3,435 base pairs and shows 89 base pairs *copT2-M2* intergenic region. Genes *copM2* and *copA2* overlapped in 11 base pairs. *copT2* was not fully isolated.

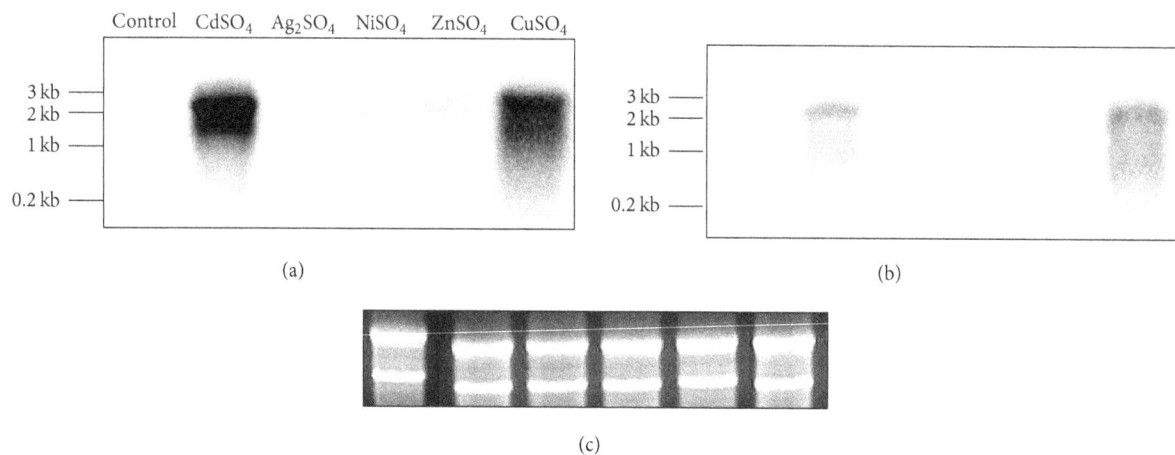

(a)

(b)

(c)

FIGURE 6: Northern blot analysis to determine the expression of *S. metallicus copA1* (a) and *copA2* (b) putative genes in response to various metals. *S. metallicus* total RNA was extracted from cultures grown either in the absence (control) or the presence of $CuSO_4$ (50 mM), Ag_2SO_4 (0.08 mM), $NiSO_4$ (15 mM), $ZnSO_4$ (50 mM), and $CdSO_4$ (2 mM). P^{32}-radioactivelly labelled DNA fragments annealing to *copA1* and *copA2* sequences, respectively, were used as probes. (c) shows rRNAs as total RNA loading control.

On the other hand, CopM1 (71 aa) and CopM2 (56 aa), sharing 66% of identity, present the proposed metal binding domain TRASH (trafficking, resistance, and sensing of heavy metals) [10]. Additionally, when analyzing the putative transcriptional regulators, the presence of a C-terminal TRASH domain in both CopT1 and CopT2 was also found. CopT1 sequence also contains an entire N-terminal helix-turn-helix (HTH) motif (not shown) that resembles the DNA-binding motifs of prokaryotic transcriptional regulators, such as Lrp-like proteins [67]. The HTH motif was only partially identified in CopT2 since the 5′ region of *copT1* gene has not been yet fully isolated.

3.5. Transcriptional Analysis of S. metallicus Cop Genes. To get insight into the role of the two *cop* loci, the expression of the corresponding genes was analyzed under various conditions. *S. metallicus* was first grown in presence of different heavy metals (Cu, Zn, Cd, Ni o Ag) at a given concentration that did not affect growth kinetics [30, 42]. Subsequently, total RNA was isolated from late exponentially grown cultures and Northern blot experiments were carried out in order to determine the expression of both *copA1* and *copA2* genes. As depicted in Figure 6, transcription of both *copA1* and *copA2* messengers were specifically induced in the presence of Cu and Cd ions. This gene expression pattern

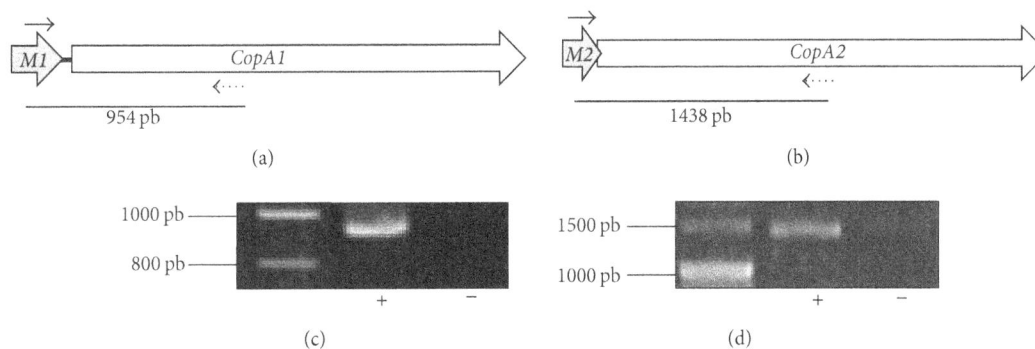

FIGURE 7: Cotranscription analysis of *copMA1* and *copMA2* genes. cDNA was synthesized with a reverse primer (dotted arrows) hybridizing toward the 3' end of either *copA1* (a) or *copA2* (b). *S. metallicus* total RNA was extracted at the late exponential phase from a culture growing in presence of 20 mM Cu. PCR amplifications were carried out with these cDNAs and each corresponding primer pair (black arrows) as listed in Table 1. (c) and (d) show RT-PCR products obtained for the *copMA1* and *copMA2* intergenic regions, respectively. Reverse transcriptase reactions with (+) and without (−) the Improm II reverse transcriptase enzyme were carried out in order to exclude amplification due to genomic DNA contamination. Expected sizes (in base pairs) for the corresponding PCR products are given in (a) and (b).

is in good agreement with what has been described for *S. solfataricus*, where the bicistronic *copMA* transcript levels were also found to be up-regulated in response to both Cu and Cd [19].

Northern blot analysis showed that the sizes seen for both *copA* transcripts (~2.5 Kb) did not correspond exactly with the individual genes length (each of about 2.25 Kb) (Figure 6). This strongly suggested that the transcripts could also include the respective *copM* gene located upstream of *copA* in each *cop* loci. To confirm this assumption, a co-transcription experiment was done (Figure 7) in which the cDNAs were obtained by using RNA extracted from a culture grown in the presence of 20 mM Cu and using a reverse primer hybridizing with *copA1* (Figure 7(a)) or *copA2* (Figure 7(b)) gene sequence, respectively. PCR amplifications were carried out by using the corresponding cDNAs as templates and each pair of primers lying in adjacent genes (*copM*). The presence of an amplicon of the expected size in each case indicated the adjacent genes were part of polycistronic messengers (Figure 7). These results clearly show that in *S. metallicus* each couple of *copMA* genes was expressed in the form of transcriptional units, as reported in *F. acidarmanus* and *S. solfataricus* [18, 19]. The coexpression of gene pair's *copMA* may suggest a coordinated and dependent function for the respective encoded proteins. In the bacterium *E. hirae*, the metallochaperone CopZ (CopM in *S. metallicus*) fulfills a pivotal role in the mechanism of Cu homeostasis [45]. Thus, it was shown that this protein interacts directly with the Cu-ATPase CopB (CopA in *S. metallicus*), handing the Cu for subsequent removal. In this context, one might expect that proteins CopM1 and CopM2 from *S. metallicus* have a similar role to that described for *E. hirae*. Furthermore, quantitative RT-PCR experiments were carried out in order to determine differential expression of the *cop* genes in response to different Cu concentrations. The expression of *copA1*, *copA2*, *copT1* and *copT2* was tested relative to the transcript levels of *rRNA 16S* gene since its levels were not significantly affected

in all assessed conditions (data not shown). As shown in Figure 8(a), *copA2* and *copA1* transcript levels were found to be concomitantly increased with the increasing Cu concentrations present in the medium. Higher transcripts levels of *copA1* were seen in all tested conditions when compared with *copA2* levels. *copA1* transcript levels were found to be 32.5-fold up-regulated when comparing 50 mM Cu condition versuscontrol (absence of metal), whereas *copA2* transcript levels showed an increment of only 17.5-fold (Figure 8(a)). The finding that Cu-ATPases mRNA levels were significantly increased in response to Cu ions exposure suggests that the transport system may operate for Cu efflux in *S. metallicus*.

Moreover, *copA2* and *copA1* gene expression was quantified when *S. metallicus* was grown using chalcopyrite ($CuFeS_2$) ore as an oxidizable substrate (Figure 8(a)). Mineral oxidation mediated by the microorganism generates a progressive increase in Cu ions concentration in the medium. To find out whether the amount of solubilized Cu induced the expression of the Cu-ATPases genes, total RNA was extracted from a *S. metallicus* culture grown to late exponential phase and in the presence of 1% $CuFeS_2$. It was clear that an up-regulation of the Cu-ATPases encoding genes also took place when *S. metallicus* was grown in the presence of $CuFeS_2$ (Figure 8(a)). *copA1* transcript levels increased 14-fold compared with the control condition in the absence of Cu, while *copA2* gene expression showed an increase of 7.6 fold. Along with this, by means of atomic absorption spectrometry (AAS) analysis, an overall amount of solubilized Cu ions (Cu^{2+}/Cu^{1+}) of 14.4±2.1 mM was determined to be present in the medium, indicating that *copA2* and *copA1* gene expression was most likely in response to the Cu present in the culture due to $CuFeS_2$ microbial-solubilization.

The finding that *copA1* was highly expressed compared with *copA2* in all tested conditions may suggest a possible physiological hierarchy between the two ATPases when overcoming either Cu or Cd stress. In this regard, by means of a genetic approach it was demonstrated in *S. solfataricus*

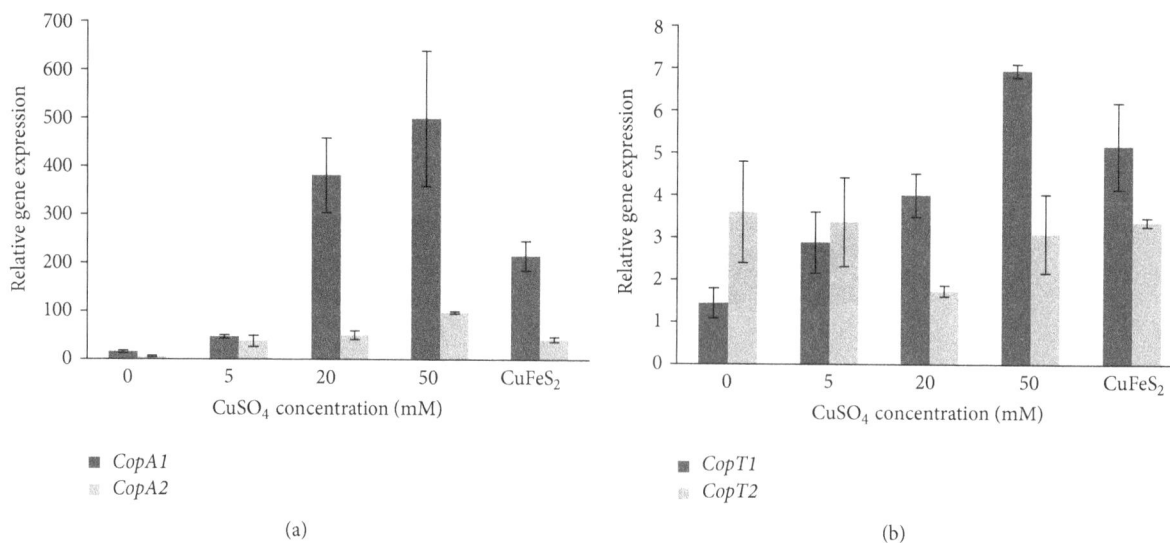

FIGURE 8: Relative expression levels of *S. metallicus* genes *copA1* and *copA2* (a) *copT1* and *copT2* (b). Expression of *S. metallicus cop* genes wasassessed by qRT-PCR and normalized against 16S rRNA gene expression in individual cultures. Mean values and standard deviations are from analyses of three independent cultures grown in the presence of 0, 5, 20, 50 mM Cu or in 1% of a chalcopyrite ore ($CuFeS_2$) concentrate.

that while CopA was an effective Cu efflux transporter at low and high Cu concentrations, the other Cu-ATPase (CopB) only appeared to be a low-affinity Cu export ATPase [68]. Moreover, by using a *M. sedula copA* deletion mutant it was demonstrated that this strain compromised metal resistance and consequently abolished chalcopyrite oxidation [22], highlighting the role of Cu detoxification mechanisms during a given bioleaching process. Our attempts to show the functionality of CopA1 and CopA2 from *S. metallicus* by using *E. coli* as a heterologous host were not successful. Apparently, the overexpression of these ORFs had a toxic effect on *E. coli* that compromised its viability. Although gene disruption tools have not been yet developed for *S. metallicus* the functional role of both *copTMA* locimight be further studied by cross-species complementation of a copper sensitive *S. solfataricus copR* mutant as it was described by Maezato et al. [22]. It will be of great interest to establish in future studies the possible functionality of the isolated putative transporters from *S. metallicus*.

Likewise, quantitative RT-PCR experiments showed that *copT1* transcript levels increased concomitantly with increasing Cu concentrations, whereas *copT2* showed relatively similar transcripts levels, most likely indicating a constitutive expression profile (Figure 8(b)). Furthermore, whereas *copT1* transcript levels were found to be increased 3.6 fold in $CuFeS_2$ grown cultures, *copT2* levels remained unchanged in comparison with the control in the absence of Cu (Figure 8(b)). The results obtained for *S. metallicus copT2* gene expression are consistent with those reported for *S. solfataricus*. CopT transcriptional regulator has been proposed to function as a repressor in *S. solfataricus*, showing a constitutive expression that in the presence of Cu loses its affinity for the promoter region of *copMA* allowing the expression of this polycistron [19]. In contrast, the increased transcripts levels of both *copA1* and *copT1* concomitant with higher Cu concentrations in the

environment suggest that the putative transcriptional regulator CopT1 may act by activating both *copMA1* and probably itself (Figure 8). Regarding this, it was recently reported that CopR (CopT in *S. metallicus*) from *S. solfataricus* strain 98/2 acts as an activator of *copT* (*copM* in *S. metallicus*) and *copA* expression [69]. Nevertheless, additional experiments would be required to test this possibility in *S. metallicus*.

4. Concluding Remarks

We previously reported that *S. metallicus* resists extremely high Cu concentrations, which was mediated to some extent by the use of a possible metal resistance system based on inorganic polyphosphate hydrolysis and consequently Cu-PO_4^{2-} efflux [15, 29, 30]. Here we have addressed the question whether this microorganism coded for other determinants that might help to explain its high heavy metal tolerance. As the genomic sequence of this microorganism is not yet available, we jointly used CODEHOP-PCR and genome walking approaches and were able to establish the occurrence of two nonidentical homologous *cop* loci into the genome of *S. metallicus* (*cop1* and *cop2*). Each *cop* locus codes for an archaeal transcriptional regulator (CopT), a metal-binding chaperone (CopM) and a Cu-transporting P-type ATPase (CopA). High levels of the polycistronic mRNAscopMA of each *cop* locus were observed after treatment with either Cu or Cd, suggesting that the encoded ATPases efflux heavy metals out in order to detoxify the intracellular environment. Altogether, previous reports and the results obtained in this study allow us to suggest that some key elements that may explain the high resistance to Cu in *S. metallicus* is the duplication of the Cu resistance *cop* genes, a defensive response to stress and a polyP-based accumulation mechanism. In the case of Cd, although some similar stress responses were

Molecular Characterization of Copper and Cadmium Resistance Determinants in the Biomining
Thermoacidophilic Archaeon Sulfolobus metallicus

95

observed, whether comparable Cu-responsive elements also participate in Cd responses remains to be seen.

Conflict of Interests

The authors declare that the research was done in the absence of any commercial or financial relationships that could be construed as a potential conflict of interests.

Acknowledgments

This work was supported by Grant no. FONDECYT 1110214 and in part by ICM P-05-001-F project.

References

[1] A. Schippers, "Microorganisms involved in bioleaching and nucleic acid-based molecular methods for their identification and quantification," in *Microbial Processing of Metal Sulfides*, R. E. Donati and W. Sand, Eds., pp. 3–33, Elsevier, Berlin, Germany, 2007.

[2] H. R. Watling, "The bioleaching of sulphide minerals with emphasis on copper sulphides: a review," *Hydrometallurgy*, vol. 84, no. 1-2, pp. 81–108, 2006.

[3] D. E. Rawlings, D. Dew, and C. Du Plessis, "Biomineralization of metal-containing ores and concentrates," *Trends in Biotechnology*, vol. 21, no. 1, pp. 38–44, 2003.

[4] L. Valenzuela, A. Chi, S. Beard et al., "Genomics, metagenomics and proteomics in biomining microorganisms," *Biotechnology Advances*, vol. 24, no. 2, pp. 197–211, 2006.

[5] C. A. Jerez, "Bioleaching and biomining for the industrial recovery of metals," in *Comprehensive Biotechnology*, M. Moo-Young, Ed., vol. 3, pp. 717–729, Elsevier, Amstaerdam, The Netherlands, 2nd edition, 2011.

[6] P. R. Norris and J. P. Owen, "Mineral sulphide oxidation by enrichment cultures of novel thermoacidophilic bacteria," *FEMS Microbiology Reviews*, vol. 11, no. 1–3, pp. 51–56, 1993.

[7] D. E. Rawlings and D. B. Johnson, "The microbiology of biomining: development and optimization of mineral-oxidizing microbial consortia," *Microbiology*, vol. 153, no. 2, pp. 315–324, 2007.

[8] P. R. Norris, N. P. Burton, and N. A. M. Foulis, "Acidophiles in bioreactor mineral processing," *Extremophiles*, vol. 4, no. 2, pp. 71–76, 2000.

[9] M. R. Bruins, S. Kapil, and F. W. Oehme, "Microbial resistance to metals in the environment," *Ecotoxicology and Environmental Safety*, vol. 45, no. 3, pp. 198–207, 2000.

[10] T. J. G. Ettema, M. A. Huynen, W. M. De Vos, and J. Van Der Oost, "TRASH: a novel metal-binding domain predicted to be involved in heavy-metal sensing, trafficking and resistance," *Trends in Biochemical Sciences*, vol. 28, no. 4, pp. 170–173, 2003.

[11] M. Dopson, C. Baker-Austin, P. R. Koppineedi, and P. L. Bond, "Growth in sulfidic mineral environments: metal resistance mechanisms in acidophilic micro-organisms," *Microbiology*, vol. 149, no. 8, pp. 1959–1970, 2003.

[12] S. Franke and C. Rensing, "Acidophiles. Mechanisms to tolerate metal and acid toxicity," in *Physiology and Biochemistry of Extremophiles*, C. Gerday and N. Glansdorff, Eds., pp. 271–278, ASM Press, Washington, DC, USA, 2007.

[13] C. A. Navarro, L. H. Orellana, C. Mauriaca, and C. A. Jerez, "Transcriptional and functional studies of *Acidithiobacillus ferrooxidans* genes related to survival in the presence of copper," *Applied and Environmental Microbiology*, vol. 75, no. 19, pp. 6102–6109, 2009.

[14] A. Orell, C. A. Navarro, R. Arancibia, J. C. Mobarec, and C. A. Jerez, "Life in blue: copper resistance mechanisms of bacteria and Archaea used in industrial biomining of minerals," *Biotechnology Advances*, vol. 28, no. 6, pp. 839–848, 2010.

[15] A. Orell, C. A. Navarro, M. Rivero, J. S. Aguilar, and C. A. Jerez, "Inorganic polyphosphates in extremophiles and their posible functions," *Extremophiles*, vol. 16, no. 4, pp. 573–583, 2012.

[16] C. Baker-Austin, M. Dopson, M. Wexler et al., "Extreme arsenic resistance by the acidophilic archaeon "Ferroplasma acidarmanus" Fer1," *Extremophiles*, vol. 11, no. 3, pp. 425–434, 2007.

[17] E. Pedone, S. Bartolucci, and G. Fiorentino, "Sensing and adapting to environmental stress: the archaeal tactic," *Frontiers in Bioscience*, vol. 9, pp. 2909–2926, 2004.

[18] C. Baker-Austin, M. Dopson, M. Wexler, R. G. Sawers, and P. L. Bond, "Molecular insight into extreme copper resistance in the extremophilic archaeon "Ferroplasma acidarmanus" Fer1," *Microbiology*, vol. 151, no. 8, pp. 2637–2646, 2005.

[19] T. J. G. Ettema, A. B. Brinkman, P. P. Lamers, N. G. Kornet, W. M. de Vos, and J. van der Oost, "Molecular characterization of a conserved archaeal copper resistance (cop) gene cluster and its copper-responsive regulator in *Sulfolobus solfataricus* P2," *Microbiology*, vol. 152, no. 7, pp. 1969–1979, 2006.

[20] A. A. Villafane, Y. Voskoboynik, M. Cuebas, I. Ruhl, and E. Bini, "Response to excess copper in the hyperthermophile *Sulfolobus solfataricus* strain 98/2," *Biochemical and Biophysical Research Communications*, vol. 385, no. 1, pp. 67–71, 2009.

[21] K. S. Auernik, Y. Maezato, P. H. Blum, and R. M. Kelly, "The genome sequence of the metal-mobilizing, extremely thermoacidophilic archaeon *Metallosphaera sedula* provides insights into bioleaching-associated metabolism," *Applied and Environmental Microbiology*, vol. 74, no. 3, pp. 682–692, 2008.

[22] Y. Maezato, T. Johnson, S. McCarthy, K. Dana, and P. Blum, "Metal resistance and lithoautotrophy in the extreme thermoacidophile *Metalosphaera sedula*," *Journal of Bacteriology*, vol. 194, no. 24, pp. 6856–6863, 2012.

[23] D. H. Nies, "Microbial heavy-metal resistance," *Applied Microbiology and Biotechnology*, vol. 51, no. 6, pp. 730–750, 1999.

[24] K. Vido, D. Spector, G. Lagniel, S. Lopez, M. B. Toledano, and J. Labarre, "A proteome analysis of the cadmium response in *Saccharomyces cerevisiae*," *The Journal of Biological Chemistry*, vol. 276, no. 11, pp. 8469–8474, 2001.

[25] G. Bertin and D. Averbeck, "Cadmium: cellular effects, modifications of biomolecules, modulation of DNA repair and genotoxic consequences (a review)," *Biochimie*, vol. 88, no. 11, pp. 1549–1559, 2006.

[26] M. H. Joe, S. W. Jung, S. H. Im et al., "Genome-wide response of *Deinococcus radiodurans* on cadmium toxicity," *Journal of Microbiology and Biotechnology*, vol. 21, no. 4, pp. 438–447, 2011.

[27] A. Lagorce, A. Fourçans, M. Dutertre, B. Bouyssiere, Y. Zivanovic, and F. Confalonieri, "Genome-wide transcriptional response of the archeon *Thermococcus gammatolerans* to cadmium," *PLOS ONE*, vol. 7, no. 7, Article ID e41935, 2012.

[28] A. Kornberg, N. N. Rao, and D. Ault-Riché, "Inorganic polyphosphate: a molecule of many functions," *Annual Review of Biochemistry*, vol. 68, pp. 89–125, 1999.

[29] J. D. Keasling, "Regulation of intracellular toxic metals and other cations by hydrolysis of polyphosphate," *Annals of the New York Academy of Sciences*, vol. 829, pp. 242–249, 1997.

[30] F. Remonsellez, A. Orell, and C. A. Jerez, "Copper tolerance of the thermoacidophilic archaeon *Sulfolobus metallicus*: possible role of polyphosphate metabolism," *Microbiology*, vol. 152, no. 1, pp. 59–66, 2006.

[31] V. Hatzimanikatis, L. H. Choe, and K. H. Lee, "Proteomics: theoretical and experimental considerations," *Biotechnology Progress*, vol. 15, no. 3, pp. 312–318, 1999.

[32] L. H. Choe and K. H. Lee, "A comparison of three commercially available isoelectric focusing units for proteome analysis: the Multiphor, the IPGphor and the Protean IEF cell," *Electrophoresis*, vol. 21, no. 5, pp. 993–1000, 2000.

[33] U. K. Laemmli, "Cleavage of structural proteins during the assembly of the head of bacteriophage T4," *Nature*, vol. 227, no. 5259, pp. 680–685, 1970.

[34] A. Shevchenko, M. Wilm, O. Vorm, and M. Mann, "Mass spectrometric sequencing of proteins from silver-stained poly-acrylamide gels," *Analytical Chemistry*, vol. 68, no. 5, pp. 850–858, 1996.

[35] P. Giavalisco, E. Nordhoff, T. Kreitler et al., "Proteome analysis of Arabidopsis thaliana by two-dimensional gel electrophoresis and matrix-assisted laser desorption/ionisation-time of flight mass spectrometry," *Proteomics*, vol. 5, no. 7, pp. 1902–1913, 2005.

[36] T. M. Rose, E. R. Schultz, J. G. Henikoff, S. Pietrokovski, C. M. McCallum, and S. Henikoff, "Consensus-degenerate hybrid oligonucleotide primers for amplification of distantly related sequences," *Nucleic Acids Research*, vol. 26, no. 7, pp. 1628–1635, 1998.

[37] T. M. Rose, J. G. Henikoff, and S. Henikoff, "CODEHOP (Consensus-Degenerate Hybrid Oligonucleotide Primer) PCR primer design," *Nucleic Acids Research*, vol. 31, no. 13, pp. 3763–3766, 2003.

[38] J. P. Acevedo, F. Reyes, L. P. Parra, O. Salazar, B. A. Andrews, and J. A. Asenjo, "Cloning of complete genes for novel hydrolytic enzymes from Antarctic sea water bacteria by use of an improved genome walking technique," *Journal of Biotechnology*, vol. 133, no. 3, pp. 277–286, 2008.

[39] S. Bathe and P. R. Norris, "Ferrous iron- and sulfur-induced genes in *Sulfolobus metallicus*," *Applied and Environmental Microbiology*, vol. 73, no. 8, pp. 2491–2497, 2007.

[40] J. Sambrook, E. F. Fritsch, and T. Maniatis, *Molecular Cloning: A Laboratory Manual*, Cold Spring Harbor Press, New York, NY, USA, 1989.

[41] K. W. Miller, S. Sass Risanico, and J. B. Risatti, "Differential tolerance of Sulfolobus strains to transition metals," *FEMS Microbiology Letters*, vol. 93, no. 1, pp. 69–73, 1992.

[42] G. Huber, C. Spinnler, A. Gambacorta, and K. O. Stetter, "Met-allosphaera sedula gen. and sp. nov. represents a new genus of aerobic, metal-mobilizing, thermoacidophilic archaebacteria," *Systematic and Applied Microbiology*, vol. 12, pp. 38–47, 1989.

[43] J. Llanos, C. Capasso, E. Parisi, D. Prieur, and C. Jeanthon, "Susceptibility to heavy metals and cadmium accumulation in aerobic and anaerobic thermophilic microorganisms isolated from deep-sea hydrothermal vents," *Current Microbiology*, vol. 41, no. 3, pp. 201–205, 2000.

[44] A. Kaur, M. Pan, M. Meislin, M. T. Facciotti, R. El-Gewely, and N. S. Baliga, "A systems view of haloarchaeal strategies to withstand stress from transition metals," *Genome Research*, vol. 16, no. 7, pp. 841–854, 2006.

[45] M. Solioz and J. V. Stoyanov, "Copper homeostasis in *Entero-coccus hirae*," *FEMS Microbiology Reviews*, vol. 27, no. 2-3, pp. 183–195, 2003.

[46] S. Sitthisak, K. Howieson, C. Amezola, and R. K. Jayaswal, "Characterization of a multicopper oxidase gene from *Staphy-lococcus aureus*," *Applied and Environmental Microbiology*, vol. 71, no. 9, pp. 5650–5653, 2005.

[47] L. Rodríguez-Montelongo, S. I. Volentini, R. N. Farías, E. M. Massa, and V. A. Rapisarda, "The Cu(II)-reductase NADH dehydrogenase-2 of *Escherichia coli* improves the bacterial growth in extreme copper concentrations and increases the resistance to the damage caused by copper and hydroperoxide," *Archives of Biochemistry and Biophysics*, vol. 451, no. 1, pp. 1–7, 2006.

[48] E. Williams, T. M. Lowe, J. Savas, and J. DiRuggiero, "Microar-ray analysis of the hyperthermophilic archaeon *Pyrococcus furiosus* exposed to gamma irradiation," *Extremophiles*, vol. 11, no. 1, pp. 19–29, 2007.

[49] L. M. Gaetke and C. K. Chow, "Copper toxicity, oxidative stress, and antioxidant nutrients," *Toxicology*, vol. 189, no. 1-2, pp. 147–163, 2003.

[50] M. J. Davies, "The oxidative environment and protein damage," *Biochimica et Biophysica Acta*, vol. 1703, no. 2, pp. 93–109, 2005.

[51] J. R. Thompson, J. K. Bell, J. Bratt, G. A. Grant, and L. J. Banaszak, "Vmax regulation through domain and subunit changes. The active form of phosphoglycerate dehydrogenase," *Biochemistry*, vol. 44, no. 15, pp. 5763–5773, 2005.

[52] V. Consalvi, R. Chiaraluce, L. Politi, A. Gambacorta, M. De Rosa, and R. Scandurra, "Glutamate dehydrogenase from the thermoacidophilic archaebacterium *Sulfolobus solfataricus*," *European Journal of Biochemistry*, vol. 196, no. 2, pp. 459–467, 1991.

[53] A. J. L. Macario, M. Lange, B. K. Ahring, and E. Conway De Macario, "Stress genes and proteins in the archaea," *Microbiol-ogy and Molecular Biology Reviews*, vol. 63, no. 4, pp. 923–967, 1999.

[54] J. A. Maupin-Furlow, M. A. Gil, M. A. Humbard et al., "Archaeal proteasomes and other regulatory proteases," *Current Opinion in Microbiology*, vol. 8, no. 6, pp. 720–728, 2005.

[55] N. Sun, F. Beck, R. Wilhelm Knispel et al., "Proteomics analysis of *Thermoplasma acidophilum* with a focus on protein com-plexes," *Molecular and Cellular Proteomics*, vol. 6, no. 3, pp. 492–502, 2007.

[56] L. S. Madding, J. K. Michel, K. R. Shockley et al., "Role of the β1 subunit in the function and stability of the 20S proteasome in the hyperthermophilic Archaeon *Pyrococcus furiosus*," *Journal of Bacteriology*, vol. 189, no. 2, pp. 583–590, 2007.

[57] R. Shibata, Y. Bessho, A. Shinkai et al., "Crystal structure and RNA-binding analysis of the archaeal transcription factor NusA," *Biochemical and Biophysical Research Communications*, vol. 355, no. 1, pp. 122–128, 2007.

[58] W. Bae, B. Xia, M. Inouye, and K. Severinov, "*Escherichia coli* CspA-family RNA chaperones are transcription antitermina-tors," *Proceedings of the National Academy of Sciences of the United States of America*, vol. 97, no. 14, pp. 7784–7789, 2000.

[59] R. J. Ram, N. C. VerBerkmoes, M. P. Thelen et al., "Microbi-ology: community proteomics of a natural microbial biofilm," *Science*, vol. 308, no. 5730, pp. 1915–1920, 2005.

[60] I. Noël-Georis, T. Vallaeys, R. Chauvaux et al., "Global analysis of the *Ralstonia metallidurans* proteome: prelude for the large-scale study of heavy metal response," *Proteomics*, vol. 4, no. 1, pp. 151–179, 2004.

Molecular Characterization of Copper and Cadmium Resistance Determinants in the Biomining
Thermoacidophilic Archaeon Sulfolobus metallicus

97

[61] M. T. M. Novo, A. C. Da Silva, R. Moreto et al., "Thiobacillus ferrooxidans response to copper and other heavy metals: growth,
protein synthesis and protein phosphorylation," *Antonie van
Leeuwenhoek*, vol. 77, no. 2, pp. 187–195, 2000.

[62] Q. She, R. K. Singh, F. Confalonieri et al., "The complete genome
of the crenarchaeon *Sulfolobus solfataricus* P2," *Proceedings of
the National Academy of Sciences of the United States of America*,
vol. 98, no. 14, pp. 7835–7840, 2001.

[63] R. P. Anderson and J. R. Roth, "Tandem genetic duplications in
phage and bacteria," *Annual Review of Microbiology*, vol. 31, pp.
473–505, 1977.

[64] D. Gevers, K. Vandepoele, C. Simillion, and Y. Van De Peer,
"Gene duplication and biased functional retention of paralogs
in bacterial genomes," *Trends in Microbiology*, vol. 12, no. 4, pp.
148–154, 2004.

[65] A. B. Reams and E. L. Neidle, "Selection for gene clustering by
tandem duplication," *Annual Review of Microbiology*, vol. 58, pp.
119–142, 2004.

[66] L. H. Orellana and C. A. Jerez, "A genomic island provides
Acidithiobacillus ferrooxidans ATCC 53993 additional copper
resistance: a possible competitive advantage," *Applied Microbiology and Biotechnology*, vol. 92, no. 4, pp. 761–767, 2011.

[67] A. B. Brinkman, T. J. G. Ettema, W. M. de Vos, and J. van der
Oost, "The Lrp family of transcriptional regulators," *Molecular
Microbiology*, vol. 48, no. 2, pp. 287–294, 2003.

[68] C. Vollmecke, S. L. Drees, J. Reimann, S. V. Albers, and M.
Lubben, "The ATPases CopA and CopB both contribute to
copper resistance of the thermoacidophilic archaeon *Sulfolobus
solfataricus*," *Microbiology*, vol. 158, no. 6, pp. 1622–1633, 2012.

[69] A. Villafane, Y. Voskoboynik, I. Ruhl et al., "CopR of *Sulfolobus
solfataricus* represents a novel class of archaeal-specific copper-
responsive activators of transcription," *Microbiology*, vol. 157,
no. 10, pp. 2808–2817, 2011.

10

Temporal and Spatial Coexistence of Archaeal and Bacterial *amoA* Genes and Gene Transcripts in Lake Lucerne

Elisabeth W. Vissers,[1] Flavio S. Anselmetti,[2,3] Paul L. E. Bodelier,[1] Gerard Muyzer,[1,4] Christa Schleper,[5] Maria Tourna,[6] and Hendrikus J. Laanbroek[1,7]

[1] *Department of Microbial Ecology, Netherlands Institute of Ecology (NIOO-KNAW), Droevendaalsesteeg 10, 6708 PB Wageningen, The Netherlands*
[2] *Swiss Federal Institute of Aquatic Science and Technology (Eawag), Überlandstrasse 133, 8600 Dübendorf, Switzerland*
[3] *Institute of Geological Sciences and Oeschger Centre for Climate Change Research, University of Bern, Zähringerstrasse 25, 3012 Bern, Switzerland*
[4] *Department of Aquatic Microbiology, University of Amsterdam, Science Park 904, 1098 XH Amsterdam, The Netherlands*
[5] *Department of Genetics in Ecology, University of Vienna, Althanstrasse 14, 1090 Vienna, Austria*
[6] *AgResearch Ltd., Ruakura Centre, East Street Private Bag 3115, Hamilton 3240, New Zealand*
[7] *Institute of Environmental Biology, Utrecht University, Padualaan 8, 3584 CH Utrecht, The Netherlands*

Correspondence should be addressed to Elisabeth W. Vissers; liesbeth.vissers@gmail.com

Academic Editor: Hans-Peter Klenk

Despite their crucial role in the nitrogen cycle, freshwater ecosystems are relatively rarely studied for active ammonia oxidizers (AO). This study of Lake Lucerne determined the abundance of both *amoA* genes and gene transcripts of ammonia-oxidizing archaea (AOA) and bacteria (AOB) over a period of 16 months, shedding more light on the role of both AO in a deep, alpine lake environment. At the surface, at 42 m water depth, and in the water layer immediately above the sediment, AOA generally outnumbered AOB. However, in the surface water during summer stratification, when both AO were low in abundance, AOB were more numerous than AOA. Temporal distribution patterns of AOA and AOB were comparable. Higher abundances of *amoA* gene transcripts were observed at the onset and end of summer stratification. In summer, archaeal *amoA* genes and transcripts correlated negatively with temperature and conductivity. Concentrations of ammonium and oxygen did not vary enough to explain the *amoA* gene and transcript dynamics. The observed herbivorous zooplankton may have caused a hidden flux of mineralized ammonium and a change in abundance of genes and transcripts. At the surface, AO might have been repressed during summer stratification due to nutrient limitation caused by active phytoplankton.

1. Introduction

Nitrogen cycling is one of the major biogeochemical processes on Earth. The discovery of novel nitrogen-converting pathways in the past decades [1] has shown the lack of knowledge we had and still have on global nitrogen cycling. Additionally, intensified use of fertilizers and nitrogenous precipitation derived from industry and traffic has led to large changes in the N-cycle in many ecosystems [2]. A major recent discovery in relation to the nitrification process was the role of Archaea in ammonia oxidation [3–5]. This

notion has led to a great interest in the presence of ammonia-oxidizing archaea and bacteria in many ecosystems, often determined by the occurrence of archaeal and bacterial *amoA* genes (e.g., [6, 7]). In most analyses, the presence of archaeal *amoA* genes outnumbered those of bacteria by orders of magnitudes. What this means for the relative activities of both groups has only been investigated in a few environmental studies [8, 9].

The ecological importance of AOA and AOB has been determined in several studies; the relative abundance of AOA and AOB in soils is thought to be influenced mainly by pH [10, 11], temperature [12], and ammonium [13, 14], while in marine systems, next to ammonium [15], oxygen concentrations are expected to play a major role in the presence and abundance of AOA and AOB [16, 17]. However, studies comprising this type of analyses in relation to the occurrence of AOA and AOB in freshwater systems lag behind those related to terrestrial and marine studies.

The ecology of nitrifying bacteria in lakes is well described throughout the years (e.g., [18–21]), but the mutual presence of AOA and AOB was recorded only in some lakes and only at one time point. Lehours et al. [22, 23] found a different archaeal and bacterial community in oxic and permanent anoxic parts of monomictic Lake Pavin. In the sediment of the hypertrophic Lake Taihu, archaea dominated the prokaryotic community, likely due to the low oxygen conditions; no archaea could be detected in the water column [24–26]. In high-altitude Tibetan lakes, salinity influenced the abundance and community composition of AOA, which outnumbered AOB [27].

A first freshwater interannual analysis of Archaea showed the presence of a high diversity of thaumarchaeota (formerly thought to be part of the crenarchaeota phylum) in sulfurous karstic Lake Vilar, but only on the basis of the presence of the 16S rRNA gene [28]. These authors observed differences in richness distribution and seasonality, but no clear correlations were obtained when multivariate statistical analyses were carried out. No temporal comparison of both AOA and AOB in freshwater ecosystems has been made to date.

Here we present a temporal and spatial study of the abundance of the *amoA* genes and the *amoA* gene transcripts as indicators of the presence and the status of activity, respectively, of AOB and AOA in the oligomictic Lake Lucerne. This lake, with high thaumarchaeota-specific crenarchaeol concentrations [29] and relatively high amounts of nitrogen [30], was expected to present a good site for studying the ecology of ammonia oxidizers (AO). The AOA and AOB have a similar temporal distribution pattern, though the AOA outnumber the AOB gene abundance at 42 m water depth and water just above the sediment. In the surface water the AO gene numbers were lower in the summer months, at which time the AOB outnumber the AOA, and a negative correlation of AOA with temperature and conductivity is found.

2. Materials and Methods

2.1. Location Description. Lake Lucerne is a perialpine lake located in Central Switzerland (47°N, 8°E; 434 m a.s.l) at the northern alpine front, with a catchment area of 2124 km^2. It covers an area of 116 km^2, contains seven basins, and is fed by four major alpine rivers (Reuss, Muota, Engelberger Aa, and Sarner Aa providing ~80% of the lakes total water supply (109 m^3/s)) [31] with a 3.4-year residence time. As an oligomictic lake, a complete overturn occurs on average every six years. Sampling was done in the Kreuztrichter basin, one of the subbasins of Lake Lucerne, situated in the relatively open, western part of the lake.

2.2. Determination of Environmental Factors. Conductivity, temperature, oxygen, and pH were measured at the sampling location throughout the water column with a CTD scanner.

The concentrations of ammonium, nitrate, and dissolved organic nitrogen (DON) were measured on a SEAL-QuAAtro autoanalyzer (Seal, Norderstedt, Germany). Detection limits were 0.16 μmol for ammonium, 0.10 μmol for nitrate, and 2 μmol for DON. The concentration of dissolved organic carbon (DOC) was determined with a Formacs DOC analyzer (Skalar, Breda, The Netherlands) with a detection limit of 20 μmol.

2.3. Sampling. Lake water was collected and filtered from the water surface (t = top, 0 m depth), the middle of the water column (m = middle, 42 m depth), and at the bottom, just above the sediment (b = bottom, varying from 72 m to 101 m depth due to slight location changes at different sampling times and the bathymetry at the sampling point in the Kreuztrichter basin) from January 2008 to April 2009. One sample was taken at each depth every month. Depending on the load of suspended particles, 1 to 3 liters of lake water were filtered. Samples for RNA analysis were frozen in a transportable liquid nitrogen freezer directly after filtration and stored at −80°C.

2.4. Nucleic Acids Extractions. DNA was extracted as described previously [32]. In brief, cells were lysed by bead-beating followed by a phenol-cholorform-isoamyl alcohol extraction. The DNA was precipitated and dissolved in 100 mL of molecular biology grade water (Sigma-Aldrich, St. Louis, MO, USA). After extraction, the DNA was purified on a Wizard column (Promega, Madison, WI, USA) and the quantity of DNA was determined spectrophotometrically using a Nanodrop ND-1000 spectrophotometer (Nanodrop Technology, Wilmington, DE, USA).

RNA was extracted with an adjusted protocol of Culley et al. [33], in which one mL of Trizol was added to a tube containing half of a 47 mm 0.2 μm pore-size membrane filter, over which a known amount of water was filtered (1.5 to 2 liters depending on the amount of suspended material) and followed by subsequent bead-beating and RNA isolation steps. RNA was purified from DNA using the Ambion Turbo DNA-free kit (Applied Biosystems, Austin, TX, USA) twice on each sample (as described by the manufacturer). DNA contamination was tested by performing PCR on the samples with primer sets F357 and R518 [34] for the 16S rRNA gene of bacteria.

The BioRad iScript kit with random hexamers (Bio-Rad Laboratories Inc., Hercules, CA, USA) was used to perform reverse transcriptase cDNA production.

2.5. Plankton Measurements. Abundances of planktonic organisms were determined by microscopy in a monthly monitor of a mixed sample of the upper 20 m of the Kreuztrichter basin and were kindly provided for this study by Dr. Hans-Rudolf Bürgi (Eawag).

A principal component analysis on the presence of phyto- and zooplankton was made, in which the explanatory power

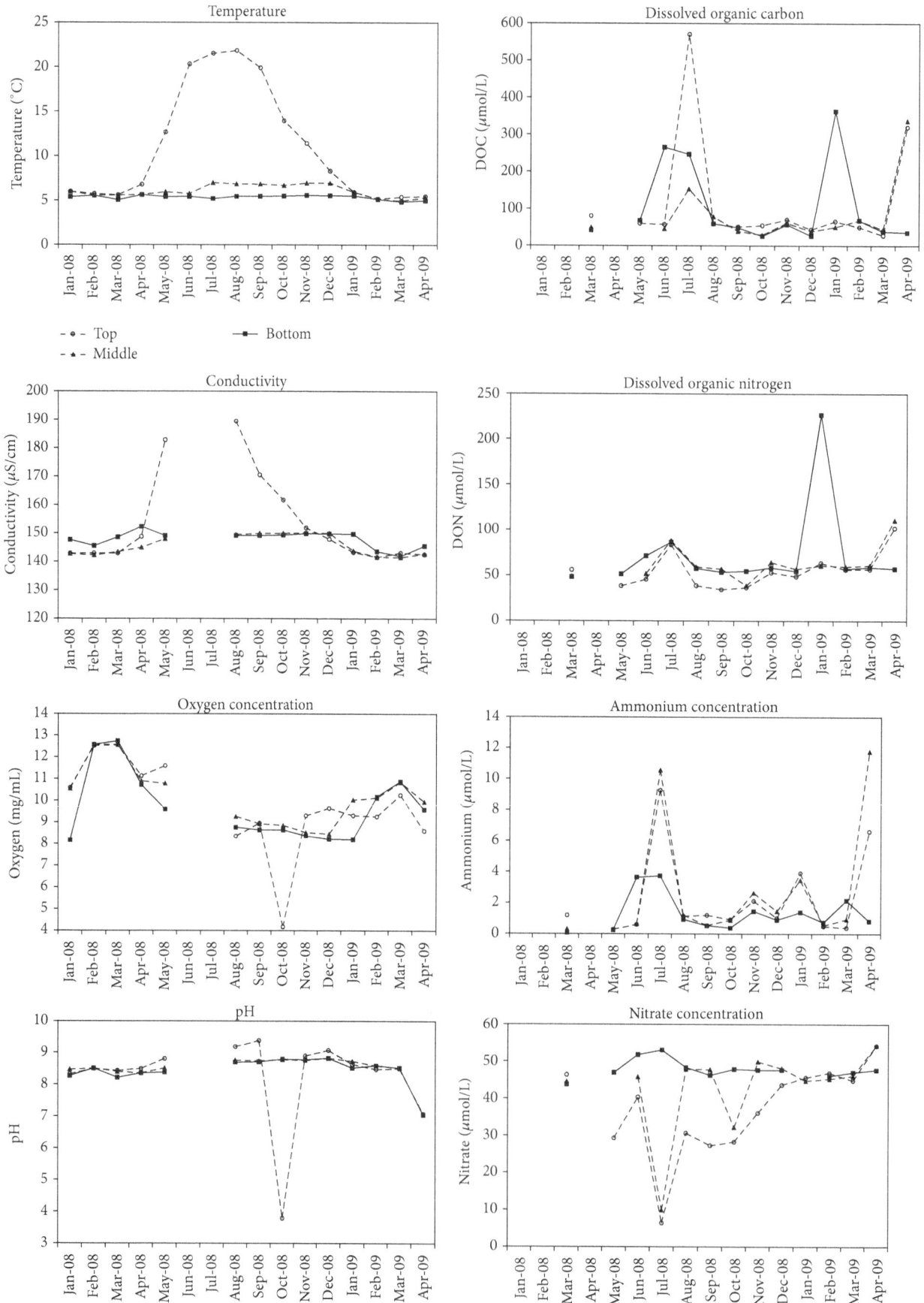

FIGURE 1: Continued.

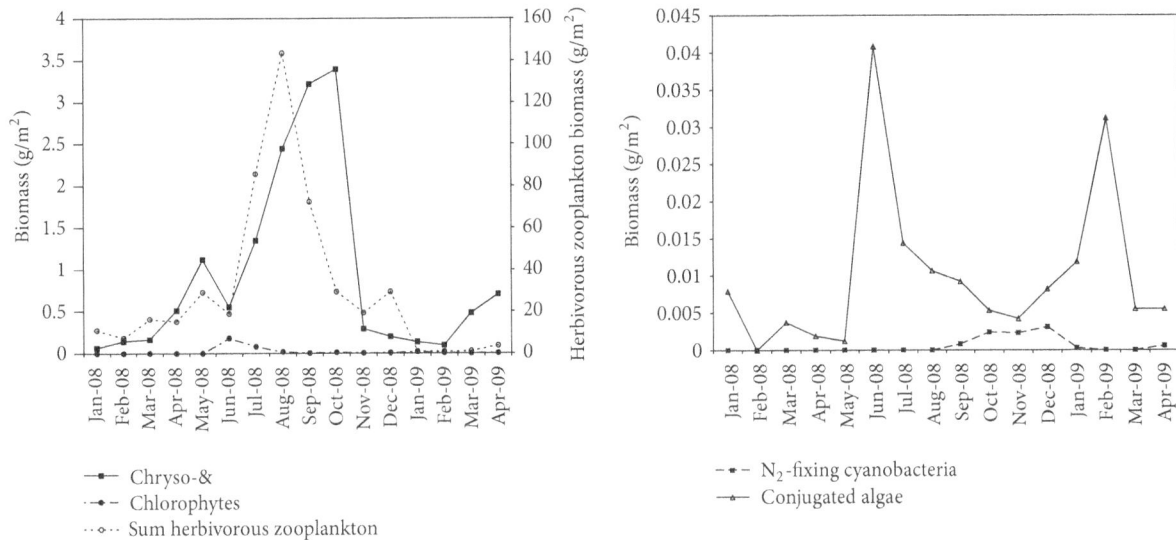

- ■ Chryso-&
- ▲ Chlorophytes
- ○ Sum herbivorous zooplankton

- ● N₂-fixing cyanobacteria
- ▲ Conjugated algae

FIGURE 1: Temporal distribution of environmental factors at three water depths in Lake Lucerne. The single drop in pH and oxygen concentration in the surface water in October 2008 is expected to be caused by a failure of the equipment as such low pH values and oxygen concentrations are not observed in Lake Lucerne.

of the abundance of these organisms on the AOA and AOB *amoA* gene abundances and diversities was established.

2.6. Clone Library Construction and Sequencing. Clone libraries of archaeal *amoA* genes were made of the water samples taken in December by the use of the pGEM-T Vector system (Promega, Madison, WI, USA). Hundred clones were processed and analyzed per water depth. Selected clones were sequenced with their amplification primers (Macrogen Inc., Republic of Korea) (Supplementary Table 2 of the Supplementary Material available online at http://dx.doi .org/10.1155/2013/289478).

2.7. Quantitative PCR of Archaeal and Bacterial amoA Genes. qPCR of archaeal and bacterial *amoA* genes was performed in a 20 μL mixture of 10 μL iQTM SYBR Green Supermix (Bio-Rad), 1 μM of forward and reverse primers, and 0.2 mg mL-1 BSA. For archaeal standards, serial dilutions of the linearized soil fosmid clone 54d9 were used. For bacterial standards, a serial dilution of the linearized plasmid (pCR4-TOPO, Invitrogen) containing the *amoA* gene of *Nitrosomonas europaea* was used. For the archaeal *amoA* gene the forward primer 104(L)F (5′-GCAGGWGAYTACATYTTCTA-3′) was designed after the alignment of soil, marine, and freshwater clone sequences [14] and modified including and favoring clone sequences obtained from archaeal *amoA* genes found in Lake Lucerne sampled in December 2008 (Supplementary Table 2). Thus, the primer should be considered specific for *amoA* gene sequences dominating this lake. Amplifications were performed in Realplex (Mastercycler ep realplex, Eppendorf). Melting curve analyses were performed at the end of every qPCR run to confirm the amplification of the target products only, followed by standard agarose gel electrophoresis for affirmation. The following qPCR-program

was used for both analyses—initial denaturation: 95°C for 15 minutes followed by 40 cycles of 95°C for 15 seconds, 55°C for 30 seconds, and 72°C for 40 seconds.

2.8. Statistical Analysis. Statistical analysis was performed using the Statistica 9 program (Statsoft Inc., Tulsa, OK, USA). The gene abundance was log-transformed to create normal distributions. A table of Spearman rank-order correlations of all variables was subsequently produced. A multiple-regression analysis and principal component analysis on the presence of phyto- and zooplankton and chemical compounds were made, in which the explanatory power of the concentrations of these compounds and organisms on the AOA and AOB *amoA* gene abundances and diversities was established.

3. Results and Discussion

3.1. Environmental Parameters. The oligotrophic nature of Lake Lucerne is reflected by an oxygenated water column with generally low nutrient levels, but with relatively high nitrogen concentrations in the form of nitrate (on average 63 μmol/L) (Figure 1).

During our sixteen-month study, pH and oxygen did not vary at the different sampling depths of Lake Lucerne. More dynamic were the conductivity and temperature in the lake, especially in the surface water.

In July 2008 and April 2009, DOC and DON showed a peak at all depths, while in December 2008, DOC and DON peaked strongly in the water above the sediment, suggesting a more active decomposing microbial community at these times.

Ammonium concentrations were mostly around the detection level of 0.16 μM but showed a peak in the surface

TABLE 1: One-way ANOVA on the differences between the means of community characteristics of ammonia-oxidizing archaea and bacteria determined for summer and winter months, respectively. The difference is significant when $F_{measured} < F_{critical}$ and $F_{critical} = 4.8443357$. Significant differences are shown in bold. Individual data have been presented by Vissers et al. (in press).

Parameter	Water depth	$F_{measured}$	P
Log archaeal 16S	Surface	**5.048039**	**0.04615**
	−42 m	0.118078	0.73761
	Above sediment	0.708741	0.41779
Log archaeal *amoA*	Surface	**7.093356**	**0.02205**
	−42 m	0.061664	0.82351
	Above sediment	0.052174	0.41779
Log bacterial *amoA*	Surface	0.020078	0.88988
	−42 m	1.228674	0.29131
	Above sediment	0.622908	0.44665
Number of archaeal 16S rRNA DGGE bands	Surface	2.678394	0.12998
	−42 m	0.206905	0.65805
	Above sediment	0.151504	0.70453
Number of archaeal *amoA* DGGE bands	Surface	**29.27228**	**0.00021**
	−42 m	2.873572	0.11813
	Above sediment	2.925275	0.11522

water and at 42 m when nitrate showed a minimum. The opposing fluctuations of ammonium and nitrate concentrations may suggest that ammonia oxidation plays a role in Lake Lucerne, which is confirmed by low AOA and AOB abundances in the periods with high concentrations of ammonium and low concentrations of nitrate and vice versa (Figures 1 and 2).

3.2. AOA and AOB amoA Gene Numbers. The increase and decrease of AOA and AOB *amoA* gene abundances showed similar patterns among the sixteen monthly collected samples at all three depths, indicating that AOA and AOB are generally displaying similar population dynamics (Figure 2). This observation is supported by significant ($P < 0.05$) and positive Spearman rank-order correlations between the gene copy numbers (Supplementary Table 1).

An increase in abundance of both AOA and AOB was observed in March (surface) and April (deeper waters) 2008, with the onset of summer stratification in the water column of Lake Lucerne, and an increase in AO was again observed in December 2008 when the water layers mixed again. During the period of water stratification, the numbers of AOA at the surface declined more than those of AOB leading to a lower percentage of the total AO of the first one. This period of lower AO numbers and AOB dominance at the surface of the lake coincided with relatively warm water and a higher conductivity (Figure 1). When comparing the gene copy numbers obtained in the summer stratification period, that is, from June till September, for which ANOVA pointed to a different temperature compared with the rest of the sampling period, it appears that the means of the archaeal gene copy numbers obtained in these two periods were only significantly different in the surface water (Table 1). With bacterial gene copy numbers, no significant differences between the means were observed throughout the water

column. Water depth did also not significantly affect the AOB *amoA* gene abundance in the water column of the lake. In contrast, the AOA *amoA* gene abundance increased from the surface to the deeper water layers, giving rise to an increasing AOA/AOB ratio with depth, which is also observed in other aquatic systems [35–37].

We observed (Figure 2) and confirmed by one-way ANOVA that AOA in the surface water behaved differently from the AOA in the deeper waters ($P < 0.005$), which was not observed for AOB ($P < 0.6$). This all suggests that the low AOA/AOB ratio at the surface water is caused by an environment in which different AOA dynamics or even communities occur compared to waters at greater depth.

The most striking result of our temporal study was the generally similar behavior of the archaeal and bacterial ammonia-oxidizing communities through time, suggesting a situation in which AOA and AOB cooccur rather than compete for nutrients.

3.3. AOA and AOB amoA Gene Transcript Numbers. On the cDNA level, the differences between the two domains were even less pronounced (Figure 2, right panels). The transcripts of the *amoA* genes also showed mutual temporal dynamics and higher abundances in the water column at the onset and end of summer stratification, except in the middle of the water column, where the transcripts were most abundant during summer stratification. Higher gene transcript numbers at moments before and after stratification are likely due to mixing of the water column and subsequent increased nutrient availability leading to higher metabolic activities [38–40].

Generally, an increased *amoA* cDNA level was observed a month before or at the same time of a rise in *amoA* genes, suggesting a higher ammonia-oxidizing activity when cells started to multiply (Figure 2). This was, however, less clear for

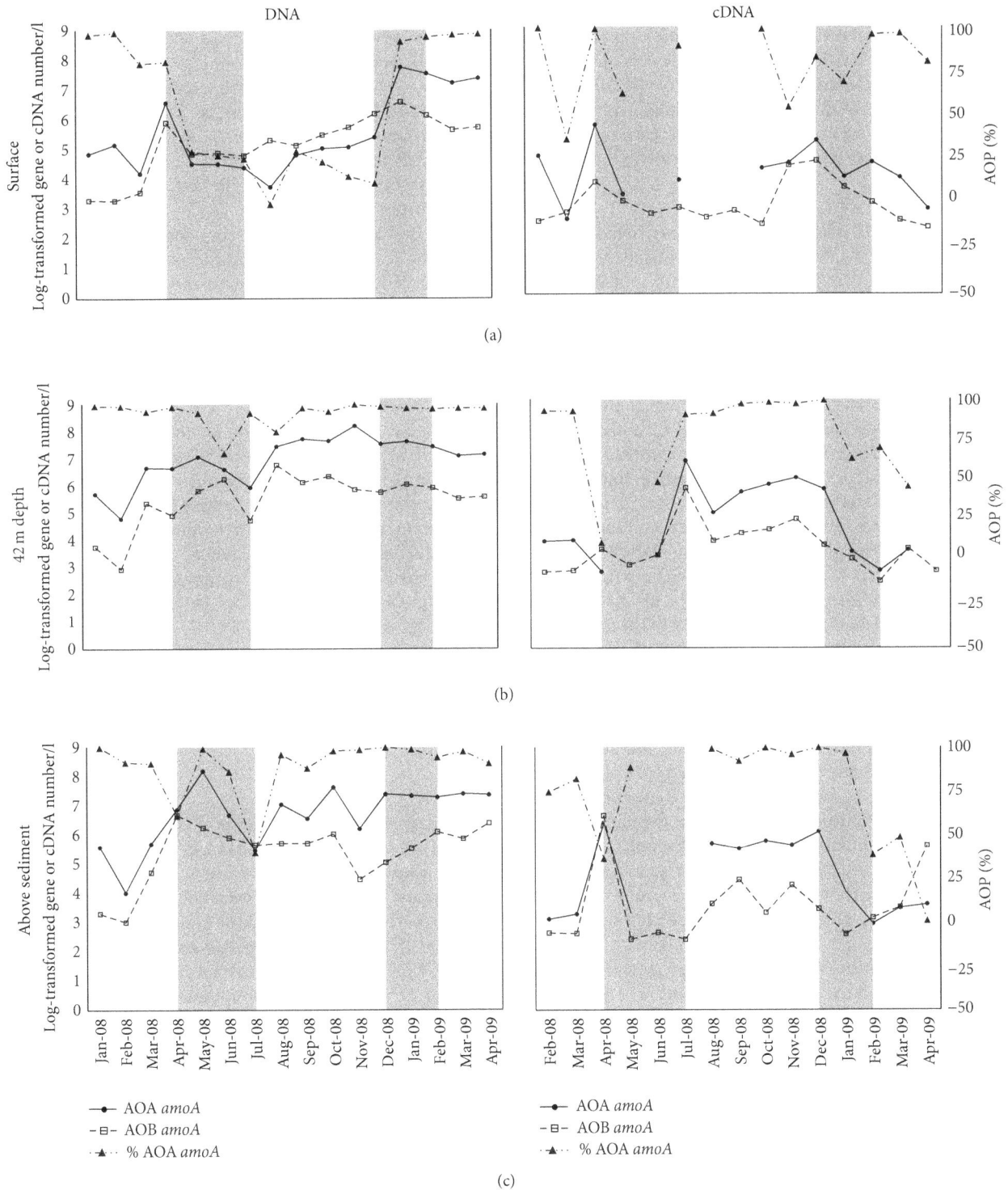

FIGURE 2: Temporal distribution of ammonia-oxidizing archaea (AOA) (solid lines, circles) and ammonia-oxidizing bacteria (AOB) (broken lines, squares), *amoA* gene abundances, and the archaeal percentage of the total *amoA* genes (broken line, triangles), all determined in three different layers in the water column of Lake Lucerne. In the left panels the DNA gene abundances are shown, on the right the cDNA abundances. Periods of mixing of the water layers are depicted by grey rectangles. Gene abundances were obtained by taking the average of three replicated qPCR analyses. Standard deviations of the replicates are indicated by error bars.

the surface layer of the water column, where cDNA was even below the detection limit in the months in which the numbers of the *amoA* gene of AOB exceeded those of the AOA. Hence, not only cell numbers of AOA were lower then, but also the transcription activity was undetectable for AOA. In the surface water in December 2008, however, when the AOA outnumbered the AOB once more, the amount of archaeal *amoA*-related cDNA had the highest increase rate, as one would expect at moments of population growth.

3.4. AOA and AOB amoA Genes and Transcripts in relation to Environmental Factors. Different environmental factors correlated to AOA and AOB *amoA* genes and transcripts throughout seasons and depths, as is shown by Spearman rank-order correlation analysis (Supplementary Table 1) and supported and visualized by PCA analysis (Figure 3). The main environmental factors influencing the AOA populations in previous studies, that is, pH, ammonium concentration, and oxygen availability, showed little dynamics in our study site; hence little influence on the AO gene and transcript abundances could be assigned to these factors. Additionally, the factors that showed the strongest explanatory power in our study, that is, temperature and conductivity, were constant throughout the season in the deeper water layers, opposite to the changes observed for the surface water. When considering all water depths of Lake Lucerne, conductivity explained 53% of the variance in the distribution of AOA. Conductivity was also of great influence on AO dispersal in Tibetan lakes [27], where lake biochemistry seemed to shape the archaeal community rather than historic events.

Conductivity in the Kreuztrichter basin was described to be affected by processes that are connected to phytoplankton dynamics, such as carbon assimilation, calcite precipitation, sedimentation, and decomposition in the hypolimnion [41]. A change in conductivity therefore may reflect a change in local nutrient availability due to phytoplankton activity, which probably affects the dynamics of AOA and AOB, though each in a specific manner as revealed by ANOVA (Table 1).

The concentration of ammonium, the expected substrate, was mostly around the detection limit and no relation with the transcript abundance of the functional gene for ammonia-oxidation could be found. The nitrate concentration in Lake Lucerne is expected to change by biochemical cycling only, as the inflow of fresh water is limited and originates from other basins of the Lake, rather than from the surrounding catchment. However, nitrate, the endproduct of nitrification, did not correlate with bacterial *amoA* genes or gene transcript abundances neither with archaeal *amoA* transcripts. Nitrate did however correlate with archaeal *amoA* gene abundance, but only in the surface water. To date the comparisons of AOA and AOB ammonium uptake kinetics are based on a limited number of pure culture experiments, and so far it is unknown if AOA and AOB in natural environments behave similarly. AOA were found to thrive at low nutrient concentrations [42] and showed growth until ammonium concentrations fell below the detection level (i.e., 10 nM),

which is a 100-fold lower than the threshold concentration for AOB (1 μM at near neutral pH) [15]. In accordance with these findings, ammonium was generally around the detection limit in the waters of our study site, where AOB only reached low cell numbers (Figures 1 and 2) and were outnumbered by AOA by 1 or 2 orders of magnitude difference in gene abundance in the deeper waters. Also in the North Sea, a similar temporal dynamic of AOA and AOB was observed with AOA outnumbering AOB by 1 or 2 orders of magnitude [3], suggesting this might be more common in aquatic environments.

In the surface water the abundance of AOB was higher than that of AOA during summer stratification when temperature and conductivity increased (Figures 1 and 2); this is due to a negative correlation of AOA with conductivity and temperature, rather than a positive correlation of AOB with these factors. However, temperature and conductivity correlated positively with cDNA derived from archaeal and bacterial *amoA* in the deeper layers, although for the bacterial cDNA only at 42 m depth. Apparently, temperature and conductivity stimulated the transcription activity of the ammonia oxidizers in the deeper layers, but not in the surface water. Hence, some other factor must have been responsible for the relative increase of AOB in relation to AOA in the surface layer during summer stratification.

It has been suggested that oxygen influences the composition of AOB communities [43] and low oxygen levels may offer a niche for AOA [16, 17, 44–46]. However, since the concentration of oxygen varied only little at the different water depths of the well-oxygenated water column of Lake Lucerne, oxygen is not likely to be a selective environmental factor with respect to the presence of AOA and AOB in lake Lucerne.

3.5. Correlation of AO Genes and Gene Transcript Numbers to the Presence of Other Plankton. AOA *amoA* genes and gene transcripts in deeper waters, as well as AOB *amoA* transcripts throughout the water column, correlated to numbers of herbivorous zooplankton and N_2-fixing cyanobacteria (Figure 4). These plankton groups may supply AOA and AOB directly or indirectly with extra ammonium from mineralization of organic nitrogen compounds. Correlations with herbivorous and mixotrophic zooplankton were found in all water depths. A possible explanation for increasing amounts of *amoA* transcripts might be the increase of activity during grazing. It has been shown in ammonia-limited chemostats containing pure cultures of AOB and heterotrophic bacteria that grazing by a flagellate lowered the number of ammonia-oxidizing cells present in the culture, but increased at the same time the oxidation rate per cell [47]. AOB cells have a higher amount of mRNA ready for ammonia oxidation at moments before growth is observed, which possibly causes the AOB population to recover faster after predation, while the AOA population needs more time to recover from phagotrophy.

In the surface water, a negative correlation was observed between AOA gene and gene transcript numbers on one side, and the numbers of conjugate algae and chrysophytes on the

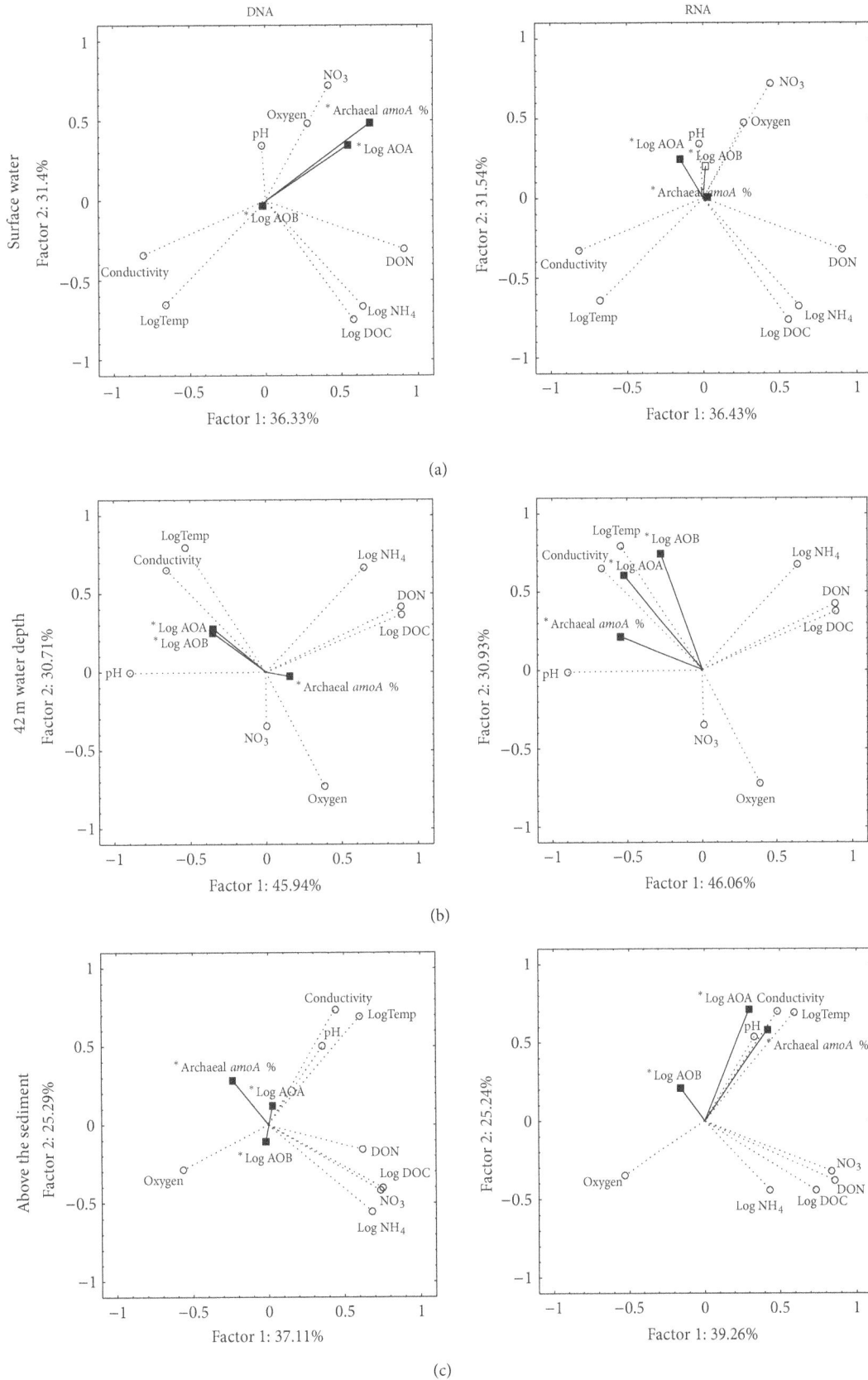

FIGURE 3: Principal component analysis of ammonia oxidizers DNA (left) and RNA (right) and environmental factors and nutrients in the surface water (above), 42 m water depth (middle), and water just above the sediment (below). A principal component analysis on chemical compounds was made, in which the explanatory power of the concentrations of these compounds on the AOA and AOB amoA gene abundances and diversities was established. Statistical analysis was performed using the Statistica 9 program (Statsoft Inc., Tulsa, OK, USA).

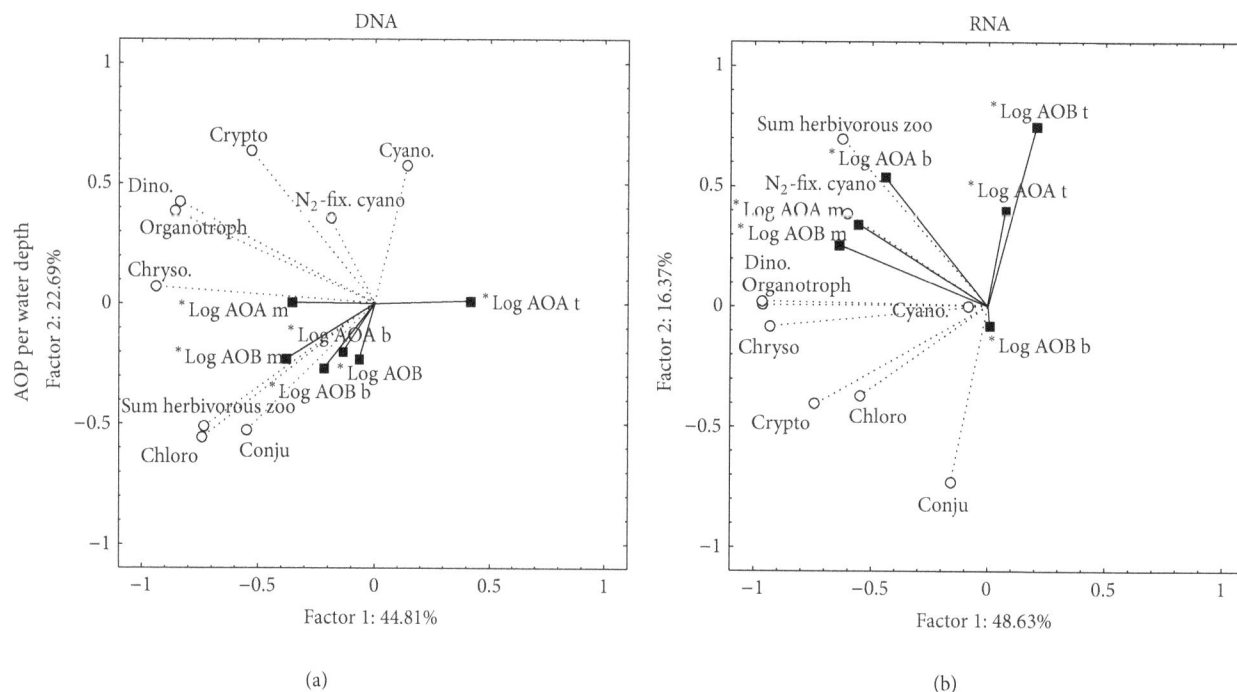

FIGURE 4: Principal component analysis of ammonia oxidizers DNA (a) and RNA (b) and other planktonic microorganisms abundance in the Kreuztrichter basin, Lake Lucerne. Ammonia oxidizers' abundances are printed per water depth on the representation of the abundance of the other planktonic organisms observed throughout the upper 20 m of the water column.

other side. Chrysophytes are described to be mixotrophs as they obtain energy either from light or by feeding on decaying or living cells [48]. This predation could cause the decline of the archaeal and bacterial cell numbers in the surface water during summer stratification, when the chrysophyte bloom was observed.

An explanation for the lower numbers of AO in the surface water may be found in surface-related factors such as a competition with phototrophic microorganisms for nutrients and CO_2 or an inhibition by light. The community of AOB in the surface water is apparently less affected than the AOA by these factors from May till December 2008. Outside this period of summer stratification the negative factors for the AOA in the surface layer seem to be less severe, which might lead again to their dominance. More research is required to elucidate this differential effect of surface water factors on AOA and AOB.

4. Conclusions

The low availability of ammonium in the lake throughout the year may favor AOA over AOB leading to larger population sizes of the first group [49]. Although with different amplitudes, AOA and AOB followed more or less the same temporal changes throughout the water column. Assuming that they have to compete for the same resource, a similarity in community dynamics between archaeal and bacterial ammonia-oxidizing microorganisms is not expected. Even chaotic behavior of pelagic populations makes such a similarity in temporal dynamics not likely [50]. This either means that the amount of ammonium was not limiting or that

AOA can utilize other resources next to ammonium, as has been suggested by [51]. Increase in gene and gene transcript abundance cooccurred with the mixing of the water column before and after summer stratification in the lake, which may indicate a rapid response to changing conditions such as ammonium availability. In Lake Lucerne, ammonium levels were mostly very low. However, ammonium could be available as a nutrient for AOA and AOB by direct local production, which was supported by the observation that AOB and AOA in the deeper waters correlated to herbivorous zooplankton, which make ammonium available by their grazing activity. In the surface water, UV inhibition as well as predation and competition for nutrients and CO_2 by zooplankton may have influenced the population size of the AOP negatively. In addition, not only the size of the AOA community based on both the abundance of the *amoA* gene and that of the 16S rRNA gene was significantly affected in the surface layer by factors prevailing during the period of summer stratification, but also the diversity of the dominant strains as appearing from DGGE profiling of the *amoA* gene [52] was significantly affected in this period (Table 1).

Conflict of Interests

The authors do not have any conflict of interests with the content of the paper.

Acknowledgments

This is publication number DW-2013-1001 of the Darwin Center for Biogeosciences, which partially funded this paper.

The authors thank Cornelia Blaga of Utrecht University, Alois Zwyssig, and Michael Schurter both of Eawag Kastanienbaum for their help during the extensive field campaign. Their gratitude is expressed to Eawag for sharing their planktonic data of the Chrueztrichter basin with them. Additional financial support for the field campaign was obtained from the Schure-Beijerinck Popping fund and Darwin Center for Biogeosciences, publication number 5409 of the Netherlands Institute of Ecology (NIOO-KNAW).

References

[1] M. S. M. Jetten, "The microbial nitrogen cycle," *Environmental Microbiology*, vol. 10, no. 11, pp. 2903–2909, 2008.

[2] N. Gruber and J. N. Galloway, "An Earth-system perspective of the global nitrogen cycle," *Nature*, vol. 451, no. 7176, pp. 293–296, 2008.

[3] C. Wuchter, B. Abbas, M. J. L. Coolen et al., "Archaeal nitrification in the ocean," *Proceedings of the National Academy of Sciences of the United States of America*, vol. 103, pp. 12317–12322, 2006.

[4] M. Könneke, A. E. Bernhard, J. R. De La Torre, C. B. Walker, J. B. Waterbury, and D. A. Stahl, "Isolation of an autotrophic ammonia-oxidizing marine archaeon," *Nature*, vol. 437, no. 7058, pp. 543–546, 2005.

[5] A. H. Treusch, S. Leininger, A. Kietzin, S. C. Schuster, H. P. Klenk, and C. Schleper, "Novel genes for nitrite reductase and Amo-related proteins indicate a role of uncultivated mesophilic crenarchaeota in nitrogen cycling," *Environmental Microbiology*, vol. 7, no. 12, pp. 1985–1995, 2005.

[6] C. A. Francis, K. J. Roberts, J. M. Beman, A. E. Santoro, and B. B. Oakley, "Ubiquity and diversity of ammonia-oxidizing archaea in water columns and sediments of the ocean," *Proceedings of the National Academy of Sciences of the United States of America*, vol. 102, no. 41, pp. 14683–14688, 2005.

[7] J. H. Rotthauwe, K. P. Witzel, and W. Liesack, "The ammonia monooxygenase structural gene *amoA* as a functional marker: molecular fine-scale analysis of natural ammonia-oxidizing populations," *Applied and Environmental Microbiology*, vol. 63, no. 12, pp. 4704–4712, 1997.

[8] G. W. Nicol, S. Leininger, C. Schleper, and J. I. Prosser, "The influence of soil pH on the diversity, abundance and transcriptional activity of ammonia oxidizing archaea and bacteria," *Environmental Microbiology*, vol. 10, no. 11, pp. 2966–2978, 2008.

[9] H. J. Di, K. C. Cameron, J. P. Shen et al., "Nitrification driven by bacteria and not archaea in nitrogen-rich grassland soils," *Nature Geoscience*, vol. 2, no. 9, pp. 621–624, 2009.

[10] G. W. Nicol, G. Webster, L. A. Glover, and J. I. Prosser, "Differential response of archaeal and bacterial communities to nitrogen inputs and pH changes in upland pasture rhizosphere soil," *Environmental Microbiology*, vol. 6, no. 8, pp. 861–867, 2004.

[11] J.-Z. He, J.-P. Shen, L.-M. Zhang et al., "Quantitative analyses of the abundance and composition of ammonia-oxidizing bacteria and ammonia-oxidizing archaea of a Chinese upland red soil under long-term fertilization practices," *Environmental Microbiology*, vol. 9, no. 9, pp. 2364–2374, 2007.

[12] M. Tourna, T. E. Freitag, G. W. Nicol, and J. I. Prosser, "Growth, activity and temperature responses of ammonia-oxidizing archaea and bacteria in soil microcosms," *Environmental Microbiology*, vol. 10, no. 5, pp. 1357–1364, 2008.

[13] L. E. Lehtovirta-Morley, K. Stoecker, A. Vilcinskas, J. I. Prosser, and G. W. Nicol, "Cultivation of an obligate acidophilic ammonia oxidizer from a nitrifying acid soil," *Proceedings of the National Academy of Sciences of the United States of America*, vol. 108, pp. 15892–15897, 2011.

[14] M. Tourna, M. Stieglmeier, A. Spang et al., "Nitrososphaera viennensis, an ammonia oxidizing archaeon from soil," *Proceedings of the National Academy of Sciences of the United States of America*, vol. 108, no. 20, pp. 8420–8425, 2011.

[15] W. Martens-Habbena, P. M. Berube, H. Urakawa, J. R. De La Torre, and D. A. Stahl, "Ammonia oxidation kinetics determine niche separation of nitrifying Archaea and Bacteria," *Nature*, vol. 461, no. 7266, pp. 976–979, 2009.

[16] P. Lam, G. Lavik, M. M. Jensen et al., "Revising the nitrogen cycle in the Peruvian oxygen minimum zone," *Proceedings of the National Academy of Sciences of the United States of America*, vol. 106, no. 12, pp. 4752–4757, 2009.

[17] M. J. L. Coolen, B. Abbas, J. Van Bleijswijk et al., "Putative ammonia-oxidizing Crenarchaeota in suboxic waters of the Black Sea: a basin-wide ecological study using 16S ribosomal and functional genes and membrane lipids," *Environmental Microbiology*, vol. 9, no. 4, pp. 1001–1016, 2007.

[18] H. J. Laanbroek and A. Bollmann, "Nitrification in inland water," in *Nitrification*, B. B. Ward, M. G. Klotz, and D. J. Arp, Eds., pp. 385–404, ASM Press, Washington, DC, USA, 2011.

[19] M. Coci, P. L. E. Bodelier, and H. J. Laanbroek, "Epiphyton as a niche for ammonia-oxidizing bacteria: detailed comparison with benthic and pelagic compartments in shallow freshwater lakes," *Applied and Environmental Microbiology*, vol. 74, no. 7, pp. 1963–1971, 2008.

[20] C. B. Whitby, J. R. Saunders, R. W. Pickup, and A. J. McCarthy, "A comparison of ammonia-oxidiser populations in eutrophic and oligotrophic basins of a large freshwater lake," *Antonie van Leeuwenhoek, International Journal of General and Molecular Microbiology*, vol. 79, no. 2, pp. 179–188, 2001.

[21] W. F. Vincent and M. T. Downes, "Nitrate accumulation in aerobic hypolimnia—relative importance of benthic and planktonic nitrifiers in an oligotrophic lake," *Applied and Environmental Microbiology*, vol. 42, pp. 565–573, 1981.

[22] A. C. Lehours, C. Bardot, A. Thenot, D. Debroas, and G. Fonty, "Anaerobic microbial communities in Lake Pavin, a unique meromictic lake in France," *Applied and Environmental Microbiology*, vol. 71, no. 11, pp. 7389–7400, 2005.

[23] A. C. Lehours, P. Evans, C. Bardot, K. Joblin, and F. Gérard, "Phylogenetic diversity of archaea and bacteria in the anoxic zone of a meromictic lake (Lake Pavin, France)," *Applied and Environmental Microbiology*, vol. 73, no. 6, pp. 2016–2019, 2007.

[24] L. Liu, Y. Peng, X. Zheng, L. Xiao, and L. Yang, "Vertical structure of bacterial and archaeal communities within the sediment of a eutrophic lake as revealed by culture-independent methods," *Journal of Freshwater Ecology*, vol. 25, no. 4, pp. 565–573, 2010.

[25] Y. Wu, Y. Xiang, J. Wang, J. Zhong, J. He, and Q. L. Wu, "Heterogeneity of archaeal and bacterial ammonia-oxidizing communities in Lake Taihu, China," *Environmental Microbiology Reports*, vol. 2, no. 4, pp. 569–576, 2010.

[26] W. Ye, X. Liu, S. Lin et al., "The vertical distribution of bacterial and archaeal communities in the water and sediment of Lake Taihu," *FEMS Microbiology Ecology*, vol. 70, no. 2, pp. 263–276, 2009.

[27] A. Hu, T. Yao, N. Jiao, Y. Liu, Z. Yang, and X. Liu, "Community structures of ammonia-oxidising archaea and bacteria in high-altitude lakes on the Tibetan Plateau," *Freshwater Biology*, vol. 55, no. 11, pp. 2375–2390, 2010.

[28] M. Llirós, E. O. Casamayor, and C. Borrego, "High archaeal richness in the water column of a freshwater sulfurous karstic lake along an interannual study," *FEMS Microbiology Ecology*, vol. 66, no. 2, pp. 331–342, 2008.

[29] C. I. Blaga, G. J. Reichart, O. Heiri, and J. S. Sinninghe Damsté, "Tetraether membrane lipid distributions in water-column particulate matter and sediments: a study of 47 European lakes along a north-south transect," *Journal of Paleolimnology*, vol. 41, no. 3, pp. 523–540, 2009.

[30] H. R. Bürgi and P. Stadelmann, "Alteration of phytoplankton structure in Lake Lucerne due to trophic conditions," *Aquatic Ecosystem Health*, vol. 5, pp. 45–49, 2002.

[31] M. Schnellmann, F. S. Anselmetti, D. Giardini, J. A. McKenzie, and S. N. Ward, "Prehistoric earthquake history revealed by lacustrine slump deposits," *Geology*, vol. 30, pp. 1131–1134, 2002.

[32] E. W. Vissers, P. L. E. Bodelier, G. Muyzer, and H. J. Laanbroek, "A nested PCR approach for improved recovery of archaeal 16S rRNA gene fragments from freshwater samples," *FEMS Microbiology Letters*, vol. 298, no. 2, pp. 193–198, 2009.

[33] D. E. Culley, W. P. Kovacik, F. J. Brockman, and W. Zhang, "Optimization of RNA isolation from the archaebacterium Methanosarcina barkeri and validation for oligonucleotide microarray analysis," *Journal of Microbiological Methods*, vol. 67, no. 1, pp. 36–43, 2006.

[34] G. Muyzer, E. C. De Waal, and A. G. Uitterlinden, "Profiling of complex microbial populations by denaturing gradient gel electrophoresis analysis of polymerase chain reaction-amplified genes coding for 16S rRNA," *Applied and Environmental Microbiology*, vol. 59, no. 3, pp. 695–700, 1993.

[35] D. L. Kirchman, H. Elifantz, A. I. Dittel, R. R. Malmstrom, and M. T. Cottrell, "Standing stocks and activity of Archaea and Bacteria in the western Arctic Ocean," *Limnology and Oceanography*, vol. 52, no. 2, pp. 495–507, 2007.

[36] C. Callieri, G. Corno, E. Caravati, S. Rasconi, M. Contesini, and R. Bertoni, "Bacteria, Archaea, and Crenarchaeota in the epilimnion and hypolimnion of a deep holo-oligomictic lake," *Applied and Environmental Microbiology*, vol. 75, no. 22, pp. 7298–7300, 2009.

[37] C. Tamburini, M. Garel, B. Al Ali et al., "Distribution and activity of Bacteria and Archaea in the different water masses of the Tyrrhenian Sea," *Deep-Sea Research Part II*, vol. 56, no. 11-12, pp. 700–712, 2009.

[38] M. Winder, "Photosynthetic picoplankton dynamics in Lake Tahoe: temporal and spatial niche partitioning among prokaryotic and eukaryotic cells," *Journal of Plankton Research*, vol. 31, no. 11, pp. 1307–1320, 2009.

[39] R. Naiman, J. J. Magnuson, D. M. McKnight, and J. A. Stanford, *The Freshwater Imperative: A Research Agenda*, Island Press, Washington, DC, USA, 1995.

[40] S. Fietz, G. Kobanova, L. Izmesteva, and A. Nicklisch, "Regional, vertical and seasonal distribution of phytoplankton and photosynthetic pigments in Lake Baikal," *Journal of Plankton Research*, vol. 27, no. 8, pp. 793–810, 2005.

[41] H. Bührer and H. Ambühl, "Lake Lucerne, Switzerland, a long term study of 1961–1992," *Aquatic Sciences*, vol. 63, no. 4, pp. 432–456, 2001.

[42] T. H. Erguder, N. Boon, L. Wittebolle, M. Marzorati, and W. Verstraete, "Environmental factors shaping the ecological niches of ammonia-oxidizing archaea," *FEMS Microbiology Reviews*, vol. 33, no. 5, pp. 855–869, 2009.

[43] A. Bollmann and H. J. Laanbroek, "Influence of oxygen partial pressure and salinity on the community composition of ammonia-oxidizing bacteria in the Schelde estuary," *Aquatic Microbial Ecology*, vol. 28, no. 3, pp. 239–247, 2002.

[44] J. M. Beman, B. N. Popp, and C. A. Francis, "Molecular and biogeochemical evidence for ammonia oxidation by marine Crenarchaeota in the Gulf of California," *ISME Journal*, vol. 2, no. 4, pp. 429–453, 2008.

[45] P. Lam, M. M. Jensen, G. Lavik et al., "Linking crenarchaeal and bacterial nitrification to anammox in the Black Sea," *Proceedings of the National Academy of Sciences of the United States of America*, vol. 104, no. 17, pp. 7104–7109, 2007.

[46] J. Yan, S. C. M. Haaijer, H. J. M. Op den Camp et al., "Mimicking the oxygen minimum zones: stimulating interaction of aerobic archaeal and anaerobic bacterial ammonia oxidizers in a laboratory-scale model system," *Environmental Microbiology*, vol. 14, pp. 3146–3158, 2012.

[47] F. J. M. Verhagen and H. J. Laanbroek, "Competition for ammonium between nitrifying and heterotrophic bacteria in dual energy-limited chemostats," *Applied and Environmental Microbiology*, vol. 57, no. 11, pp. 3255–3263, 1991.

[48] D. A. Holen and M. E. Boraas, "Mixotrophy in chrysophytes," in *Chrysophyte Algae Ecology, Phylogeny and Development*, pp. 119–140, Cambridge University Press, 1995.

[49] C. Schleper and G. W. Nicol, "Ammonia-oxidising archaea—physiology, ecology and evolution," in *Advances in Microbial Physiology*, vol. 57, pp. 1–41, Academic Press Ltd-Elsevier Science, London, UK, 2010.

[50] J. Huisman and F. J. Weissing, "Biodiversity of plankton by species oscillations and chaos," *Nature*, vol. 402, no. 6760, pp. 407–410, 1999.

[51] P. C. Blainey, A. C. Mosier, A. Potanina, C. A. Francis, and S. R. Quake, "Genome of a low-salinity ammonia-oxidizing archaeon determined by single-cell and metagenomic analysis," *PLoS ONE*, vol. 6, no. 2, Article ID e16626, 2011.

[52] E. W. Vissers, C. I. Blaga, P. L. E. Bodelier et al., "Seasonal and vertical distribution of putative ammonia-oxidizing thaumarchaeotal communities in an oligotrophic lake," *FEMS Microbiology Ecology*, vol. 83, pp. 515–526, 2013.

11

Acetate Activation in *Methanosaeta thermophila*: Characterization of the Key Enzymes Pyrophosphatase and Acetyl-CoA Synthetase

Stefanie Berger, Cornelia Welte, and Uwe Deppenmeier

Institute for Microbiology and Biotechnology, University of Bonn, Meckenheimer Allee 168, 53115 Bonn, Germany

Correspondence should be addressed to Uwe Deppenmeier, udeppen@uni-bonn.de

Academic Editor: Francesca Paradisi

The thermophilic methanogen *Methanosaeta thermophila* uses acetate as sole substrate for methanogenesis. It was proposed that the acetate activation reaction that is needed to feed acetate into the methanogenic pathway requires the hydrolysis of two ATP, whereas the acetate activation reaction in *Methanosarcina sp.* is known to require only one ATP. As these organisms live at the thermodynamic limit that sustains life, the acetate activation reaction in *Mt. thermophila* seems too costly and was thus reevaluated. It was found that of the putative acetate activation enzymes one gene encoding an AMP-forming acetyl-CoA synthetase was highly expressed. The corresponding enzyme was purified and characterized in detail. It catalyzed the ATP-dependent formation of acetyl-CoA, AMP, and pyrophosphate (PP$_i$) and was only moderately inhibited by PP$_i$. The breakdown of PP$_i$ was performed by a soluble pyrophosphatase. This enzyme was also purified and characterized. The pyrophosphatase hydrolyzed the major part of PP$_i$ ($K_M = 0.27 \pm 0.05$ mM) that was produced in the acetate activation reaction. Activity was not inhibited by nucleotides or PP$_i$. However, it cannot be excluded that other PP$_i$-dependent enzymes take advantage of the remaining PP$_i$ and contribute to the energy balance of the cell.

1. Introduction

Methanogenic archaea are of high ecological importance as they are responsible for closure of the global carbon cycle and production of the greenhouse gases CO_2 and methane [1–3]. They are also an integral part of biogas reactors and contribute to the production of the combustible gas methane that is a source of renewable energy [4, 5]. Methanogenic archaea use end products of anaerobic bacterial degradation processes like H_2/CO_2 and acetate as substrates for growth. It is estimated that about two thirds of the methane produced by methanogenic archaea on earth derives from acetate degradation [6]. But despite its high abundance only two genera are able to use acetate as substrate for methanogenesis, namely, *Methanosarcina* and *Methanosaeta*. While *Methanosarcina* species are metabolically versatile, members of the genus *Methanosaeta* are specialized on acetate utilization. This is reflected in a very high affinity for the substrate. For growth, a minimal concentration of only

7–$70\,\mu$M is needed [7]. Therefore, *Methanosaeta* species prevail over members of the genus *Methanosarcina* in low acetate environments frequently encountered in natural habitats. Important biotechnological habitats are biogas facilities [8–12], where *Methanosaeta* species are of special importance for reactor performance and stability [12, 13].

In acetate-degrading (aceticlastic) methanogenesis, acetate first has to be activated at the expense of ATP. This reaction can be catalysed by the high activity but low affinity acetate kinase/phosphotransacetylase (AK/PTA) system that is used by *Methanosarcina* sp. [14, 15] or by the low-activity but high-affinity AMP-dependent acetyl-CoA-synthetases (ACS) [16–18]. While the AK/PTA system generates ADP, P$_i$ and acetyl-CoA from ATP, CoA, and acetate [15, 19, 20], the ACS converts ATP, CoA, and acetate to acetyl-CoA, AMP and pyrophosphate (PP$_i$) [16, 18]. In the first step of aceticlastic methanogenesis, acetyl-CoA is cleaved into its methyl and carbonyl moiety by the action of a CO dehydrogenase/acetyl-CoA synthase. In the course of this reaction, the carbonyl group is oxidized to CO_2 and electrons are transferred to ferredoxin [21–23]. The methyl group is donated to the methanogenic cofactor tetrahydrosarcinapterin and subsequently transferred to coenzyme M (CoM) by a membrane bound Na^+ translocating methyltransferase. Reduction of the methyl group to methane with coenzyme B as electron donor leads to the formation of the so-called heterodisulfide (CoM-S-S-CoB). Only recently we demonstrated that *Methanosaeta (Mt.) thermophila* uses the heterodisulfide as terminal electron acceptor in an anaerobic respiratory chain with reduced ferredoxin as the sole electron donor [24]. However, the way this organism conserves energy is not yet fully understood. It can be estimated that the amount of ions translocated over the cytoplasmic membrane in the course of aceticlastic methanogenesis could be sufficient for the phosphorylation of two ADP molecules. Yet AMP-dependent acetyl-CoA synthetase and soluble pyrophosphatase (PPiase) activities could be demonstrated for the closely related *Mt. concilii* [16, 18, 25]. Taking non-energy coupled hydrolysis of pyrophosphate into account, two ATP equivalents are consumed in the course of the acetate activation reaction. According to this model, the obligate aceticlastic methanogen *Mt. thermophila* is not able to conserve energy during methanogenesis. To clarify this contradiction, the acetate activation reaction in *Mt. thermophila* was reevaluated by gene expression analysis and characterization of ACS and PPiase.

2. Materials and Methods

2.1. Materials. All chemicals and reagents were purchased from Sigma-Aldrich (Munich, Germany) or Carl Roth GmbH (Karlsruhe, Germany). Restriction endonucleases, T4 DNA ligase, Taq DNA polymerase, and PCR reagents were purchased from Fermentas (St. Leon-Rot, Germany). Phusion DNA polymerase was purchased from New England Biolabs (Frankfurt am Main, Germany). Oligonucleotides were synthesized by Eurofins (Ebersberg, Germany).

2.2. Bioinformatics. For Blast analyses, the respective tool on NCBI (http://www.ncbi.nlm.nih.gov/) was used. For the batch Blast analysis, those proteins that had a threshold E-value $< e^{-40}$ were referred to as homologous. The programs PsiPred (http://bioinf.cs.ucl.ac.uk/psipred/) and InterPro (http://www.ebi.ac.uk/interpro/) were utilized for bioinformatic analyses of CBS domains.

2.3. qRT-PCR. Total RNA from *Mt. thermophila* DSM 6194 was isolated by TRI Reagent extraction. 250 mL cultures were grown anaerobically to the mid- to late- exponential growth phase in DSMZ medium 387 at 55°C with 50 mM sodium acetate. The cultures were filled into centrifuge tubes in an anaerobic hood and were quick-chilled by shaking in an ice-cold ethanol bath (−70°C) for 5 min. Afterwards, cells were harvested under anaerobic conditions by centrifugation (11000 ×g, 25 min, 4°C). Cell pellets were resuspended in 5 mL TRI Reagent and lysed via a freeze-thaw treatment at −70°C overnight. Total RNA was extracted according

to the manufacturer's instructions (Ambion, Darmstadt, Germany). Preparations were treated with DNAse I to reduce DNA contaminations. Cleaning and concentration of RNA were achieved using the SurePrep RNA Cleanup and Concentration kit (Fisher Scientific, Schwerte, Germany). RNA purity was quantified spectrophotometrically by examining the 260 nm/280 nm ratio as well as by denaturing agarose gel electrophoresis.

Primers for qRT-PCR were designed using the Primer3 software (http://frodo.wi.mit.edu/primer3/input.htm). For the highly homologous ACS genes, the least homologous areas were used as templates to guarantee specificity of the primers. The genes encoding glyceraldehyde-3-phosphate dehydrogenase (GAP-DH, *mthe_0701*) and ribosomal protein S3P (*mthe_1722*) were chosen as reference genes. Sequences of the primers used can be seen from Table 1.

PCR reactions were performed according to the manufacturer's instructions (http://www1.qiagen.com) with on average 250 ng of RNA. The QuantiTect SYBR Green RT-PCR kit (Qiagen, Hilden, Germany) and the iCycler (Bio-Rad, Munich, Germany) were used for labeling and quantification, respectively. For data analysis, the Bio-Rad iCycler software was used. Each PCR product gave a single narrow peak in the melting curve analysis. A relative value (ΔC_t) for the initial target concentration in each reaction was determined by subtracting C_t values of the reference genes from those of the genes of interest. By subtracting ΔC_t values, comparisons among the genes of interest could be accomplished. In addition, negative-control assays were included that were not incubated with reverse transcriptase. These assays contained only traces of DNA that were not removed by DNase treatment. The C_t values of the negative controls were analyzed and were at least five cycles higher than the assays with reverse transcriptase treatment.

2.4. Cloning into Expression Vectors. Genes from *Mt. thermophila* were amplified from chromosomal DNA extracted with CTAB [26]. Restriction endonuclease sites were inserted by PCR; Primers had the following sequences (recognition sites for restriction endonucleases are underlined): *mthe_0236* for ATGGTA<u>GGTCTC</u>AAA-TGGCAGATAATATCTATGTGGTCGGG, *mthe_0236* rev ATGGTA<u>GGTCTC</u>AGCGCTCTTCTTGAATGCGGA-CTCGAGC, *mthe_1194* for ATGGTA<u>ACCTGC</u>ATTAGC-GCCGCTGAGACTGCAAAGACTGCTG, *mthe_1194* rev ATGGTA<u>ACCTGC</u>ATTATATCAGACTATGAGCGG-GATGTTCTCG. For cloning of the pyrophosphatase gene (*mthe_0236*), Eco31I was used, for cloning of the AMP-dependent ACS gene (*mthe_1194*) BveI. Amplicons were cut and ligated into pASK-IBA3 or pASK-IBA5 (IBA GmbH, Göttingen, Germany) to produce pASK-*mthe0236*-3 and pASK-*mthe1194*-5, respectively. Both vectors contained plasmid encoded ribosomal binding sites and a Strep-tag II either C-terminal (pASK-IBA3) or N-terminal (pASK-IBA5). The constructs were confirmed by sequencing and transformed into *E. coli* [27].

2.5. Protein Overproduction and Purification. Overproduction of proteins was performed in *E. coli* BL21 (DE3)

Acetate Activation in Methanosaeta thermophila: Characterization of the Key Enzymes Pyrophosphatase and
Acetyl-CoA Synthetase

111

TABLE 1: Gene number, function of corresponding protein and primers used for amplification of genes analyzed by qRT-PCR.

Function	Gene number	Primer sequence
AMP-dependent ACS	mthe_1194	for CCAGTGGATCATCGAGTA
		rev CAGAAATCGAGGTAGTTC
	mthe_1195	for TAAGGAGCTTGCTGAGAA
		rev CAGAACTCTATGTAGTGG
	mthe_1196	for TCGAAGGCGTATGCTGAC
		rev CGCCTCGTCAGCCTGCTT
	mthe_1413	for CAGGCGCGCTCCGCGAG
		rev GGCCTTTATCGGGATAGG
ADP-dependent ACS	mthe_0554	for TATCATTGGGGTTACAAG
		rev CAGAGATGGGTATTGATC
PPiase	mthe_0236	for GCCAGCATGTATGAGCTG
		rev CATGTGGGTGACTTGAAT
GAP-DH	mthe_0701	for CTATGCCGTTGCTGTGAA
		rev TTGGCGGTGCATTTATCT
ribosomal protein S3P	mthe_1722	for GTTCGTCATGATTGGCAC
		rev CCCCTTCTGGAGCTTATC
intergenic region	Between mthe_1194 and mthe_1195	for GCGGTCAACCTATTTTATTT
		rev TTACATACCTCCATTCATCT
intergenic region	Between mthe_1195 and mthe_1196	for AACGTCCGCAATTTTTATTT
		rev CTGCCTCCAGCCCATCCCG

including the plasmid pLysS (Novagen/Merck, Darmstadt, Germany). Cells were grown on modified maximal induction medium [28] with 3.2% [w/v] tryptone, 2% [w/v] yeast extract, and additions of M9 salts as well as 0.1 mM $CaCl_2$, 1 mM $MgSO_4$ and 1 μM ammonium iron(III) citrate. Ampicillin (100 μg mL^{-1}) and chloramphenicol (25 μg mL^{-1}) were added for plasmid maintenance. Cultures were grown aerobically at 37°C to an OD_{600} of 0.6; protein production was induced by addition of anhydrotetracyclin (200 ng mL^{-1}). Cells were allowed to grow for another 3-4 hours, harvested by centrifugation (11000 ×g, 10 min) and lysed by sonication. Protein purification by Strep-tactin affinity chromatography was performed aerobically according to the manufacturer's instructions (IBA GmbH, Göttingen, Germany). The purified protein was stored at –70°C.

2.6. Protein Visualisation. SDS-PAGE was done according to Laemmli [29] with a 5% [w/v] polyacrylamide stacking gel and a 12.5% [w/v] slab gel. Samples were diluted in sample loading buffer (2% [w/v] SDS, 5% [v/v] β-mercaptoethanol, 50% [v/v] glycerol, 20% [v/v] collecting buffer (0.625 M Tris-HCl pH 6.8), 0.001% [w/v] bromophenol blue), boiled for 5 min at 95°C and loaded to the gel. Molecular masses were calculated by comparison to a molecular mass standard (Fermentas, St. Leon-Rot, Germany). Proteins were visualized by silver staining [30].

2.7. Gel Filtration Chromatography. For gel filtration chromatography a Hi Load 16/60 Superdex 75 prep grade column (GE Healthcare, Munich, Germany) was employed. Calibration was done using the kit for molecular weights,

29000–700000 for gel filtration chromatography (Sigma–Aldrich, Munich, Germany) according to the manufacturer's instructions. For determination of the void volume Blue Dextran was employed. The K_{av} was calculated according to

$$K_{av} = \frac{(v_e - v_o)}{(v_c - v_o)}, \qquad (1)$$

v_e being the elution volume, v_o the void volume and v_c the column volumn. K_{av} was plotted against the decadal logarithm of the molecular weight of the proteins used for calibration, and the resulting curve was used for molecular mass determination. Averaged 1.5 mg of the soluble pyrophosphatase were loaded and run in 40 mM Tris-HCl pH 8, 150 mM NaCl, and 1 mM $MnCl_2$ at a rate of 0.5 mL min^{-1}.

2.8. Enzyme Assays. Assay mixtures for Mthe_0236 routinely contained 200 μL total volume with 40 mM Tris-HCl pH 8, 5 mM $MgCl_2$ and 1 mM PP_i. For measuring the manganese containing enzyme, the protein preparation was preincubated for 5 min at room temperature in 40 mM Tris-HCl pH 8, 5 mM $MgCl_2$ and 1 mM $MnCl_2$ prior to the measurement. For measuring inhibitory effects of nucleotides, either 750 μM AMP or 5 μM ADP were included. For measuring the effect of phosphate between 0 and 1.5 mM, KH_2PO_4 were added. The activity of the pyrophosphatase was determined with a discontinuous assay so samples were taken at different time points and the content of the reaction product orthophosphate was measured (modified after Saheki et al. [31]). Values were compared to standard curves (0–2 mM P_i). To run more reactions in parallel, tests were performed in 96-well plates. Therefore, 10 μL of sample from the assay mixture were stopped with $2\,\mu L$ 10% [w/v] trichloroacetic acid. $150\,\mu L$ of molybdate reagent (15 mM $(NH_4)_6Mo_7O_{24}$, 70 mM zinc acetate, pH 5.0 with HCl) were added as well as $50\,\mu L$ of 10% [w/v] ascorbic acid (pH 5.0 with NaOH). After incubation at 30°C for 15 min, absorption at 850 nm was measured with the Nanoquant Infinite M200 (Tecan, Männedorf, Switzerland). One unit was defined as μmol PP_i hydrolyzed min^{-1}.

For measuring the activity of the AMP-dependent ACS (Mthe_1194) two different methods were employed. Temperature stability, the K_M value for acetate, and inhibition by PP_i were measured via auxiliary enzymes according to a method modified after Meng et al. [32] (Table 2). In this method production of AMP by Mthe_1194 was coupled to NADH consumption that was followed photometrically at 340 nm. In a standard 1 mL assay 50 mM HEPES pH 7.5, 5 mM $MgCl_2$, 3 mM phosphoenolpyruvate, 1 mM CoA, 2.5 mM ATP, 1 mM DTT, 20 mM sodium acetate, and 0.15 mM NADH were included. Reaction temperature was set to 55°C. Auxiliary enzymes were sufficiently stable at this temperature, and the amounts of auxiliary enzymes (5.7 U myokinase, 2.3 U pyruvate kinase, 2.1 U lactate dehydrogenase) were not rate limiting. The extinction coefficient of NADH was $6.22\,mM^{-1}\,cm^{-1}$. One unit was defined as one μmol of acetate consumed per min that was equal to two μmols of NADH consumed per min.

The K_M values for ATP and CoA (reaction volume 3.5 mL) as well as substrate specificity (reaction volume 2 mL) were determined by using a discontinuous assay. At different time points $380\,\mu L$ samples were taken and the content of PP_i was measured according to a method modified after Kuang et al. [33]. The reaction in the samples was stopped with $380\,\mu L$ 12% TCA [w/v] and $100\,\mu L$ molybdate reagent (2.5% [w/v] $(NH_4)_6Mo_7O_{24}$ in 5 N H_2SO_4), $100\,\mu L$ 0.5 M β-mercaptoethanol and $40\,\mu L$ Eikonogen reagent were added for detection of PP_i. The Eikonogen reagent was prepared by dissolving 0.25 g Na_2SO_3, 14.65 g $KHSO_3$ and 0.25 g 1-amino-2-naphthol-4-sulfonic acid in 100 mL hot water. The solution was cooled down and filtered before use. The reaction mixture for PP_i analysis was incubated for 15 min at 37°C and the absorption at 580 nm measured. Quantification was done using standard curves (0–0.5 mM PP_i). One unit was defined as μmol acetate depleted per min.

3. Results

3.1. Comparison of Genomes of Methanosaeta thermophila and Methanosarcina mazei.
The recently completed genome sequence of Mt. thermophila [23] indicated that the majority of the core steps of aceticlastic methanogenesis are similar in comparison to the genus Methanosarcina, but striking differences have been discovered in electron transfer reactions and energy conservation apparatus. These findings led us to a detailed and comprehensive comparison of proteins. A batch Blast analysis of all amino acid sequences from Ms. mazei against Mt. thermophila and vice versa was performed. In summary, there were about 900 proteins identified that were present in Ms. mazei and Mt. thermophila (not shown). Among the homologs are enzymes that participate in the

TABLE 2: Activity measurement of the acetyl-CoA synthetase via auxiliary enzymes. The decrease of the absorption of NADH was tracked photometrically at 340 nm.

Enzyme	Reaction catalyzed
Acetyl-CoA synthetase	acetate + ATP + CoA $\rightleftharpoons$ acetyl-CoA + PP_i + AMP
Myokinase	AMP + ATP $\rightleftharpoons$ 2 ADP
Pyruvate kinase	ADP + PEP $\rightleftharpoons$ pyruvate + ATP
Lactate dehydrogenase	pyruvate + NADH $\rightleftharpoons$ lactate + NAD^+

central part of aceticlastic methanogenesis and proteins involved in DNA replication, transcription, and translation. Taking into account that the genome of Mt. thermophila codes for 1698 proteins, about 800 proteins found in Mt. thermophila had no counterpart in Ms. mazei. On the other hand, Ms. mazei contains 3371 genes indicating that about 2500 proteins can be produced in Ms. mazei that are not found in Mt. thermophila.

A detailed inspection of the genome of Mt. thermophila revealed that the respiratory chain is simpler in comparison to Methanosarcina species and is composed only of the $F_{420}H_2$ dehydrogenase and the heterodisulfide reductase. There are no genes on the chromosome that encode hydrogenases (neither F_{420}-reducing hydrogenase (Frh) and F_{420}-nonreducing hydrogenase (Vho), nor Ech hydrogenase) [23] or the Rnf complex (encoding a membrane-bound enzyme able to oxidize reduced ferredoxin) [23]. Also membrane fractions of Mt. thermophila were shown not to exhibit any hydrogenase activity [24]. In addition, there is no membrane-bound pyrophosphatase and the electron input module of the $F_{420}H_2$ dehydrogenase FpoF [34] is also missing. Furthermore, genes for acetate kinase and phosphotransacetylase are absent. In contrast to this limited equipment, Mt. thermophila possesses four genes encoding acetyl-CoA synthetases (ACS) [23]. No homologs to these four genes are found in Methanosarcina species. There was no evidence for a membrane-bound pyrophosphatase that could couple the hydrolysis of pyrophosphate to ion extrusion [35] and thus contribute to energy conservation. Instead, a single soluble type II pyrophosphatase was identified (mthe_0236) [23].

3.2. Characterization of the Pyrophosphatase.
The current hypothesis of the acetate activating reaction in Methanosaeta species is that pyrophosphate, produced in the course of acetyl-CoA formation, is hydrolyzed by a pyrophosphatase [25]. However, from our knowledge of the energy conserving system of these organisms it is evident that at least part the energy from the pyrophosphate bond has to be conserved. Therefore, the soluble pyrophosphatase from Mt. thermophila was characterized with respect to gene expression and enzyme activity.

The transcript level of the gene encoding the soluble type II pyrophosphatase was analyzed by qRT-PCR experiments. The number of transcripts was three- to four fold higher than that of the reference genes encoding GAP-DH and

FIGURE 1: SDS-PAGE analysis of purified soluble pyrophosphatase (Mthe_0236) and AMP-dependent ACS (Mthe_1194). Enzymes were purified by Strep-tactin affinity chromatography. M: molecular mass marker (PAGE Ruler prestained protein ladder, Fermentas, St. Leon-Rot, Germany), lane 1: Mthe_0236 0.5 μg, lane 2: Mthe_1194 1 μg.

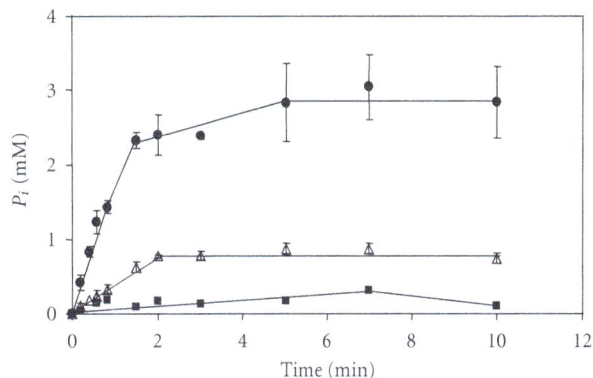

FIGURE 2: Activity measurement of the soluble type II pyrophosphatase (Mthe_0236). Assays contained 40 mM Tris-HCl pH 8 with 5 mM MgCl$_2$ and 1.25 μg enzyme/mL. (•) activity measurement after 5 min preincubation with 1 mM MnCl$_2$, (Δ) activity without preincubation, and (■) control without Mthe_0236.

ribosomal protein S3P (Figure S1). Since at least the gene encoding the S3P protein has to expressed in high amounts for efficient ribosome formation, it is evident that the soluble pyrophosphatase mRNA exists in great copy numbers in cells of *Mt. thermophila*.

Furthermore, the soluble type II pyrophosphatase was found to contain a single CBS domain situated near the N-terminus that could have regulatory effects triggered by binding of ligands such as AMP and ADP [36–38]. Consequently, the pyrophosphatase from *Mt. thermophila* could be potentially inhibited under low-energy conditions (low ATP/ADP ratio) enabling the cell to take advantage of the phosphate group transfer potential of pyrophosphate.

Blast analyses revealed that CBS domains are rarely found in pyrophosphatases of methanogenic archaea. They could only be identified in soluble pyrophosphatases from species of *Methanosaeta*, *Methanocaldococcus*, and *Methanotorris*. However, biochemical data exists only for the pyrophosphatase from *Mt. concilii* [25]. Thus, functionality of CBS domains in pyrophosphatases of methanogenic archaea has not yet been shown. To evaluate the kinetic parameters of the pyrophosphatase from *Mt. thermophila*, the gene *mthe_0236* was cloned into an expression vector and the respective protein was overproduced in *E. coli*. A single band was detected at 35 kDa on the SDS gel after Streptactin affinity purification (Figure 1). This was in accordance with the predicted molecular mass of 35 kDa. The native conformation of the pyrophosphatase was assayed by gel

filtration chromatography. The molecular mass of the native enzyme was 71.4 ± 5 kDa. Thus, the native pyrophosphatase was a homodimer. Crystal structures of the soluble type II pyrophosphatases from *Bacillus subtilis*, *Streptococcus gordonii*, and *Streptococcus mutans* revealed that these enzymes are also homodimers in their native conformation [39, 40].

Kinetic analysis showed that the v$_{max}$ of the enzyme was 157 ± 33 U/mg with Mg^{2+} and 726 ± 40 U/mg with Mn^{2+} as metal cofactor (Figure 2). The K$_M$-value for PP$_i$ was measured with Mn^{2+} as metal ion in the catalytic center and was found to be 0.27 ± 0.05 mM. As indicated earlier, the presence of the CBS domain pair pointed towards a possible regulation of enzyme activity by nucleotides. However, our experiments demonstrated that the pyrophosphatase was not inhibited by nucleotides or its end product phosphate: neither addition of 750 μM AMP or 5 μM ADP nor addition of up to 1.5 mM phosphate led to a reduced reaction rate. Hence, the results indicate that the single CBS domain found in the soluble pyrophosphatase of *Mt. thermophila* is not involved in the regulation of enzyme activity. In contrast, two CBS domains (referred to as Bateman domain [38]) were identified in the membrane-bound pyrophosphatase from the bacterium *Moorella thermoacetica* and inhibition by adenine nucleotides was demonstrated [41].

3.3. Characterization of the ACS. The genome of *Mt. thermophila* contains four genes encoding putative AMP-dependent ACS enzymes, three of which are tandemly positioned (*mthe_1194–mthe_1196*). The gene encoding the fourth putative ACS is located elsewhere as a single gene (*mthe_1413*). Additionally, we identified a gene encoding a putative ADP-dependent acetyl-CoA-synthetase (*mthe_0554*).

To unravel which of the ACS enzymes catalyzes the acetate activation reaction *in vivo*, the transcript amount of the respective genes was investigated. qRT-PCR experiments demonstrated that *mthe_1194* is the most abundantly expressed of the ACS genes (see Figure S1 in Supplementary

Material available online at doi: 10.1155/2012/315153). The transcript level of *mthe_1194* was 2.6 and 2.0 fold higher in comparison to the *gap* gene and the gene encoding the S3P protein, respectively. In contrast, the other ACS encoding genes *mthe_1195* and *mthe_1196* showed 23-fold and 37-fold reduced transcript concentrations compared to *mthe1194*, respectively. Expression of the single gene *mthe_1413* was only slightly lower than expression of *mthe_1194*, whereas the mRNA content of the putative ADP-dependent acetyl-CoA-synthetase, *mthe_0554*, was about 4000-fold lower under the chosen growth conditions and was near to the detection limit of our assays.

The question arose whether the ACS encoding genes *mthe_1194-1196* are organized in one operon. A closer inspection indicated that the genes are separated by at least 300 bp that contain potential transcriptional starting elements (TATA and BRE boxes). Therefore, the organization of this gene cluster was further analyzed by qRT-PCR using primers pairs that bridged the intergenic regions starting from the beginning and the end of the *acs* genes. With this technique, we could detect mRNA that covered the intergenic regions between the genes. The results of qRT-PCR clearly indicated that the intergenic region between *mthe_1195* and *mthe_1196* was transcribed to the same extent as the genes themselves, indicating that *mthe_1195* and *mthe_1196* were transcribed together (see Figure S1). For the intergenic region between *mthe 1194* and *mthe_1195* no transcript could be detected. Hence, it is highly possible that *mthe_1194* represented a single transcriptional unit.

As the ACS encoded by *mthe_1194* was found to be the most abundantly expressed acetate activation enzyme, it was overexpressed in *E. coli* and the corresponding protein was purified via Strep-tactin affinity chromatography. SDS-PAGE and silver staining revealed a single band at approximately 75 kDa, which was in accordance with the predicted molecular mass (Figure 1).

Enzymatic measurements revealed that Mthe_1194 is a thermostable enzyme since 85% of the original activity was retained after incubation at 55°C for 30 min (Figure 3). The optimal growth temperature of *Mt. thermophila* is 55°C and was thus chosen as standard temperature for all enzymatic measurements. In an assay that coupled the formation of AMP to the oxidation of NADH via auxiliary enzymes (see Section 2) a maximal activity of 21.7 U/mg was measured with a K_M value for acetate at 0.4 mM. An alternate assay utilizing the detection of the pyrophosphate resulted in a maximal activity of 28 U/mg. This test was also used to determine K_M values for ATP and CoA, which were found to be 20 μM and 14.5 μM, respectively. To differentiate whether the ACS Mthe_1194 was involved only in the activation of acetate in energy metabolism or also in the metabolism of fatty acids, the substrate spectrum was tested. As expected from an enzyme involved in energy metabolism, the enzyme specifically converted acetate to the corresponding thioester. A reaction was also observed with propionate but the specific activity was only 1% compared to the reactivity with acetate. Butyrate did not serve as a substrate for the ACS. It was observed that AMP, ADP, ATP or P_i did not inhibit Mthe_1194. In contrast, addition of the final product PP_i

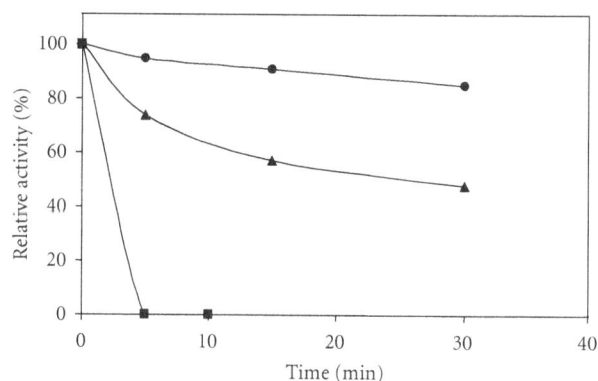

FIGURE 3: Temperature stability of the AMP-dependent ACS Mthe_1194. (•) Incubation at 55°C, (▲) incubation at 75°C, and (■) incubation at 92°C. Enzyme activity was measured in the NADH consumption assay with auxiliary enzymes at 55°C.

led to inhibition of enzyme activity. Addition of 0.25 mM pyrophosphate resulted in 50% reduction of the reaction rate.

A central question of the acetate activation reaction in *Mt. thermophila* is whether the energy that is released by the hydrolysis of ATP to AMP and pyrophosphate (2) is sufficient to drive the activation reaction or whether it is necessary to additionally hydrolyse the pyrophosphate to two inorganic phosphates (3) [42]:

$$ATP \longrightarrow AMP + PP_i \qquad \Delta G'_0 = -31\,kJ/mol\ [42] \qquad (2)$$

$$PP_i \longrightarrow 2\,P_i \qquad \Delta G'_0 = -20\,kJ/mol\ [42] \qquad (3)$$

Therefore, the activation energy of acetyl-CoA formation by Mthe_1194 was determined. For this purpose, the reaction rate between 20 and 92°C was measured and the natural logarithm of the specific activity plotted against the reciprocal value of the absolute temperature. The activation energy was calculated by using (4) and was 30 kJ/mol.

$$E_a = -m \cdot R, \qquad (4)$$

where R is the universal gas constant and m is the slope ($R = 8.314\,J\,mol^{-1}\,K^{-1}$, $m = -3535,5\,K^{-1}$).

4. Discussion

Methanogenic archaea performing aceticlastic methanogenesis are living at the thermodynamic limit as the free energy change of this reaction is only −36 kJ/mol. The first step of acetate breakdown is acetate activation. It was proposed that this step differs in the two genera that are able to grow on acetate, *Methanosarcina* and *Methanosaeta*. For *Methanosarcina* sp., it is well established that the acetate kinase/phosphotransacetylase system is used for acetate activation [43, 44]. It is of bacterial origin and was acquired by *Methanosarcina* sp. by lateral gene transfer [45]. In this pathway acetate is activated to acetyl phosphate with concomitant hydrolysis of ATP to ADP and phosphate. In the subsequent step, acetyl phosphate is transformed into acetyl-CoA without further expense of ATP. In total, one ATP equivalent is hydrolyzed. For *Methanosaeta* sp., however, the acetate activation reaction is more ambiguous. The genome sequences of *Mt. concilii* and *Mt. thermophila* indicate that the acetate kinase/phosphotransacetylase enzyme system is absent in these organisms [23, 46]. In addition, these enzyme activities could not be found in cell extract of *Mt. concilii* [17]. Instead, it was proposed that an AMP-dependent acetyl-CoA synthetase should catalyze this reaction [18]. Jetten et al. purified one of these enzymes from *Mt. concilii* [18] that converts acetate to acetyl-CoA and thereby hydrolyzes ATP to AMP and PP_i. Together with the activity of a soluble pyrophosphatase that was purified by the same group [25] this mode of activation requires the hydrolysis of two ATP equivalents. However, the anaerobic respiratory chain of *Methanosaeta sp.* is purported to be incapable of supporting the generation of more than two ATP molecules from one acetate molecule [24]. Hence, it is rather intriguing how these organisms generate metabolic energy for growth. To overcome this contradiction, we reevaluated the acetate activation reaction in *Mt. thermophila*.

In the genome of *Mt. thermophila*, five different putative ACS enzymes are encoded, four are annotated as AMP-dependent and one as ADP-dependent. So far ADP-dependent acetyl-CoA-synthetases have never been shown to work in the direction of acetyl-CoA formation *in vivo*. Nevertheless, this possibility was considered due to its energetic benefit to the cell. However, qRT-PCR experiments clearly demonstrated that the respective gene is not expressed during the exponential growth phase. Therefore, and because the acetate kinase/phosphotransacetylase system is missing, acetate activation in *Mt. thermophila* is probably catalyzed by an AMP-dependent ACS. It could be shown that one of the four genes encoding AMP-dependent ACS, *mthe_1194*, was the most abundantly expressed. Therefore, the corresponding protein was overproduced and characterized. Involvement in energy metabolism was verified by the fact that acetate is by far the best substrate, which could also be demonstrated for the ACS enzyme from *Methanothermobacter thermoautotrophicus* [47]. Also the K_M value for ATP was low, which means that acetate activation by Mthe_1194 is possible even under low-energy conditions. Inhibition by AMP has been shown but did not occur in this case [18, 48]. Instead PP_i that is the other reaction product and has also been shown to inhibit ACS enzymes [18, 48] could reduce the reaction rate by 50% at a concentration of 0.25 mM. Thus, accumulation of excess acetyl-CoA along with ATP consumption is avoided in the cytoplasm of *Mt. thermophila*.

The close relative *Mt. concilii* contains five genes encoding putative AMP-dependent ACS enzymes [46]. An ACS from *Mt. concilii* was previously purified from cell extracts and characterized and showed similar enzyme properties to the ACS characterized in this study [18]. However, in light of the recent genome sequencing, it is not certain which of the five ACS isozymes was purified from *Mt. concilii*, or if a mixture of the five highly homologous enzymes (≥58% identity) was obtained.

The finding that PP_i is generated during the acetate activation reaction led to the question if the energy released

during hydrolysis of PP_i is dissipated as heat or if it is (at least in part) used for energy conservation. As indicated above, the genome of *Mt. thermophila* contains only one pyrophosphatase gene, coding for a soluble type II pyrophosphatase (Mthe_0236). We heterologously overproduced the enzyme in *E. coli* and the biochemical characterization indicated that the enzyme indeed possessed the characteristics of a soluble type II pyrophosphatase.

The result of gel filtration chromatography showed that the pyrophosphatase from *Mt. thermophila* is active as a homodimer. In contrast, the pyrophosphatase purified from *Mt. concilii* was found to be a heterotetramer by gel filtration chromatography and SDS-PAGE analysis [25]. However, there is only one gene encoding a soluble type II pyrophosphatase in the genome of *Mt. concilii* [46] that appears as a single transcription unit and is not part of an operon structure. It is tempting to speculate that in *Mt. concilii* a posttranslational modification takes place and two forms of the protein are produced or the smaller protein is a result of a proteolytic cleavage as a first stage of degradation in a normal turn-over process. A native conformation with three to four subunits was found for the type II soluble pyrophosphatase from *Methanocaldococcus jannaschii* [49]. In contrast, soluble type II pyrophosphatase purified from bacteria are made of a single subunits and form homodimers [39, 40]. Hence, the pyrophosphatase from *Mt. thermophila* resembles bacterial enzymes with respect to subunit composition.

Pyrophosphate is formed in enzymatic reactions of various metabolic pathways (e.g., DNA, RNA, and protein biosynthesis) and is supposed to be subsequently hydrolyzed by pyrophosphatases to shift the overall reaction equilibrium towards product formation. However, this view may be too restrictive because a considerable amount of metabolic energy is lost and released as heat. Instead, it might be possible that some of the energy of the PP_i anhydride bond could be conserved. For example, by coupling the hydrolysis of PP_i to the phosphorylation of cellular compounds thereby forming energy-rich intermediates for biosynthesis. Such enzymes have already been detected in many organisms, such as PP_i-dependent phosphofructokinases from the sulphur-reducing archaeon *Thermoproteus tenax* [50], bacteria like *Methylococcus capsulatus* [51], *Methylomicrobium alcaliphilum* [52], *Borrelia burgdorferi* [53], and the protozoan *Entamoeba histolytica* [54]. Another prominent example is the pyruvate phosphate dikinase, catalyzing the reversible reaction between pyruvate, ATP and phosphate to phosphoenolpyruvate, AMP, and pyrophosphate. Among others it has been found in *T. tenax* [55], *Bacteroides symbiosis* [56], and *Microbispora rosea* [57]. Genes encoding PP_i-dependent kinases were not yet annotated in the genome of *Mt. thermophila*. However, the deduced amino acid sequence from gene *mthe_1637* revealed a low but significant homology (e-value of $2 \times e^{-40}$) to the pyruvate phosphate dikinase from *T. tenax*.

The question whether PP_i is completely hydrolyzed by the pyrophosphatase in *Mt. thermophila* or whether part of the energy-rich molecule is used for phosphorylation reactions is not clear and will be examined in the future.

However, the kinetic parameters of the pyrophosphatase from *Mt. thermophila* are intriguing. The K_M values for PP_i of the above-mentioned pyrophosphate-scavenging enzymes generally range between 0.2 and 0.015 mM and are thus below the K_M of 0.3 mM of the pyrophosphatase described here. That means that these could take advantage of part of the pyrophosphate that is released during the process of aceticlastic methanogenesis and thus contribute to the generation of high group transfer potential intermediates that would subsequently contribute to energy conservation.

In summary, it could be shown that the acetate activation reaction in *Mt. thermophila* requires two ATP equivalents per molecule of acetate. It cannot be excluded that part of PP_i generated in this process might be used by an unknown enzyme to transfer phosphate groups to an intermediary metabolite. Further investigation into the energy conservation mechanisms of *Methanosaeta* sp. is needed to understand how these organisms that live close to the thermodynamic limits of life can thrive.

Acknowledgments

The authors thank Elisabeth Schwab for technical assistance and Paul Schweiger for critical reading of the manuscript. This work was supported by the Deutsche Forschungsgemeinschaft (DE488/10-1).

References

[1] M. A. K. Khalil and R. A. Rasmussen, "Global emissions of methane during the last several centuries," *Chemosphere*, vol. 29, no. 5, pp. 833–842, 1994.

[2] R. Conrad and M. Klose, "Anaerobic conversion of carbon dioxide to methane, acetate and propionate on washed rice roots," *FEMS Microbiology Ecology*, vol. 30, no. 2, pp. 147–155, 1999.

[3] D. S. Reay, "Sinking methane," *Biologist*, vol. 50, no. 1, pp. 15–19, 2003.

[4] D. R. Kashyap, K. S. Dadhich, and S. K. Sharma, "Biomethanation under psychrophilic conditions: a review," *Bioresource Technology*, vol. 87, no. 2, pp. 147–153, 2003.

[5] Q. Li, L. Li, T. Rejtar, D. J. Lessner, B. L. Karger, and J. G. Ferry, "Electron transport in the pathway of acetate conversion to methane in the marine archaeon *Methanosarcina acetivorans*," *Journal of Bacteriology*, vol. 188, no. 2, pp. 702–710, 2006.

[6] J. G. Ferry and D. J. Lessner, "Methanogenesis in marine sediments," *Annals of the New York Academy of Sciences*, vol. 1125, pp. 147–157, 2008.

[7] M. S. M. Jetten, A. J. M. Stams, and A. J. B. Zehnder, "Methanogenesis from acetate: a comparison of the acetate metabolism in *Methanothrix soehngenii* and *Methanosarcina* spp," *FEMS Microbiology Letters*, vol. 88, no. 3-4, pp. 181–197, 1992.

[8] I. H. Franke-Whittle, B. A. Knapp, J. Fuchs, R. Kaufmann, and H. Insam, "Application of COMPOCHIP microarray to investigate the bacterial communities of different composts," *Microbial Ecology*, vol. 57, no. 3, pp. 510–521, 2009.

[9] C. Lee, J. Kim, K. Hwang, V. O'Flaherty, and S. Hwang, "Quantitative analysis of methanogenic community dynamics in three anaerobic batch digesters treating different wastewaters," *Water Research*, vol. 43, no. 1, pp. 157–165, 2009.

[10] M. Lee, T. Hidaka, W. Hagiwara, and H. Tsuno, "Comparative performance and microbial diversity of hyperthermophilic and thermophilic co-digestion of kitchen garbage and excess sludge," *Bioresource Technology*, vol. 100, no. 2, pp. 578–585, 2009.

[11] S. G. Shin, S. Lee, C. Lee, K. Hwang, and S. Hwang, "Qualitative and quantitative assessment of microbial community in batch anaerobic digestion of secondary sludge," *Bioresource Technology*, vol. 101, no. 24, pp. 9461–9470, 2010.

[12] S. Supaphol, S. N. Jenkins, P. Intomo, I. S. Waite, and A. G. O'Donnell, "Microbial community dynamics in mesophilic anaerobic co-digestion of mixed waste," *Bioresource Technology*, vol. 102, no. 5, pp. 4021–4027, 2011.

[13] D. Karakashev, D. J. Batstone, and I. Angelidaki, "Influence of environmental conditions on methanogenic compositions in anaerobic biogas reactors," *Applied and Environmental Microbiology*, vol. 71, no. 1, pp. 331–338, 2005.

[14] D. J. Aceti and J. G. Ferry, "Purification and characterization of acetate kinase from acetate-grown *Methanosarcina thermophila*. Evidence for regulation of synthesis," *Journal of Biological Chemistry*, vol. 263, no. 30, pp. 15444–15448, 1988.

[15] L. L. Lundie and J. G. Ferry, "Activation of acetate by *Methanosarcina thermophila*. Purification and characterization of phosphotransacetylase," *Journal of Biological Chemistry*, vol. 264, no. 31, pp. 18392–18396, 1989.

[16] H. P. E. Kohler and A. J. B. Zehnder, "Carbon monoxide dehydrogenase and acetate thiokinase in *Methanothrix soehngenii*," *FEMS Microbiology Letters*, vol. 21, no. 3, pp. 287–292, 1984.

[17] P. Pellerin, B. Gruson, G. Prensier, G. Albagnac, and P. Debeire, "Glycogen in *Methanothrix*," *Archives of Microbiology*, vol. 146, no. 4, pp. 377–381, 1987.

[18] M. S. M. Jetten, A. J. M. Stams, and A. J. B. Zehnder, "Isolation and characterization of acetyl-coenzyme A synthetase from *Methanothrix soehngenii*," *Journal of Bacteriology*, vol. 171, no. 10, pp. 5430–5435, 1989.

[19] M. T. Latimer and J. G. Ferry, "Cloning, sequence analysis, and hyperexpression of the genes encoding phosphotransacetylase and acetate kinase from *Methanosarcina thermophila*," *Journal of Bacteriology*, vol. 175, no. 21, pp. 6822–6829, 1993.

[20] J. G. Ferry, "Methane from acetate," *Journal of Bacteriology*, vol. 174, no. 17, pp. 5489–5495, 1992.

[21] K. C. Terlesky and J. G. Ferry, "Ferredoxin requirement for electron transport from the carbon monoxide dehydrogenase complex to a membrane-bound hydrogenase in acetate-grown *Methanosarcina thermophila*," *Journal of Biological Chemistry*, vol. 263, no. 9, pp. 4075–4079, 1988.

[22] G. W. J. Allen and S. H. Zinder, "Methanogenesis from acetate by cell-free extracts of the thermophilic acetotrophic methanogen *Methanothrix thermophila* CALS-1," *Archives of Microbiology*, vol. 166, no. 4, pp. 275–281, 1996.

[23] K. S. Smith and C. Ingram-Smith, "*Methanosaeta*, the forgotten methanogen?" *Trends in Microbiology*, vol. 15, no. 4, pp. 150–155, 2007.

[24] C. Welte and U. Deppenmeier, "Membrane-bound electron transport in *Methanosaeta thermophila*," *Journal of Bacteriology*, vol. 193, no. 11, pp. 2868–2870, 2011.

[25] M. S. M. Jetten, T. J. Fluit, A. J. M. Stams, and A. J. B. Zehnder, "A fluoride-insensitive inorganic pyrophosphatase isolated from *Methanothrix soehngenii*," *Archives of Microbiology*, vol. 157, no. 3, pp. 284–289, 1992.

[26] F. Ausubel, R. Brent, and R. Kingston, *Current Protocols in Molecular Biology*, John Wiley & Sons, Brooklyn, NY, USA, 1987.

[27] D. Hanahan, "Studies on transformation of *Escherichia coli* with plasmids," *Journal of Molecular Biology*, vol. 166, no. 4, pp. 557–580, 1983.

[28] J. E. Mott, R. A. Grant, Y. S. Ho, and T. Platt, "Maximizing gene expression from plasmid vectors containing the λ P(L) promoter: strategies for overproducing transcription termination factor ρ," *Proceedings of the National Academy of Sciences of the United States of America*, vol. 82, no. 1, pp. 88–92, 1985.

[29] U. K. Laemmli, "Cleavage of structural proteins during the assembly of the head of bacteriophage T4," *Nature*, vol. 227, no. 5259, pp. 680–685, 1970.

[30] H. Blum, H. Beier, and H. J. Gross, "Improved silver staining of plant proteins, RNA and DNA in polyacrylamide gels," *Electrophoresis*, vol. 8, no. 2, pp. 93–99, 1987.

[31] S. Saheki, A. Takeda, and T. Shimazu, "Assay of inorganic phosphate in the mild pH range, suitable for measurement of glycogen phosphorylase activity," *Analytical Biochemistry*, vol. 148, no. 2, pp. 277–281, 1985.

[32] Y. Meng, C. Ingram-Smith, L. L. Cooper, and K. S. Smith, "Characterization of an archaeal medium-chain acyl coenzyme A synthetase from *Methanosarcina acetivorans*," *Journal of Bacteriology*, vol. 192, no. 22, pp. 5982–5990, 2010.

[33] Y. Kuang, N. Salem, F. Wang, S. J. Schomisch, V. Chandramouli, and Z. Lee, "A colorimetric assay method to measure acetyl-CoA synthetase activity: application to woodchuck model of hepatitis virus-induced hepatocellular carcinoma," *Journal of Biochemical and Biophysical Methods*, vol. 70, no. 4, pp. 649–655, 2007.

[34] H. Brüggemann, F. Falinski, and U. Deppenmeier, "Structure of the $F_{420}H_2$:quinone oxidoreductase of *Archaeoglobus fulgidus* identification and overproduction of the $F_{420}H_2$-oxidizing subunit," *European Journal of Biochemistry*, vol. 267, no. 18, pp. 5810–5814, 2000.

[35] S. Bäumer, S. Lentes, G. Gottschalk, and U. Deppenmeier, "Identification and analysis of proton-translocating pyrophosphatases in the methanogenic archaeon *Methansarcina mazei*," *Archaea*, vol. 1, no. 1, pp. 1–7, 2002.

[36] J. W. Scott, S. A. Hawley, K. A. Green et al., "CBS domains form energy-sensing modules whose binding of adenosine ligands is disrupted by disease mutations," *The Journal of Clinical Investigation*, vol. 113, no. 2, pp. 274–284, 2004.

[37] S. Meyer, S. Savaresi, I. C. Forster, and R. Dutzler, "Nucleotide recognition by the cytoplasmic domain of the human chloride transporter ClC-5," *Nature Structural & Molecular Biology*, vol. 14, no. 1, pp. 60–67, 2007.

[38] B. E. Kemp, "Bateman domains and adenosine derivatives form a binding contract," *The Journal of Clinical Investigation*, vol. 113, no. 2, pp. 182–184, 2004.

[39] S. Ahn, A. J. Milner, K. Fütterer et al., "The "open" and "closed" structures of the type-C inorganic pyrophosphatases from *Bacillus subtilis* and *Streptococcus gordonii*," *Journal of Molecular Biology*, vol. 313, no. 4, pp. 797–811, 2001.

[40] M. C. Merckel, I. P. Fabrichniy, A. Salminen et al., "Crystal structure of *Streptococcus mutans* pyrophosphatase: a new fold for an old mechanism," *Structure*, vol. 9, no. 4, pp. 289–297, 2001.

[41] J. Jämsen, H. Tuominen, A. Salminen et al., "A CBS domain-containing pyrophosphatase of *Moorella thermoacetica* is regulated by adenine nucleotides," *Biochemical Journal*, vol. 408, no. 3, pp. 327–333, 2007.

[42] J. M. Davies, R. J. Poole, P. A. Rea, and D. Sanders, "Potassium transport into plant vacuoles energized directly by a proton-pumping inorganic pyrophosphatase," *Proceedings of the National Academy of Sciences of the United States of America*, vol. 89, no. 24, pp. 11701–11705, 1992.

[43] P. E. Jablonski, A. A. DiMarco, T. A. Bobik, M. C. Cabell, and J. G. Ferry, "Protein content and enzyme activities in methanol- and acetate-grown *Methanosarcina thermophila*," *Journal of Bacteriology*, vol. 172, no. 3, pp. 1271–1275, 1990.

[44] K. Singh-Wissmann and J. G. Ferry, "Transcriptional regulation of the phosphotransacetylase-encoding and acetate kinase-encoding genes (pta and ack) from *Methanosarcina thermophila*," *Journal of Bacteriology*, vol. 177, no. 7, pp. 1699–1702, 1995.

[45] U. Deppenmeier, A. Johann, T. Hartsch et al., "The genome of *Methanosarcina mazei*: evidence for lateral gene transfer between bacteria and archaea," *Journal of Molecular Microbiology and Biotechnology*, vol. 4, no. 4, pp. 453–461, 2002.

[46] R. D. Barber, L. Zhang, M. Harnack et al., "Complete genome sequence of *Methanosaeta concilii*, a specialist in aceticlastic methanogenesis," *Journal of Bacteriology*, vol. 193, no. 14, pp. 3668–3669, 2011.

[47] C. Ingram-Smith, B. I. Woods, and K. S. Smith, "Characterization of the acyl substrate binding pocket of acetyl-CoA synthetase," *Biochemistry*, vol. 45, no. 38, pp. 11482–11490, 2006.

[48] Y. L. Teh and S. H. Zinder, "Acetyl-coenzyme A synthetase in the thermophilic, acetate-utilizing methanogen *Methanothrix* sp. strain CALS-1," *FEMS Microbiology Letters*, vol. 98, no. 1–3, pp. 1–7, 1992.

[49] N. J. Kuhn, A. Wadeson, S. Ward, and T. W. Young, "*Methanococcus jannaschii* ORF mj0608 codes for a class C inorganic pyrophosphatase protected by Co^{2+} or Mn^{2+} ions against fluoride inhibition," *Archives of Biochemistry and Biophysics*, vol. 379, no. 2, pp. 292–298, 2000.

[50] B. Siebers and R. Hensel, "Pyrophosphate-dependent phosphofructokinase from *Thermoproteus tenax*," *Methods in Enzymology*, vol. 331, pp. 54–62, 2001.

[51] A. S. Reshetnikov, O. N. Rozova, V. N. Khmelenina et al., "Characterization of the pyrophosphate-dependent 6-phosphofructokinase from *Methylococcus capsulatus* Bath," *FEMS Microbiology Letters*, vol. 288, no. 2, pp. 202–210, 2008.

[52] V. N. Khmelenina, O. N. Rozova, and Y. A. Trotsenko, "Characterization of the recombinant pyrophosphate-dependent 6-phosphofructokinases from *Methylomicrobium alcaliphilum* 20Z and *Methylococcus capsulatus* Bath," *Methods in Enzymology*, vol. 495, pp. 1–14, 2011.

[53] Z. Deng, D. Roberts, X. Wang, and R. G. Kemp, "Expression, characterization, and crystallization of the pyrophosphate-dependent phosphofructo-1-kinase of *Borrelia burgdorferi*," *Archives of Biochemistry and Biophysics*, vol. 371, no. 2, pp. 326–331, 1999.

[54] I. Bruchhaus, T. Jacobs, M. Denart, and E. Tannich, "Pyrophosphate-dependent phosphofructokinase of *Entamoeba histolytica*: molecular cloning, recombinant expression and inhibition by pyrophosphate analogues," *Biochemical Journal*, vol. 316, no. 1, pp. 57–63, 1996.

[55] B. Tjaden, A. Plagens, C. Dörr, B. Siebers, and R. Hensel, "Phosphoenolpyruvate synthetase and pyruvate, phosphate dikinase of *Thermoproteus tenax*: key pieces in the puzzle of archaeal carbohydrate metabolism," *Molecular Microbiology*, vol. 60, no. 2, pp. 287–298, 2006.

[56] R. E. Reeves, "Pyruvate,phosphate dikinase from *Bacteroides symbiosus*," *Biochemical Journal*, vol. 125, no. 2, pp. 531–539, 1971.

[57] N. Eisaki, H. Tatsumi, S. Murakami, and T. Horiuchi, "Pyru-
     vate phosphate dikinase from a thermophilic actinomyces
     Microbispora rosea subsp. aerata: purification, characterization
     and molecular cloning of the gene," *Biochimica et Biophysica
     Acta*, vol. 1431, no. 2, pp. 363–373, 1999.

12

Virus-Host and CRISPR Dynamics in Archaea-Dominated Hypersaline Lake Tyrrell, Victoria, Australia

Joanne B. Emerson,[1,2] Karen Andrade,[3] Brian C. Thomas,[1] Anders Norman,[1,4]
Eric E. Allen,[5,6] Karla B. Heidelberg,[7] and Jillian F. Banfield[1,3]

[1] Department of Earth and Planetary Science, University of California, Berkeley, 307 McCone Hall, Berkeley, CA 94720-4767, USA
[2] Cooperative Institute for Research in Environmental Sciences, University of Colorado, Boulder, CO, USA
[3] Department of Environmental Science, Policy, and Management, University of California, Berkeley, 54 Mulford Hall,
  Berkeley, CA 94720, USA
[4] Department of Biology, University of Copenhagen, Copenhagen, Denmark
[5] Marine Biology Research Division, Scripps Institution of Oceanography, La Jolla, CA, USA
[6] Division of Biological Sciences, University of California, San Diego, La Jolla, CA 92093-0202, USA
[7] Department of Biological Sciences, University of Southern California, Los Angeles, CA 90089, USA

Correspondence should be addressed to Joanne B. Emerson; jemerson@berkeley.edu

Academic Editor: Yoshizumi Ishino

The study of natural archaeal assemblages requires community context, namely, a concurrent assessment of the dynamics of archaeal, bacterial, and viral populations. Here, we use filter size-resolved metagenomic analyses to report the dynamics of 101 archaeal and bacterial OTUs and 140 viral populations across 17 samples collected over different timescales from 2007–2010 from Australian hypersaline Lake Tyrrell (LT). All samples were dominated by Archaea (75–95%). Archaeal, bacterial, and viral populations were found to be dynamic on timescales of months to years, and different viral assemblages were present in planktonic, relative to host-associated (active and provirus) size fractions. Analyses of clustered regularly interspaced short palindromic repeat (CRISPR) regions indicate that both rare and abundant viruses were targeted, primarily by lower abundance hosts. Although very few spacers had hits to the NCBI nr database or to the 140 LT viral populations, 21% had hits to unassembled LT viral concentrate reads. This suggests local adaptation to LT-specific viruses and/or undersampling of haloviral assemblages in public databases, along with successful CRISPR-mediated maintenance of viral populations at abundances low enough to preclude genomic assembly. This is the first metagenomic report evaluating widespread archaeal dynamics at the population level on short timescales in a hypersaline system.

1. Introduction

As the most abundant and ubiquitous biological entities, viruses influence host mortality and community structure, food web dynamics, and geochemical cycles [1, 2]. In order to better characterize the potential influence that viruses have on archaeal evolution and ecology, it is important to understand the coupled dynamics of viruses and their archaeal hosts in natural systems. Although previous studies have demonstrated dynamics in virus-host populations, most of these studies have focused on bacterial hosts, often restricted to targeted groups of virus-host pairs, and little is known about archaeal virus-host dynamics in natural systems.

Community-scale virus-host analyses have often been based on low-resolution measurements of the whole community, relying on techniques such as denaturing gradient gel electrophoresis (DGGE), pulsed-field gel electrophoresis (PFGE), and microscopic counts (e.g., [3–5]). One exception is a study that examined viral and microbial dynamics through single read-based metagenomic analyses in four aquatic environments, including an archaea-dominated hypersaline crystallizer pond [6]. In that work, it was proposed that microorganisms and viruses persisted over time at the level of individual taxa (species) but were highly dynamic at the genotype (strain) level. However, in a reanalysis of some of those data by our group using metagenomic assembly, we concluded that viruses were actually dynamic at the population (taxon) level in that system [7]. This result suggests that further analyses are necessary to determine whether archaeal populations tend to be dynamic or stable in hypersaline systems on short timescales.

Of the relatively few metagenomic analyses of virus-host dynamics that have been reported, several have considered the clustered regularly interspaced short palindromic repeat (CRISPR) system, which provides an opportunity to study hosts' responses to viral predation and to link viruses to hosts [8–12]. The CRISPR system is a genomic region in nearly all archaea and some bacteria, and CRISPRs (at least in all systems that have been biochemically characterized to date) have been shown to confer adaptive immunity to viruses and/or other mobile genetic elements through nucleotide sequence identity between the host CRISPR system and invading nucleic acids [13, 14]. The hallmarks of CRISPR regions are spacers, generally derived from foreign nucleic acids, including plasmid and viral DNA, and short palindromic repeat sequences between each spacer (reviewed in [15, 16]). Different strains of the same species can have highly divergent CRISPR regions (e.g., [17]). A highly genomically resolved study of virus-host dynamics in an archaea-dominated acid-mine drainage system demonstrated that only the most recently acquired CRISPR spacers matched coexisting viruses and showed that viruses rapidly recombined to evade CRISPR targeting, indicating that community stability was achieved by the rapid coevolution of host resistance to viruses and viral resistance to the host CRISPR system [9]. Archaeal CRISPR dynamics have also been investigated in *Sulfolobus islandicus* populations [18, 19], indicating clear biogeography of viral populations and adaptation of CRISPR sequences to local viral populations. Whether similar dynamics occur in archaea-dominated hypersaline systems is not well understood.

Previously, our group tracked the dynamics of 35 viral populations in eight viral concentrates (representing the 30 kDa–0.1 μm size fraction) collected during three summers from archaea-dominated hypersaline Lake Tyrrell (LT), Victoria, Australia [7]. We demonstrated that viruses in the LT system were generally stable on the timescale of days and dynamic over years. In this study, we sought to expand our analyses to include LT viruses from metagenomic libraries generated from 0.1, 0.8, and 3.0 μm filters, potentially including proviruses, viruses larger than 0.1 μm, actively infecting viruses, and viruses otherwise retained on the filters. This allowed us to increase the temporal and spatial scope of our study to 17 samples, including four winter samples from which viral concentrate DNA was not sequenced. To give context to the current and previous viral analyses from LT and to test the prevailing theory that microbial taxa are stable at the species level in archaea-dominated hypersaline systems (presented in [6]), in this study we also characterized archaeal and bacterial dynamics through 16S rRNA gene analyses, and we used CRISPR analyses to assist in the interpretation of the results.

2. Materials and Methods

2.1. Sample Collection and Preparation. Sample collection, DNA extraction, and sequencing methods have been described previously [7, 20, 21]. Briefly, 10 L surface water samples were collected from LT and sequentially filtered through 20, 3.0, 0.8, and 0.1 μm filters. Post-0.1 μm filtrates were concentrated through tangential flow filtration and retained for viral DNA extraction. Viral concentrates and 0.1, 0.8, and 3.0 μm filters were retained for each sample, and sequencing was undertaken for different size ranges, depending on the sample (Table 1). Sample names include the month (J for January, A for August), year, site (A or B, ~300 m apart), and time point if the sample was part of a days-scale time series (e.g., t1, t2, etc.). Where necessary, the size fraction is also indicated in the sample name (3.0, 0.8, or 0.1 for filter size in μm, or VC for viral concentrate). Sites A and B are isolated pools in the summer (January samples) but continuous with the lake in the winter (August samples). GPS coordinates for sites A and B are 35° 19′ 09.6″ S, 142° 47′ 59.7″ E and 35° 19′ 18.71″ S, 142° 48′ 4.23″ E, respectively.

2.2. Recovery of 140 Viral Contigs >10 kb and Detection of Viruses in Each Sample. In addition to the 35 LT viral and virus-like (meaning virus or plasmid) populations that we described previously [7], we incorporated all contigs >10 kb from a new IDBA UD [22] assembly of the six Illumina-sequenced viral concentrate samples (default parameters). We also sought to include as many assembled viral sequences as possible from libraries from larger size-fraction filters. To do that, we first attempted to assemble reads from all Illumina-sequenced filter samples and reads from at least one 0.8 μm filter per sample (regardless of sequencing type), using IDBA UD with default parameters [22] for Illumina-sequenced samples and gsAssembler [23] with default parameters for 454-sequenced samples. The assemblies were generally fragmentary, and, as such, no viral contigs larger than 10 kb were identified from most assemblies. However, we recovered viral contigs >10 kb from metagenomic assemblies of all 0.1 μm filters from January 2010, identified through annotation that could be confidently assigned to viruses or proviruses (e.g., contigs that included viral capsid proteins, tail proteins, and/or terminases). Annotation parameters were the same as those described previously [7]. We used BLASTn to identify duplicate viral sequences across assemblies and samples, and we removed any contigs that shared >2 kb at >95% nt identity (all contigs that were removed actually shared ≥99% nt identity because no contigs shared >2 kb at 88–98% nt identity). The remaining 140 viruses share up to 2 kb at up to 87% nt identity, with some smaller shared regions at higher identity.

To determine whether or not a given virus was present in a given sample, we used the 140 viral contig sequences >10 kb as references for fragment recruitment (i.e., recruitment of Illumina sequencing reads or equivalent 100 bp read

TABLE 1: Sequencing and sample information, all filter sizes and viral concentrates.

Sample	Date	Time	T (°C)	pH	TDS (wt%)	Filter size	Sequencing technology	Library type(s)	Reads
J2007At1	Jan. 23, 2007	15:00	22	7.2	31	0.1	Sanger	8–10 kb	650566
						0.8	Sanger, Illumina	fosmid, 8–10 kb, and 100 cycles SR	22192398
						VC	Illumina	100 cycles PE	2436330
J2007At2	Jan. 25, 2007	15:00	28	7.1	31	0.1	Sanger	8–10 kb	905142
						0.8	Sanger	fosmid, 8–10 kb	832880
						VC	Illumina	100 cycles PE	7330099
A2007At1	Aug. 7, 2007	14:00	24	nm	25	0.1	454	SR	5558982
						0.8	454	SR	5333532
						3	Illumina	100 cycles SR	1297810
A2007At2	Aug. 9, 2007	10:30	23	nm	25	0.1	Sanger	8–10 kb	12609766
						0.8	Sanger	8–10 kb	746394
						3	Illumina	100 cycles SR	3949427
A2008At1	Aug. 11, 2008	11:00	12	7.2	25	3	Illumina	100 cycles SR	15786056
A2008At2	Aug. 12, 2008	10:45	11	nm	25	0.8	454	SR and PE	10359280
J2009At1	Jan. 3, 2009	11:45	20	7	28	0.1	454	SR and PE	6303283
						0.8	454	SR	5920276
J2009At2	Jan. 7, 2009	15:00	27	6.9	27	0.1	454	SR	5519946
						0.8	454	SR	6181544
J2009Bt1	Jan. 5, 2009	7:21	18	6.9	24	0.8	454, Illumina	SR, 100 cycles SR	6844436
J2009Bt2	Jan. 5, 2009	12:37	30	7.1	26	0.8	454	SR	7372159
J2009Bt3	Jan. 5, 2009	18:00	36	7	27	0.8	454	SR	7546428
J2009B*	Jan. 5, 2009					VC	Illumina	100 cycles PE	19567468
J2010Bt1	Jan. 7, 2010	7:45	20	7.2	32	0.1	Illumina	100 cycles PE	13213244
						VC	454	SR	2373021
J2010Bt2	Jan. 7, 2010	20:00	32	7.3	36	0.1	Illumina	100 cycles PE	27300634
						3	Illumina	100 cycles PE	23375315
						VC	Illumina	100 cycles PE	3312787
J2010Bt3	Jan. 8, 2010	8:00	21	7.2	34	0.1	Illumina	100 cycles PE	38287968
						0.8	Illumina	100 cycles PE	13808599
						VC	454	SR	2243916
J2010Bt3.5**	Jan. 9, 2010	16:15	45	7.1	27	0.1	Illumina	100 cycles PE	21747692
J2010Bt4**	Jan. 10, 2010	12:50	33	7.2	32	3	Illumina	100 cycles PE	15465664
						VC	Illumina	100 cycles PE	9610233
J2010A	Jan. 10, 2010	12:50	37	7.1	35	0.1	Illumina	100 cycles PE	52520328
						3	Illumina	100 cycles PE	6854358
						VC	Illumina	100 cycles PE	9268384

VC: viral concentrate.
Nm: not measured.
PE: paired-end sequencing.
SR: single-read sequencing.
*VC from J2009B is a pool of DNA from three viral concentrates collected throughout a single day.
**To maintain consistent sample naming with previous publications, we are retaining sample names that were based on consecutive viral concentrate samples. Sample 3.5 was actually collected fourth in the series and sample 4 was fifth.

fragments from other sequencing technologies, as described in [7]), using the Burrows-Wheeler Aligner (bwa) with default parameters [24]. We required at least 1x read coverage across at least 50% of a given reference contig sequence for detection. For hierarchical clustering (Pearson correlation, average linkage clustering), the number of reads that mapped to a given viral contig sequence in a given sample was normalized by the length of the viral sequence and the number of reads in the sample, as described previously [7].

2.3. Generation of 16S rRNA Gene Data and Calculations of Host Relative Abundance. To generate a reference database of 16S rRNA gene sequences, we used the EMIRGE algorithm [25] to reconstruct near full-length 16S rRNA genes from Illumina metagenomic data. All 0.1 and 0.8 μm filters, from which DNA was Illumina sequenced and from which viral concentrate DNA was also sequenced from the same sample, underwent EMIRGE analysis in order to generate a reference 16S rRNA gene database for the LT system. The following filter sample metagenomes were EMIRGE analyzed: 2007Atl (0.8 μm), 2010Bt3 (0.8 μm), 2010Btl (0.1 μm), 2010Bt2 (0.1 μm), 2010Bt3 (0.1 μm), and 2010A (0.1 μm). We clustered all EMIRGE-generated 16S rRNA gene sequences at 97% nt identity, using UCLUST [26], resulting in a database of 101 16S rRNA gene sequences. Taxonomy was assigned to these OTUs, using the SILVA Incremental Aligner (SINA) [27] on the SILVA website [28, 29]. Using bwa with default parameters [24], we mapped metagenomic reads (split into 100 bp lengths to limit biases associated with different sequencing technologies, as described above and in [7]) from all filter samples to the reference database of 101 OTUs, in order to generate relative abundance estimates for each OTU across samples. For a given OTU to be detected in a given library, we required $\geq 1x$ coverage across $\geq 70\%$ of the length of the EMIRGE-generated 16S rRNA gene sequence. In order to account for differences in sequencing throughput, we used relative abundances; that is, we estimated the percent abundance of each OTU in each sample as the number of reads that mapped to that OTU divided by the total number of reads that mapped to any OTU in that sample times 100. Those relative abundances were used for hierarchical clustering (Pearson correlation, average linkage clustering) and appear in Table S1 (see Supplementary Material available online at http://dx.doi.org/10.1155/2013/370871).

These sequences (140 viral contigs >10 kb and 101 16S rRNA gene sequences) have been submitted to NCBI under BioProject accession no. PRJNA81851.

2.4. Correlation with CRISPR Spacers. We used Crass [30] to identify clustered regularly interspaced short palindromic repeat (CRISPR) repeat and spacer sequences in each sample. For this analysis, we considered all sequenced filter sizes for a given water sample together (i.e., 0.1, 0.8, and 3.0 μm filters). Using BLASTn with an *E*-value cutoff of $1e - 10$, we assessed the number of CRISPR spacers from each sample that matched (1) any of the 140 viral contig sequences >10 kb, (2) reads from viral concentrates collected from the same sample (applicable only to the eight samples from which viral concentrates were sequenced), and (3) reads from viral concentrates collected from other samples.

3. Results and Discussion

3.1. Relative Abundances of Viral Populations across Size Fractions. Contigs >10 kb from 105 new viruses were reconstructed, increasing the number of genomically characterized viruses from Lake Tyrrell (LT), Victoria, Australia, from 35 to 140. The 140 contigs, including seven previously reported complete genomes, range in size from 10,050 to 93,283 bp. We analyzed the relative abundances of these 140 viral genotypes in metagenomic libraries from viral concentrates and filter size fractions (0.1–0.8, 0.8–3.0, and 3.0–20 μm) across time and between locations in LT. As few as 42 and as many as 116

viruses were detected in a given sample (any size fraction), and more viruses were detected in samples from which viral concentrates were sequenced. Though we acknowledge that a comparison of 16S rRNA gene microbial OTUs to >10 kb viral contig OTUs is an imperfect proxy for comparing the diversity of these groups, we find approximately 10 viral populations in the viral concentrate size fraction per host OTU (any filter size fraction) in most samples, suggesting that planktonic virus diversity and host diversity scale approximately with abundance, which has previously been established to be approximately 10 : 1 in most environments (e.g., [31]).

In an attempt to distinguish among free (planktonic) viruses physically trapped on filters and host-associated viruses (i.e., active viruses and/or proviruses), we assumed that the 0.8 and 3.0 μm filter pore sizes would be too large to retain significant numbers of viral particles and should predominantly include host-associated viruses. We assumed that the 0.1 μm filters could retain both host-associated and planktonic viruses and that the viral concentrates (30 kDa–0.1 μm size fraction) should generally exclude host cells and be dominated by planktonic viruses. Therefore, we considered viral detection in libraries from three filter size groups separately: (1) viral concentrates, (2) 0.1 μm filters, and (3) 0.8 and/or 3.0 μm filters. For five samples from which at least one library from each of those groups was sequenced, the number of viruses detected only in the viral concentrates was always higher than the number detected only on filters (Figure 1(a)). That trend is robust to the addition of two more samples, from which libraries from viral concentrates and at least one 0.8 and/or 3.0 μm filter sample were sequenced (Figure 1(b)). Although we detected more unique viruses in the viral concentrates, the ratio of total viruses detected on 0.8 or 3.0 μm filters, relative to total viruses detected in the viral concentrates, tended to be approximately equal (Figure 1(c)), meaning that the richness of viruses in viral concentrates tended to be similar to viral richness in the host-associated size fractions.

We used hierarchical clustering (Figure 2(a)) to determine whether patterns in the relative abundances of the 140 viruses (i.e., the 140 viral contigs >10 kb) could be observed, according to season, filter size, and/or sample site. No universal patterns were observed, but there was some clustering according to filter type and/or for samples collected from the same site during the same week. Specifically, all four viral concentrate samples from January 2010, site B clustered together, and they clustered more closely with all of the 0.1 μm filters from the same time series than with other viral concentrate samples. Two additional viral concentrate samples collected during the same season but at different sites and years (J2007Atl and J2009B) complete a larger cluster for that group, indicating similarity among planktonic viral fractions, relative to viruses on larger filter sizes. Interestingly, the remaining two viral concentrate samples (the only sample collected in January 2010 from site A and another sample from site A collected in January 2007) clustered separately from each other and from the rest of the dataset. Overall, this suggests that viral concentrates represent a different part of the viral community than is sampled by other size fractions,

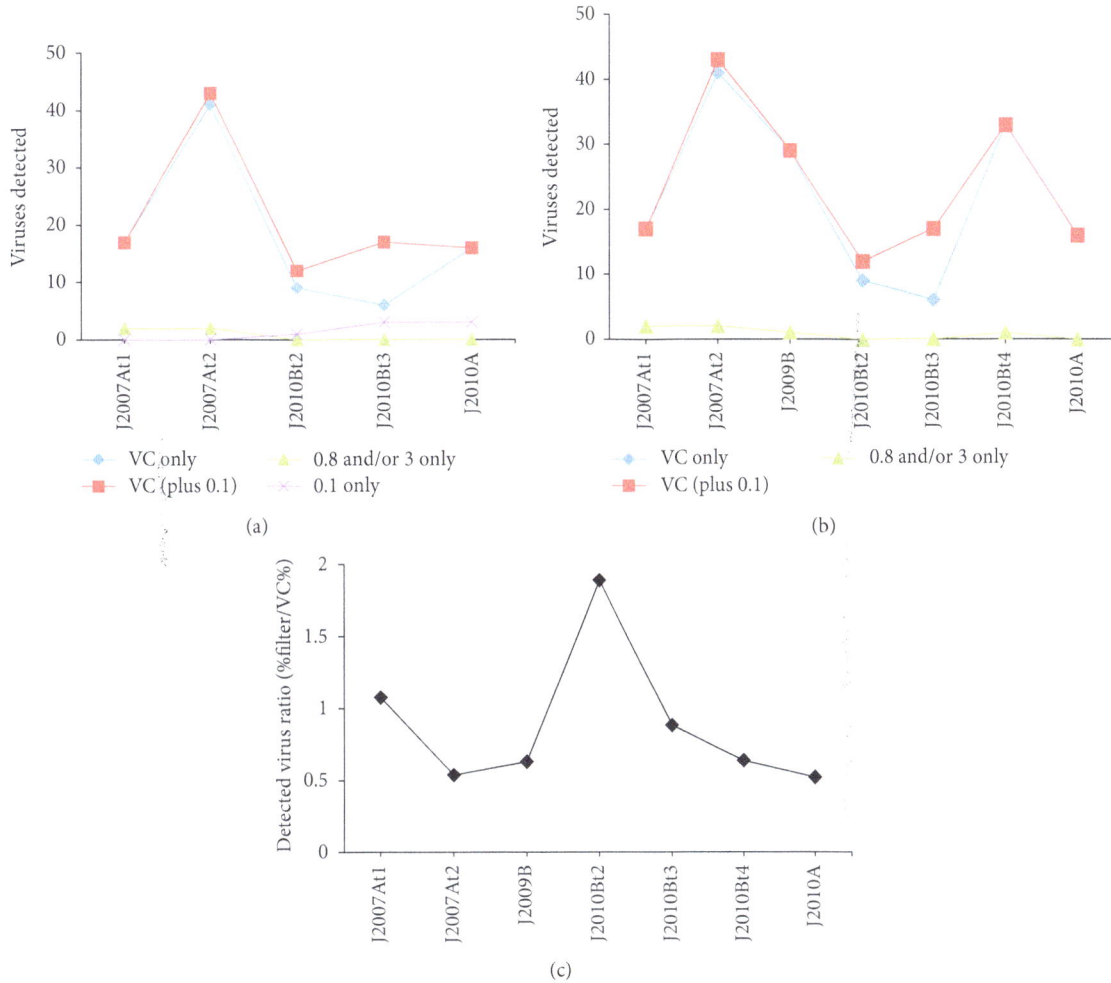

(a)

(b)

(c)

FIGURE 1

particularly fractions >0.8 μm. Interestingly, this is despite the similarity in richness described above (i.e., the number of viral OTUs remains relatively similar across samples, but the composition differs in viral concentrates, relative to host-associated size fractions).

Although the most obvious trend in the viral assemblage hierarchical clustering analysis was the separation of viral concentrates from other filter sizes described above, some clustering was observed for 0.1, 0.8, and 3.0 μm filter size fractions. Specifically, all four filter samples collected from January 2009 site A cluster together, as do all of the 0.8 and 3.0 μm filters from January 2010 site B, suggesting stability of the active viral and/or proviral assemblages over four days in both cases. Although the 0.1 and 0.8 μm filter samples collected in August 2007 from site A time 1 cluster together, they are separate from filters of the same size collected two days later, suggesting a shift in the active viral assemblage over days or possibly the induction of proviruses on that timescale. Together, these data suggest that, although some turnover in active viruses and/or proviruses was observed, most active viruses and/or proviruses were stable within the archaea-dominated LT system over days.

3.2. Archaeal and Bacterial OTUs. We also characterized the relative abundances of potential host OTUs across time and between sites in the LT system. Of 101 total archaeal and bacterial 16S rRNA gene OTUs at 97% nt identity, 29 were detected at 5% abundance or higher on any filter in any sample. For easier visualization, Figure 3 shows only those 29 OTUs, and it includes only samples from which at least five of those OTUs were detected.

In the filter size-resolved plots (Figures 3(a)–3(c)), it is clear that even the most abundant archaeal groups change significantly over time and space, in terms of both presence/absence and relative abundance. For example, the relative proportions of Halorubrum-like and Haloquadratum-like OTUs tend to differ significantly across samples, from almost exclusively Halorubrum-like organisms (e.g., in the August 2007 time series) to almost exclusively Haloquadratum-like organisms (e.g., in the January 2009, site A, time series and in January 2010, site A). Some changes in the relative abundances of these groups can even be observed over hours on 0.8 μm filters from the January 2009, site B, time series (Figure 3(b)). A significant change in temperature was measured between the first and second samples of the January

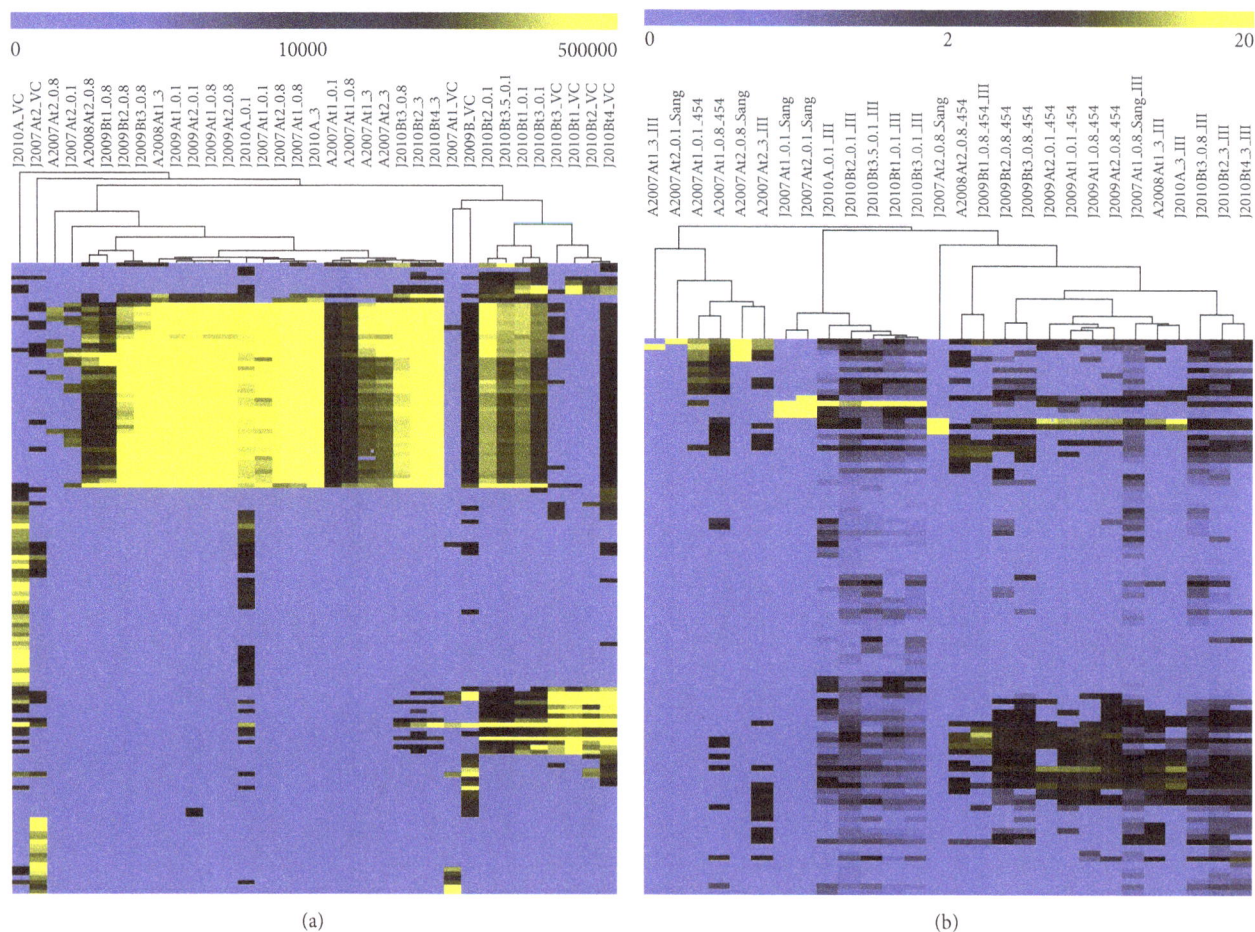

(a)   (b)

FIGURE 2

2009, site B, time series (Table 1), and we hypothesize that this shift in community structure may mark a response to the temperature change. With the exception of the January 2010, site B, time series, the diversity and relative abundance of *Haloquadratum*-like and *Halorubrum*-like organisms are generally anticorrelated, suggesting that these archaea may compete for a similar niche in the LT system. The observation of a relatively low diversity and abundance of *Haloquadratum*-like organisms in some samples (i.e., A2007At1-t2, A2008At2, and J2009Bt1) suggests that *Haloquadratum* species are more dynamic in hypersaline systems than has been previously appreciated [32].

In addition to trends within organism types, specific OTUs also exhibited interesting dynamics. For example, OTU *Haloquadratum walsbyi* 2, which belongs to a species generally considered to be among the most abundant organisms in hypersaline lakes [32], is at relatively low abundance or not detected on 0.8 and 3.0 μm filters from August 2007, site A, and January 2010, site B, though it is at high abundance in that size fraction at site A in January 2009 and January 2010. Interestingly, that OTU also appears dynamic on a days scale in August 2008 at site A, appearing at high abundance on the 3.0 μm filter at time 1 but not detected on the 0.8 μm filter at time 2. Three OTUs, including two related to *Salinibacter*,

were more abundant in the August 2007, site A, time series (all filter sizes) than in any other sample. The Nanohaloarchaeon, Candidatus *Nanosalina* [21], was detected at reasonably high abundance on all 0.1 μm filters from January 2010 sites A and B but was less abundant at other sites and times (Figure 3(a)). Consistent with the small cell size for that organism, it and the other abundant Nanohaloarchaeal OTU, Candidatus *Nanosalinarum,* were found at low abundance or not detected on filters larger than 0.1 μm.

Overall, analyses of the 29 most abundant archaeal and bacterial OTUs indicate dynamics at the population level across time and space, particularly on months-to-years timescales, with some dynamics indicated over hours to days. These results indicate dynamics in the most abundant archaeal and bacterial populations at the taxon level, in contrast to a previous study, which predicted that the most abundant microbial taxa were stable over timescales of weeks to one month in a hypersaline crystallizer pond near San Diego (SD), CA, USA [6]. That study was based on taxonomic affiliations predicted from ~100 bp metagenomic reads. It is possible that differences in sampling timescales, geochemistry, and/or community composition could result in true differences in microbial dynamics between these two systems. Also, LT microbial eukaryotic communities are

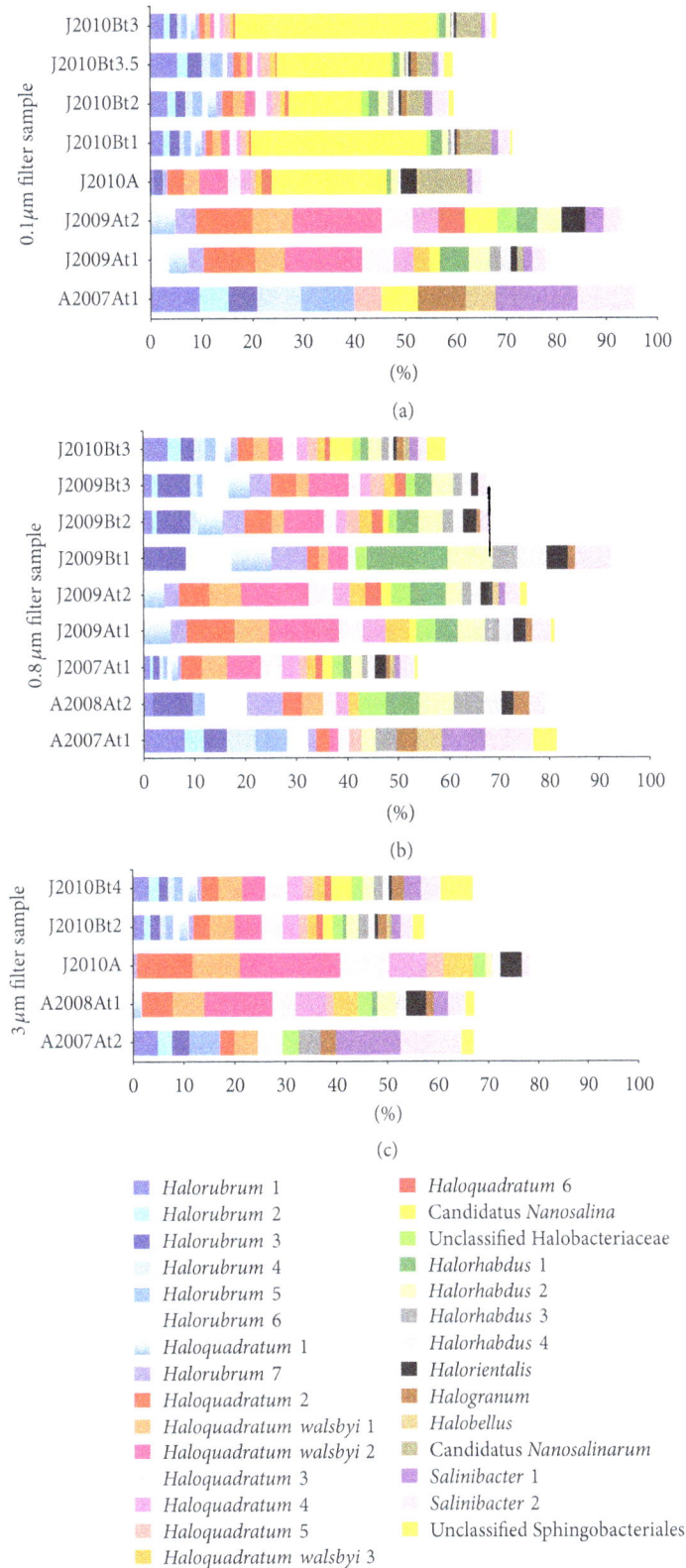

Figure 3

dominated by a predatory *Colpodella* sp., and it is unclear what effect top-down grazing may have on shifts in bacterial and archaeal community structure [20]. However, the SD study also suggested stability of viral populations in that system, and we previously demonstrated through a reanalysis of the SD data that the most abundant viral populations from that study were in fact dynamic [8], so it is possible that different sequencing and analytical methods would have revealed dynamics in archaeal and bacterial populations at the SD site as well. Unfortunately, a reanalysis of the existing SD data using the methods in this study is not possible, due to the incompatibility of 454 sequencing reads with the EMIRGE algorithm, which is designed to reconstruct near-complete 16S rRNA genes from paired-end Illumina metagenomic data.

Based on the relative abundances of the 101 archaeal and bacterial OTUs (including lower abundance LT organisms), we used hierarchical clustering to determine whether similarities in overall archaeal and bacterial community structure would group LT samples according to season, sample type, or filter size (Figure 2(b)). In general, archaeal and bacterial communities sampled from the same location over days clustered more closely than did viral assemblages from the same samples (see above and Figure 2(a)), indicating greater stability of host populations than viral populations over days. The only samples for which no within-time series (i.e., days scale) clustering was observed for host communities were samples from August 2008, site A. For that time series, DNA from different filter sizes was sequenced from each sample, and different sequencing technologies were used (454 and Illumina), so it is possible that similarities between the samples exist but were masked by different methodologies. However, interestingly, all samples and filter sizes from the August 2007, site A, time series clustered together, including DNA from 0.1, 0.8, and 3.0 μm size fractions sequenced by all three technologies (Sanger, 454, and Illumina). This indicates a distinct community at site A in August 2007, which was stable over days and robust to methodological differences in library construction and sequencing.

In some cases, the archaeal and bacterial communities on the 0.1 μm filters clustered together and separately from other filters from the same time point, reflecting enrichment for Nanohaloarchaea. Specifically, both 0.1 μm filters from January 2007, site A, cluster together, and they belong to a larger cluster that includes all 0.1 μm filters from January 2010, site B. The prevalence of Nanohaloarchaea in these samples but not in others is evident in Figure 3(a), which is based on the 29 most abundant OTUs described above. In all other cases, the 0.1 μm filters clustered with larger filters from the same time series. For the January 2009, site B, single-day time series, from which only DNA from 0.8 μm filters (and a viral concentrate) was sequenced, the morning sample clusters separately from the afternoon and evening samples, which cluster together, indicating a shift in community structure over a single day. This is consistent with the shift in relative abundance of *Halorubrum*-like and *Haloquadratum*-like organisms from that time series (Figure 3) and may be related to a temperature shift, as suggested above.

Interestingly, a seasonal trend was not indicated in either the analysis of the 29 most abundant OTUs or in the 101 OTU hierarchical clustering analysis (seasonal trends in viral populations would be difficult to infer, as no viral concentrates were sequenced from winter samples). Samples from site A in August 2007 have quite different archaeal and bacterial community structures from samples at the same site in August 2008, and January samples from site A are also fairly distinct year to year. Although some samples from site B, collected in January 2009 and January 2010, might appear to be the exception (Figure 3(b)), there was as much of a shift in community structure over hours in January 2009 at site B as there was across samples collected over years.

As with any metagenomic study, we cannot say for certain to what extent abundance in metagenomic libraries reflects true abundances. However, we have taken care to avoid biases that are often associated with sequencing-based community analyses. Specifically, by focusing on 16S rRNA gene sequences from metagenomic data, we have avoided PCR amplification biases in our estimates of archaeal and bacterial dynamics. Similarly, we were able to generate enough viral concentrate metagenomic DNA for sequencing without multiple-displacement amplification (MDA), which is known to generate significant biases, especially in viral metagenomes (discussed in [33, 34]).

3.3. CRISPR Analyses. Using the Crass algorithm [30], we identified 549 unique repeats and 8,095 unique spacer sequences from clustered regularly interspaced short palindromic repeat (CRISPR) regions in the LT dataset. Apart from a single sample from January 2009, site A, the only libraries with spacers that matched any of the 140 complete and near-complete viral genomes were those from samples from which viral concentrates were also sequenced (Table 2). This may indicate that different planktonic viral assemblages existed in LT at time points from which viral concentrates were not sequenced, which would be consistent with the viral population dynamics observed across the sequenced viral concentrates [7]. Notably, no spacers from LT August metagenomes matched any of the 140 viruses, consistent with a possible seasonal shift in viral community structure (no viral concentrates were sequenced from August samples), as has been observed in marine systems (e.g., [35]). However, a concomitant seasonal shift in host community structure was not supported, so it is also possible that the August samples could harbor different planktonic viral assemblages each year.

The only spacer matches to any of the seven previously described complete LT viral and virus-like genomes [7] were to LTV2 (76,716 bp) and LTVLE3 (71,341 bp) in libraries from three samples from the January 2010, site B, time series (Table 2). Of the seven viral concentrate genomes, those two achieved the highest abundance by approximately one order of magnitude [7], and they were at their most abundant in the time series from which spacers targeting them were detected. This indicates that LT CRISPRs can actively target abundant, coexisting viruses, consistent with previous observations of CRISPR sampling of coexisting viruses in an acid-mine drainage system [9, 11]. However, it should be noted that the vast majority of spacers do not match any of the 140 viral

TABLE 2: CRISPR analyses by sample.

Sample	Unique repeats	Unique spacers	Hits to 140 viruses	Viral contig match(es)	Predicted host
J2007At1	67	681	1	scaffold_16	
J2007At2	82	633	1	Contig999004	
A2007At1	30	141	0		
A2007At2	64	554	0		
A2008At1	20	275	0		
A2008At2	23	121	0		
J2009At1	41	163	1	scaffold_117	
J2009At2	40	173	0		
J2009B	29	114	0		
J2010Bt1	22	501	0		
J2010Bt2	79	1798	4	LTV2	
				LTV2	
				LTVLE3	
				Contig999004	
J2010Bt3	58	1171	8	LTVLE3	
				Contig1100059	Natronomonas-like
				Contig998975	
				scaffold_55	
				scaffold_29	
				LTVLE3	
				LTVLE3	Nanohaloarchaea
				LTVLE3	Nanohaloarchaea
J2010Bt3.5	60	930	4	LTV2	
				LTVLE3	Nanohaloarchaea
				Contig999004	
				Contig998975	
J2010Bt4	43	853	1	Contig999004	
J2010A	20	340	0		

contigs > 10 kb, which we assume to be among the most abundant viruses because they assembled significantly. This suggests that most CRISPRs target rare viruses. An alternative explanation might be a preponderance of inactive, vestige CRISPR regions, but if that were the case, then we would expect to see the same CRISPR spacers duplicated across samples (i.e., CRISPR regions would be clonally inherited over multiple generations, as opposed to actively integrating new spacers on subgenerational timescales). Given that only 353 spacer sequences were duplicated across samples, relative to 8,095 unique spacer sequences in the dataset, we infer either that most CRISPR regions were highly dynamic and/or that we detected different CRISPR regions over time, due to shifts in host community structure. Regardless, there is a high diversity of spacers in the LT system, consistent with the high diversity of viruses.

Of the CRISPR regions that had matches to the 140 LT viral contigs, only five had repeats with matches to assemblies from 0.1, 0.8, and/or 3.0 μm filter sequences (a match would be indicative of the host organism), and none matched any of the 12 complete and near-complete bacterial and archaeal genomes assembled from the LT January 2007 site A time series [36]. Three matches to the January 2010 site B assemblies (Andrade et al., unpublished data) were to contigs that could only be identified at the order level of Halobacteriales, based on best BLAST hits throughout the contigs, and one repeat matched a contig predicted to belong to a Natronomonas-like organism (Table 2). The most interesting CRISPR repeat sequence match was to a contig from a likely Nanohaloarchaeon, based on the taxonomy of best BLAST hits throughout the contig. A spacer associated with that repeat matched LTVLE3, a virus or plasmid described in [7], strongly suggesting that LTVLE3 is a virus or plasmid of the Nanohaloarchaea. Interestingly, while LTVLE3 was only at high abundance in viral concentrates from the January 2010 site B four-day time series, from which the only spacers targeting it were identified, the most abundant Nanohaloarchaeon, Candidatus *Nanosalina* [21], was detected in most LT samples and is as abundant at site A as it is at site B (isolated pools 300 m apart) in 2010 (Figure 3(a)). This suggests that the Nanohaloarchaea at site B may have been adapted locally to the presence of abundant LTVLE3 at site B in January 2010.

BLAST searches of all 8,095 LT CRISPR spacers and 549 repeat sequences revealed essentially no hits to the NCBI nr and environmental databases, and no hits were detected for

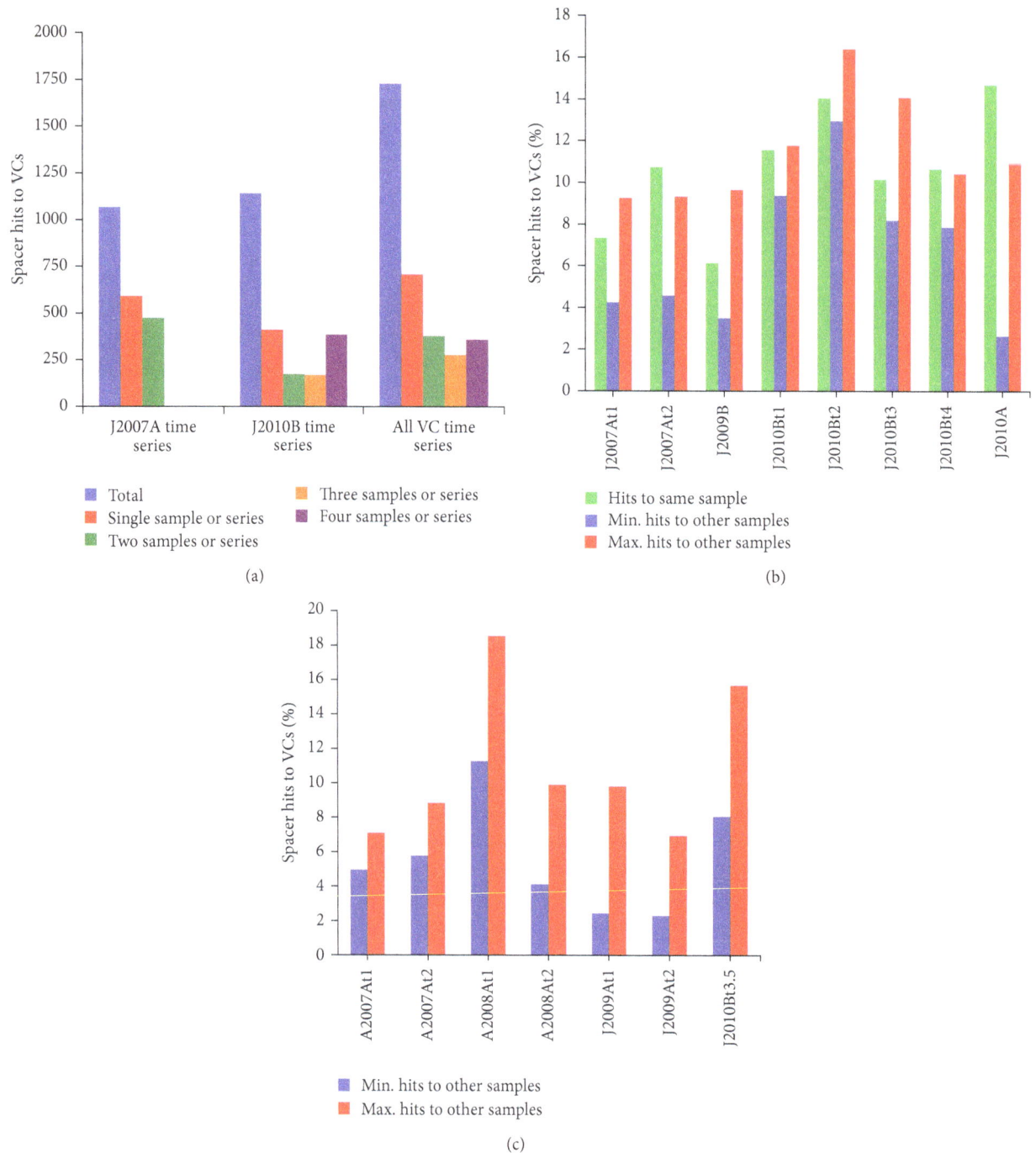

(a)

(b)

(c)

FIGURE 4

any of the repeat sequences. Three spacers had 100% matches *Homo sapiens*, indicating that they are likely either to be erroneous spacers or representative of mobile genetic elements that are not present in public databases. Overall, these data demonstrate that the LT CRISPR spacers target mobile genetic elements that have not been previously identified, suggesting that LT archaea may be highly adapted to local viral predation, though we cannot rule out the possibility of

public database bias (i.e., perhaps LT CRISPRs could target viruses from other locations that are not represented in public databases). In support of local adaptation, despite the relatively small number of spacer matches to the 140 LT viral contigs >10 kb, 1,729 spacers (21%) had matches to unassembled LT viral concentrate reads. This may indicate selection for CRISPR spacers that target locally adapted, potentially coexisting viruses, consistent with previously observed local

CRISPR adaptation to coexisting viruses in an acid-mine drainage system and in geographically distinct sludge bioreactors [8, 9]. Because these matches are to unassembled reads, we infer that most targeted LT viruses are at relatively low abundance, at least at the times and locations sampled by this study (with the exception of LTV2 and LTVLE3, as described above). This may indicate successful CRISPR-mediated maintenance of viral populations at abundances low enough to preclude genomic assembly.

In terms of the timescales on which CRISPR spacers are retained and match coexisting, low-abundance viruses in the LT system, we found several lines of evidence that support the stability of many CRISPR spacers and low-abundance viruses over all timescales sampled by this study (up to three years). LT spacers tended to match reads from multiple viral concentrate metagenomes collected over the course of days almost as often as (January 2007, site A) or more often than (January 2010, site B) they matched reads from a single viral concentrate metagenome from the same time series (Figure 4(a)). This suggests that many spacers and their targeted, low-abundance viruses are stable over days. On the timescale of years, spacers and their targeted coexisting viruses were found most often across more than one of the four sites and times sampled by viral concentrates (January 2007, site A; January 2009, site B; January 2010, site A; January 2010, site B). Similarly, in all eight samples from which viral concentrates were sequenced, nearly as many (and often more) spacers had hits to viral concentrate reads from other samples as matched viral concentrate reads from the same sample (Figure 4(b)). A similar number of matches to the eight viral concentrates were observed for samples from which viral concentrates were not sequenced (Figure 4(c)). Overall, these results demonstrate that CRISPR regions generally retain spacer targets against low-abundance viruses on timescales of 1–3 years. We also infer that low-abundance viruses are likely more stable in the LT system than their highly dynamic, higher abundance counterparts.

4. Conclusions

Both virus (140 contigs >10 kb) and host (101 16S rRNA gene OTUs) assemblage structures were highly dynamic over time and space in the archaea-dominated LT system, particularly over months to years. However, archaeal and bacterial populations were generally more stable than viral populations, and lower abundance viruses were inferred to be more stable than abundant viruses. Filter size-resolved analyses of viral populations revealed different viral assemblages in the planktonic and host-associated (i.e., active and provirus) fractions, suggesting that a higher diversity of viruses may have the potential to infect than are actively infecting at any given time. Consistent with that hypothesis, CRISPR analyses revealed persistent targeting of lower abundance viral populations, along with a high diversity of spacer sequences. However, interestingly, the most abundant viruses (~2% of the viral community, relative to ≤0.1% for most viruses, [7]) were also targeted by CRISPRs, indicating that archaeal hosts strike a balance between protection against persistent, low-abundance viruses and those viruses potentially abundant enough to effect catastrophic changes in host community structure.

Acknowledgments

Funding for this work was provided by the National Science Foundation Award 0626526 and DE-FG02-07ER64505 from the Department of Energy. Thanks to Cheetham Salt Works (Victoria, Australia) for permission to collect samples; John Moreau, Jochen Brocks, and Mike Dyall-Smith for assistance in the field; Matt Lewis and the J. Craig Venter Institute (JCVI) for library construction and sequencing; Shannon Williamson and Doug Fadrosh (JCVI) for training Joanne B. Emerson in virus-related laboratory techniques; Connor Skennerton and Gene Tyson for early access to and assistance with the Crass program; Christine Sun for helpful discussions; and two anonymous reviewers for constructive comments that improved the paper.

References

[1] F. Rohwer, D. Prangishvili, and D. Lindell, "Roles of viruses in the environment," *Environmental Microbiology*, vol. 11, no. 11, pp. 2771–2774, 2009.

[2] C. A. Suttle, "Marine viruses—major players in the global ecosystem," *Nature Reviews Microbiology*, vol. 5, no. 10, pp. 801–812, 2007.

[3] K. Porter, B. E. Russ, and M. L. Dyall-Smith, "Virus-host interactions in salt lakes," *Current Opinion in Microbiology*, vol. 10, no. 4, pp. 418–424, 2007.

[4] R.-A. Sandaa and A. Larsen, "Seasonal variations in virus-host populations in Norwegian coastal waters: focusing on the cyanophage community infecting marine *Synechococcus* spp," *Applied and Environmental Microbiology*, vol. 72, no. 7, pp. 4610–4618, 2006.

[5] S. M. Short and C. A. Suttle, "Temporal dynamics of natural communities of marine algal viruses and eukaryotes," *Aquatic Microbial Ecology*, vol. 32, no. 2, pp. 107–119, 2003.

[6] B. Rodriguez-Brito, L. Li, L. Wegley et al., "Viral and microbial community dynamics in four aquatic environments," *ISME Journal*, vol. 4, no. 6, pp. 739–751, 2010.

[7] J. B. Emerson, B. C. Thomas, K. Andrade, E. E. Allen, K. B. Heidelberg, and J. F. Banfield, "Dynamic viral populations in hypersaline systems as revealed by metagenomic assembly," *Applied and Environmental Microbiology*, vol. 78, pp. 6309–6320, 2012.

[8] V. Kunin, S. He, F. Warnecke et al., "A bacterial metapopulation adapts locally to phage predation despite global dispersal," *Genome Research*, vol. 18, no. 2, pp. 293–297, 2008.

[9] A. F. Andersson and J. F. Banfield, "Virus population dynamics and acquired virus resistance in natural microbial communities," *Science*, vol. 320, no. 5879, pp. 1047–1050, 2008.

[10] R. E. Anderson, W. J. Brazelton, and J. A. Baross, "Using CRISPRs as a metagenomic tool to identify microbial hosts of a diffuse flow hydrothermal vent viral assemblage," *FEMS Microbiology Ecology*, vol. 77, no. 1, pp. 120–133, 2011.

[11] G. W. Tyson and J. F. Banfield, "Rapidly evolving CRISPRs implicated in acquired resistance of microorganisms to viruses," *Environmental Microbiology*, vol. 10, no. 1, pp. 200–207, 2008.

[12] J. F. Heidelberg, W. C. Nelson, T. Schoenfeld, and D. Bhaya, "Germ warfare in a microbial mat community: CRISPRs provide insights into the co-evolution of host and viral genomes," *PLoS ONE*, vol. 4, no. 1, Article ID e4169, 2009.

[13] R. Barrangou, C. Fremaux, H. Deveau et al., "CRISPR provides acquired resistance against viruses in prokaryotes," *Science*, vol. 315, no. 5819, pp. 1709–1712, 2007.

[14] F. J. M. Mojica, C. Díez-Villaseñor, J. García-Martínez, and E. Soria, "Intervening sequences of regularly spaced prokaryotic repeats derive from foreign genetic elements," *Journal of Molecular Evolution*, vol. 60, no. 2, pp. 174–182, 2005.

[15] R. Sorek, V. Kunin, and P. Hugenholtz, "CRISPR—a widespread system that provides acquired resistance against phages in bacteria and archaea," *Nature Reviews Microbiology*, vol. 6, no. 3, pp. 181–186, 2008.

[16] B. Wiedenheft, S. H. Sternberg, and J. A. Doudna, "RNA-guided genetic silencing systems in bacteria and archaea," *Nature*, vol. 482, no. 7385, pp. 331–338, 2012.

[17] R. T. DeBoy, E. F. Mongodin, J. B. Emerson, and K. E. Nelson, "Chromosome evolution in the Thermotogales: large-scale inversions and strain diversification of CRISPR sequences," *Journal of Bacteriology*, vol. 188, no. 7, pp. 2364–2374, 2006.

[18] N. L. Held, A. Herrera, H. C. Quiroz, and R. J. Whitaker, "CRISPR associated diversity within a population of Sulfolobus islandicus," *PLoS ONE*, vol. 5, no. 9, Article ID e12988, 2010.

[19] N. L. Held and R. J. Whitaker, "Viral biogeography revealed by signatures in Sulfolobus islandicus genomes," *Environmental Microbiology*, vol. 11, no. 2, pp. 457–466, 2009.

[20] K. B. Heidelberg, W. C. Nelson, J. B. Holm, N. Eisenkolb, K. Andrade, and J. B. Emerson, "Characterization of eukaryotic microbial diversity in hypersaline Lake Tyrrell, Australia," *Frontiers in Microbiology*, vol. 4, Article ID 115, 2013.

[21] P. Narasingarao, S. Podell, J. A. Ugalde et al., "De novo metagenomic assembly reveals abundant novel major lineage of Archaea in hypersaline microbial communities," *ISME Journal*, vol. 6, no. 1, pp. 81–93, 2012.

[22] Y. Peng, H. C. M. Leung, S. M. Yiu, and F. Y. L. Chin, "IDBA-UD: a de novo assembler for single-cell and metagenomic sequencing data with highly uneven depth," *Bioinformatics*, vol. 28, no. 11, pp. 1420–1428, 2012.

[23] M. Margulies, M. Egholm, W. E. Altman et al., "Genome sequencing in microfabricated high-density picolitre reactors," *Nature*, vol. 437, pp. 376–380, 2005.

[24] H. Li and R. Durbin, "Fast and accurate short read alignment with Burrows-Wheeler transform," *Bioinformatics*, vol. 25, no. 14, pp. 1754–1760, 2009.

[25] C. S. Miller, B. J. Baker, B. C. Thomas, S. W. Singer, and J. F. Banfield, "EMIRGE: reconstruction of full-length ribosomal genes from microbial community short read sequencing data," *Genome Biology*, vol. 12, no. 5, article R44, 2011.

[26] R. C. Edgar, "Search and clustering orders of magnitude faster than BLAST," *Bioinformatics*, vol. 26, no. 19, pp. 2460–2461, 2010.

[27] E. Pruesse, J. Peplies, and F. O. Glockner, "SINA: accurate high-throughput multiple sequence alignment of ribosomal RNA genes," *Bioinformatics*, vol. 28, pp. 1823–1829, 2012.

[28] E. Pruesse, C. Quast, K. Knittel et al., "SILVA: a comprehensive online resource for quality checked and aligned ribosomal RNA sequence data compatible with ARB," *Nucleic Acids Research*, vol. 35, no. 21, pp. 7188–7196, 2007.

[29] C. Quast, E. Pruesse, P. Yilmaz et al., "The SILVA ribosomal RNA gene database project: improved data processing and web-based tools," *Nucleic Acids Research*, vol. 41, pp. D590–D596, 2013.

[30] C. T. Skennerton, M. Imelfort, and G. W. Tyson, "Crass: identification and reconstruction of CRISPR from unassembled metagenomic data," *Nucleic Acids Research*, vol. 41, no. 10, p. e105, 2013.

[31] R. Danovaro, C. Corinaldesi, A. Dell'Anno et al., "Marine viruses and global climate change," *FEMS Microbiology Reviews*, vol. 35, no. 6, pp. 993–1034, 2011.

[32] M. L. Dyall-Smith, F. Pfeiffer, K. Klee et al., "Haloquadratum walsbyi: limited diversity in a global pond," *PLoS ONE*, vol. 6, no. 6, Article ID e20968, 2011.

[33] M. B. Duhaime and M. B. Sullivan, "Ocean viruses: rigorously evaluating the metagenomic sample-to-sequence pipeline," *Virology*, vol. 434, pp. 181–186, 2012.

[34] M. B. Duhaime, L. Deng, B. T. Poulos, and M. B. Sullivan, "Towards quantitative metagenomics of wild viruses and other ultra-low concentration DNA samples: a rigorous assessment and optimization of the linker amplification method," *Environmental Microbiology*, vol. 14, pp. 2526–2537, 2012.

[35] R. J. Parsons, M. Breitbart, M. W. Lomas, and C. A. Carlson, "Ocean time-series reveals recurring seasonal patterns of virioplankton dynamics in the northwestern Sargasso Sea," *ISME Journal*, vol. 6, no. 2, pp. 273–284, 2012.

[36] S. Podell, J. A. Ugalde, P. Narasingarao, J. F. Banfield, K. B. Heidelberg, and E. E. Allen, "Assembly-driven community genomics of a hypersaline microbial ecosystem," *PLoS ONE*, vol. 8, Article ID e61692, 2013.

13

Lipids of Archaeal Viruses

Elina Roine and Dennis H. Bamford

Department of Biosciences and Institute of Biotechnology, University of Helsinki, P.O. Box 56,
Viikinkaari 5, 00014 Helsinki, Finland

Correspondence should be addressed to Elina Roine, elina.roine@helsinki.fi

Academic Editor: Angela Corcelli

Archaeal viruses represent one of the least known territory of the viral universe and even less is known about their lipids. Based on the current knowledge, however, it seems that, as in other viruses, archaeal viral lipids are mostly incorporated into membranes that reside either as outer envelopes or membranes inside an icosahedral capsid. Mechanisms for the membrane acquisition seem to be similar to those of viruses infecting other host organisms. There are indications that also some proteins of archaeal viruses are lipid modified. Further studies on the characterization of lipids in archaeal viruses as well as on their role in virion assembly and infectivity require not only highly purified viral material but also, for example, constant evaluation of the adaptablity of emerging technologies for their analysis. Biological membranes contain proteins and membranes of archaeal viruses are not an exception. Archaeal viruses as relatively simple systems can be used as excellent tools for studying the lipid protein interactions in archaeal membranes.

1. Introduction

Viruses are obligate parasites. Their hallmark is the virion, an infectious particle made of proteins and encapsidating the viral genome. Many viruses, however, also contain lipids as essential components of the virion [1]. The majority of viral lipids are found in membranes, but viral proteins can also be modified with lipids [2, 3].

1.1. Membrane Containing Viruses in the Viral Universe. Membrane containing viruses can roughly be divided into two subclasses [1]. The first subclass contains viruses in which the membrane, also called an envelope, is the outermost layer of the viral particle. In the second class of viruses, the membrane is underneath the usually icosahedral protein capsid. Few viruses contain both the inner membrane as well as an envelope [1]. Lipid membranes of viruses have evolved into essential components of virions that in many cases seem to be involved in the initial stages of infection [4–6]. The majority of membrane containing viruses infect animals both vertebrate and invertebrate that do not have a cell wall surrounding the cytoplasmic membrane. For other host organisms such as plants and prokaryotes there are much

fewer membrane containing viruses known [1]. Usually the cells of these organisms are covered with a cell wall. By far the majority of known viruses that infect prokaryotes, that is, bacteria (bacteriophages), and archaea (archaeal viruses) belong to the order *Caudovirales*, the tailed viruses (Figure 1) [1, 7]. These viruses are made of the icosahedrally organized head and a helical tail. Tailed viruses do not usually contain a membrane, although there are some early reports of tailed mycobacteriophages containing lipids [8, 9]. Viral proteins can also be modified with lipids [3], and there are some indications that proteins of archaeal viruses may also contain lipid modifications [10]. Since very little is known about the lipid modifications of archaeal virus proteins, this paper will concentrate mostly on the membrane lipids of archaeal viruses.

1.2. How Do Viruses Obtain Membranes? Viral-encoded genes possibly involved in lipid modifications have been found in large eukaryotic viruses such as Mimivirus [11] and *Paramecium bursaria* Chlorella virus 1 (PBCV-1) [12]. In prokaryotic viruses, however, no genes encoding components for lipid metabolism have been recognized, but the membranes of prokaryotic viruses are mostly obtained from

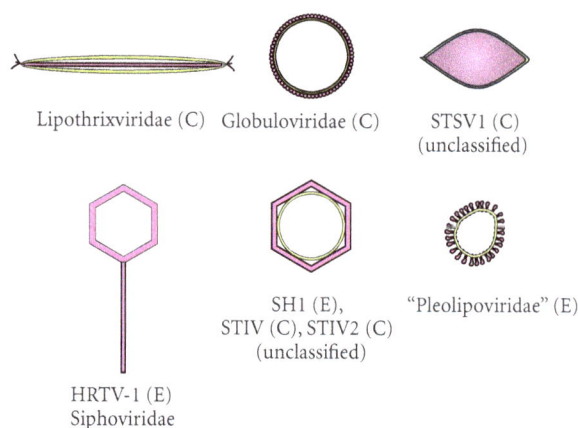

FIGURE 1: Schematic representation of the currently known lipid containing archaeal viruses (C = viruses infecting crenarchaeal hosts, E = viruses infecting euryarchaeal hosts). As a comparison, an archaeal virus devoid of a membrane [32] is also shown. Membrane is illustrated as a yellow layer either inside or outside of the protein capsid depicted in purple. The viral particles are not drawn in scale.

the host cytoplasmic membranes [13, 14]. Enveloped viruses obtain the membrane during budding, that is, egress of the viral particles from the cells without disturbing the cell membrane integrity [15]. The inner membrane of prokaryotic viruses is presumed to be obtained from specific patches of host cytoplasmic membrane containing viral membrane proteins and mechanistically analogous to the formation of clathrin coated pits [6, 16–18]. Consequently, enveloped viruses often exit the cells without lysis, whereas the viruses containing a membrane inside the capsid usually lyse the cells. At least one exception to this can be noted. Prokaryotic lipid containing virus $\phi6$ contains an envelope, but its infection cycle ends in lysis of the host cells [19, 20].

As mentioned above, viral membranes are often involved in the initial stages of infection. This is especially true for the enveloped viruses where the proteins responsible for host recognition (spikes or fusion proteins) are usually incorporated in the envelope. At some point during the often multiphase entry process, the viral envelope fuses with a host membrane releasing the contents into the cell [4, 5]. Among viruses that contain the membrane inside the capsid, the involvement of the membrane in the entry has been shown for the bacteriophage PRD1. After the receptor recognition, the protein rich membrane forms a tubular structure through which the DNA enters the cell cytoplasm [21–23]. Such tubular structures, however, are not formed by all prokaryotic icosahedral viruses containing the membrane inside the capsid [24–28]. For the bacteriophage PM2 that infects the marine bacterium *Pseudoalteromonas*, fusion of the viral inner membrane with the host outer membrane was suggested [29]. Similarly, fusion of the *Sulfolobus* turreted icosahedral virus (STIV) membrane with the cytoplasmic membrane of *S. solfataricus* was suggested [30]. In addition to the function in viral entry, the inner membrane of viruses act, together with the viral membrane proteins, as the scaffold for capsid protein assembly [18, 26, 31].

2. Analysis of Viral Membranes

How do we know if a membrane is part of the viral structure? Chloroform treatment can be used as the first step in screening for viral membranes: the infectivity of the virus is usually considerably reduced if the virions contain a membrane [32–34]. Chloroform treatment can, however, also abolish the infectivity of virions that have not been reported to contain lipids [35] and therefore further studies are always required. Low buoyant density is also an indicator of the lipid membrane in the virions [6, 36]. Sudan Black B can be used to stain the polyacrylamide gel containing separated virion proteins and lipids [10, 37]. Although Sudan Black B is not entirely specific for lipids, positive staining is an indication of the presence of lipid membranes in highly purified viral material and also shows if some viral proteins are putatively lipid modified [10]. Further proof for the presence of a lipid membrane and analyses of its different components can be obtained by techniques also used for the analyses of the membrane lipids of the host cells, for example, thin layer chromatography (TLC), mass spectrometry (e.g., electrospray ionization, ESI-MS), and nuclear magnetic resonance (NMR) [38, 39]. Lipids must be obtained from highly purified viral material [10, 40, 41] or from distinct dissociation components of the virion [28, 42] as it is often difficult to separate virions from membrane vesicles of host origin.

3. Archaeal Lipids

Since the membrane lipids of archaeal viruses are derived from the host lipid pool, analysis of the host lipids is an important part of the lipid analysis of their viruses. Archaeal lipids are known to be drastically different from the ones of bacterial and eukaryotic membranes: instead of lipids based on diacylglycerol the most common core lipid of archaeal phospholipids is the diether of diphytanylglycerol [43, 44]. Archaeal lipids can be divided according to the two major kingdoms of Archaea. As a crude generalization, one can say that the haloarchaeal cell membranes consist mostly of bilayer-forming diether lipids, whereas membranes of archaeal thermophilic organisms are largely composed of tetraether lipids that form monolayer membranes [38, 45, 46]. As in other organisms, phospholipids are the major components of archaeal membranes. In halophilic Archaea, approximately 10% of total lipids are neutral lipids such as bacterioruberin [38]. The major core structure of haloarchaeal lipids consists of archaeol, a 2,3-di-O-phytanyl-sn-glycerol with C_{20} isoprenoid chains [38, 43]. One of the major lipids in extremely thermophilic archaea such as *Sulfolobus* sp. is the macrocyclic tetraether lipid caldarchaeol [46–48]. The composition of lipid membranes is modified according to the environmental conditions in all organisms, and Archaea are not an exception [38, 45, 47, 49].

Some archaeal proteins are known to be modified by isoprenoid derivatives [2, 60–62], and structural analysis revealed a diphytanylglyceryl methyl thioether lipid of one modified protein [61]. Modification of the *Haloferax volcanii*

S-layer protein with a lipid of unknown structure was shown to be crucial to the maturation of the protein [62].

4. Archaeal Membrane Containing Viruses

Our knowledge of archaeal viruses is scarce, but even less is known about their lipids. The known archaeal membrane containing viruses are listed in Table 1. Especially crenarchaeal viruses that infect thermophilic or hyperthermophilic hosts are difficult to produce in amounts high enough for closer analysis of their membrane lipids by traditional methods (e.g., [58]).

4.1. Crenarchaeal Viruses. The presence of lipid membranes have been reported for icosahedral crenarchaeal viruses STIV and *Sulfolobus* turreted icosahedral virus 2 (STIV2), for filamentous viruses *Acidianus* filamentous virus 1 (AFV-1), *Sulfolobus islandicus* filamentous virus (SIFV), and *Thermoproteus tenax* virus 1 (TTV1), for spindle-shaped *Sulfolobus tengchongensis* spindle-shaped virus 1 (STSV1), and for spherical virus *Pyrobaculum* spherical virus (PSV; Table 1). The *Acidianus* bottle-shaped virus (ABV) virions were reported to contain a 9 nm thick envelope [63]. The lipid nature of this envelope, however, has not been reported and the estimated thickness of 9 nm is more than that of usual membranes of archaea [28, 58, 64, 65].

The case of spindle-shaped viruses such as *Sulfolobus* spindle-shaped virus 1 (SSV1) of family *Fuselloviridae* is interesting, because the virions have a buoyant density that is in the same range as in those virions containing a membrane (1.24 g/cm^3 in CsCl), and the virions are sensitive to chloroform [66]. It has been reported that "10% of the SSV-1 virion envelope consists of host lipids" [67], but no further membrane studies have been conducted. This all may suggest that some other type of lipid component than a lipid membrane is present.

The situation among the members of the family *Lipothrixviridae* is also confusing, because these viruses are defined as rod-shaped viruses containing an envelope. The family is further divided into genera *Alpha-*, *Beta-*, *Gamma-*, and *Deltalipothrixvirus* according to the specific structures involved in the host attachment located in the virion ends [68]. The envelope is reported to consist of viral proteins and host derived lipids [68]. The presence of a lipid envelope has been shown for alphalipothrixvirus TTV1 [51], betalipothrixvirus SIFV, and gammalipothrixvirus AFV-1 [53]. However, no evidence for a lipid membrane in the type species of *Deltalipothrixvirus* genus, the *Acidianus* filamentous virus 2 (AFV-2), could be found [69].

Further analysis using thin layer chromatography (TLC) has been reported for AFV1 [53], SIFV [54], STSV1 [59], and PSV [50]. The lipid composition of STIV was analysed using ESI-MS [70]. In conclusion, it could be shown that in general the lipids of crenarchaeal viruses were obtained from the host lipid pool, but some lipid species were found to be quantitatively and qualitatively different from the host lipids [50, 53, 54]. Although viral lipids are considered to be derived from the host membrane lipids, it was suggested that they derived from host lipids by modification [53] and possibly by virus encoded enzyme apparatus [50]. Since no such enzyme apparatus has been described in prokaryotic let alone archaeal viruses, the more probable explanation for the differences at the moment is a strong selection for some minor lipid species of the host. Recent study on the assembly of STIV using cryo-electron tomography suggests that the viral membrane is derived *de novo* in the host cell and not as a result of a membrane invagination [31, 71]. This would, at least in theory, allow the possibility that there is viral enzymatic machinery responsible for the lipid modification. The comparative lipid analysis of STIV and its host *S. solfataricus* showed, however, that the viral lipids consisted of a subpopulation of the host lipids but in different proportions [70].

4.2. Euryarchaeal Viruses. Among euryarchaeal viruses, the icosahedral SH1 and *Haloarcula hispanica* icosahedral virus 2 (HHIV-2) virions contain an inner membrane [40, 42, 72] and the pleomorphic viruses contain a membrane envelope [10, 34, 36, 41].

SH1 was the first icosahedral virus characterized among haloarchaea [33]. Inside the rather complex protein capsid, there is a lipid membrane enclosing the approximately 31 kb linear double stranded (ds) DNA genome [28, 40, 42]. The major protein component of the membrane is the approximately 10 kDa VP12, one of the major structural proteins of the SH1 virion [42]. Although there are no detailed studies reporting the assembly steps of SH1, similarities in virion structure to PRD1 suggests a similar assembly pathway [28, 40, 42]. Therefore, it is likely that the viral capsid and the inner membrane are assembled with the help of the membrane proteins and the genome is packaged into these empty particles (procapsid) before the lysis of the cells [28, 33, 40, 42, 73]. Mass spectrometric analysis of the SH1 lipids revealed major archaeal phospholipid species of phosphatidylglycerol (PG), the methyl ester of phosphatidylglycerophosphate (PGP-Me), and phosphatidylglycerol sulfate (PGS). The proportion of PGP-Me, however, was higher in SH1 than in its host *Haloarcula hispanica* [40]. Quantitative dissociation studies of SH1 allowed the separation of the virion into fractions of soluble capsid proteins and lipid core particle (LC) which consisted of the same phospholipid classes and in the same proportions as the intact virions confirming the presence of the inner membrane [42]. Sudan Black B staining was used to show the presence of lipids in the highly purified HHIV-2 virions [72]. Cryo-electron microscopy (cryo-EM) and image reconstruction of SH1 particles show that as in PRD1 [74] and STIV [25] the inner membrane of SH1 follows the shape of the capsid and the membrane is highly curved at the fivefold vertices where there is a clear transmembrane complex probably containing VP2 protein [28].

Haloarchaeal pleomorphic viruses is a newly characterized group of viruses with relatively simple virion architecture [10, 34, 36, 41, 56, 73, 75]. The genome (single stranded or double stranded DNA) is enclosed in a membrane vesicle derived from the host membrane [10, 34, 41].

TABLE 1: Currently known membrane containing archaeal viruses, exit strategy, and presence of lipid envelope or inner membrane.

Family or Genus[a]	Type species/example of species/species lipids studied	Exit strategy	Lipids	References
Globuloviridae (C)	*Pyrobaculum* spherical virus, PSV	No lysis detected	Lipid envelope	[50]
Lipothrixviridae (C)				
Genus *Alphalipothrixvirus*	*Thermoproteus tenax* virus 1, TTV1	Lysis	Lipid envelope	[51, 52]
Genus *Betalipothrixvirus*	*Acidianus* filamentous virus 1, AFV1	No lysis detected	Lipid envelope	[53]
Genus *Gammalipothrixvirus*	*Sulfolobus islandicus* filamentous virus 1, SIFV1	No lysis detected	Lipid envelope	[54]
Genus *Salterprovirus* (E)	His2[b]	No lysis detected	Lipid envelope	[10, 55]
"Pleolipoviridae" (E)[b]	*Halorubrum* pleomorphic virus 1, HRPV-1	No lysis detected	Lipid envelope	[34, 41]
	Halorubrum pleomorphic virus 2, HRPV-2	No lysis detected	Lipid envelope	[10, 56]
	Halorubrum pleomorphic virus 3, HRPV-3	No lysis detected	Lipid envelope	[10, 56]
	Halorubrum pleomorphic virus 6, HRPV-6	No lysis detected	Lipid envelope	[10]
	Haloarcula hispanica pleomorphic virus 1, HHPV-1	No lysis detected	Lipid envelope	[36]
Unclassified	*Sulfolobus* turreted icosahedral virus, STIV (C)	Lysis	Inner membrane	[25, 57]
	Sulfolobus turreted icosahedral virus 2, STIV2 (C)	Lysis	Inner membrane	[58]
	Sulfolobus tengchongensis spindle-shaped virus 1 (C)	No lysis detected	Lipid envelope	[59]
	SH1 (E)	Lysis	Inner membrane	[33, 40, 42]

[a] Host domain: B: bacteria, C: crenarchaea, E: euryarchaea.
[b] His2 has been suggested to belong to the new family *Pleolipoviridae*. The approval of the suggested new family is pending at the ICTV.

There are two major structural proteins, the larger proteins (approximately 50 kDa in size) are mostly exposed and C-terminally anchored to the membrane [10, 41]. This larger protein is N-glycosylated in HRPV-1 [41, 76], and in HGPV-1 it stains with Sudan Black B [10] suggesting a lipid modification. The smaller structural proteins (approximately 10 to 14.5 kDa) are predicted to contain several transmembrane domains [10, 34, 41]. New progeny viruses are released from the infected cell without lysis [10, 34, 36]. Thus, the viral envelope is most probably acquired by budding from the sites of host cytoplasmic membrane containing the viral membrane proteins and the genome [10, 34, 36]. The detailed sequence of events and the viral and host proteins involved will be the subject of future studies. Currently, the group of haloarchaeal pleomorphic viruses consists of seven members: *Halorubrum* pleomorphic viruses 1, 2, 3 and 6 (HRPV-1, HRPV-2, HRPV-3, and HRPV-6), respectively [10, 34, 56], *Haloarcula hispanica* pleomorphic virus 1 (HHPV-1) [36], and *Halogeometricum* sp. pleomorphic virus 1 (HGPV-1) [56]. In addition, His2 [55], the second member of genus *Salterprovirus*, is suggested to belong to the pleomorphic viruses [10, 75]. Lipid analysis by TLC or mass spectrometry of the highly purified viral material suggests that the composition of lipids was similar to that of their hosts [10, 34, 41]. The lipids of viruses infecting *Halorubrum* sp. hosts consisted mostly of the archaeal forms of PG, PGP-Me, and PGS, whereas in *Halogeometricum* sp. the PGS was missing both in the host lipids as well as in the lipids of HGPV-1 [10, 41, 77]. Sudan Black B staining of the HGPV-1 and His2 proteins showed that some of the major structural proteins may also be lipid modified [10].

Studies on lipid containing haloarchaeal viruses of different morphotypes have also allowed the comparison of the differences in the proportions of incorporated lipids. For example, the isolation and characterization of the *Haloarcula hispanica* pleomorphic virus 1 (HHPV-1) [36] allowed to compare the differences of lipid composition between an icosahedral membrane containing virus SH1and the enveloped, pleomorphic HHPV-1 that infect the same host, *Har. hispanica* (Figure 2). The comparison showed that the lipid composition of the pleomorphic virus HHPV-1 envelope was more similar to the lipids of the host membrane than those of SH1 (Figure 2) [36]. This may be explained by the constraints that the inner membrane curvature poses on the selection of lipids in SH1 and consequently suggests that SH1 is able to selectively acquire lipids from the host membrane [36, 73]. Different lipids are known to have different shapes and therefore can be found in different positions in the curved membrane [14, 78]. It is known that different membrane proteins attract different types of lipids [79], and it would be very interesting to determine which viral proteins are involved in this process.

5. Concluding Remarks

Research on lipid containing archaeal viruses is still in its infancy. The presence of lipids and characterization of their nature has been shown for some archaeal viruses [10, 25, 34, 36, 40–42, 50, 51, 53, 54, 58, 59, 70]. Deeper understanding of their role in virus biology is largely still missing. Partly this problem can be assigned to an inability to produce enough material of high enough purity. Partly this problem is due to missing techniques comparable to those developed for lipid research of bacteria and eukaryotes. Lipid research, as many other fields of research, benefits from a thorough characterization of the systems studied. Characterization of the virus life cycle and studying its different steps using the cutting edge technologies in electron microscopy, for example, complements the information obtained using biochemical and genomic methods [25, 28, 31, 40, 42].

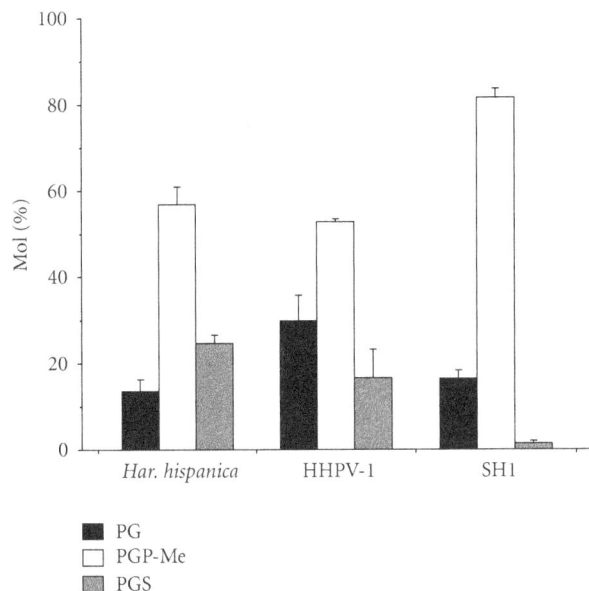

FIGURE 2: Comparison of the phospholipid compositions of *Har. hispanica*, HHPV-1 and SH1. Concentrations are expressed as the mol% of the total phospholipids. Only phospholipids representing more than 1% of the total are shown. Error bars represent standard deviations of data from at least three independent experiments. Copyright American Society for Microbiology, [36].

The examples set by crystallization of the whole virions of membrane containing bacteriophages PRD1 [74, 80] and PM2 [26] show how valuable different perspectives on lipid membranes can be. Studies on the finding of the proposed viral-encoded genes involved in novel lipid modifications is hampered by the fact that a high amount of predicted gene content in archaeal viruses do not have homologues in the data bases. A more systematic approach of cloning and expression analyses of genes as well as crystallization of the gene products could be used in screening for the functions of interest.

Archaeal lipids are unique in terms of their chemical, physical, structural, and biological properties. Not only can they be admired in their complexity and variability, but as material adjusted to extreme conditions, they can be considered unique for biotechnological applications designed for extreme conditions. The archaeosomes made of one of the major phospholipid of haloarchaeal membranes, the archaeal form of the methyl ester of phophatidylglycerophosphate (PGP-Me), for example, have been shown to be superior in terms of stability and low permeability in high salt conditions [81]. Similar findings were reported for the performance of archaeosomes made of thermophilic lipids in a wide range of temperatures [65]. Although the lipids of archaeal viruses are obtained from host lipids, they can be present in different proportions and the mechanisms for the selection must be driven by viral components. The simplicity of many membrane containing archaeal viruses can be exploited in studying the mechanisms of protein-lipid interplay in archaeal membranes.

Acknowledgments

This work was supported by the Helsinki University three year grant (2010–2012) to E. Roine and Academy Professor funding (Academy of Finland) grants 256197 and 256518 to D. H. Bamford.

References

[1] A. M. Q. King, M. J. Adams, E. B. Carstens, and E. J. Lefkowitz, *Virus Taxonomy, Ninth Report of the International Committee on Taxonomy of Viruses*, Elsevier, Oxford, UK, 2011.

[2] J. Eichler and M. W. W. Adams, "Posttranslational protein modification in *Archaea*," *Microbiology and Molecular Biology Reviews*, vol. 69, no. 3, pp. 393–425, 2005.

[3] D. E. Hruby and C. A. Franke, "Viral acylproteins: greasing the wheels of assembly," *Trends in Microbiology*, vol. 1, no. 1, pp. 20–25, 1993.

[4] M. M. Poranen, R. Daugelavičius, and D. H. Bamford, "Common principles in viral entry," *Annual Review of Microbiology*, vol. 56, pp. 521–538, 2002.

[5] A. E. Smith and A. Helenius, "How viruses enter animal cells," *Science*, vol. 304, no. 5668, pp. 237–242, 2004.

[6] H. M. Oksanen, M. M. Poranen, and D. H. Bamford, "Bacteriophages: lipid-containing," in *Encyclopedia of Life Sciences (ELS)*, John Wiley & Sons, Chichester, UK, 2010.

[7] H. W. Ackermann, "5500 Phages examined in the electron microscope," *Archives of Virology*, vol. 152, no. 2, pp. 227–243, 2007.

[8] B. L. Soloff, T. A. Rado, B. E. Henry II, and J. H. Bates, "Biochemical and morphological characterization of mycobacteriophage R1," *Journal of Virology*, vol. 25, no. 1, pp. 253–262, 1978.

[9] M. L. Gope and K. P. Gopinathan, "Presence of lipids in mycobacteriophage I3," *Journal of General Virology*, vol. 59, no. 1, pp. 131–138, 1982.

[10] M. K. Pietilä, N. S. Atanasova, V. Manole et al., "Virion architecture unifies globally distributed pleolipoviruses infecting halophilic *Archaea*," *Journal of Virology*, vol. 86, no. 9, pp. 5067–5079, 2012.

[11] J. M. Claverie, C. Abergel, and H. Ogata, "Mimivirus," in *Current Topics in Microbiology and Immunology*, J. L. Van Etten, Ed., vol. 328, pp. 89–121, 2009.

[12] W. H. Wilson, J. L. Van Etten, and M. J. Allen, "The *Phycodnaviridae*: the story how tiny giants rule the world," in *Current Topics in Microbiology and Immunology*, J. L. Van Etten, Ed., pp. 1–42, 2009.

[13] G. J. Brewer, "Control of membrane morphogenesis in bacteriophage," *International Review of Cytology*, vol. 68, pp. 53–96, 1980.

[14] S. Laurinavičius, *Phospholipids of lipid-containing bacteriophages and their transbilayer distribution. [Ph.D. thesis]*, University of Helsinki, Helsinki, Finland, 2008.

[15] H. Garoff, R. Hewson, and D. J. E. Opstelten, "Virus maturation by budding," *Microbiology and Molecular Biology Reviews*, vol. 62, no. 4, pp. 1171–1190, 1998.

[16] L. Mindich, D. Bamford, T. McGraw, and G. Mackenzie, "Assembly of bacteriophage PRD1: particle formation with wild-type and mutant viruses," *Journal of Virology*, vol. 44, no. 3, pp. 1021–1030, 1982.

[17] D. H. Bamford, J. Caldentey, and J. K. Bamford, "Bacteriophage PRD1: a broad host range dsDNA tectivirus with an

internal membrane," *Advances in Virus Research*, vol. 45, pp. 281–319, 1995.

[18] P. S. Rydman, J. K. H. Bamford, and D. H. Bamford, "A minor capsid protein P30 is essential for bacteriophage PRD1 capsid assembly," *Journal of Molecular Biology*, vol. 313, no. 4, pp. 785–795, 2001.

[19] A. K. Vidaver, R. K. Koski, and J. L. Van Etten, "Bacteriophage φ6: a lipid containing virus of *Pseudomonas phaseolicola*," *Journal of Virology*, vol. 11, no. 5, pp. 799–805, 1973.

[20] L. Mindich and J. Lehman, "Cell wall lysin as a component of the bacteriophage φ6 virion," *Journal of Virology*, vol. 30, no. 2, pp. 489–496, 1979.

[21] K. H. Lundström, D. H. Bamford, E. T. Palva, and K. Lounatmaa, "Lipid-containing bacteriophage PR4: structure and life cycle," *Journal of General Virology*, vol. 43, no. 3, pp. 583–592, 1979.

[22] D. Bamford and L. Mindich, "Structure of the lipid-containing bacteriophage PRD1: disruption of wild-type and nonsense mutant phage particles with guanidine hydrochloride," *Journal of Virology*, vol. 44, no. 3, pp. 1031–1038, 1982.

[23] A. M. Grahn, R. Daugelavičius, and D. H. Bamford, "Sequential model of phage PRD1 DNA delivery: active involvement of the viral membrane," *Molecular Microbiology*, vol. 46, no. 5, pp. 1199–1209, 2002.

[24] H. M. Kivelä, R. Daugelavičius, R. H. Hankkio, J. K. H. Bamford, and D. H. Bamford, "Penetration of membrane-containing double-stranded-DNA bacteriophage PM2 into *Pseudoalteromonas* hosts," *Journal of Bacteriology*, vol. 186, no. 16, pp. 5342–5354, 2004.

[25] R. Khayat, L. Tang, E. T. Larson, C. M. Lawrence, M. Young, and J. E. Johnson, "Structure of an archaeal virus capsid protein reveals a common ancestry to eukaryotic and bacterial viruses," *Proceedings of the National Academy of Sciences of the United States of America*, vol. 102, no. 52, pp. 18944–18949, 2005.

[26] N. G. Abrescia, J. M. Grimes, H. M. Kivelä et al., "Insights into virus evolution and membrane biogenesis from the structure of the marine lipid-containing bacteriophage PM2," *Molecular cell*, vol. 31, no. 5, pp. 749–761, 2008.

[27] S. T. Jaatinen, L. J. Happonen, P. Laurinmäki, S. J. Butcher, and D. H. Bamford, "Biochemical and structural characterisation of membrane-containing icosahedral dsDNA bacteriophages infecting thermophilic *Thermus thermophilus*," *Virology*, vol. 379, no. 1, pp. 10–19, 2008.

[28] H. T. Jäälinoja, E. Roine, P. Laurinmäki, H. M. Kivelä, D. H. Bamford, and S. J. Butcher, "Structure and host-cell interaction of SH1, a membrane-containing, halophilic euryarchaeal virus," *Proceedings of the National Academy of Sciences of the United States of America*, vol. 105, no. 23, pp. 8008–8013, 2008.

[29] V. Cvirkaitė-Krupovič, M. Krupovič, R. Daugelavičius, and D. H. Bamford, "Calcium ion-dependent entry of the membrane-containing bacteriophage PM2 into its *Pseudoalteromonas* host," *Virology*, vol. 405, no. 1, pp. 120–128, 2010.

[30] R. Khayat, C. Y. Fu, A. C. Ortmann, M. J. Young, and J. E. Johnson, "The architecture and chemical stability of the archaeal *Sulfolobus* turreted icosahedral virus," *Journal of Virology*, vol. 84, no. 18, pp. 9575–9583, 2010.

[31] C. Y. Fu, K. Wang, L. Gan et al., "In vivo assembly of an archaeal virus studied with whole-cell electron cryotomography," *Structure*, vol. 18, no. 12, pp. 1579–1586, 2010.

[32] P. Kukkaro and D. H. Bamford, "Virus-host interactions in environments with a wide range of ionic strengths," *Environmental Microbiology Reports*, vol. 1, no. 1, pp. 71–77, 2009.

[33] K. Porter, P. Kukkaro, J. K. H. Bamford et al., "SH1: a novel, spherical halovirus isolated from an Australian hypersaline lake," *Virology*, vol. 335, no. 1, pp. 22–33, 2005.

[34] M. K. Pietilä, E. Roine, L. Paulin, N. Kalkkinen, and D. H. Bamford, "An ssDNA virus infecting *Archaea*: a new lineage of viruses with a membrane envelope," *Molecular Microbiology*, vol. 72, no. 2, pp. 307–319, 2009.

[35] M. L. Dyall-Smith, "Genus *Salterprovirus*," in *Virus Taxonomy, Ninth Report of the International Committee on Taxonomy of Viruses*, A. M. Q. King, M. J. Adams, E. B. Carstens, and E. J. Lefkowitz, Eds., pp. 183–186, ElsevierOxford, UK, 2011.

[36] E. Roine, P. Kukkaro, L. Paulin et al., "New, closely related haloarchaeal viral elements with different nucleic acid types," *Journal of Virology*, vol. 84, no. 7, pp. 3682–3689, 2010.

[37] J. P. Prat, J. N. Lamy, and J. D. Weill, "Staining of lipoproteins after electrophoresis in polyacrylamide gel," *Bulletin de la Société de Chimie Biologique*, vol. 51, no. 9, article 1367, 1969.

[38] A. Corcelli and S. Lobasso, "Characterization of lipids of halophilic *Archaea*," in *Methods in Microbiology: Extremophiles*, F. A. Rainey and A. Oren, Eds., vol. 35, pp. 585–613, Elsevier, New York, NY, USA, 2006.

[39] M. S. da Costa, M. F. Nobre, and R. Wait, "Analysis of lipids from extremophilic bacteria," in *Methods in Microbiology: Extremophiles*, F. A. Rainey and A. Oren, Eds., vol. 35, pp. 127–159, Elsevier, New York, NY, USA, 2006.

[40] D. H. Bamford, J. J. Ravantti, G. Rönnholm et al., "Constituents of SH1, a novel lipid-containing virus infecting the halophilic euryarchaeon *Haloarcula hispanica*," *Journal of Virology*, vol. 79, no. 14, pp. 9097–9107, 2005.

[41] M. K. Pietilä, S. Laurinavičius, J. Sund, E. Roine, and D. H. Bamford, "The single-stranded DNA genome of novel archaeal virus *Halorubrum pleomorphic* virus 1 is enclosed in the envelope decorated with glycoprotein spikes," *Journal of Virology*, vol. 84, no. 2, pp. 788–798, 2010.

[42] H. M. Kivelä, E. Roine, P. Kukkaro, S. Laurinavičius, P. Somerharju, and D. H. Bamford, "Quantitative dissociation of archaeal virus SH1 reveals distinct capsid proteins and a lipid core," *Virology*, vol. 356, no. 1-2, pp. 4–11, 2006.

[43] M. Kates, "The phytanyl ether-linked polar lipids and isoprenoid neutral lipids of extremely halophilic bacteria," *Progress in the Chemistry of Fats and Other Lipids*, vol. 15, no. 4, pp. 301–342, 1977.

[44] G. D. Sprott, "Structures of archaebacterial membrane lipids," *Journal of Bioenergetics and Biomembranes*, vol. 24, no. 6, pp. 555–566, 1992.

[45] S. V. Albers, W. N. Konings, and A. J. M. Driessen, "Membranes of thermophiles and other extremophiles," in *Methods in Microbiology: Extremophiles*, F. A. Rainey and A. Oren, Eds., vol. 35, pp. 161–171, Elsevier, New York, NY, USA, 2006.

[46] Y. Boucher, "Lipids: biosynthesis, function, and evolution," in *Archaea: Molecular and Cellular Biology*, R. Cavicchioli, Ed., pp. 341–353, ASM Press, Washington, DC, USA, 2007.

[47] A. Gliozzi, R. Rolandi, M. de Rosa, and A. Gambacorta, "Monolayer black membranes from bipolar lipids of archaebacteria and their temperature-induced structural changes," *The Journal of Membrane Biology*, vol. 75, no. 1, pp. 45–56, 1983.

[48] B. Nicolaus, A. Trincone, E. Esposito, M. R. Vaccaro, A. Gambacorta, and M. de Rosa, "Calditol tetraether lipids of the

archaebacterium *Sulfolobus solfataricus*. Biosynthetic studies," *Biochemical Journal*, vol. 266, no. 3, pp. 785–791, 1990.

[49] A. Corcelli, "The cardiolipin analogues of *Archaea*," *Biochimica et Biophysica Acta*, vol. 1788, no. 10, pp. 2101–2106, 2009.

[50] M. Häring, X. Peng, K. Brügger et al., "Morphology and genome organization of the virus PSV of the hyperthermophilic archaeal genera *Pyrobaculum* and *Thermoproteus*: a novel virus family, the *Globuloviridae*," *Virology*, vol. 323, no. 2, pp. 233–242, 2004.

[51] M. Rettenberger, *Das Virus TTV1 des extreme thermophilen Schwefel-Archaebacteriums Thermoproteus tenax: Zusammensetzung und Structur [Ph.D. thesis]*, Ludwig-Maximillians-Universität, Munich, Germany, 1990.

[52] D. Janekovic, S. Wunderl, and I. Holz, "TTV1, TTV2 and TTV3, a family of viruses of the extremely thermophilic, anaerobic, sulfur reducing archaebacterium *Thermoproteus tenax*," *Molecular and General Genetics*, vol. 192, no. 1-2, pp. 39–45, 1983.

[53] M. Bettstetter, X. Peng, R. A. Garrett, and D. Prangishvili, "AFV1, a novel virus infecting hyperthermophilic *Archaea* of the genus *Acidianus*," *Virology*, vol. 315, no. 1, pp. 68–79, 2003.

[54] H. P. Arnold, W. Zillig, U. Ziese et al., "A novel lipothrixvirus, SIFV, of the extremely thermophilic crenarchaeon *Sulfolobus*," *Virology*, vol. 267, no. 2, pp. 252–266, 2000.

[55] C. Bath, T. Cukalac, K. Porter, and M. L. Dyall-Smith, "His1 and His2 are distantly related, spindle-shaped haloviruses belonging to the novel virus group, *Salterprovirus*," *Virology*, vol. 350, no. 1, pp. 228–239, 2006.

[56] N. S. Atanasova, E. Roine, A. Oren, D. H. Bamford, and H. M. Oksanen, "Global network of specific virus-host interactions in hypersaline environments," *Environmental Microbiology*, vol. 14, no. 2, pp. 426–440, 2012.

[57] G. Rice, L. Tang, K. Stedman et al., "The structure of a thermophilic archaeal virus shows a double-stranded DNA viral capsid type that spans all domains of life," *Proceedings of the National Academy of Sciences of the United States of America*, vol. 101, no. 20, pp. 7716–7720, 2004.

[58] L. J. Happonen, P. Redder, X. Peng, L. J. Reigstad, D. Prangishvili, and S. J. Butcher, "Familial relationships in hyperthermo- and acidophilic archaeal viruses," *Journal of Virology*, vol. 84, no. 9, pp. 4747–4754, 2010.

[59] X. Xiang, L. Chen, X. Huang, Y. Luo, Q. She, and L. Huang, "*Sulfolobus tengchongensis* spindle-shaped virus STSV1: virus-host interactions and genomic features," *Journal of Virology*, vol. 79, no. 14, pp. 8677–8686, 2005.

[60] H. Sagami, A. Kikuchi, K. Ogura, K. Fushihara, and T. Nishino, "Novel isoprenoid modified proteins in *Halobacteria*," *Biochemical and Biophysical Research Communications*, vol. 203, no. 2, pp. 972–978, 1994.

[61] H. Sagami, A. Kikuchi, and K. Ogura, "A novel type of protein modification by isoprenoid-derived materials. Diphytanylglycerylated proteins in *Halobacteria*," *The Journal of Biological Chemistry*, vol. 270, no. 25, pp. 14851–14854, 1995.

[62] Z. Konrad and J. Eichler, "Lipid modification of proteins in *Archaea*: attachment of a mevalonic acid-based lipid moiety to the surface-layer glycoprotein of *Haloferax volcanii* follows protein translocation," *Biochemical Journal*, vol. 366, no. 3, pp. 959–964, 2002.

[63] M. Häring, R. Rachel, X. Peng, R. A. Garrett, and D. Prangishvili, "Viral diversity in hot springs of Pozzuoli, Italy, and characterization of a unique archaeal virus, *Acidianus* bottle-shaped virus, from a new family, the *Ampullaviridae*," *Journal of Virology*, vol. 79, no. 15, pp. 9904–9911, 2005.

[64] S. Paula, A. G. Volkov, A. N. Van Hoek, T. H. Haines, and D. W. Deamer, "Permeation of protons, potassium ions, and small polar molecules through phospholipid bilayers as a function of membrane thickness," *Biophysical Journal*, vol. 70, no. 1, pp. 339–348, 1996.

[65] Y. Zhai, P. L. Chong, L. J. Taylor et al., "Physical properties of archaeal tetraether lipid membranes as revealed by differential scanning and pressure perturbation calorimetry, molecular acoustics, and neutron reflectometry: effects of pressure and cell growth temperature," *Langmuir*, vol. 28, no. 11, pp. 5211–5217, 2012.

[66] W. D. Reiter, W. Zillig, and P. Palm, "Archaebacterial viruses," *Advances in Virus Research*, vol. 34, pp. 143–188, 1988.

[67] D. Prangishvili, "Family *Fuselloviridae*," in *Virus Taxonomy, Ninth Report of the International Committee on Taxonomy of Viruses*, A. M. Q. King, M. J. Adams, E. B. Carstens, and E. J. Lefkowitz, Eds., pp. 183–186, Elsevier, Oxford, UK, 2011.

[68] D. Prangishvili, "Family *Lipothrixviridae*," in *Virus Taxonomy, Ninth Report of the International Committee on Taxonomy of Viruses*, A. M. Q. King, M. J. Adams, E. B. Carstens, and E. J. Lefkowitz, Eds., pp. 211–221, Elsevier, Oxford, UK, 2011.

[69] M. Häring, G. Vestergaard, K. Brügger, R. Rachel, R. A. Garrett, and D. Prangishvili, "Structure and genome organization of AFV2, a novel archaeal lipothrixvirus with unusual terminal and core structures," *Journal of Bacteriology*, vol. 187, no. 11, pp. 3855–3858, 2005.

[70] W. S. A. Maaty, A. C. Ortmann, M. Dlakić et al., "Characterization of the archaeal thermophile *Sulfolobus* turreted icosahedral virus validates an evolutionary link among double-stranded DNA viruses from all domains of life," *Journal of Virology*, vol. 80, no. 15, pp. 7625–7635, 2006.

[71] C. Y. Fu and J. E. Johnson, "Structure and cell biology of archaeal virus STIV," *Current Opinion in Virology*, vol. 2, no. 2, pp. 122–127, 2012.

[72] S. T. Jaakkola, R. K. Penttinen, S. T. Vilén et al., "Closely related archaeal *Haloarcula hispanica* icosahedral viruses HHIV-2 and SH1 have nonhomologous genes encoding host recognition functions," *Journal of Virology*, vol. 86, no. 9, pp. 4734–4742, 2012.

[73] E. Roine and H. M. Oksanen, "Viruses from the hypersaline environment," in *Halophiles and Hypersaline Environments: Current Research and Future Trends*, Ventosa, A. Oren, and Y. Ma, Eds., pp. 153–172, Springer, Berlin, Germany, 2011.

[74] J. J. B. Cockburn, N. G. A. Abrescia, J. M. Grimes et al., "Membrane structure and interactions with protein and DNA in bacteriophage PRD1," *Nature*, vol. 432, no. 7013, pp. 122–125, 2004.

[75] A. Senčilo, L. Paulin, S. Kellner, M. Helm, and E. Roine, "Related haloarchaeal pleomorphic viruses contain different genome types," *Nucleic Acids Research*, vol. 40, no. 12, pp. 5523–5534, 2012.

[76] L. Kandiba, O. Aitio, J. Helin et al., "Diversity in prokaryotic glycosylation: an archaeal-derived N-linked glycan contains legionaminic acid," *Molecular Microbiology*, vol. 84, no. 3, pp. 578–593, 2012.

[77] R. Montalvo-Rodríguez, R. H. Vreeland, A. Oren, M. Kessel, C. Betancourt, and J. López-Garriga, "*Halogeometricum borinquense* gen. nov., sp. nov., a novel halophilic archaeon from Puerto Rico," *International Journal of Systematic Bacteriology*, vol. 48, no. 4, pp. 1305–1312, 1998.

[78] I. R. Cooke and M. Deserno, "Coupling between lipid shape and membrane curvature," *Biophysical Journal*, vol. 91, no. 2, pp. 487–495, 2006.

[79] L. Adamian, H. Naveed, and J. Liang, "Lipid-binding surfaces of membrane proteins: evidence from evolutionary and structural analysis," *Biochimica et Biophysica Acta*, vol. 1808, no. 4, pp. 1092–1102, 2011.

[80] N. G. A. Abrescia, J. J. B. Cockburn, J. M. Grimes et al., "Insights into assembly from structural analysis of bacterio-phage PRD1," *Nature*, vol. 432, no. 7013, pp. 68–74, 2004.

[81] B. Tenchov, E. M. Vescio, G. D. Sprott, M. L. Zeidel, and J. C. Mathai, "Salt tolerance of archaeal extremely halophilic lipid membranes," *The Journal of Biological Chemistry*, vol. 281, no. 15, pp. 10016–10023, 2006.

14

New Strategy for a Suitable Fast Stabilization of the Biomethanization Performance

L. A. Fernández-Güelfo,[1] C. J. Álvarez-Gallego,[1]
D. Sales Márquez,[2] and L. I. Romero García[1]

[1] Department of Chemical Engineering and Food Technology, Faculty of Science, University of Cadiz,
  Cadiz, 11510 Puerto Real, Spain
[2] Department of Environmental Technologies, Faculty of Marine and Environmental Sciences, University of Cadiz,
  Cadiz, 11510 Puerto Real, Spain

Correspondence should be addressed to L. A. Fernández-Güelfo, alberto.fdezguelfo@uca.es

Academic Editor: Michael Hoppert

The start-up strategies for thermophilic anaerobic reactors usually consist of an initial mesophilic stage (35°C), with an approximate duration of 185 days, and a subsequent thermophilic stage (55°C), which normally requires around 60 days to achieve the system stabilizatio. During the first 8–10 days of the mesophilic stage, the reactor is not fed so that the inoculum, which is generally a mesophilic anaerobic sludge, may be adapted to the organic solid waste. Between mesophilic and thermophilic conditions the reactor is still not fed in an effort to prevent possible imbalances in the proces. As a consequence, the start-up and stabilization of the biomethanization performance described in the literature require, at least, around 245 days. In this sense, a new strategy for the start-up and stabilization phases is presented in this study. This approach allows an important reduction in the overall time necessary for these stages in an anaerobic continuous stirred tank reactor (CSTR) operated at thermophilic-dry conditions for treating the organic fraction of the municipal solid waste (OFMSW): 60 days versus 245 days of conventional strategies. The new strategy uses modified SEBAC technology to adapt an inoculum to the OFMSW and the operational conditions prior to seeding the CSTR.

1. Introduction

The organic fraction of municipal solid wastes (OFMSW) has been commonly treated by means of anaerobic digestion (AD) [1–3]. Among the main advantages of this biological process, the low energy consumption and sludge generation and the high hydrogen and/or methane productions must be highlighted; however, the main disadvantage is its slowness.

In order to avoid this important inconvenience and to accelerate the process with regard to the mesophilic-wet conditions (35°C, 5–10% total solids concentration), AD may be operated at thermophilic-dry (55°C, 30% total solids concentration) conditions. At these new conditions the hydrolysis phase is faster and more effective and, therefore, the overall rate of the process is improved. On the other hand, it must be noted that for the start-up and stabilization of the biomethanization process in continuous stirred tank reactor

(CSTR) for anaerobic biodegrading of OFMSW is necessary operational times extremely long.

As it is reported by the authors Bolzonella et al. [4] and Michaud et al. [5], the strategy generally employed to start-up and stabilization of thermophilic anaerobic digesters consists of two stages.

(1) A mesophilic stage (35°C) of about 185 days of operation. During the first 8–10 days the reactor is not fed in order to the inoculum (generally anaerobic mesophilic sewage sludge) may be adapted to the waste.

(2) A thermophilic stage (55°C) with an approximate duration of 60 days at least. In addition, during the mesophilic-thermophilic transition the reactor is not fed to prevent destabilization episodes.

Following this, the strategies reported in the literature have an approximate duration of about 245 days at least. This fact is mainly because the inoculum is not adapted to the type of waste and/or the operational conditions.

With specific reference to the full-scale industrial application of the AD processes, at first, because of the historical background, more reactors adopting the wet processes (<10% dry solids in the reactor) were applied; since then, dry digestion (more than 25% dry solids in the feed) has prevailed because of the reduced volume of reactors and wastewater production. Most applied technologies for dry processes are Dranco, Valorga, Linde, and Kompogas, all working in the range 30–40% of total solids in the reactor feeding [6].

About the conventional ranges of temperature of the AD processes, a total treatment capacity for solid waste organics, excluding the tonnage used for sewage sludge and manures, evolved from 122,000 ton per year in 1990 to 1,037,000 ton available or under construction by the last decade in 53 plants across Europe, an increase by 750%. Both mesophilic and thermophilic technologies have been proven, with about 38% of capacity being operated at thermophilic temperatures. All digestion plants were initially operated at mesophilic temperatures. The first thermophilic plants were dry fermentation plants and came online in 1992 and 1993. The capacity of mesophilic operation increased by 350,000 ton during 1994 through 1999, while thermophilic capacity increased by 280,000 ton or 70,000 ton and 56,000 ton per year, respectively. During some years, more mesophilic plants are added while during other years more thermophilic capacity is constructed. No clear trend can be observed. It can be expected that the increase will be level for both temperature ranges, even though more suppliers are starting to provide thermophilic digestion. Thermophilic operation was developed later but has been established as a reliable and accepted mode of fermentation. It provides the added benefit of treating the waste at higher temperatures and thereby increasing pathogen kill-off during the anaerobic phase. The added amount of heat does not seem to stop companies operating thermophilically, as higher gas production yields and rates are being claimed by various suppliers [7].

For all the above reasons, an inoculum adapted to the waste (OFMSW) and the operational conditions (thermophilic-dry AD) was obtained by mean of the modified sequencing batch anaerobic composting (SEBAC) technology. This technology and its modifications are fully detailed in the literature [8–11] and have been successfully employed to develop the AD of OFMSW with acceptable conversions in only 30 days.

The modification of the SEBAC technology used in this research is based on the interconnection of two anaerobic digesters (reactors A and B). Daily, reactors A and B are fed by means of the recirculation of the leachate generated from the other reactor. In this way the fresh organic waste (reactor A) to be digested is inoculated through recirculation of the leachate or effluent from the reactor containing the digested waste (reactor B), while the leachate generated by the reactor with fresh waste (reactor A) is recirculated to the reactor with the digested waste (reactor B). In this way a flow of

microorganisms is established to the undigested waste and of organic material to the digested waste [12].

Based on all the stated above and in order to reduce the long-time periods required for the start-up and stabilization of the biomethanization in a CSTR operated at thermophilic-dry conditions for treating OFMSW, two main goals may be defined in this work which are as follows.

(i) To obtain a suitable inoculum quickly by means of the modified SEBAC technology commented previously. This inoculum will be adapted to the OFMSW and the operational conditions typical of the thermophilic-dry AD (55°C and 30% total solids concentration).

(ii) To achieve stable biomethanization performance for the thermophilic-dry AD of OFMSW in a CSTR using the inoculum obtained by the modified SEBAC technology.

2. Materials and Methods

2.1. Modified SEBAC Technology. This system consists of the interconnection of two anaerobic 25 L-reactors (Figure 1), operating under thermophilic-dry conditions (55°C and 30% total solids concentration).

(i) Reactor A contains alternate layers of source selected OFMSW and pig manure with a total solids concentration of 30%. The pig manure accelerates the colonization of the OFMSW since it is a potential source of anaerobic microorganisms.

(ii) Reactor B generally contains a stabilized waste previously degraded by anaerobic digestion, with a high concentration of viable and active microorganisms [13]. In this study, anaerobic mesophilic sludge, from the anaerobic digesters of a full-scale plant for treating sewage sludge, was used.

Daily, the leachate produced in reactor A was exchanged to reactor B and an equivalent quantity of sludge from reactor B was added to reactor A. This procedure causes a flow of microorganisms from B to A and a flow of organic matter from reactor A to B. Reactors do not require agitation and the time required for effective start-up is around 30 days. The composition of the modified SEBAC reactors is shown in Table 1.

2.2. Continuous Stirred Tank Reactor (CSTR). The CSTR was initially loaded with 1.5 kg of milled and dry synthetic OFMSW (90% in total solids concentration). The moisture was adjusted using the inoculum obtained by means of the modified SEBAC technology described above. Concretely, the inoculums consisted of a 1 : 1 v/v mixture [14–16] of thermophilic sludge and leachate. In this sense, 4 litres of inoculum (2 litres of sludge + 2 litres of leachate) were required to add moisture to the synthetic OFMSW.

The compositions of the different wastes used in this study are given in Table 2. It must be noted that the synthetic OFMSW was prepared based on the nutritional requirements of the main populations of microorganisms involved in the

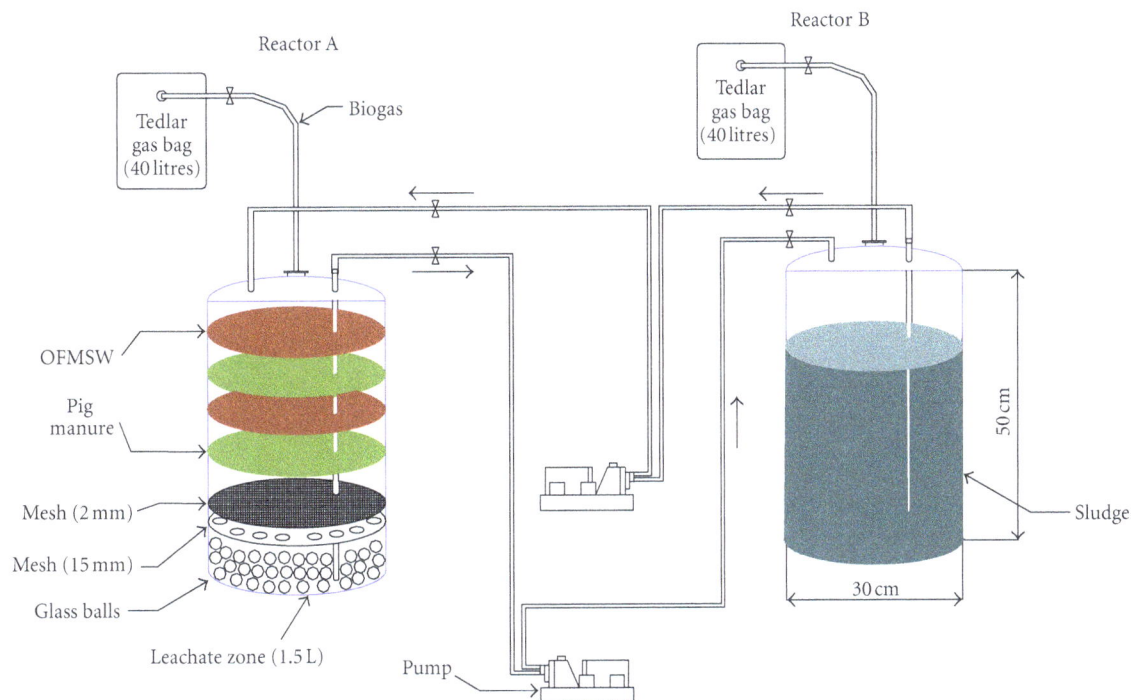

FIGURE 1: Flow of leachate in the modified SEBAC reactors.

TABLE 1: Composition of the modified SEBAC reactors.

	Wastes	Layers	Weight/layer (kg)
Reactor A	OFMSW	2	1
	Pig manure	2	1.5
Reactor B	Sludge	—	21

AD [17]. This type of feed avoids the problem of high variations in the composition of the source selected OFMSW. This aspect is important in order to determine an accurate efficiency for the process.

2.3. Processing of the OFMSW. Control of the total solid concentration of the feed is necessary to obtain a suitable level of performance for the dry AD. Therefore, pretreatment of the OFMSW samples was necessary to adjust them to the required optimum values. In this study, the samples were dried at 55°C for 48 hours and then at ambient temperature for 72 hours until final moisture content of 10% was achieved. The dried OFMSW was milled until a particle size of approximately 1 cm was obtained and, finally, the moisture was adjusted to 70–75% (25–30% in total solids concentration, which is characteristic of dry AD) with tap water, leachate from garbage, sludge, or combinations of these.

2.4. Analytical Techniques. For the control of the reactors, the following parameters were determined: the volume and composition of the biogas (H_2, O_2, N_2, CH_4, and CO_2), volatile fatty acids (VFA), total solids (TS), suspended total solids (STS), total volatile solids (TVS), suspended volatile solid (SVS), alkalinity, pH, dissolved organic carbon (DOC), ammonium, chemical oxygen demand (COD), and density. The analytical techniques were performed according to procedures described by Álvarez-Gallego [18].

3. Results and Discussion

3.1. Inoculum Preparation through Modified SEBAC Technology. The daily volume of leachate exchanged between the two modified SEBAC reactors must be between 5 and 10% of the initial volume of OFMSW for digestion [8]. For this reason, Reactor A controls the start-up phase, which is the *rate limiting step* of the process [19]. The theoretical calculation indicates that the volume of leachate that should be exchanged is approximately 600 mL. From the fourth day of operation until the conclusion of the experiment (hundredth day), the flow of leachate between the two reactors was maintained at 600 mL.

The minimum time required to obtain a suitable inoculum through modified SEBAC technology may be determined from the accumulated methane production curves generated in reactors A and B. As can be seen in Figure 2, the curves of both reactors present the maximum slope in 30 days of operation, more pronounced in reactor B (sludge). This fact indicates an exponential growth of the methanogenic Archaea in the system and, therefore, if the inoculum is taken in this moment, it will present a high methanogenic activity. Thus, the sludge from day 30 can be considered as a viable inoculum for the biomethanization of OFMSW at thermophilic-dry conditions. Finally, the initial and final compositions of wastes at the end of the assay are shown in Table 3.

TABLE 2: Composition of the wastes used to start-up the CSTR.

Parameter	Leachate inoculum	Sewage sludge inoculum	OFMSW	OFMSW/Inoculum mixture
pH	8.62	8.35	7.78	8.70
Density (kg/m^3)	980	985	750	1116
Alkalinity ($gCaCO_3/L$)	21.78	16.54	4.29	5.14
Ammonium (gNH_3-N/L)	26.88	14.56	1.68	2.8
Total Nitrogen	25.66 gNH_3-N/L	21.46 gNH_3-N/L	207.2 gNH_3-N/kg	72.8 gNH_3-N/kg
gTSS/L	14.46	20.46	—	—
gVSS/L	10.73	9.16	—	—
gTS/g sample	—	—	0.90	0.31
gTVS/g sample	—	—	0.71	0.25
Total carbon (mg/g)	80.78	35.27	112.6	65.07
Total inorganic carbon (mg/g)	2.07	0.96	0.29	0.30
Total organic carbon (mg/g)	78.41	34.31	112.3	64.75
Acidity (mgAcH/L)	12403	17353	1440	356

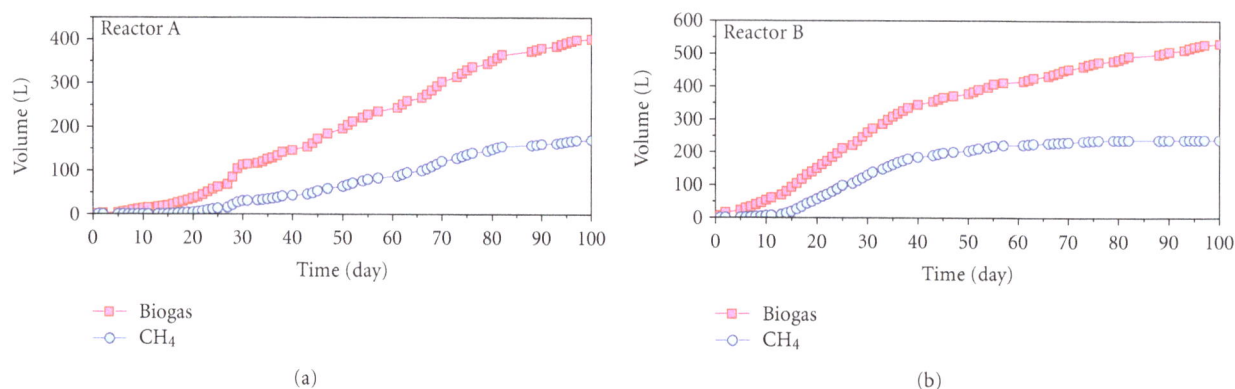

FIGURE 2: Accumulation of biogas and methane productions in the modified SEBAC reactors.

3.2. Start-Up and Stabilization of the Biomethanization Process.

A continuously stirred tank reactor (CSTR) was started-up under thermophilic-dry conditions and a series of four solid retention times (SRT) was carried out in order to study the effect of the added organic loading rate (OLR_0) on the biomethanization performance at semicontinuous regime of feeding.

Along the four consecutive stages, the OLR_0 (expressed as mgDOC/L·d and mgTVS/L·d) was increased and it was maintained constant in each SRT. The SRT, OLR_0, and operation time data for each stage are shown in Table 4.

In the first stage the OLR imposed was relatively low (0.704 gDOC/L·day) in order to check if the system evolved appropriately. The results obtained in the first 14 days were favourable and, therefore, the OLR was increased to 0.805 gDOC/L·day. The OLR_0 used at stage 1 was different to the values reported in the literature. Bolzonella et al. [4] carried out start-up phase studies in the mesophilic range with an extremely low OLR—less than 0.16 gDOC/L·day— for approximately 40 days. It must be highlighted that, in this study, the start-up phase was carried out using the SEBAC inoculum, which had been previously adapted to the waste and operational conditions. This fact allows that the reactor may be operated at higher OLR.

3.2.1. Study of the Gas Productions.

As can be seen from Figure 3, the biogas generated in stage 1 is not useful and this stage may be considered as a latency period in which the hydrolysis and colonization of the waste takes place. During stage 2, the specific methane yield reaches its maximum average values of 1.11 LCH_4/gDOC degraded and 0.51 LCH_4/gTVS degraded due to the biodegradation of the VFA accumulated in the previous stage. Finally, during stages 3 and 4, the methane yield coefficient stays constant at around 0.91 LCH_4/gDOC and 0.1 LCH_4/gTVS respectively, indicating stable biomethanization performance in the system.

On the other hand, the average specific methane yield in term of COD reaches the value of 0.42 LCH_4/gCOD in stage 2 and 0.34 LCH_4/gCOD in stages 3 and 4. However, in stage 1 the average specific methane yield is practically zero, 0.01 LCH_4/gCOD. In accordance with Bushwell and Mueller [20], the stoichiometric value for methane generation is 0.35 LCH_4/gCOD, which indicates that in stages 3 and 4 the reactor is working with a methane yield coefficient very close to the theoretical maximum. However, the value obtained for the 35-day SRT (stage 2) is higher than the theoretical maximum. This discrepancy is due to the fact that in this SRT, in addition to the degradation of the OLR added, the

TABLE 3: Initial and final compositions of the wastes.

Parameters	OFMSW		Pig manure		Sludge	
	Initial	Final	Initial	Final	Initial	Final
Density (kg/m³)	600	850	1200	1000	900	1000
Total solids (g/kg)	878	173	586	80	42	26.6
Total volatile solids (g/kg)	700.4	85	464.1	60	15	2.6
Suspended total solids (g/L)	0.5	7.9	3.9	5	20.2	11.4
Suspended volatile solids (g/L)	3.6	6.9	3.5	3.4	7.7	7.4
pH	0.2	8.1	7.1	8.4	8.3	8.35
Alkalinity (gCaCO₃/L)	7.6	8.4	50	75.4	20.1	16.5
Chemical oxygen demand (mgO₂/L)	112000	41558	14814	6509	10527	25526

TABLE 4: Initial organic loading rate (OLR$_0$) for each SRT.

Stage	SRT (day)	Operation time (day)	OLR$_0$ gDOC/L·day	gTVS/L·day
1	40	14	0.704	4.42
2	35	17	0.805	5.07
3	30	25	0.940	5.92
4	25	50	1.123	7.50

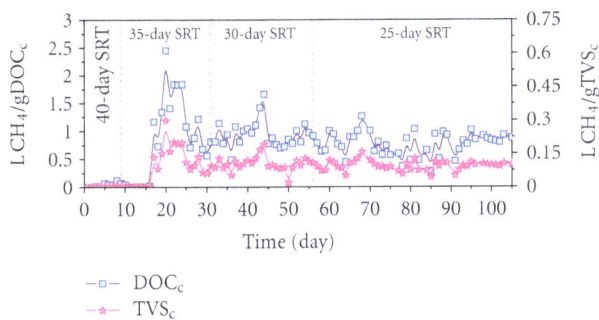

FIGURE 3: Evolution of the specific methane yield expressed as LCH$_4$/gDOC$_c$ and LCH$_4$/gTVS$_c$.

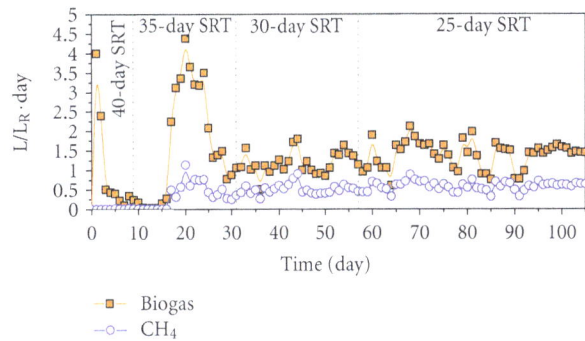

FIGURE 4: Daily biogas and methane productions expressed as L/L$_{Reactor}$·day.

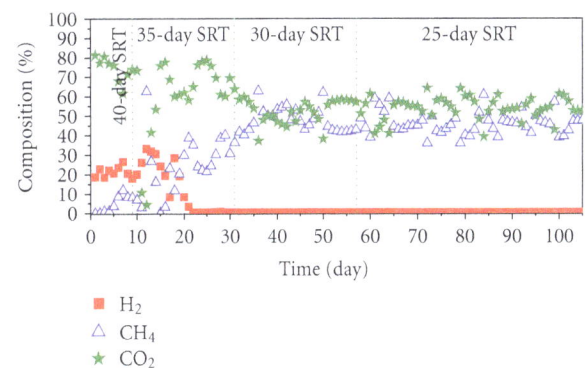

FIGURE 5: Biogas composition expressed as percentage.

transformation of the organic matter accumulated in the system during the previous 40-day SRT takes place. As a consequence, the methane yield coefficient obtained is higher than its theoretical value.

As far as the daily biogas generation is concerned, during the first 3 days of stage 1 a significant level of production was observed due to the hydrolysis of the waste (Figure 4). The composition observed in this period is usual for hydrolytic phase: H$_2$ (20%) and CO$_2$ (80%), see Figure 5.

During the hydrolysis phase complex molecules are transformed into other simpler products, without methane production. For this reason, in Figure 5, stage 1 has been considered as a latency phase.

However, in stage 2, H$_2$ levels drop to zero due to the methanogenic activity while CH$_4$ and CO$_2$ converge at around 50%, which is typical behaviour of a stable biomethanization process. In this phase, the intermediate products generated in the hydrolysis are converted into CO$_2$ and CH$_4$ by the methanogenic Archaea and as consequence the daily

average biogas production reaches its maximum average value of 1.834 L$_{Biogas}$/L$_{Reactor}$·day an average stable methane production of 0.55 LCH$_4$/L$_{Reactor}$·day after 15 days of operation time. About the last value, 0.55 LCH$_4$/L$_{Reactor}$·day, it is higher than the result reported by Fernández et al. [21] in their dry AD studies of OFMSW using the same technology (CSTR) with similar SRTs. In that work, sewage sludge anaerobically digested at mesophilic regime of temperature coming from a full-scale WWTP was used as inoculums to start-up a CSTR. In that case, the system reaches a stable methane production of 0.48 LCH$_4$/L$_R$·day after 25 days of operation time and at 30-day SRT. In this work, the system reaches a stable methane production of 0.55 LCH$_4$/L$_R$·day after 15 days of operation time and at

35-day SRT (very positive result since higher SRTs are associated with low methane yields). Therefore, the start-up period is decreased about 40% with this strategy and, in addition, the methane production is improved around 15% although the SRT is higher (35 versus 30 days), and, therefore, the methane production should be lower.

In stage 3 the daily average biogas production decreases to $1.138 \, L_{Biogas}/L_{Reactor} \cdot day$ since most of the initial waste which was loaded into the reactor has been degraded. Nevertheless, the composition of the biogas is stabilized with values of CO_2 and CH_4 at around 50%, which indicates that a balance between the different microbial populations involved in the digestion has been reached in the reactor.

Finally, in stage 4 the daily average biogas production increase to $1.768 \, L_{Biogas}/L_{Reactor} \cdot d$ due to the OLR_0 increased, with a composition in CO_2 and CH_4 of 55 and 45%, respectively (Figure 5).

3.2.2. VFA Evolution. As can be seen in Figure 6, from day 90 at 35-day SRT, the total VFA and butyric and acetic acids concentrations reach very stable values of around 8000, 2500, and 400 mg/L, respectively. It must be noted that in this specific period (from 90 to 105 days), the stabilization of VFA concentrations matches with a stable specific methane yield, expressed in terms of DOC and TVS (Figure 3), and with a stable biogas and methane productions (Figure 4). In addition, removal percentages of 56% for TS, 89% for TVS, and 63% for DOC were observed. These values have been compared with literature values [8, 22] and they confirm that the biomethanization operates efficiently at stable conditions. Hence, it is possible to reach stable biomethanization performance in SRT that is appropriate for full-scale plants (25 days) in approximately 90 days.

4. Conclusions

As a general conclusion, a successful strategy for the start-up and stabilization phase of the biomethanization process of OFMSW in a CSTR operated at thermophilic-dry conditions has been developed. The new strategy allows stable operation in a reduced time versus other literature protocols. Taking into account the above main conclusion, the following specific conclusions may be established.

(1) In the first stage, a thermophilic anaerobic inoculum adapted to the OFMSW must be obtained by means of the modified SEBAC technology. This inoculum is used in the second stage to inoculate the CSTR. The semicontinuous reactor must be subsequently fed with milled OFMSW in a high SRT (40 days). When the system is stabilized, the SRT imposed can be progressively diminished until reaching 25-day SRT.

(2) The results obtained from the modified SEBAC reactors indicate that an incubation period of approximately 30 days is necessary to obtain an appropriate inoculum. From the day 30 of operation the system reaches a high biogas production with a high methane percentage.

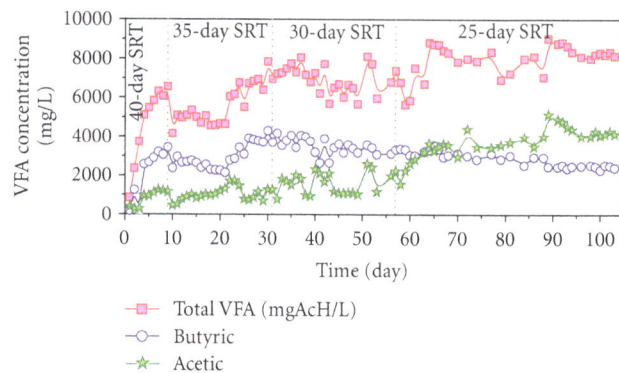

Figure 6: Evolution of total VFA, butyric, and acetic acids.

(3) The semicontinuous reactor can be inoculated with a 1 : 1 mixture of the effluents (leachate of OFMSW and sludge) from the modified SEBAC reactors and a period for the acclimatization of microorganisms is not required prior to feeding the system.

(4) A high retention time must be initially imposed (40 days) to avoid irreversible distortions in the process. The SRT is subsequently reduced progressively, which is associated with an increase in the OLR_0, until the required operational conditions are reached. The stabilization of the system, for an SRT of 35 days (OLR_0 of 0.805 gDOC/L·d), requires 30 days of operation. Under these conditions the maximum average methane productions is reached, $1.834 \, L_{Biogas}/L_{Reactor} \cdot day$.

(5) For a successful start-up of the system, it is necessary a period of 60 days at least (30 days to obtain the inoculum and 30 days to stabilize the system at 35-day SRT. In addition, it is possible to reach stable operation in times that are appropriate for industrial operation (25-day SRT) in approximately 90 days. These data compare favourably with literature results for similar reactors, where the start-up periods are higher than 245 days.

In summary, the preparation of an inoculum adapted to the solid waste and operational conditions by means of modified SEBAC procedure enables us to reduce the time necessary to start-up and stabilization of a CSTR for the thermophilic-dry AD of OFMSW by a factor of four (60 days versus 245) with respect to conventional strategies reported in the literature.

Acknowledgments

This work was supported by the Ministerio de Ciencia e Innovación of Spain (Project CTM2010-17654), the Consejería de Innovación, Ciencia y Empresa of the Junta de Andalucía, Spain (Project P07-TEP-02472), the European Regional Development Fund (ERDF), and the Ministerio de Educación y Ciencia of Spain (Project NovEDAR_Consolider CSD2007-00055).

References

[1] K. F. Fannin, J. R. Conrad, V. J. Srivastava, D. E. Jerger, and D. P. Chynoweth, "Anaerobic processes," *Journal of the Water Pollution Control Federation*, vol. 55, no. 6, pp. 623–632, 1983.

[2] T. Akao, E. Mizuki, H. Saito, S. Okumura, and S. Murao, "The methane fermentation of *Citrus unshu* peel pretreated with fungus enzymes," *Bioresource Technology*, vol. 41, no. 1, pp. 35–39, 1992.

[3] K. K. Moorhead and R. A. Nordstedt, "Batch anaerobic digestion of water hyacinth: effects of particle size, plant nitrogen content, and inoculum volume," *Bioresource Technology*, vol. 44, no. 1, pp. 71–76, 1993.

[4] D. Bolzonella, L. Innocenti, P. Pavan, P. Traverso, and F. Cecchi, "Semi-dry thermophilic anaerobic digestion of the organic fraction of municipal solid waste: focusing on the start-up phase," *Bioresource Technology*, vol. 86, no. 2, pp. 123–129, 2003.

[5] S. Michaud, N. Bernet, P. Buffière, M. Roustan, and R. Moletta, "Methane yield as a monitoring parameter for the start-up of anaerobic fixed film reactors," *Water Research*, vol. 36, no. 5, pp. 1385–1391, 2002.

[6] D. Bolzonella, P. Pavan, S. Mace, and F. Cecchi, "Dry anaerobic digestion of differently sorted organic municipal solid waste: a full-scale experience," *Water Science and Technology*, vol. 53, no. 8, pp. 23–32, 2006.

[7] L. de Baere, "Anaerobic digestion of solid waste: state-of-the-art," *Water Science and Technology*, vol. 41, no. 3, pp. 283–290, 2000.

[8] T. Forster-Carneiro, L. A. Fernández, M. Pérez, L. I. Romero, and C. J. Álvarez, "Optimization of sebac start-up phase of municipal solid waste anaerobic digestion," *Chemical and Biochemical Engineering Quarterly*, vol. 18, no. 4, pp. 429–439, 2004.

[9] T. Forster-Carneiro, M. Pérez, L. I. Romero, and D. Sales, "Dry-thermophilic anaerobic digestion of organic fraction of the municipal solid waste: focusing on the inoculum sources," *Bioresource Technology*, vol. 98, no. 17, pp. 3195–3203, 2007.

[10] D. P. Chynoweth and R. LeGrand, "Apparatus and method for sequential batch anaerobic composting of high–solids organics feedstocks," Unites States Patent Number 5269634, University of Florida, 1993.

[11] D. P. Chynoweth, G. Bosch, J. F. K. Earle, R. Legrand, and K. Liu, "A novel process for anaerobic composting of municipal solid waste," *Applied Biochemistry and Biotechnology*, vol. 28-29, no. 1, pp. 421–432, 1991.

[12] L. T. Angenent, S. Sung, and L. Raskin, "Methanogenic population dynamics during start-up of a full-scale anaerobic sequencing batch reactor treating swine waste," *Water Research*, vol. 36, no. 18, pp. 4648–4654, 2002.

[13] S. Chugh, D. P. Chynoweth, W. Clarke, P. Pullammanappallil, and V. Rudolph, "Degradation of unsorted municipal solid waste by a leach-bed process," *Bioresource Technology*, vol. 69, no. 2, pp. 103–115, 1999.

[14] M. Kim, Y. H. Ahn, and R. E. Speece, "Comparative process stability and efficiency of anaerobic digestion; mesophilic versus thermophilic," *Water Research*, vol. 36, no. 17, pp. 4369–4385, 2002.

[15] C. Y. Lin and Y. S. Lee, "Effect of thermal and chemical pretreatments on anaerobic ammonium removal in treating septage using the UASB system," *Bioresource Technology*, vol. 83, no. 3, pp. 259–261, 2002.

[16] Q. Wang, M. Kuninobu, H. I. Ogawa, and Y. Kato, "Degradation of volatile fatty acids in highly efficient anaerobic digestion," *Biomass and Bioenergy*, vol. 16, no. 6, pp. 407–416, 1999.

[17] D. J. Martin, L. G. A. Potts, and A. Reeves, "Small-scale simulation of waste degradation in landfills," *Biotechnology Letters*, vol. 19, no. 7, pp. 683–685, 1997.

[18] C. J. Álvarez-Gallego, *Ensayo de diferentes procedimientos para el arranque de un proceso de co-digestión anaerobia seca de OFMSW and lodos de depuradora en rango termofílico [Ph.D. thesis]*, Universidad de Cádiz, 2005.

[19] T. E. Lai, A. Nopharatana, P. C. Pullammanappallil, and W. P. Clarke, "Cellulolytic activity in leachate during leach-bed anaerobic digestion of municipal solid waste," *Bioresource Technology*, vol. 80, no. 3, pp. 205–210, 2001.

[20] A. M. Buswell and H. F. Mueller, "Mechanism of methane fermentation," *Industrial and Engineering Chemistry*, vol. 44, no. 3, pp. 550–552, 1952.

[21] J. Fernández, M. Pérez, and L. I. Romero, "Effect of substrate concentration on dry mesophilic anaerobic digestion of organic fraction of municipal solid waste (OFMSW)," *Bioresource Technology*, vol. 99, no. 14, pp. 6075–6080, 2008.

[22] A. Davidsson, C. Gruvberger, T. H. Christensen, T. L. Hansen, and J. L. C. Jansen, "Methane yield in source-sorted organic fraction of municipal solid waste," *Waste Management*, vol. 27, no. 3, pp. 406–414, 2007.

15

Archaeal Phospholipid Biosynthetic Pathway Reconstructed in *Escherichia coli*

Takeru Yokoi, Keisuke Isobe, Tohru Yoshimura, and Hisashi Hemmi

Department of Applied Molecular Bioscience, Graduate School of Bioagricultural Sciences, Nagoya University, Furo-cho, Chikusa-ku, Nagoya, Aichi 460-8601, Japan

Correspondence should be addressed to Hisashi Hemmi, hhemmi@agr.nagoya-u.ac.jp

Academic Editor: Yosuke Koga

A part of the biosynthetic pathway of archaeal membrane lipids, comprised of 4 archaeal enzymes, was reconstructed in the cells of *Escherichia coli*. The genes of the enzymes were cloned from a mesophilic methanogen, *Methanosarcina acetivorans*, and the activity of each enzyme was confirmed using recombinant proteins. *In vitro* radioassay showed that the 4 enzymes are sufficient to synthesize an intermediate of archaeal membrane lipid biosynthesis, that is, 2,3-di-*O*-geranylgeranyl-*sn*-glycerol-1-phosphate, from precursors that can be produced endogenously in *E. coli*. Introduction of the 4 genes into *E. coli* resulted in the production of archaeal-type lipids. Detailed liquid chromatography/electron spray ionization-mass spectrometry analyses showed that they are metabolites from the expected intermediate, that is, 2,3-di-*O*-geranylgeranyl-*sn*-glycerol and 2,3-di-*O*-geranylgeranyl-*sn*-glycerol-1-phosphoglycerol. The metabolic processes, that is, dephosphorylation and glycerol modification, are likely catalyzed by endogenous enzymes of *E. coli*.

1. Introduction

Archaeal membrane lipids are very specific to the organisms in the domain Archaea and have structures that are distinct from those of bacterial/eukaryotic lipids [1–3]. Although they are, essentially, analogues of glycerolipids from bacteria or eukaryotes, they have specific structural features as follows: (1) hydrocarbon chains of archaeal lipids are multiply-branched isoprenoids typically derived from (all-*E*) geranylgeranyl diphosphate (GGPP), while linear acyl groups are general in bacterial/eukaryotic lipids; (2) the isoprenoid chains are linked with the glycerol moiety with ether bonds, while ester bonds are general in bacterial/eukaryotic lipids; (3) the glycerol moiety of archaeal lipids is derived from *sn*-glycerol-1-phosphate (G-1-P), which is the enantiomer of *sn*-glycerol-3-phosphate, the precursor for bacterial/eukaryotic glycerolipids; (4) dimerization of membrane lipids by the formation of carbon-carbon bonds between the ω-terminals of hydrocarbon chains, which generates macrocyclic structures such as caldarchaeol-type lipids with a typically 72-membered ring, is often observed in thermophilic and methanogenic archaea.

These characteristics affect the properties of membranes formed with the lipids. In general, the permeability of membranes composed of archaeal lipids is lower than that of membranes that consist of bacterial/eukaryotic lipids [4, 5]. Moreover, the structural differences between archaeal and bacterial/eukaryotic lipids are believed to cause their black-and-white distribution between these domains without exception (the "lipid divide") [6, 7]. This hypothesis is based on the idea that a membrane composed of both archaeal- and bacterial/eukaryotic-type lipids is disadvantageous to the organism, compared with membranes composed of one type. Although this hypothesis is attractive, no proof of it has been reported so far. To obtain proof of this hypothesis, two lines of experiments can be designed. One is to compare the physical properties of artificial membranes prepared with the archaeal- and/or bacterial/eukaryotic-type lipids. A few studies of this type have been done [8, 9]. Shimada and Yamagishi [9] recently reported that hybrid liposomes constructed from both archaeal- and bacterial-type lipids were generally more stable (impermeable) than they had expected. Based on these results, they concluded that the common ancestor of life (and the origin of eukaryotes supposedly formed by the fusion of archaea- and bacteria-like cells) might have had such hybrid lipid membranes, but they did not explain how the lipid divide occurred. The other, more straightforward line of experiments is to generate an organism that synthesizes both archaeal- and bacterial/eukaryotic-type membrane lipids and therefore has hybrid membranes. If the hybrid membranes are disadvantageous for the organism because of properties such as stability, permeability, and fluidity, the organism may lower viability or become susceptible to stresses such as heat and osmotic shock. Lai et al. recently reported the construction of such an organism, although they did not determine its phenotypes [10]. In their study, phospholipid biosynthetic genes from a hyperthermophilic archaeon *Archaeoglobus fulgidus* were introduced into *Escherichia coli*. The authors demonstrated the synthesis of precursors for archaeal membrane lipids, that is, 3-*O*-geranylgeranyl-*sn*-glycerol-1-phosphate (GGGP) and 2,3-di-*O*-geranylgeranyl-*sn*-glycerol-1-phosphate (DGGGP), in the recombinant *E. coli*, based on the detection of corresponding alcohols from the lipid extract from the cells after phosphatase treatment. However, it was still unclear whether the archaeal-type lipids produced in the cells actually acted as the structural components of a membrane bilayer, because the authors did not show the intact structures of the lipids. Moreover, they did not describe the level of production of the archaeal-type lipids, which is also important to the evaluation of their effects on the membranes of *E. coli*.

In the present study, we reconstructed a part of the biosynthetic pathway of an archaeal phospholipid (Figure 1), which consisted of G-1-P dehydrogenase, GGPP synthase, GGGP synthase, and DGGGP synthase from a mesophilic methanogenic archaeon, *Methanosarcina acetivorans*, in *E. coli*. These enzymes can synthesize DGGGP from the endogenous precursors of isoprenoid in *E. coli*, that is, (all-*E*) farnesyl diphosphate (FPP), isopentenyl diphosphate (IPP), and dihydroxyacetone phosphate (DHAP). In addition, the enzymes from the mesophile were expected to have optimal activities at the growth temperature of *E. coli*, which would lead to high-level production of the archaeal phospholipid precursor and its derivatives. We evaluated the total amount and intact structures of the archaeal-type lipids extracted from the cells by liquid chromatography/electron spray ionization-mass spectrometry (LC/ESI-MS) analysis and showed that DGGGP was metabolized by enzymes endogenous to *E. coli*.

2. Materials and Methods

2.1. Materials. LKC-18F precoated, reversed-phase, thin-layer chromatography plates were purchased from Whatman, UK. FPP was donated by Drs. Kyozo Ogura and Tanetoshi Koyama, Tohoku University. [1-^{14}C]IPP was purchased from American Radiolabeled Chemicals, USA. All other chemicals were of analytical grade.

2.2. General Procedures. Restriction enzyme digestions, transformations, and other standard molecular biological

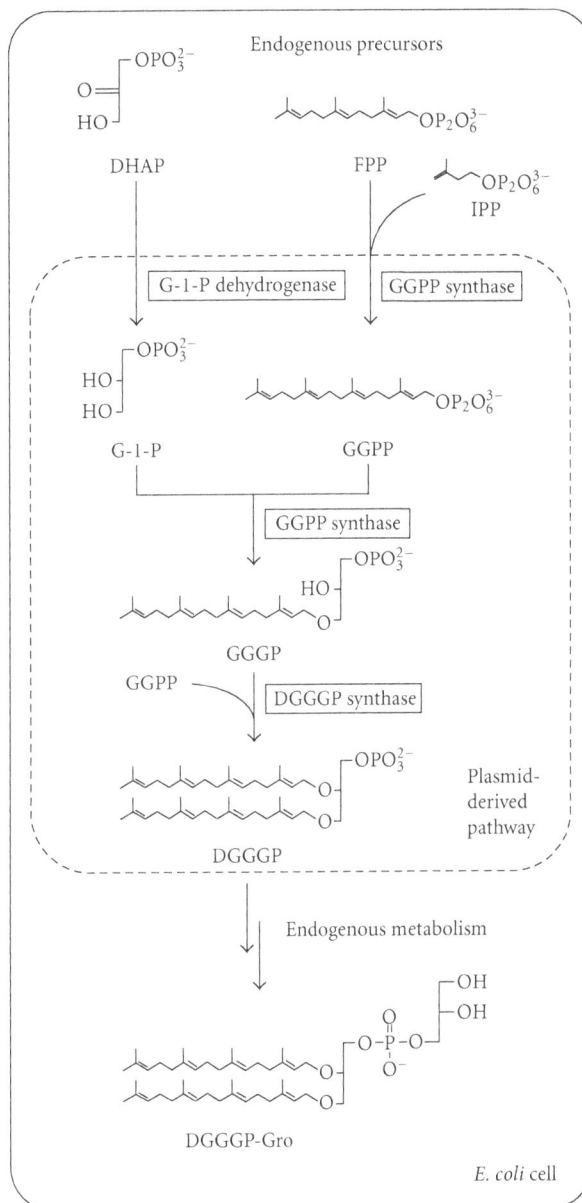

FIGURE 1: Biosynthetic pathway of archaeal-type lipids reconstructed in *E. coli*.

techniques were carried out as described by Sambrook et al. [11].

2.3. Cultivation of M. acetivorans. The *M. acetivorans* C2A (JCM 12185) strain was provided by the Japan Collection of Microorganisms (JCM), RIKEN BRC, through the Natural Bio-Resource Project of the MEXT, Japan. The mesophilic methanogenic archaeon was cultivated in a JCM 385 *Methanosarcina acetivorans* medium at 37°C and harvested at the log phase.

2.4. Construction of Plasmids Containing Phospholipid Biosynthetic Genes from M. acetivorans. The genome of *M. acetivorans* was extracted from the cells using a DNA extraction kit,

TABLE 1: Primers used for plasmid construction.

Primers	Sequences (restriction enzymes that recognize the underlined sites)
For the construction of pBAD-MA0606	
ma0606fw	GTAAAGAATTCAGATATAAGGAAATAGATGTGATGCTTATGATGCTTAT (EcoRI)
ma0606rv	GGTATTTCTAGATTGTATCCTTATTTTTCAGTATTCCCTTGCAATCA (XbaI)
For the construction of pBAD-MA3969	
ma3969fw	TTATATAGCTAGCTATTAAAAATAAGGATAATTAATGCAGGTGGAAGCACACCT (NheI)
ma3969rv	GAAATGTCGACGATATATCTCCTTTTATTTTTAGCTTTTTATAGCTGATA (SalI)
For the construction of pBAD-MA0961	
ma0961fw	ATAGAATTCAAGAAGATTATAATGTCTGCCGGAATAC (EcoRI)
ma0961rv	GATTCTAGATCATACACCGGCAATGAAAG (XbaI)
For the construction of pBAD-MA3686	
ma3686fw	CTATTGAGCTCAAATAAAAGGAGATATATCATGAAATTGACCATCAATA (SacI)
ma3686rv	ATATTGGTACCATCTATTTCCTTATATCTTCAACTTATGACCTTTGTGA (KpnI)
For the construction of pBAD-ALB2 by amplification of ma0961	
alb2fw	ACAATCTAGAGTCGAAGGAAGATTATAATGTCTGCCGGAATAC
alb2rv	ATGCCTGCAGGTCGACTCATACACCGGCAATGAAAG (SalI)
For the construction of pBAD-ALB3 by amplification of ma3969	
alb3fw	CGGTGTATGAGTCGAAAGGAGTAATTAATGCAGGTGGAAGCACACCT
alb3rv	ATGCCTGCAGGTCGACTTAGCTTTTTATAGCTGATA (SalI)
For the construction of pBAD-ALB4 by amplification of ma3686	
alb4fw	AAAAAGCTAAGTCGAAAGGAGATATATCATGAAATTGACCATCAATA
alb4rv	ATGCCTGCAGGTCGACTCAACTTATGACCTTTGTGA (SalI)

ISOPLANT II (Nippon Gene). Each of the hypothetical genes for archaeal phospholipid biosynthesis, that is, *MA3686*, *MA0606*, *MA3969*, and *MA0961*, was amplified using the primers shown in Table 1, using the genome of *M. acetivorans* as a template, and using KOD DNA polymerase (Toyobo, Japan). The amplified DNA fragment was digested by restriction enzymes that recognize the sites in the primers and then inserted into the pBAD18 vector cut with the same restriction enzymes to construct the plasmid for expression of each archaeal enzyme, that is, pBAD-MA3686, pBAD-MA0606, pBAD-MA3969, and pBAD-MA0961.

For the construction of plasmids for expression of multiple archaeal genes, an In-Fusion Advantage PCR cloning kit (Takara, Japan) was used according to the manufacturer's instructions. The *MA0961* gene was amplified using the primers shown in Table 1 and pBAD-MA0961 as a template. By the action of the In-Fusion enzyme, the amplified fragment was inserted into the plasmid pBAD-MA0606, which had been digested with *Sal*I, to construct the plasmid pBAD-ALB2. Next, the *MA3969* gene, which was amplified using the primers in Table 1 and pBAD-MA3969 as a template, was inserted into pBAD-ALB2 digested with *Sal*I to construct pBAD-ALB3. The plasmid was then digested with *Sal*I, and the *MA3686* gene, amplified using the primers in Table 1 and pBAD-MA3686 as a template, was inserted to construct pBAD-ALB4.

2.5. Recombinant Expression of the Archaeal Enzymes. *E. coli* Top10, transformed with each plasmid containing a homologous gene for archaeal phospholipid biosynthesis, that is, pBAD-MA0606, pBAD-MA3969, pBAD-MA0961, pBAD-MA3686, or pBAD-ALB4, was cultivated at 37°C in 250 mL LB medium supplemented with 100 mg/L ampicillin. When the optical density at 660 nm of the culture reached 0.5, then 0.02% of L-arabinose was added for induction. After an additional 16 h incubation, the cells were harvested and disrupted by sonication in 5 mL of 100 mM 3-(N-morpholino)propanesulfonic acid (MOPS)-NaOH buffer, pH 7.0. The homogenates were centrifuged at 24,000 g for 30 min to recover the supernatants as a crude extract, which was used for enzyme assay.

2.6. In Vitro Assay for the Biosynthesis of Phospholipid Precursors. The assay mixture for prenyltransferases contained, in a final volume of $200\,\mu$L 0.2 nmol of [1-^{14}C]IPP (2.04 GBq/mmol), 1 nmol of FPP, $2.0\,\mu$mol of $MgCl_2$, $20\,\mu$mol of MOPS-NaOH, pH 7.0, and suitable volumes of the crude extracts from *E. coli* containing pBAD-MA0606, pBAD-MA3969, or pBAD-MA0961. α-Glycerophosphate (racemic mixture) was added only to the mixtures containing MA3969.

The assay mixture for G-1-P dehydrogenase contained, in a final volume of $200 \mu L$, 0.2 nmol of $[1-^{14}C]IPP$ (2.04 GBq/mmol), 1 nmol of FPP, $2.0 \mu mol$ of $MgCl_2$, $20 \mu mol$ of MOPS-NaOH, pH 7.0, and suitable volumes of the crude extracts from *E. coli* containing pBAD-MA0606, pBAD-MA3969, or pBAD-MA3686. If needed, 200 nmol of α-glycerophosphate or DHAP was added to the mixture.

In a final volume of $200 \mu L$, the assay mixture for the 4 archaeal enzymes simultaneously expressed in *E. coli* contained, 0.2 nmol of $[1-^{14}C]IPP$ (2.04 GBq/mmol), 1 nmol of FPP, 200 nmol of DHAP, $2.0 \mu mol$ of $MgCl_2$, $20 \mu mol$ of MOPS-NaOH, pH 7.0, and a suitable volume of the crude extract from *E. coli* containing pBAD-ALB4.

After incubation at $37°C$ for 30 min, the reaction was stopped by chilling in an ice bath. $200 \mu L$ of water saturated with NaCl was added to the mixture, and then the products were extracted with $600 \mu L$ of 1-butanol saturated with NaCl-saturated water. They were treated with acid phosphatase according to the method of Fujii et al. [12], and the hydrolysates were extracted with *n*-pentane and analyzed by reversed-phase, thin-layer chromatography (TLC) using a precoated plate LKC-18F developed with acetone/H_2O (9:1). The distribution of radioactivity was detected using a BAS2000 bioimaging analyzer (Fujifilm, Japan). The authentic samples were prepared as described in our previous reports, using the enzymes from *Sufolobus acidocaldarius* and *S. solfataricus* [13].

2.7. Lipid Isolation from E. coli Harboring pBAD-ALB4. Lipid was extracted from 2 g of wet cells of *E. coli* harboring pBAD-ALB4, cultured as described above, except that induction with l-arabinose was performed for 18 h. The cells were dissolved with 15 mL of 1-butanol/75 mM ammonium water/ethanol (4:5:11). The mixture was heated to $70°C$ and shaken vigorously for 1 min. It was heated again at $70°C$ for 20 min and shaken vigorously again for 1 min. After cooling to room temperature, the mixture was centrifuged at $1,000 g$ for 10 min. The supernatant was recovered and dried under a stream of nitrogen at $55°C$. The dried residue was then dissolved with 7.2 mL of 1-butanol/methanol/0.5 M acetate buffer, pH 4.6 (3:10:5). Lipids in the mixture were extracted with 3 mL *n*-pentane and dried under a stream of nitrogen at $55°C$. The dried residue was then redissolved in 1 mL of methanol/2-propanol (1:1).

2.8. LC/ESI-MS Analysis. ESI-MS was performed with an Esquire 3000 ion trap system (Bruker Daltonics, USA). MS-parameters used were as follows: sheath gas, N_2 of 30 psi; dry gas, N_2 of $7.0 L·min^{-1}$ at $320°C$; scanning range, 50–1,000 *m/z*; scan speed, 13,000 $m/z·sec^{-1}$; ion charge control target, 50,000 or 20,000; maximum accumulation time, 100 ms; averages, 10; rolling averaging, 2. The system was equipped with an Agilent 1100 Series HPLC system (Agilent Technologies, USA) using UV detection at 210 nm and COS-MOSIL Packed Column $5C_{18}$-AR-II (2.0×150 mm, Nacalai, Japan). The mobile phase consisted of methanol/$100 mg·L^{-1}$ sodium acetate (9:1) or methanol/$120 mg·L^{-1}$ potassium acetate (9:1). The flow rate was $0.2 mL·min^{-1}$.

2.9. Sodium Periodate Treatment. For sodium periodate treatment of 2,3-di-*O*-geranylgeranyl-*sn*-glycero-1-phosphoglycerol (DGGGP-Gro), the peak fraction from HPLC contained about 1 nmol of the mixture of DGGGP-Gro and 2,3-di-*O*-geranylgeranyl-*sn*-glycerol (DGGGOH), and $1 \mu mol$ sodium periodate was added to 1 mL of 1-butanol/75 mM ammonium water/ethanol (4:5:11). The mixture was reacted at $25°C$ for 1 h in the dark. The reaction was stopped by adding $1.5 \mu mol$ of glycerol. After 15 min, the product was extracted by 1 mL of *n*-pentane and dried with N_2. The dried residue was dissolved with $100 \mu L$ of methanol/2-propanol (1:1) and analyzed by LC/ESI-MS.

3. Results and Discussion

We first searched for and cloned the genes from a mesophilic, methanogenic archaeon, *M. acetivorans*, which encoded the closest homologues of the enzymes involved in the biosynthesis of archaeal membrane lipids. The homologue of G-1-P dehydrogenase, which shows 59% sequential identity with G-1-P dehydrogenase from *Methanothermobacter thermautotrophicus* [14], is encoded in the gene *MA3686*. The closest homologue of GGPP synthase, with 39% identity with the enzyme from *S. acidocaldarius* [15], is encoded in *MA0606*. The GGGP synthase homologue with 57% identity with the enzyme from *M. thermautotrophicus* [16] is encoded in *MA3969*. The closest homologue of DGGGP synthase, with 31% identity with the enzyme from *S. solfataricus* [13], is encoded in *MA0961*. Each of the genes, *MA3686*, *MA0606*, *MA3969*, and *MA0961*, was recombinantly expressed in *E. coli*. The cells of *E. coli* were disrupted and centrifuged to recover the supernatant as the crude extract. Then the enzyme activity in the crude extract was confirmed by radio-TLC assay. As shown in Figure 2(a), incubation of the crude extract from *E. coli* expressing *MA0606* with FPP and $[^{14}C]IPP$ yielded a radiolabeled hydrophobic product, and treatment of the product with acid phosphatase produced a compound that comigrated with authentic (all-*E*) geranylgeraniol on a reversed-phase TLC plate ($R_f = 0.60$). Addition of the crude extract from *E. coli* expressing *MA3969* and α-glycerophosphate to the reaction mixture resulted in the movement of the radiolabeled spot on TLC. The new spot ($R_f = 0.68$) comigrated with authentic 3-*O*-geranylgeranyl-*sn*-glycerol (GGGOH). The movement did not occur in the absence of α-glycerophosphate (data not shown). When the crude extract from *E. coli* expressing *MA0961* was additionally mixed, new radiolabeled spots ($R_f = 0.34$ and 0.12) emerged on TLC, accompanied by diminishing radioactivity of the other spot. The spot with an R_f of 0.34 comigrated with authentic DGGGOH. These results indicate that *MA0606*, *MA3969*, and *MA0961* encode, as expected from their homologies, GGPP synthase, GGGP synthase, and DGGGP synthase, respectively. The spot with an R_f of 0.12 was considered to have originated from an unknown modification of DGGGP catalyzed by enzymes contained in the crude extracts. To confirm G-1-P dehydrogenase activity in the crude extract of *E. coli* expressing *MA3686*, the extract was incubated with the crude extracts containing

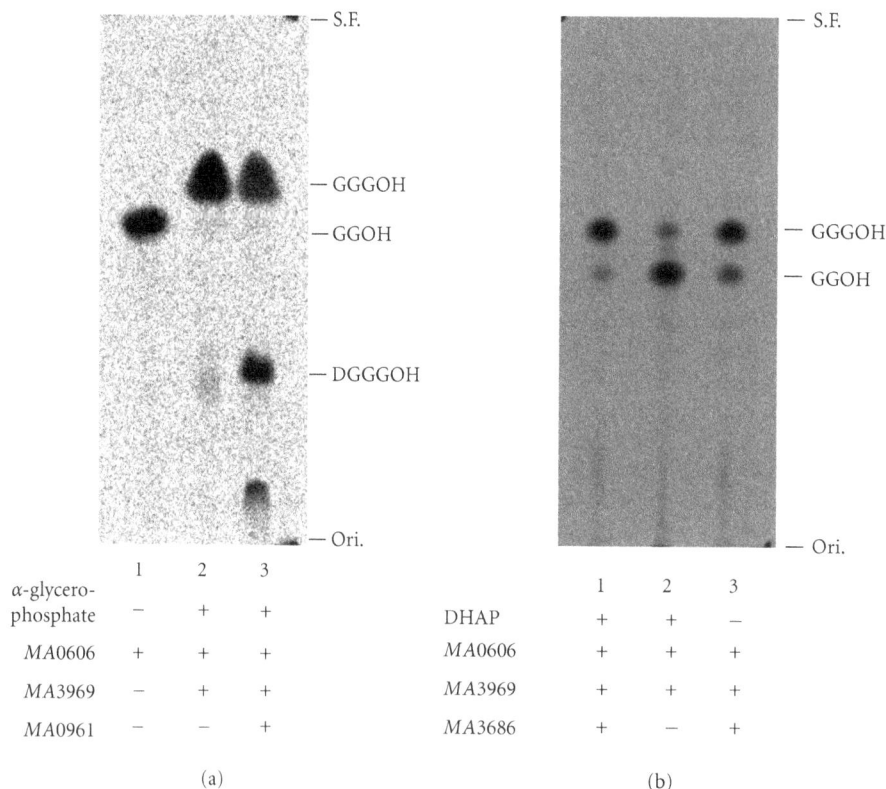

FIGURE 2: *In vitro* assay of the archaeal enzymes for phospholipid biosynthesis. (a) Thin-layer radiochromatogram of the dephosphorylated products from the reactions with recombinant *M. acetivorans* GGPP synthase, GGGP synthase, and/or DGGGP synthase. The enzyme assays were performed using FPP, [^{14}C]IPP, and α-glycerophosphate as the substrates. (b) Thin-layer radiochromatogram of the products from the reaction with recombinant *M. acetivorans* G-1-P dehydrogenase, coupled with GGPP synthase and GGGP synthase. FPP, [^{14}C]IPP, and DHAP were used as the substrates. S.F., solvent front; Ori., origin.

M. acetivorans GGPP synthase and GGGP synthase, DHAP, FPP, and [^{14}C]IPP. The hydrophobic product was extracted, treated with phosphatase, and analyzed by TLC, giving a main spot that comigrated with GGGOH (Figure 2(b)). After removal of the crude extract of *E. coli* expressing *MA3686* from the reaction mixture, the GGGOH spot became thinner, and a spot that comigrated with geranylgeraniol became the major spot. This result shows that *MA3686* encodes G-1-P dehydrogenase. In contrast, removal of DHAP from the mixture did not change the TLC profile of the products, suggesting that a sufficient amount of DHAP existed in the reaction mixture, which contained cell extracts from *E. coli*. It is noteworthy that a small amount of GGGOH appears to be synthesized even in the absence of *M. acetivorans* G-1-P dehydrogenase. It is possible that the enzyme has only low affinity for *sn*-glycerol-3-phosphate, as has been reported with G-1-P-specific archaeal homologues [17–19].

We next constructed a plasmid vector containing the 4 archaeal genes, which formed an artificial operon in the order *MA0606-MA0961-MA3969-MA3686*, to reconstruct the biosynthetic pathway of archaeal phospholipid in *E. coli*. The activities of the enzymes were confirmed by *in vitro* radio-TLC assay. The cell extract from recombinant *E. coli* expressing the 4 archaeal genes showed activities related to the formation of DGGGP from DHAP, IPP, and FPP *in vitro*

(Figure 3(a)), which indicated that the enzymes from *M. acetivorans*, that is, G-1-P dehydrogenase, GGPP synthase, GGGP synthase, and DGGGP synthase, were all expressed in the cells. In addition, a radioactive spot with a lower R_f value (∼0.1) was observed. This spot probably corresponded with the one with an R_f of 0.12 observed in Figure 2(a). Because these spots accompanied the formation of DGGGP and because reaction mixtures for these assays contained cell extracts from *E. coli*, they were considered to arise from an unknown derivative of DGGGP, which might be formed through endogenous metabolic pathways in *E. coli*.

Thus, we extracted lipids from the recombinant *E. coli* cells to confirm *in vivo* synthesis of the archaeal phospholipid precursors or their derivatives. The results of LC/ESI-MS analysis of the extract from *E. coli* containing pBAD-ALB4 showed a relatively broad LC peak of A$_{210}$, which eluted from the column at ∼22 min (Figure 3(b)). This peak was absent in the analysis of the extract from *E. coli* containing the parent plasmid pBAD18. Specific ion peaks with *m/z* of 659.6 and 835.6 were detected through MS analysis of the peak in the positive ion mode (Figure 3(c)). These ions had similar but slightly different peak retention times, so the smaller ion was unlikely derived from fragmentation of the larger one. The smaller ion with *m/z* of 659.6 corresponded with [DGGGOH+Na]$^+$. As shown in Figure 3(d), MS/MS

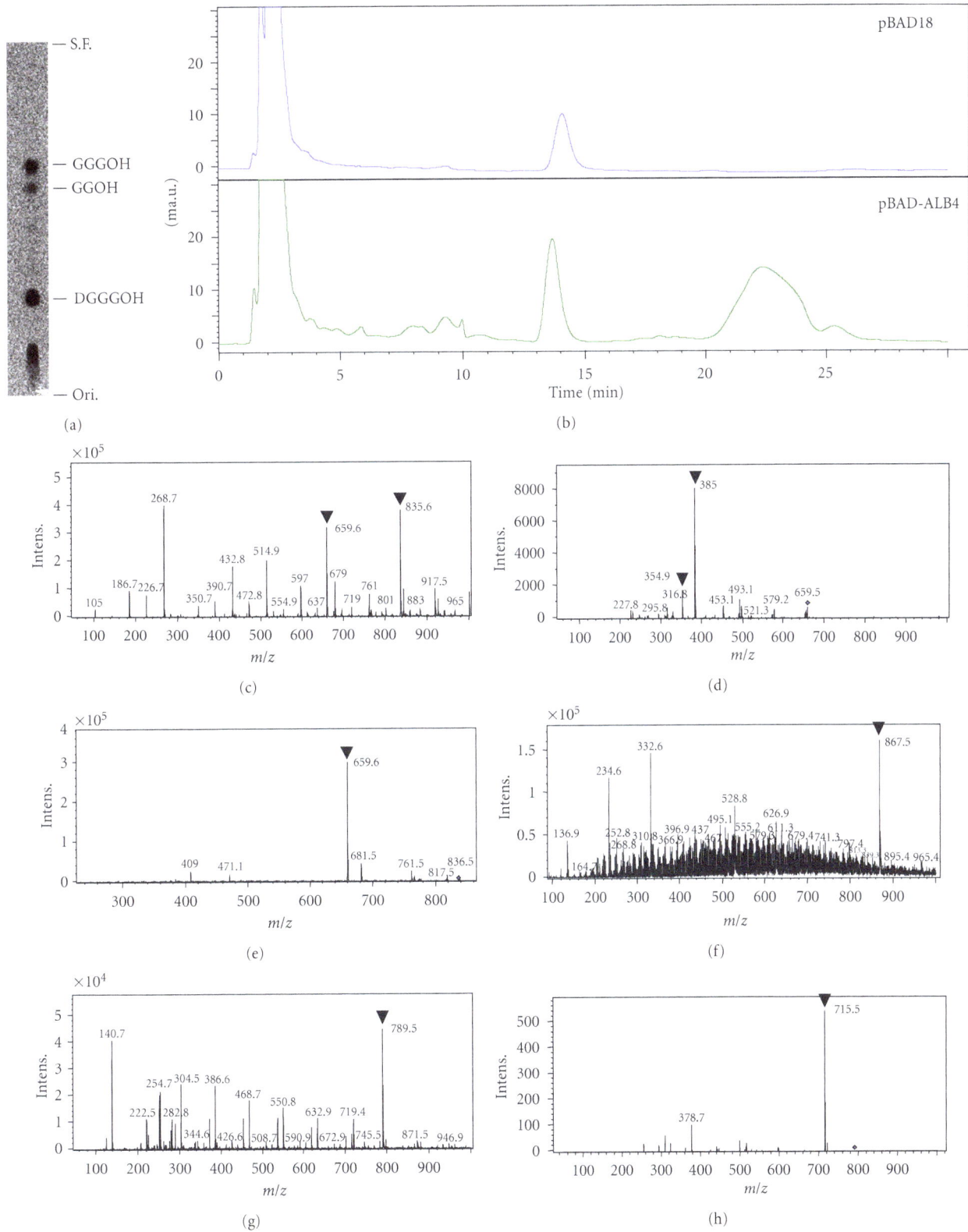

Figure 3: Radio-TLC and LC/ESI-MS analyses of the archaeal-type lipids synthesized in *E. coli*. (a) Crude extract from *E. coli* harboring pBAD-ALB4 was incubated with FPP, [¹⁴C]IPP, and DHAP. Reversed-phase radio-TLC analysis of the products was performed after dephosphorylation. S.F., solvent front; Ori., origin. (b) LC profiles of lipids extracted from *E. coli* harboring pBAD-ALB4 (lower) or its parent plasmid, pBAD18 (upper). (c) Positive ESI-MS ion spectrum of a peak in (b) around 22 min. (d) and (e) MS/MS analyses of the ions in (c), with *m/z* of 659.6 and 835.6, respectively. (f) Positive ESI-MS ion spectrum of the LC peak corresponding with that analyzed in (c). Exclusively for this analysis, the elution buffer was changed from sodium based to potassium based. (g) Negative ESI-MS ion spectrum of a peak in (b) around 22 min. (h) MS/MS analysis of the ion in (g), with an *m/z* of 789.5.

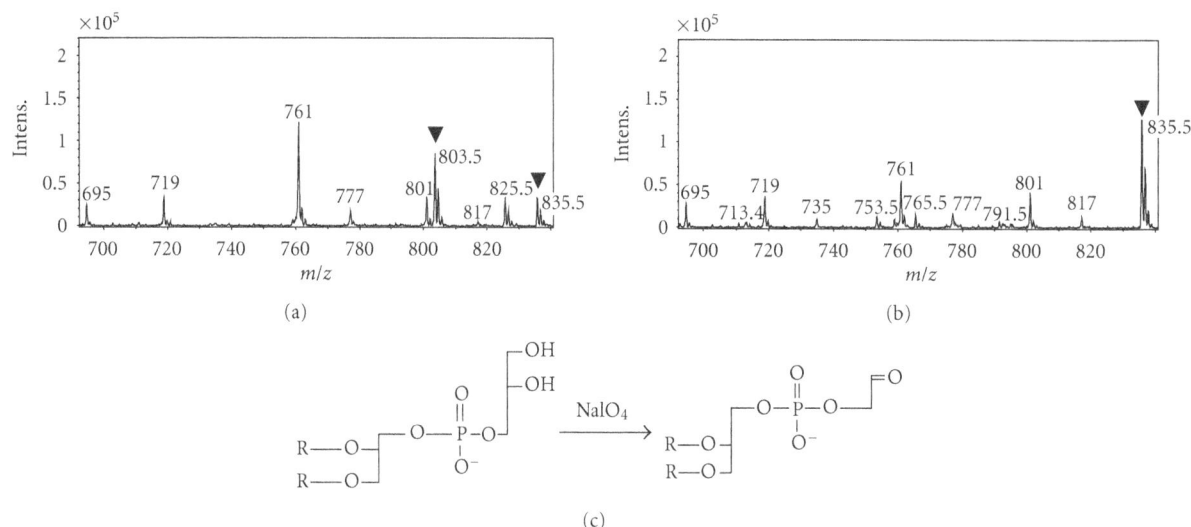

FIGURE 4: Sodium periodate treatment of DGGGP-Gro. (a) Positive ion spectrum from LC/ESI-MS analysis of the archaeal-type phospholipid postperiodate treatment and (b) preperiodate treatment. (c) The scheme of periodate treatment of DGGGP-Gro. R represents a geranylgeranyl group.

analysis of the ion gave a fragment ion with an m/z of 385.0, which corresponded with [GGGOH+Na-2H]$^+$. In addition, a smaller fragment ion with an m/z of 354.9, which corresponded with [GGGOH+Na-CH$_2$O]$^+$, was detected. The fragmentation pattern supported the idea that the peak in Figure 3(b) contained DGGGOH, which probably synthesized by the action of the exogenous archaeal enzymes and endogenous phosphatases in E. coli. On the other hand, the MS/MS analysis of the larger ion with m/z of 835.6 found a fragment ion with an m/z of 659.6, suggesting that the parent ion contained the DGGGOH structure (Figure 3(e)). The MS/MS/MS analysis of the fragment ion with an m/z of 659.6 yielded fragment ions similar to those observed in Figure 3(d) (data not shown). We therefore presumed that the ion peak with an m/z of 835.6 was derived from the cationic bisodium salt of the phosphatidylglycerol-type derivative of DGGGP (DGGGP-Gro). To confirm this idea, the elution buffer for LC/ESI-MS was changed from one containing sodium acetate to one containing potassium acetate, and the same lipid extract was analyzed. As a result, an ion with m/z of 867.5, which corresponded well with that expected for the cationic bis-potassium salt of DGGGP-Gro, was detected instead (Figure 3(f)). In addition, MS analysis of the ion shown in Figure 3(b), in the negative ion mode, yielded an ion with m/z of 789.5, which corresponded with [DGGGP-Gro]$^-$ (Figure 3(g)). MS/MS analysis of the ion showed a fragmentation ion with an m/z of 715.5, which is consistent with [DGGGP]$^-$ (Figure 3(h)).

Moreover, we recovered the LC peak in Figure 3(b), which probably contained DGGGP-Gro, and treated the phospholipid with sodium periodate to confirm the structure of the polar head group. LC/ESI-MS analysis of the treated lipid with the elution buffer containing sodium acetate gave a positive ion with an m/z of 803.5 (Figure 4(a)), which was absent in the analysis of the untreated sample

(Figure 4(b)). The emergence of this ion seemed to accompany the decline of the ion with an m/z of 835.5. The m/z of 803.5 corresponded well with the cationic bisodium salt of DGGGP modified with glycoaldehyde (2,3-di-O-geranylgeranyl-sn-glycero-1-phosphoglycoaldehyde), which had been expected as the product of the sodium periodate treatment of DGGGP-Gro (Figure 4(c)).

These results show that DGGGP, which should be synthesized from the precursors in E. coli cells by the action of the 4 exogenous archaeal enzymes, has been metabolized by endogenous E. coli enzymes to yield DGGGP-Gro. It is unclear whether the radioactive TLC spots with of ~0.1, observed in Figures 2(a) and 3(a), are derived from DGGGP-Gro. The archaeal-type phospholipid probably acts as a component of membranes in E. coli. Modification of phospholipids with glycerol is usual in E. coli, which produces phosphatidylglycerol as a major component of membrane phospholipids [20]. However, the most common phospholipid in the bacterium is phosphatidylethanolamine. The biosynthesis of these phospholipids starts from the cytidylation of phosphatidic acid, which yields CDP-diacylglycerol [21]. sn-Glycerol-3-phosphate or L-serine is then transferred to form phosphatidyl-sn-glycero-3-phosphate or phosphatidyl-L-serine, respectively. Dephosphorylation of the former intermediate yields phosphatidylglycerol, while decarboxylation of the latter yields phosphatidylethanolamine. If the formation of DGGGP-Gro proceeds through this pathway, the cytidyltransferase, sn-glycerol-3-phosphate transferase, and phosphatase of E. coli must accept the archaeal-type phospholipid as the substrate. However, the addition of CTP to the reaction mixture of the in vitro radio-TLC assay did not intensify the spot with an R_f of ~0.1 (data not shown). In contrast, the fact that DGGGP modified with ethanolamine (or serine) was not detected in the LC/ESI-MS analyses suggested that the L-serine transferase did not accept the archaeal-type substrate. In fact, *E. coli* phosphatidylserine synthase, which belongs to an enzyme superfamily different from that which includes archaeal phosphatidylserine synthases, reportedly does not accept CDP-activated DGGGOH [22]. If the cytidylation-dependent pathway does not work, which seems more likely, the inner membrane-periplasmic phosphoglyceroltransferase system [23, 24] may transfer the *sn*-1-phosphoglycerol group from the 6-(glycerophospho)-D-glucose moiety of osmoregulated periplasmic glucans "membrane derived oligosaccharides", or their lipid-linked precursors, to DGGGOH to yield DGGGP-Gro directly.

It should be noted that the growth rate of *E. coli* harboring pBAD-ALB4 was almost identical to that of *E. coli* harboring pBAD18 (data not shown). This fact suggests that the production of archaeal-type glycerolipids, which differ from endogenous bacterial ones in hydrocarbon structures and in chirality of the glycerol moiety, does not strongly affect the viability of *E. coli*. The total amount of archaeal-type lipids extracted from *E. coli* cells, which was estimated by comparing the area of the LC peak at A_{210} with that of known amounts of GGPP, was only $\sim 60\,\mu g/g$ of wet cells. In addition, the archaeal-type lipids detected in this work, that is, DGGGP-Gro and DGGGOH, still retained double bonds in their hydrocarbon chains, which are rarely found in mature archaeal lipids. Therefore, it appears to be too early to conclude that the coexistence of archaeal and bacterial lipids is not disadvantageous for the organisms.

Acknowledgments

This work was supported by grants-in-aid for scientific research from the Asahi Glass Foundation (for H. H.) and from MEXT, Japan (No. 23108531, for H. Hemmi). The authors thank Dr. Susumu Asakawa, Nagoya University, for his help with the cultivation of *M. acetivorans*. They thank Mr. Shigeyuki Kitamura, Nagoya University, for his help with LC/ESI-MS analyses.

References

[1] M. De Rosa and A. Gambacorta, "The lipids of archaebacteria," *Progress in Lipid Research*, vol. 27, no. 3, pp. 153–175, 1988.

[2] A. Gambacorta, A. Trincone, B. Nicolaus, L. Lama, and M. De Rosa, "Unique features of lipids of Archaea," *Systematic and Applied Microbiology*, vol. 16, no. 4, pp. 518–527, 1994.

[3] Y. Koga and H. Morii, "Biosynthesis of ether-type polar lipids in archaea and evolutionary considerations," *Microbiology and Molecular Biology Reviews*, vol. 71, no. 1, pp. 97–120, 2007.

[4] J. C. Mathai, G. D. Sprott, and M. L. Zeidel, "Molecular mechanisms of water and solute transport across archaebacterial lipid membranes," *Journal of Biological Chemistry*, vol. 276, no. 29, pp. 27266–27271, 2001.

[5] J. L. C. M. van de Vossenberg, A. J. M. Driessen, and W. N. Konings, "The essence of being extremophilic: the role of the unique archaeal membrane lipids," *Extremophiles*, vol. 2, no. 3, pp. 163–170, 1998.

[6] Y. Koga, "Early evolution of membrane lipids: how did the lipid divide occur?" *Journal of Molecular Evolution*, vol. 72, no. 3, pp. 274–282, 2011.

[7] G. Wächtershäuser, "From pre-cells to Eukarya—a tale of two lipids," *Molecular Microbiology*, vol. 47, no. 1, pp. 13–22, 2003.

[8] Q. Fan, A. Relini, D. Cassinadri, A. Gambacorta, and A. Gliozzi, "Stability against temperature and external agents of vesicles composed of archaeal bolaform lipids and egg PC," *Biochimica et Biophysica Acta*, vol. 1240, no. 1, pp. 83–88, 1995.

[9] H. Shimada and A. Yamagishi, "Stability of heterochiral hybrid membrane made of bacterial *sn*-G3P lipids and archaeal *sn*-G1P lipids," *Biochemistry*, vol. 50, no. 19, pp. 4114–4120, 2011.

[10] D. Lai, B. Lluncor, I. Schröder, R. P. Gunsalus, J. C. Liao, and H. G. Monbouquette, "Reconstruction of the archaeal isoprenoid ether lipid biosynthesis pathway in *Escherichia coli* through digeranylgeranylglyceryl phosphate," *Metabolic Engineering*, vol. 11, no. 3, pp. 184–191, 2009.

[11] J. Sambrook, *Molecular Cloning, A Laboratory Manual*, 2nd edition, 1989.

[12] H. Fujii, T. Koyama, and K. Ogura, "Efficient enzymatic hydrolysis of polyprenyl pyrophosphates," *Biochimica et Biophysica Acta*, vol. 712, no. 3, pp. 716–718, 1982.

[13] H. Hemmi, K. Shibuya, Y. Takahashi, T. Nakayama, and T. Nishino, "(S)-2,3-Di-O-geranylgeranylglyceryl phosphate synthase from the thermoacidophilic archaeon *Sulfolobus solfataricus*: molecular cloning and characterization of a membrane-intrinsic prenyltransferase involved in the biosynthesis of archaeal ether-linked membrane lipids," *Journal of Biological Chemistry*, vol. 279, no. 48, pp. 50197–50203, 2004.

[14] Y. Koga, T. Kyuragi, M. Nishihara, and N. Sone, "Did archaeal and bacterial cells arise independently from noncellular precursors? A hypothesis stating that the advent of membrane phospholipid with enantiomeric glycerophosphate backbones caused the separation of the two lines of descent," *Journal of Molecular Evolution*, vol. 46, no. 1, pp. 54–63, 1998.

[15] S. I. Ohnuma, M. Suzuki, and T. Nishino, "Archaebacterial ether-linked lipid biosynthetic gene. Expression cloning, sequencing, and characterization of geranylgeranyl-diphosphate synthase," *Journal of Biological Chemistry*, vol. 269, no. 20, pp. 14792–14797, 1994.

[16] T. Soderberg, A. Chen, and C. D. Poulter, "Geranylgeranylglyceryl phosphate synthase. Characterization of the recombinant enzyme from *Methanobacterium thermoautotrophicum*," *Biochemistry*, vol. 40, no. 49, pp. 14847–14854, 2001.

[17] A. Chen, D. Zhang, and C. D. Poulter, "(S)-geranylgeranylglyceryl phosphate synthase. Purification and characterization of the first pathway-specific enzyme in archaebacterial membrane lipid biosynthesis," *Journal of Biological Chemistry*, vol. 268, no. 29, pp. 21701–21705, 1993.

[18] N. Nemoto, T. Oshima, and A. Yamagishi, "Purification and characterization of geranylgeranylglyceryl phosphate synthase from a thermoacidophilic archaeon, *Thermoplasma acidophilum*," *Journal of Biochemistry*, vol. 133, no. 5, pp. 651–657, 2003.

[19] D. Zhang and C. Dale Poulter, "Biosynthesis of archaebacterial ether lipids. Formation of ether linkages by prenyltransferases," *Journal of the American Chemical Society*, vol. 115, no. 4, pp. 1270–1277, 1993.

[20] C. Miyazaki, M. Kuroda, A. Ohta, and I. Shibuya, "Genetic manipulation of membrane phospholipid composition in *Escherichia coli*: pgsA mutants defective in phosphatidylglycerol synthesis," *Proceedings of the National Academy of Sciences of the United States of America*, vol. 82, no. 22, pp. 7530–7534, 1985.

[21] J. E. Cronan and P. R. Vagelos, "Metabolism and function of the membrane phospholipids of *Escherichia coli*," *Biochimica et Biophysica Acta*, vol. 265, no. 1, pp. 25–60, 1972.

[22] H. Morii and Y. Koga, "CDP-2,3-di-*O*-geranylgeranyl-*sn*-glycerol:L-serine *O*-archaetidyltransferase (archaetidylserine synthase) in the methanogenic archaeon *Methanothermobacter thermautotrophicus*," *Journal of Bacteriology*, vol. 185, no. 4, pp. 1181–1189, 2003.

[23] B. J. Jackson, J. P. Bohin, and E. P. Kennedy, "Biosynthesis of membrane-derived oligosaccharides: characterization of *mdoB* mutants defective in phosphoglycerol transferase I activity," *Journal of Bacteriology*, vol. 160, no. 3, pp. 976–981, 1984.

[24] B. J. Jackson and E. P. Kennedy, "The biosynthesis of membrane-derived oligosaccharides. A membrane-bound phosphoglycerol transferase," *Journal of Biological Chemistry*, vol. 258, no. 4, pp. 2394–2398, 1983.

16

PH1: an Archaeovirus of *Haloarcula hispanica* Related to SH1 and HHIV-2

Kate Porter,[1] **Sen-Lin Tang,**[2] **Chung-Pin Chen,**[2] **Pei-Wen Chiang,**[2]
Mei-Jhu Hong,[2] **and Mike Dyall-Smith**[3]

[1] *Biota Holdings Limited, 10/585 Blackburn Road, Notting Hill, VIC 3168, Australia*
[2] *Biodiversity Research Center, Academia Sinica, Nankang, Taipei 115, Taiwan*
[3] *School of Biomedical Sciences, Charles Sturt University, Locked Bag 588, Wagga Wagga, NSW 2678, Australia*

Correspondence should be addressed to Mike Dyall-Smith; mike.dyallsmith@gmail.com

Academic Editor: Shaun Heaphy

Halovirus PH1 infects *Haloarcula hispanica* and was isolated from an Australian salt lake. The burst size in single-step growth conditions was 50–100 PFU/cell, but cell density did not decrease until well after the rise (4–6 hr p.i.), indicating that the virus could exit without cell lysis. Virions were round, 51 nm in diameter, displayed a layered capsid structure, and were sensitive to chloroform and lowered salt concentration. The genome is linear dsDNA, 28,064 bp in length, with 337 bp terminal repeats and terminal proteins, and could transfect haloarchaeal species belonging to five different genera. The genome is predicted to carry 49 ORFs, including those for structural proteins, several of which were identified by mass spectroscopy. The close similarity of PH1 to SH1 (74% nucleotide identity) allowed a detailed description and analysis of the differences (divergent regions) between the two genomes, including the detection of repeat-mediated deletions. The relationship of SH1-like and pleolipoviruses to previously described genomic loci of virus and plasmid-related elements (ViPREs) of haloarchaea revealed an extensive level of recombination between the known haloviruses. PH1 is a member of the same virus group as SH1 and HHIV-2, and we propose the name *halosphaerovirus* to accommodate these viruses.

1. Introduction

Viruses of *Archaea* (archaeoviruses [1]) show considerable diversity and encompass novel morphotypes not seen in *Bacteria* or *Eukarya*. Relatively few have been examined in detail, partly because of the demanding growth requirements of many extremophilic *Archaea* (particularly thermophiles), and also because genetic analysis is often technically difficult compared to bacterial systems such as *Escherichia coli*. Although viruses of thermophilic *Archaea* show the most innovative capsids and replication strategies [1], the viruses of halophilic *Archaea* (haloarchaea) are of increasing attention as new isolates are found with unexpected properties. Many of the earliest reported haloviruses, including all those described before 1998, are bacteriophage-like (*Caudovirales*) with typical head-tail capsids and linear dsDNA genomes. These include groups of related viruses, such as the ΦH-like genus (ΦH, ΦCh1, and BJ1) [2–4] and the unassigned virus group comprising of HF1 and HF2 [5]. The first

spindle-shaped halovirus, His1, was reported in 1998 [6, 7], and the first round virus, SH1, was described in 2003 [8, 9], and electron microscopic studies indicate that these morphotypes dominate in natural waters [10–12]. Over the last 5 years, there has been a wonderful increase in the number and types of described haloviruses, including further examples of SH1-like viruses (e.g., HHIV-2 [13]) and a range of His2-related viruses that are now classified as pleolipoviruses (e.g., HRPV-1 and HHPV-1 [14, 15]). One example of the biological novelty displayed by these archaeoviruses is the geometry of the SH1 capsid, which was found to be of a previously undescribed type, $T = 28$ dextro [16].

Halovirus His2 has a 16 kb dsDNA genome with terminal proteins and probably replicates via an encoded protein-primed DNA polymerase [7]. It is now known to be a pleomorphic virus, distinct from the spindle-shaped His1 [17]. When the genome sequence was first described, it was shown to be related to a cryptic plasmid (pHK2) of *Haloferax lucentense* [18, 19] and to a number of genomic loci found in several different haloarchaea. Later descriptions of haloviruses HRPV-1, HHPV-1, and several others revealed a spectrum of related viruses with very different genome structures (circular ds- and ssDNA), lengths, and replication strategies [14, 15], but they all share a similar set of capsid proteins. The mechanisms underlying the movement of the capsid genes between viruses having such different characteristics and modes of replication (protein primed versus rolling circle) remain unclear. One possibility is that such modular recombinations could be facilitated by previously described genomic loci of virus and plasmid-related elements (ViPREs) [20]. Initially, these did not appear to be simple provirus genomes in varying states of decay, but the description of pleolipoviruses with genomes much smaller than His2 and differing replication strategy can explain many of them as virus integrants. On the other hand, some of these genomic loci have expanded in length as more virus and plasmid sequences become available, and their gene homologs can be recognised in flanking sequences. For these, we retain the ViPRE epithet. From the decades of the study of bacteriophages, it is well documented that related phages frequently recombine (by both legitimate and illegitimate recombination events), giving rise to mosaic genomes with varying evolutionary histories [21, 22]. This process can be assisted by the modular arrangement of genes (e.g., capsid formation or replication) commonly found in many virus genomes [23]. The same appears true of haloviruses, such as the large recombination event evident in the comparison of halovirus HF1 and HF2 genomes [5]. More generally, there is good evidence of widespread recombination from the study of archaeal MCM helicase genes (often associated with mobile genetic elements) [24] and in the comparative genomics of archaeal caudoviruses [25].

The group of round haloviruses exemplified by SH1 has been recently expanded by the descriptions of HHIV-2 and SNJ1 [13, 26]. Members of the SH1 virus group are related by their capsid proteins but differ in genome length and replication strategy. SH1 and HHIV-2 have linear dsDNA genomes (~30.5 kb) with terminal proteins, indicative of replication by protein priming, while SNJ1 has a circular dsDNA genome of only 16.3 kb. This diversity of genome types within the same virus group closely parallels that of the pleolipoviruses described above. SH1 has been the most intensively studied, including host range, particle stability, virion structure, genome sequence, transcription mapping, transfection, and the establishment of a method of genetic manipulation [8, 13, 16, 27–31]. The aim of the current study is to describe a new member of this group, PH1. Its virological characteristics, genome sequence, and major proteins are presented and compared to other members of the SH1 virus group and to genomic loci containing related genes. For convenience, we provide the group name *halosphaerovirus* to encompass the SH1 virus group, currently consisting of SH1, PH1, HHIV-2, and SNJ1.

2. Results

2.1. Virus Isolation and Host Range.

A water sample from Pink Lake, a hypersaline lake ($32°00'$ S and $115°30'$ E) in

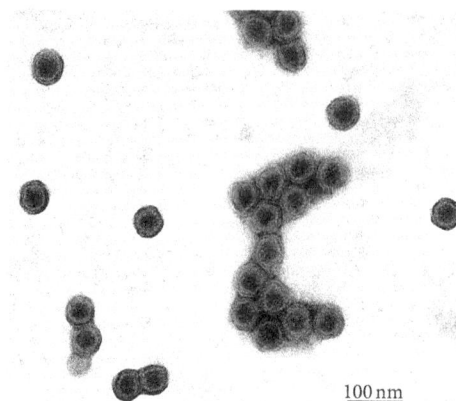

FIGURE 1: Purified PH1 virus examined by negative-stain electron microscopy. Particles were stained with 2% (w/v) uranyl acetate. Scale bar represents 100 nm.

western Australia, was screened for haloviruses by plating directly on lawns of *Har. hispanica* using overlay plates of modified growth medium (MGM) with 12% or 18% salts (w/v). A high titre of similar plaque morphology was observed (1.2×10^5 PFU/mL) on the plates with 18% (w/v) salts. A novel halovirus was isolated from one of these plaques and designated PH1 based on the source (Pink Lake) and the isolating host (*Har. hispanica*).

Plaques were fully developed after two days at 37°C using overlay plates of 18% (w/v) MGM and were 1-2 mm in diameter, clear, and with ragged edges. At 30°C, plaques took three days to develop, and, at 25°C, plaques were hazy and took seven days to develop (data not shown).

The host range of PH1 was identical to that of SH1 [8]; it was unable to plaque on lawns of 12 different species belonging to six different genera of the *Halobacteriaceae* (*Haloarcula*, *Halobacterium*, *Haloferax*, *Halorubrum*, *Haloterrigena*, and *Natrialba*) but could plaque on *Halorubrum* strain CSW 2.09.4 (described in [8]), an uncharacterized Australian isolate (Table 1).

2.2. Virus Purification and Particle Morphology.

Virus was purified by a slight modification of the method described previously for halovirus SH1 ([8] and Section 4). PH1 banded at a density of 1.29 g/mL in CsCl gradients and gave a final specific infectivity of $\sim 5 \times 10^{11}$ PFU/A_{260}. Negative-stain electron microscopy revealed spherical particles, with an average diameter of ~51 nm (Figure 1). The capsid displayed two layers, with a compact core particle of ~43 nm in diameter. This morphology is similar to that of halovirus SH1, a closely related virus (see below) that was isolated at the same time as PH1, but from a neighbouring salt lake [8].

2.3. Single-Burst and Single-Step Growth of PH1.

Around 30% of *Har. hispanica* cells could be infected by PH1, and single burst experiments [32] indicated an average burst size of 87 PFU/cell (data not shown). This value was supported by single-step growth curves, which gave burst sizes of between 50 and 100 PFU/cell. The number of infectious centres usually increased at 4–6 hr p.i., but cell lysis did not appear to begin

TABLE 1: Strains used in this study.

Species/isolate	Strain	Reference
CSW 2.09.4	Original isolate	[51]
Haloarcula hispanica	ATCC 33960	[52]
Haloarcula marismortui	ATCC 43044	[53]
"Haloarcula sinaiiensis"	ATCC 33800	[54]
Halobacterium salinarum	NCIMB 763	[55]
Haloferax gibbonsii	ATCC 33959	[52]
Haloferax lucentense	NCIMB 13854	[56]
Haloferax mediterranei	ATCC 33500	[54]
Haloferax volcanii	ATCC 29605	[57]
Halorubrum coriense	ACAM 3911	[58]
Halorubrum lacusprofundi	ACAM 34	[59]
Halorubrum saccharovorum	NCIMB 2081	[60]
Haloterrigena turkmenica	NCIMB 784	[61]
Natrialba asiatica	JCM 9576	[62]

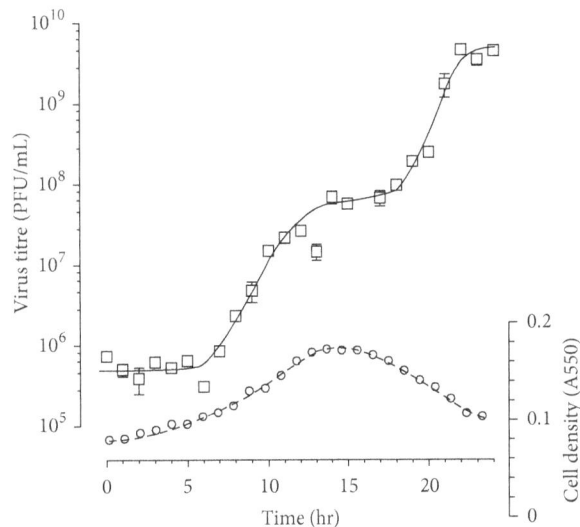

FIGURE 2: Growth curve of PH1. An early exponential culture of *Har. hispanica* in 18% (w/v) MGM was infected with virus (MOI, 50), washed to remove unbound virus, and incubated at 37°C. At regular intervals, samples were removed, the absorbance at 550 nm were measured (circles), and the number of infectious centres were determined by plaque assay (squares). Error bars represent one standard deviation of the average titre.

until well after this, usually between 14 and 24 hr p.i. A representative example of a single-step growth curve for PH1 is shown in Figure 2, where the rise begins at ~6 hr p.i., and visible cell lysis begins at ~14 hr p.i. This figure shows that a second round of growth occurs at around 18 hr p.i., reflecting the initial infection of only 30% of cells, and it is during this second round of virus release that the cell density decreases most rapidly, reaching a value of about 0.1 A_{550}, which is close to the initial cell density. In this example, the average burst size for the first growth step was 95 PFU/cell.

2.4. Virus Stability. The stability of PH1 virus was tested under several conditions (Figures 3(a)–3(d)). When stored in HVD at 4°C, the infectivity of PH1 remained unaltered for several months (data not shown). A thermal stability curve is presented in Figure 3(a) and shows that PH1 is stable up to 56°C above in which it rapidly loses titre. Particles were sensitive to a reduced salt environment (Figure 3(b)) and to chloroform (Figure 3(d)). PH1 was most stable between pH 8 and pH 9 (Figure 3(c)). In general, the stability of PH1 was similar to that described previously for SH1 [8].

2.5. PH1 Structural Proteins and Protein Complexes. The proteins of purified PH1 virus were separated by SDS-polyacrylamide gel electrophoresis, alongside the proteins of SH1 virus (Figure 4). Nine PH1 protein bands were detected, with molecular weights from 7 to 185 kDa (Figure 4(a)), and these were designated with the prefix VP and a number corresponding to the homologous protein of SH1 [8, 27] (see later). This nomenclature is consistent with that of the recently described HHIV-2 virus, a member of the same virus group [13]. The protein profile of PH1 was very similar to that of SH1 [8] and with similar relative masses of the protein bands. One notable difference was that PH1 proteins VP9 and VP10 ran closely together (calculated MWs are 16.5 and 16.7 kDa, resp.) compared to their SH1 homologs (16.5 and 16.9 kDa, resp.). It is probable, given the strong sequence similarity to SH1 (see later), that at least six additional PH1 structural proteins were unable to be visualized using our staining techniques.

The potential glycosylation of virus proteins was examined by staining similar protein gels either with a periodate-acid-Schiff (PAS) stain (GelCode Glycoprotein Staining Kit, Pierce Biotechnology, USA) or the more sensitive fluorescent stain (Pro-Q Emerald 488 Glycoprotein Gel and Blot Stain Kit, Molecular Probes, USA). No glycoproteins were detected.

The major protein bands of PH1 (asterisked in Figure 4) were excised from gels, digested with trypsin, and analysed by MALDI-TOF MS, and the results are summarized in Table 2. From these data, the structural proteins VP 1–4, 7, 9, 10, and 12 were found to be specified by ORFs 12, 24, 28, 21, 20, 27, 26 and 19, respectively (see later).

2.6. Characteristics of the PH1 Genome. Nucleic acid was extracted from purified PH1 virus preparations, treated with proteinase K and incubated with various nucleases to determine the characteristics of the genome (Table 3). The PH1 genome was sensitive to dsDNA endo- and exonucleases but not to ssDNA nuclease (mung bean) or RNase A. This indicated that the PH1 genome is linear dsDNA, with free (i.e., not covalently closed) termini. Restriction endonuclease digestions of the PH1 genome gave a length of approximately 29 kb (data not shown).

The presence of terminal bound proteins was examined using a silica binding assay [33]. Nonproteinase K-treated PH1 DNA was restricted with *Ase*I, and the four resulting fragments passed through GF/C filters under conditions where proteins bind firmly to the glass. As shown in Figure 5, the two internal *Ase*I fragments (2.3 and 7.6 kb) passed through the filter, but the right terminal fragment (17.4 kb) was bound, indicating it carried an attached protein. The left terminal fragment was too small (0.62 kb) and faintly

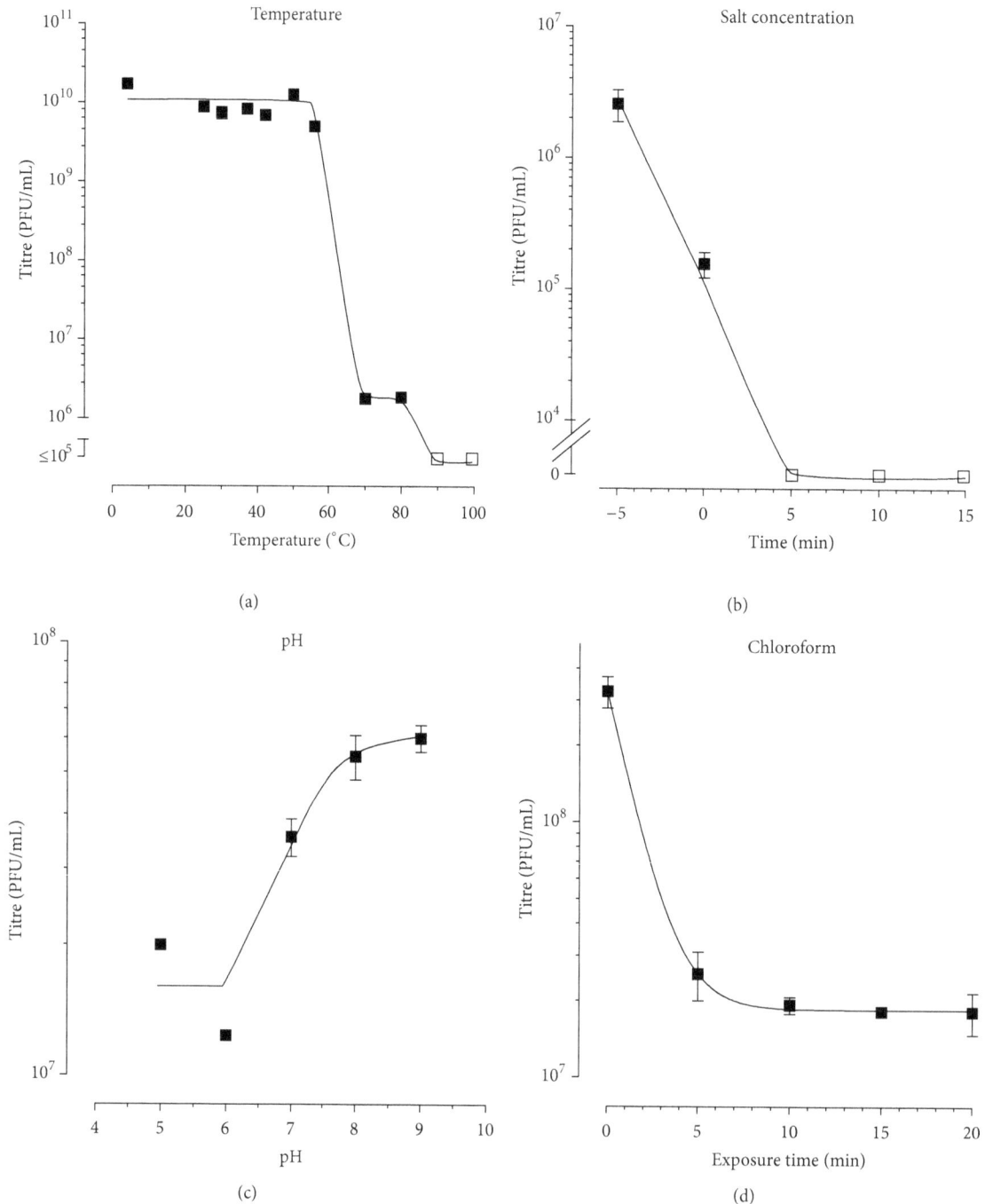

FIGURE 3: Stability of PH1 virus to various treatments and conditions. Virus preparations (infected-cell supernatants in 18% (w/v) MGM) were exposed to various conditions, after which the virus titre was determined (in duplicate) on *Har. hispanica* cells. (a) The effect of temperature. Virus was incubated for 1 hr with constant agitation at temperatures between 4 and 100°C. (b) The effect of lowered salt concentration. Virus was diluted 1 : 1,000 in double-distilled H$_2$O and incubated at room temperature, with constant agitation. Samples were removed at regular intervals. (c) The effect of pH. Virus was diluted 1 : 100 in Tris-HCl buffers at the different pHs and incubated with constant agitation for 30 min. (d) The effect of chloroform. Chloroform was mixed with virus (1 : 4 ratio) and incubated at room temperature with constant agitation. Samples were removed at regular intervals. Open square symbols indicate where virus titres were undetectable. Error bars represent one standard deviation of the average titre.

TABLE 2: PH1 virus proteins identified by mass spectroscopy of tryptic peptides.

Protein[a]	Locus tag	ORF	MW (kDa) Observed (calculated)	Matching peptide masses[b]
VP1	HhPH1_gp12	12	185 (158)	16
VP2	HhPH1_gp24	24	100 (78)	15
VP3	HhPH1_gp28	28	40 (38)	5
VP4	HhPH1_gp21	21	35 (26)	5
VP7	HhPH1_gp20	20	24 (20)	5
VP9	HhPH1_gp27	27	16 (17)	3
VP10	HhPH1_gp26	26	15 (17)	4
VP12	HhPH1_gp19	19	7 (9.9)	4

[a] Virus capsid proteins are numbered according to their similarity with SH1 capsid proteins.

[b] The number of peptide masses between 750.0 and 3,513.0 m/z identified by MALDI-TOF MS that correspond to the theoretical tryptic peptides of the predicted virus protein.

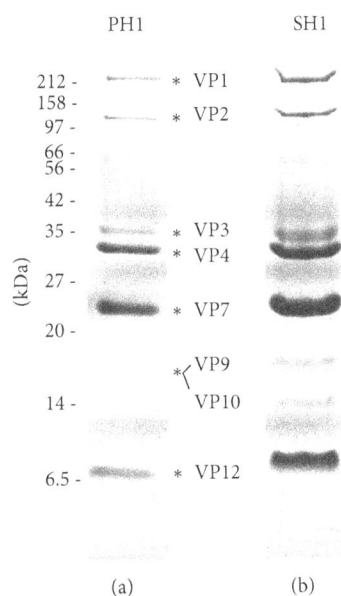

(a)   (b)

FIGURE 4: Structural proteins of halovirus PH1. Viral structural proteins were separated by SDS-PAGE on a 12% (w/v) acrylamide gel and stained with Brilliant Blue G (a). They were run in parallel with the proteins of purified SH1 virus (b). The sizes of protein standard markers are indicated on the left side (in kDa). Asterisks denote proteins bands of PH1 that were identified by mass spectroscopy. The numbering of proteins of PH1 (VP1–VP12) follows that of the SH1 homologs seen in (b) (and explained also in the text).

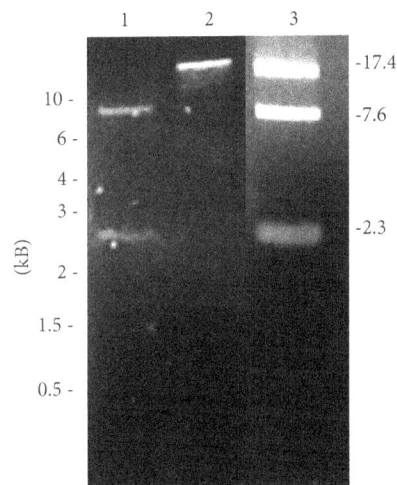

FIGURE 5: Detection of proteins bound to the termini of PH1 DNA. DNA was extracted from purified PH1 particles without proteinase K, digested with AseI, then passed through a silica filter (GF/C) under conditions where proteins would bind to the filter. Lane 1: AseI fragments that did not bind and were eluted. Lane 2: AseI fragments that bound and were eluted only after protease treatment. Lane 3: AseI digest of PH1 DNA. The positions of DNA size standards are indicated on the left side, in kb. The calculated sizes of the AseI fragments of PH1 DNA in lane 3 are indicated on the right side (also in kb). The predicted 0.62 kb PH1 AseI fragment was not detected.

staining to be detected. Further evidence was obtained by a nuclease protection assay, as previously used for SH1 DNA [31]. As shown in Table 3, exonuclease III (a 3′ exonuclease) was able to digest nonproteinase K-treated PH1 DNA, but T7 exonuclease (a 5′ exonuclease) and Bal31 (5′ and 3′ exonucleases) were unable to digest PH1 DNA unless it had been previously treated with proteinase K. This indicated that both 5′ termini of the genome had protein attached.

2.7. Sequence of the PH1 Genome. The complete PH1 genome sequence was determined (accession KC252997) using a

combination of cloned fragments, PCR, primer walking on virus DNA, and 454 whole genome sequencing (see Section 4). It was found to be 28,072 bp in length, 67.6% G + C, with inverted terminal repeat sequences (ITRs) of 337 bp. Using GLIMMER [34], manual BLAST searching at the GenBank database, and comparison to related viruses, the PH1 genome sequence was predicted to contain 49 ORFs (Table 4). Most ORFs were closely spaced or overlapping, giving a gene density of 1.74 genes/kb (average of 572 nt/gene). The cumulative AT-skew plot of the PH1 genome shown at the bottom of Figure 6 shows inflection points (circled) that are consistent with the changes in transcription direction

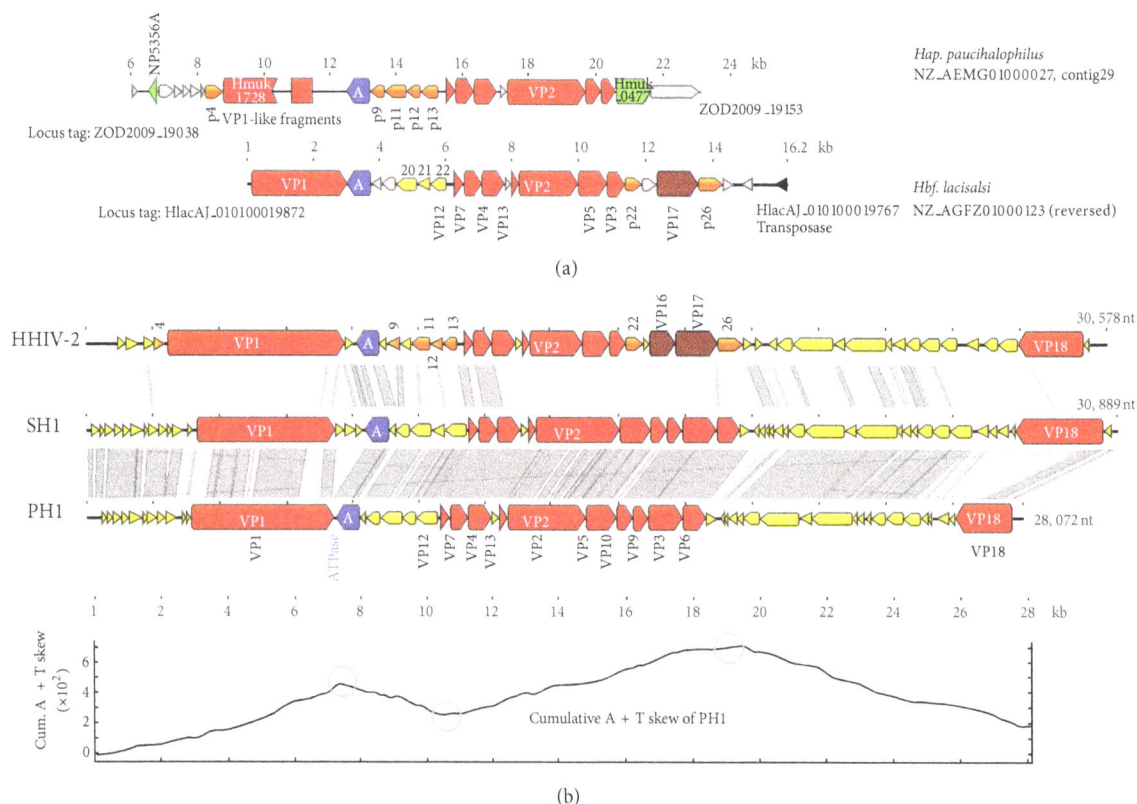

FIGURE 6: Genome alignments of haloviruses PH1, SH1, and HHIV-2, along with two related genomic loci. (a) Genomic loci of *Hap. paucihalophilus* and *Hbf. lacisalsi* that contain genes related to *halosphaeroviruses* SH1, PH1, and HHIV-2. The names and GenBank accessions for these contigs are given on the far right, and ORFs are coloured and labeled to indicate the relationships of these ORFs to those of the viruses below. The locus tag numbers for the first and last ORFs shown in each locus are given nearby their respective ORFs. In addition, grey coloured ORFs represent sequences that do not match any of the haloviruses, and green coloured ORFs represent protein sequences that are closely related to ORFs found within or very close to previously described virus/plasmid loci, so called ViPREs [20]. The scale bars shown above each contig show the position of the described region within the respective contig. (b) The three virus genomes are labeled at the left, with scale markers below (in kb) and the total length indicated at the far right. At the bottom is a cumulative AT-skew plot of the PH1 genome (http://molbiol-tools.ca/Jie_Zheng/), with inflection points circled. The grey shaded bands between the genome diagrams indicate significant nucleotide similarity (using ACT [63]). Annotated ORFs are represented by arrows, with colours indicating structural proteins (red or brown), nonstructural proteins (yellow or orange) or the packaging ATPase (blue). The names of structural protein ORFs are indicated either within the arrow (e.g., VP1) or in text nearby. The numbered, orange coloured ORFs of HHIV-2 are homologous to ORFs found in the genomic loci (probably proviruses or provirus remnants) pictured in (a).

indicated by the annotated ORFs, as has been seen in other haloviruses [7]. There are no GATC motifs in the genome, and the inverse sequence, CTAG, is strongly underrepresented with just 2 motifs (68 expected).

When compared to the related viruses SH1 and HHIV-2 (Figure 6), the PH1 genome is seen to be of similar length and is much more closely related to SH1 than to HHIV-2 (74% and 54% nt identity, resp.). Like PH1, the genomes of SH1 and HHIV-2 lack GATC motifs, and CTAG is either absent (SH1) or underrepresented (HHIV-2). The ITRs of PH1 and SH1 share 78.5% nucleotide identity, but the PH1 ITR is longer than that of SH1 (337 and 309 nt, resp.) partly because SH1 has replaced an 18 bp sequence (gtcgtgcggtttcggcgg) found at the internal end of its left-hand ITR with a sequence at the corresponding position of its right-hand ITR that shows little inverted sequence similarity. In the PH1 ITR, this sequence is retained at both ends. Whether this difference represents a recombination event or a mistake in replication of the ITRs

is unclear. An alignment of the ITRs of all three viruses identified a number of highly conserved regions (labeled 1–9 in Supplementary Figure 2, see Supplementary Material available online at http://dx.doi.org/10.1155/2013/456318). A 16 bp sequence at the termini (region 1) is conserved, consistent with the conservation seen at the termini of linear *Streptomyces* plasmids [35]. A 16 bp GC-rich sequence around the middle of the ITR (region 3) is also conserved. Shorter motifs of either C-rich (region 8) or AT-rich sequence (regions 7 and 9) are found near the internal end of the ITRs.

The grey shading between the schematic virus genomes in Figure 6(b) indicates regions of high nucleotide similarity, revealing that the differences between PH1 and SH1 are not evenly distributed but vary considerably along the length of the aligned genomes. A comparison between SH1 and HHIV-2 has been published recently [13], and since the PH1 and SH1 genomes are so similar, we will focus the following description on the differences between PH1 and SH1.

TABLE 3: Digestion of the PH1 genome by nucleases[a].

Nuclease (amount)	Proteinase K-treated		Untreated	
	Control	PH1 DNA	Control	PH1 DNA
DNase I (RNase-free) (100 U)	+[b]	+	+[b]	+
Exonuclease III (100 U)	+[b]	+	+[b]	+
Mung bean nuclease (10 U)	+[c]	−	+[c]	−
Nuclease BAL-31 (1 U)	+[b]	+	+[b]	−
RNase A (5 μg/mL)	+[d]	−	+[d]	−
T7 exonuclease (10 U)	+[b]	+	+[b]	−

[a]Extracted nucleic acid from purified PH1 virus was treated with various nucleases and then analysed by gel electrophoresis to detect whether the nuclease digested (+) or did not digest (−) the genome. Control nucleic acids used to confirm activity of the nucleases were [b]λ DNA, [c]a DNA oligonucleotide, and [d]yeast transfer RNA.

Within corresponding ORFs, there are blocks of coding sequence with low (or no) similarity. These may represent *in situ* divergence or short recombination events (e.g., indels). There are also cases where an entire ORF in one virus has no corresponding homolog in the other because of an insertion/deletion event (indel). Lastly, there are replacements, where the sequences within the corresponding regions show low or no similarity but are flanked by sequence with high similarity. The largest visible differences seen in Figure 6 are due to sequence changes within or near the long genes encoding capsid proteins VP1 and VP18. A summary of the main sequence differences between SH1 and PH1 and their locations is given in Table 5, where these divergent regions (DV) are numbered from DV1 to DV18.

The sequences of the two viruses are close enough that in many cases the mechanism for the observed differences can be inferred. For example, DV2 is likely the result of a deletion event that has removed the SH1 ORF10 gene homolog from the PH1 genome. At either end of the SH1 ORF are almost perfect direct repeats (cggcctgac/cggcatgac) that would allow repeat-mediated deletion to occur, removing the intervening sequence. Small direct repeats leading to deletions have been described previously in the comparative analysis of *Haloquadratum walsbyi* genomes [20]. DV5 appears to be an indel, where SH1 ORFs 14–16 are absent in PH1, and an inverted repeat (AGCCATG) found at each end of the SH1 divergent region may be significant in the history of this change. The proteins specified by SH1 ORFs 14–16 are presumably dispensable for PH1 (or their functions supplied by other proteins), but a homolog of SH1 ORF14 is found also in HHIV-2 (ORF 6), and a homolog of SH1 ORF15 occurs in halovirus His1 (ORF13). Replacement regions are also common and can occur within ORFs (e.g., DV3, in capsid protein VP1) or provide additional or alternative genes (e.g., DV12 and DV13). Several divergent regions occur within the gene for capsid protein VP2, a hot spot for change because of the repetitive nature of the coding sequence, which specifies a protein with runs of glycines and many heptapeptide repeats, as described previously for SH1 VP2 [27]. DV18 is a large replacement that significantly alters the predicted length of the minor capsid protein VP18 in the two viruses (865/519 aa for SH1/PH1, resp.) and also provides SH1 with 3 ORFs that are different or absent in PH1: ORFs 52, 53, and 54. For example, a homolog of SH1 ORF52 is not present in PH1 but is present in HHIV-2 (putative protein 40). While ORF48 of PH1 shows no aa sequence homology to SH1 ORFs 53 and 54, all of these predicted proteins carry CxxC motifs, suggestive of related functions, such as DNA binding activity [36].

The genes of the PH1 genome are syntenic with those of SH1, and are found in similarly oriented blocks of closely spaced or overlapping ORFs that suggest that transcription is organised into operons. The programme of transcription of the SH1 genome has been reported previously [30], and the sequences in PH1 corresponding to the six promoter regions (P1–P6) determined in SH1 showed that five of these (P1–P5) are well conserved. Only the region corresponding to SH1 P6 was poorly conserved (data not shown), but this promoter is strongly regulated in SH1 and only switches on late in infection [30].

2.8. Structural Protein Genes. The genes coding for the major virus structural proteins of PH1 (VP 1–4, 7, 9 10, and 12) were identified by MALDI-TOF (see above), and the genes for these proteins are found in the same order and approximate positions as in the SH1 genome (Figure 6(b), red ORFs). Genes coding for minor structural proteins can be deduced by sequence comparison with SH1 so that VP5 and VP6 are likely to be encoded by PH1 ORFs 25 and 29, respectively, and VP13 and VP18 by ORFs 23 and 49, respectively. This would give PH1 a total of 11 structural proteins, the same as SH1. Of these, VP7 is the most conserved between the two viruses (98% aa identity), followed by VP4, VP9, VP3, and VP12 (91%–94% identity), VP5, VP6, and VP13 (82%–88% identity), VP2 (77%), VP1 (65%), and, lastly, the least conserved protein was VP18 (31%).

2.9. Genomic Loci Related to PH1 and Other Halosphaeroviruses. Clusters of halosphaerovirus-related genes are present in the genomes of two recently sequenced haloarchaea, *Hap. paucihalophilus* and *Hbf. lacisalsi* (Figure 6(a)). Neither of these regions appears to represent complete virus genome, but they retain a number of genes that share a similar sequence and synteny with the virus genomes depicted below them (Figure 6(a)). Both loci show a mixed pattern of relatedness to SH1/PH1 and HHIV-2, as indicated by ORFs of matching colour and gene name. If these genomic loci represent provirus integrants that have decayed over time, then the relationships they show suggest not only that halosphaeroviruses are a diverse virus group but also that a significant level of recombination occurs between them, leading to mosaic gene combinations. The two ORFs coloured green in the *Hap. paucihalophilus* locus (Figure 6) are related to ORFs found within or very close to genomic loci of virus/plasmid genes found in other haloarchaea, suggesting additional links between (pro)viruses [20].

TABLE 4: ORF annotations of the halovirus PH1 genome.

ORF[a]	Position[b]	Locus tag	Length[c] (aa)	pKi	Similarity/characteristics
1	441–605	HhPH1_gp01	50	4.3	76% aa similarity to SH1 ORF2 (YP_271859)
2	602–802	HhPH1_gp02	66	10.9	80% aa similarity to SH1 ORF3 (YP_271860)
3	802–1059	HhPH1_gp03	85	3.9	85% aa similarity to SH1 ORF4 (YP_271861). 44% similarity to HHIV-2 protein 1
4	1056–1283	HhPH1_gp04	75	3.9	85% aa identity to SH1 ORF5 (YP_271862). Predicted coiled-coil region.
5	1276–1686	HhPH1_gp05	136	4.9	80% aa similarity to SH1 ORF6 (YP_271863). Two transmembrane domains. Also related to HHIV-2 protein 2 (38%, AFD02283) and to *Halobiforma lacisalsi* ORF ZP_09950018 (35%)
6	1683–1829	HhPH1_gp06	48	6.1	81% aa similarity to SH1 ORF7 (YP_271864). Predicted signal sequence (SignalP).
7	1822–2130	HhPH1_gp07	102	5.8	91% aa similarity to SH1 ORF8 (YP_271865) and 86% to HHIV-2 putative protein 3 (AFD02284). Other homologs include *Nocardia* protein pnf2110 (YP_122060.1), *Hrr. lacusprofundi* Hlac_0751 (YP_002565421.1), ORFs in actinophage VWB (AAR29707), *Frankia* (EAN11657), and gp40 of *Mycobacterium* phage Dori (AER47690)
8	2123–2383	HhPH1_gp08	86	4.2	90% aa similarity to SH1 ORF9 (YP_271866)
9	2380–2714	HhPH1_gp09	111	5.2	88% aa similarity to SH1 ORF11 (YP_271868) and also similarity to putative protein 4 of HHIV-2 (AFD02285) HlacAJ_19877 of *Hbf. lacisalsi* AJ5 (ZP_09950016)
10	2861–3004	HhPH1_gp10	47	10.4	Weak similarity to SH1 ORF12 (YP_271869) only over the N-terminal 18 residues
11	3029–3100	HhPH1_gp11	23	12.3	Not present in SH1 or HHIV-2
12	3134–7453	HhPH1_gp12	1439	4.2	*Capsid protein VP1*. 72% similarity to SH1 VP1 (ORF 13, YP_271870); HHIV-2 VP1 (AFD02286); HlacAJ_19872 *Hbf. lacisalsi* AJ5. Predicted helix-turn-helix and RuvA-like domain (InterProScan)
13	7505–8227	HhPH1_gp13	240	5.4	Predicted P-loop ATPase domain (COG0433). 92% aa similarity to SH1 ORF17 (YP_271874) and also similarity to putative ATPase of HHIV-2 (AFD02288); ATPase of *Haladapatus paucihalophilus* DX253 (ZP_08046180); HlacAJ_19867 of *Hbf. lacisalsi* (ZP_09950014)
14	8419–8228c	HhPH1_gp14	63	5.1	43% aa similarity to SH1 ORF18 (YP_271875). Alanine-rich. Central transmembrane domain (Phobius)
15	8856–8416c	HhPH1_gp15	146	4.1	92% aa similarity to SH1 ORF19 (YP_271876) and also to HHIV-2 putative protein 9 (AFD02290); HlacAJ_19857 of *Hbf. lacisalsi* AJ5; ZOD2009_19093 of *Hap. paucihalophilus*
16	8853–9500c	HhPH1_gp16	215	3.9	65% aa similarity to SH1 ORF20 (YP_271877) and also to putative protein 11 of HHIV-2 (AFD02292); ZOD2009_19098 of *Hap. paucihalophilus*; HlacAJ_19852 *Hbf. lacisalsi* AJ5. C-terminal transmembrane domain (Phobius)
17	9500–9919c	HhPH1_gp17	139	4.1	79% aa similarity to SH1 ORF21 (YP_271878) and also to putative protein 12 of HHIV-2 (AFD02293); hypothetical protein HlacAJ_19847 of *Hbf. lacisalsi* (ZP_09950010). Transmembrane domain (Phobius)
18	9923–10594c	HhPH1_gp18	223	4.3	83% aa similarity to SH1 ORF22 (YP_271879) and also to putative protein 13 of HHIV-2 (AFD02294). ZOD2009_19108 of *Hap. paucihalophilus*; HlacAJ_19842 of *Hbf. lacisalsi*. Predicted signal sequence (signalP). Contains 4 CxxC motifs.
19	10659–10943	HhPH1_gp19	94	10.4	*Capsid protein VP12* (ORF19). 91% aa similarity to SH1 VP12 (ORF23, YP_271880) and also to VP12 of HHIV-2 (AFD02295) and HlacAJ_19837 of *Hbf. lacisalsi*; ZOD2009_19113 of *Hap. paucihalophilus*. Two transmembrane domains (Phobius)
20	10960–11517	HhPH1_gp20	185	4.4	*Capsid protein VP7* (ORF20). 98% aa similarity to SH1 VP7 (ORF24, YP_271881) and also to VP7 of HHIV-2 (AFD02296); ZOD2009_19118 of *Hap. paucihalophilus*; HlacAJ_19832 of *Hbf. lacisalsi* AJ5
21	11519–12217	HhPH1_gp21	232	4.1	*Capsid protein VP4* (ORF21). 94% aa similarity to SH1 ORF25 (YP_271882) and also to VP4 of HHIV-2 (AFD02297); ZOD2009_19123 of *Hap. paucihalophilus*; HlacAJ_19827 of *Hbf. lacisalsi* AJ5
22	12233–12454	HhPH1_gp22	73	3.9	81% aa similarity to SH1 ORF26 (YP_271883) and also to putative protein 17 of HHIV-2 (AFD02298);
23	12458–12697	HhPH1_gp23	79	4.8	78% aa similarity to SH1 *capsid protein VP13* (ORF27, YP_271884) and also to VP13 of HHIV-2 (AFD02299). Predicted coil-coil domain. Predicted C-terminal transmembrane domain (Phobius).

TABLE 4: Continued.

ORF[a]	Position[b]	Locus tag	Length[c] (aa)	pKi	Similarity/characteristics
24	12701–15064	HhPH1_gp24	787	3.9	Capsid protein VP2 (ORF24). 75% aa similarity to SH1 VP2 (ORF28, YP_271885) and to VP2 of HHIV-2 (AFD02300); ZOD2009_19133 of Hap. pauchalophilus (ZP_08046189).
25	15065–15961	HhPH1_gp25	298	4.5	Putative capsid protein VP5 (ORF25). 81% aa similarity to SH1 VP5 (ORF29, YP_271886) and to HHIV-2 VP5 (AFD02301); HlacAJ_19807 of Hbf.lacisalsi A]5; ZOD2009_19133 of Hap. pauchalophilus.
26	15964–16446	HhPH1_gp26	160	4.6	73% aa similarity to SH1 capsid protein VP10 (ORF30, YP_271887) and to VP10 of HHIV-2 (AFD02302); ZOD2009_19138 of Hap. pauchalophilus; HlacAJ_19802 of Hbf. lacisalsi. Predicted C-terminal transmembrane domain (TMHMM).
27	16446–16895	HhPH1_gp27	149	4.2	Capsid protein VP9 (ORF27). 94% aa similarity to SH1 VP9 (ORF31, YP_271888) and to ZOD2009_19143 of Hap. pauchalophilus (ZP_08046191)
28	16908–17921	HhPH1_gp28	337	4.3	Capsid protein VP3 (ORF28). 91% aa similarity to SH1 VP3 (ORF32, YP_271889) and to NJ7G_2365 of Natrinema sp. J7-2
29	17928–18617	HhPH1_gp29	229	4.2	Capsid protein VP6 (ORF28). 83% aa similarity to SH1 VP6 (ORF33, YP_271890). Also to NJ7G_3394 of Natrinema sp. J7-2 and Hmuk-0476 of Halomicrobium mukohataei. Predicted carboxypeptidase regulatory-like domain (CarboxypepD_reg, pfam13620)
30	18614–18943	HhPH1_gp30	109	4.9	71% aa similarity to SH1 ORF34 (YP_271891) and to putative protein 27 of HHIV-2 (AFD02308); Predicted signal sequence (signalP)
31	19054–19176c	HhPH1_gp31	40	5.0	Two CxxC motifs. No homolog in SH1 or HHIV-2
32	19173–19289c	HhPH1_gp32	38	4.1	68% aa similarity to SH1 ORF37 (YP_271894)
33	19286–19552c	HhPH1_gp33	88	5.3	58% aa similarity to SH1 ORF39 (YP_271896) and to HHIV-2 putative protein 29 (AFD02310). Four CxxC motifs
34	19549–19731c	HhPH1_gp34	60	7.0	Contains CxxC motif. No homolog in SH1 or HHIV-2
35	19728–20219c	HhPH1_gp35	163	5.6	93% aa similarity to SH1 ORF 41 (YP_271898) and to putative protein 30 of HHIV-2 (AFD02311)
36	20216–21415c	HhPH1_gp36	399	5.0	95% aa similarity to SH1 ORF 42 (YP_271899) and to putative protein 31 of HHIV-2 (AFD02312)
37	21419–21778c	HhPH1_gp37	119	4.1	91% aa similarity to SH1 ORF43 (YP_271900) and to putative protein 32 of HHIV-2 (AFD02313)
38	21762–23006c	HhPH1_gp38	414	4.2	83% aa similarity to SH1 ORF44 (YP_271901) and to putative protein 33 of HHIV-2 (AFD02314). Predicted coil-coil and helix-turn-helix domains
39	23003–23158c	HhPH1_gp39	51	6.5	69% aa similarity to SH1 ORF45 (YP_271902)
40	23161–23370c	HhPH1_gp40	69	3.5	74% aa similarity to SH1 ORF46 (YP_271903), and 55% similarity to putative protein 35 of HHIV-2 (AFD02316)
41	23422–23586c	HhPH1_gp41	54	7.6	86% aa similarity to SH1 ORF47 (YP_271904); Contains 2 CxxC motifs, and shows similarity to protein domain family PF14206. Arginine-rich
42	23583–24023c	HhPH1_gp42		4.8	77% aa similarity to SH1 ORF48 (YP_271905) and to putative protein 36 of HHIV-2 (AFD02317); Hham1_14540 of Hcc. hamelinensis (ZP_I1271999); PhiChIp72 of Natrialba phage PhiCh1 (NP_665989). DUF4326 (pfam14216) family domain
43	24020–24538c	HhPH1_gp43	172	4.7	79% aa similarity to SH1 ORF49 (YP_271906), and 56% similarity to putative protein 37 of HHIV-2 (AFD02318)
44	24535–25053c	HhPH1_gp44	172	4.3	92% aa similarity to SH1 ORF50 (YP_271907), and 72% similarity to putative protein 38 of HHIV-2 (AFD02319)
45	25050–25277c	HhPH1_gp45	75	4.3	No homolog in SH1 or HHIV-2.
46	25274–25387c	HhPH1_gp46	37	4.8	No homolog in SH1 or HHIV-2
47	25533–25904	HhPH1_gp47	123	4.3	61% aa similarity to SH1 ORF51 (YP_271908); 66% similarity to putative protein 39 of HHIV-2 (AFD02320). Predicted COG1342 domain (DNA binding/helix-turn-helix)
48	25901–26092c	HhPH1_gp48	63	4.9	Contains CxxC motif. No homolog in SH1 or HHIV-2
49	26089–27648c	HhPH1_gp49	506	4.0	81% aa similarity to SH1 ORF55 (YP_271912) (but only in the N-terminal half). Homolog of virus structural protein VP18 of HHIV-2 (AFD02323)

[a] ORFs were predicted either by GLIMMER or by manual searching for homologs in the GenBank database.
[b] Start and end positions of ORFs are give in bp number according to the PH1 sequence deposited at GenBank (KC252997). ORFs on the complementary strand are denoted by the suffix c.
[c] Length of the predicted ORF, in number of amino acids.

TABLE 5: Main differences (divergent regions) between the SH1 and PH1 genomes.

Region[a]	SH1 start[b]	SH1 stop	Length (bp)	PH1 start	PH1 stop	Length (bp)	Comment
DV1	527	650	123	538	604	66	Replacement. Both regions have direct repeats at their borders; PH1 has two sets GACCCGGC and CGCTGC, while SH1 has CCCGAC. In SH1, this replacement region includes the C-terminal region of ORF2 and the N-terminal region of ORF3. In PH1, it covers only the C-terminal region of ORF1
DV2	2433	2588	155	2386	2387	—	RMD[c] from PH1. SH1 repeat at border is TGACCG. This removes the homolog of SH1 ORF10 from PH1
DV3	3324	3598	274	3137	3416	279	Replacement at the beginning of SH1 ORF13/PH1 ORF12 (capsid protein VP1)
DV4	6656	7128	472	6478	7085	607	Replacement near the C-terminus of SH1 ORF13/PH1 ORF12 (capsid protein VP1)
DV5	7491	8341	850	7446	7447	—	Indel that results in ORFs14-16 of SH1 not being present in PH1. SH1 ORF14 has a close homolog in HHIV-2 (ORF6), in a similar position, just after the VP1 homolog. SH1 ORF15 is a conserved protein in haloarchaea (e.g. NP_2552A, and Hmuk_2978) and halovirus His1 (ORF13). Possible inverted repeat (AGCCATG) at border of SH1 region
DV6	13705	13706	—	12818	12851	33	RMD from SH1. PH1 repeat at border is CAGCGG(g/t)G. This removes a part of capsid protein VP2 sequence from the SH1 protein
DV7	13789	13863	74	12934	12941	7	Replacement. This occurs within VP2 gene of both viruses
DV8	15142	15186	44	14220	14221	—	RMD from PH1. SH1 repeat at border is TG(t/c)CCGACGA and occurs within capsid protein VP2 gene of both viruses
DV9	15303	15317	14	14337	14338	—	RMD from PH1. SH1 repeat at border is GCCGACGA and occurs within capsid protein VP2 gene of both viruses
DV10	15504	15527	23	14529	14530	—	RMD from PH1. SH1 repeat at border is GACGA and occurs within capsid protein VP2 of both viruses
DV11	20026	20405	379	19019	19173	154	Replacement beginning at the start of SH1 ORF35/PH1 ORF31. This region is longer in SH1, where it includes ORF36, an ORF that has no homolog present in PH1
DV12	20440	20661	221	19208	19286	78	Replacement. This is unequal and includes an ORF in SH1 (ORF38) that is not present in PH1
DV13	20943	21085	142	19562	19728	166	Replacement. This covers SH1 ORF40/PH1 ORF34. The predicted proteins are not homologous
DV14	23541	23542	—	22171	22197	27	Probable RMD in SH1. PH1 repeat at border is CGTCTCGG and occurs in SH1 ORF44/PH1 ORF38
DV15	24506	24521	15	23152	23153	—	Probable RMD in PH1. SH1 repeat at border is CTCGGT and occurs near the end of SH1 ORF45/PH1 ORF39
DV16	24951	24964	13	23579	23580	—	Probable RMD in PH1. SH1 repeat at border is CGGTC and occurs in SH1 ORF47/PH1 ORF41
DV17	26449	26450	—	25050	25274	224	Probable RMD in SH1. PH1 repeat at border is TCATGCG and occurs near the start of SH1 ORF50/PH1 ORF44 and provides an extra ORF for PH1 (ORF45)
DV18	27074	29764	690	25895	26933	1038	Replacement. Left border at start of SH1 ORF51/PH1 ORF47 and extends rightwards into SH1 ORF55/PH1 ORF49 (VP18 gene). It is an unequal replacement and SH1 has two more ORFs in this region than PH1

[a]DV: Divergent regions between the genomes of SH1 and PH1.
[b]Start and stop positions refer to the GenBank sequences of the two viruses: SH1, NC_007217.1; PH1, KC252997.
[c]RMD: repeat-mediated deletion event, as described in [20].

The recently described halovirus SNJ1 carries many ORFs that have homologs adjacent to a previously described ViPRE of *Hmc. mukohataei* (from here on denoted by ViPRE$_{Hmuk1}$), a locus that contains genes related to His2 (and other pleolipoviruses) as well as to *Haloferax* plasmid pHK2 (probably a provirus) and to two small plasmids of *Hqr. walsbyi* (~6 kb, pL6A and pL6B). The latter relationship provides wider links to other viruses because one of the PL6 genes (Hqrw_6002) has homologs in some pleolipoviruses (HPRV-3 and HGPV-1), while another gene (Hqrw_6005) is related to ORF16 of the spindle-shaped halovirus His1 [15]. After adding the SNJ1 gene homologs to ViPRE$_{Hmuk1}$, it is extended significantly, and a close inspection of the flanking regions revealed a tRNA-ala gene at one end and a partial copy of the same tRNA-ala gene (next to a phage integrase gene) at the other end (supplementary Figure 1). Genes beyond these points

TABLE 6: Transfection of haloarchaea by PH1 DNA.

Species[a]	Efficiency of transfection/transformation with	
	PH1 DNA[b] (PFU/μg of DNA)	pUBP2 DNA[b,c] (CFU/μg of DNA)
Har. hispanica	$5.3 \pm 0.5 \times 10^3$	$1.4 \pm 0.8 \times 10^4$
Har. marismortui	$4.5 \pm 0.7 \times 10^3$	$1.8 \pm 0.1 \times 10^4$
"Har. sinaiiensis"	$3.2 \pm 0.6 \times 10^2$	—
Halobacterium salinarum	—	$5.4 \pm 4.2 \times 10^3$
Haloferax gibbonsii	$2.8 \pm 0.8 \times 10^3$	$4.0 \pm 2.6 \times 10^3$
Haloferax lucentense	—	$5.7 \pm 1.1 \times 10^3$
Hfx. volcanii	—	$2.1 \pm 0.7 \times 10^5$
Hrr. lacusprofundi	$7.0 \pm 2.2 \times 10^2$	$9.5 \pm 1.5 \times 10^3$
Haloterrigena turkmenica	$1.6 \pm 1.7 \times 10^2$	$2.4 \pm 0.6 \times 10^3$
Natrialba asiatica	$3.5 \pm 3.9 \times 10^2$	$3.1 \pm 1.0 \times 10^4$

[a]Only those species positive for transfection and/or plasmid transformation are shown.

[b]Rates of transfection by (nonprotease treated) PH1 DNA or transformation by plasmid pUB2 are averages of three independent experiments, each performed in duplicate (± standard deviation).

[c]Transformants were selected on plates with 2, 4, or 6 μg/mL simvastatin, depending on the strain.

— no plaques or colonies observed.

appear to be related to cellular metabolism. ViPRE$_{Hmuk1}$ is now 39379 nt in length, flanked by a 59 nt direct repeat (potential *att* sites), and includes 54 ORFs, many of which are related to known haloviruses, are virus-like (e.g., integrases, methyltransferases, transcriptional regulators, DNA methyltransferase, and DNA glycosylase), are homologs in or near other known ViPREs, or show little similarity to other known proteins. This locus also includes an ORC1/CDC6 homolog (Hmuk_0446), which could provide a replication function.

The mixture of very different virus and plasmid genes seen within ViPRE$_{Hmuk1}$ is remarkable, but it also reveals homologs found in or adjacent to genomic loci of other species, such as the gene homologs of *Har. marismortui* seen in the left end of ViPRE$_{Hmuk1}$ (supplementary Figure 1). When these genes are added to the previously described *Haloarcula* locus (ViPRE$_{Hmuk1}$), it is also extended significantly. It becomes 25,550 bp in length, encompasses genes from Hmar_2382 to Hmar_2404, and is flanked by a full tRNA-ala gene at one end and a partial copy at the other. It carries an ORC1/CDC6 homolog, an integrase, and a phiH-like repressor as well as pleolipovirus homologs.

2.10. Transfection of Haloarchaea by PH1 DNA. PH1 DNA was introduced into *Har. hispanica* cells using the PEG method [37], and the cells were screened for virus production by plaque assay. Figure 7 shows the increase in transfected cells with input (nonproteinase K-treated) virus DNA. The estimated efficiency was 5.3×10^3 PFU per μg DNA. PH1 DNA that had been treated with proteinase K or with DNAase I (RNase-free) did not produce plaques (data not shown). PH1 DNA was found to transfect six species of haloarchaea other than *Har. hispanica*, including members of the genera *Haloarcula*, *Haloferax*, *Halorubrum*, *Haloterrigena*, and *Natrialba* (Table 6). For these experiments, the virus production from transfected cells was detected by plaque assay on indicator lawns of *Har. hispanica*.

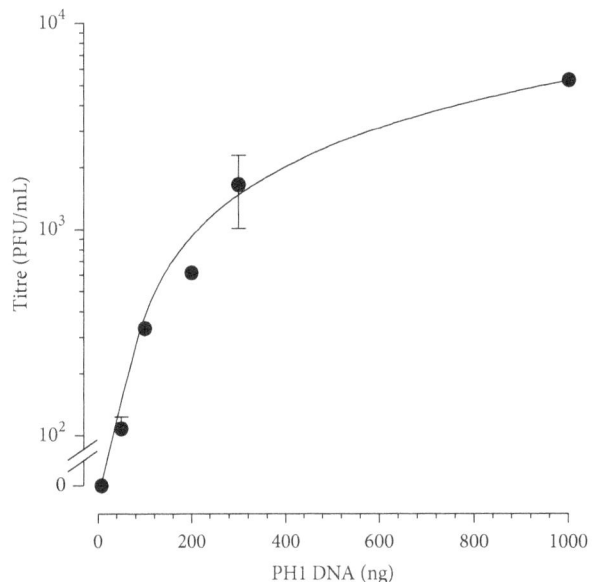

FIGURE 7: Transfection of *Har. hispanica* cells with PH1 DNA. Varying amounts of nonproteinase K-treated PH1 DNA were introduced into cells of *Har. hispanica* using the PEG method [64]. Cells were then screened for infective centres by plaque assay. Data shown is the average of three independent experiments, performed in duplicate. Error bars represent one standard deviation of the mean. If protease-treated DNA was used, no transfectant plaques were observed.

3. Discussion

PH1 is very similar in particle morphology, genome structure, and sequence to the previously described halovirus SH1 [8, 16]. Like SH1, it has long inverted terminal repeat sequences and terminal proteins, indicative of protein-primed replication. Purified PH1 particles have a relatively low buoyant density and are chloroform sensitive and show a layered capsid structure, consistent with the likely presence of an internal membrane layer, as has been shown for SH1, HHIV-2, and SNJ1 [13, 26]. The particle stability of PH1 to temperature, pH, and reduced salt was similar to SH1 [8]. The structural proteins of PH1 are very similar in sequence and relative abundance to those of SH1, so the two viruses are likely to share the same particle geometry [16].

Viruses that use protein-primed replication, such as bacteriophage Φ29, carry a viral type B DNA polymerase that can interact specifically with the proteins attached to the genomic termini and initiate strand synthesis [38]. Archaeoviruses His1 [7] and *Acidianus* bottle-shaped virus [39] probably use this mode of replication. However, a polymerase gene cannot be found in the genomes of PH1, SH1, or HHIV-2. DNA polymerases are large enzymes, and the only ORF of SH1 that was long enough to encode such an enzyme and had not previously been assigned as a structural protein was ORF55, which specified an 865 aa protein that contained no conserved domains indicative of polymerases [27]. The recent study of HHIV-2 showed that the corresponding ORF in this virus (gene 42) is a structural protein of the virion, and, by inference, this is likely also for the corresponding proteins of SH1 and PH1 (no homolog is present in SNJ1). Without a polymerase gene, these viruses must use a host enzyme, but searches for a viral type B DNA polymerase in the genome of the host, *Har. hispanica*, or in the genomes of other sequenced haloarchaea, did not find any matches. This argues strongly for a replication mechanism that is different to that exemplified by Φ29. An attractive alternative is that displayed by *Streptomyces* linear plasmids, which have 5′-terminal proteins and use a cellular polymerase for replication [40]. The terminal proteins are not used for primary replication but for end patching [41], and it has been shown that if the terminal repeat sequences are removed from these plasmids and the ends ligated, they can replicate as circular plasmids [42]. This provides a testable hypothesis for the replication of halosphaeroviruses and also offers a pathway for switching between linear (e.g., PH1) and circular (e.g., SNJ1) forms. Use of a host polymerase would also fit with the ability of SH1 and PH1 DNA to transfect many different haloarchaeal species [31] as these enzymes and their mode of action are highly conserved.

The growth characteristics of PH1 in *Har. hispanica*, both by single-cell burst and single-step growth experiments, gave values of 50–100 viruses/cell, significantly less than that of SH1 (200 viruses/cell) [8] and HHIV-2 (180 viruses/cell) [13]. The latter virus is lytic, whereas SH1, PH1, and SNJ1 start producing extracellular virus well before any decrease in cell density, indicating that virus release can occur without lysis. This has been most clearly shown with SH1, where almost 100% of cells can be infected [8]. Infection of *Har. hispanica* by PH1 was less efficient (~30% of cells), but the kinetics of virus production in single-step growth curves of SH1 and PH1 are similar, with virus production occurring over several hours rather than at a clearly defined time after infection. The two viruses are also very closely related and infect the same host species, so it is likely that they use the same scheme for cell exit. For both viruses, the cell density eventually decreases in

single-step cultures, showing that virus infection does result in cell death. This mode of exit appears to maximise the production of the virus over time, as the host cells survive for an extended period. In natural hypersaline waters, cell numbers are usually high, but growth rates are low [12], and the high incident UV (on often shallow ponds) would damage virus DNA, so reducing the half-life of released virus particles [43]. These factors may have favoured the exit strategy displayed by these viruses, improving their chance of transmission to a new host.

The motif GATC was absent in the genomes of PH1, SH1, and HHIV-2, and the motif CTAG was either absent or greatly underrepresented. All three viruses share the same host, but GATC and CTAG motifs are plentiful in the *Har. hispanica* genome (accessions CP002921 and CP002923). Avoidance of these motifs in haloviruses His1, His2, HF1, and HF2 has been reported previously [7]. Such purifying selection commonly results from host restriction enzymes, and, in *Hfx. volcanii*, it is exactly these two motifs that are targeted [44]. One *Hfx. volcanii* enzyme recognises A-methylated GATC sites [45], and another recognises un-methylated CTAG sites (which are protected by methylation in the genome). In the current study, this could explain the negative transfection results for *Hfx. volcanii*, as the PH1 genome contains two CTAG motifs. By comparison, the SH1 genome contains no CTAG motifs and is able to transfect *Hfx. volcanii* [31].

Comparison of the PH1 and SH1 genomes allowed a detailed picture of the natural variation occurring between closely related viruses, revealing likely deletion events mediated by small repeats, a process described previously in a study of *Hqr. walsbyi* and termed repeat-mediated deletion [20]. There are also many replacements, including blocks of sequence that occur within long open reading frames (such as the capsid protein genes, VP1 and VP18). VP3 and VP6 are known to form the large spikes on the external surface of the SH1 virion, and presumably one or both interact with host cell receptors [16]. PH1 and SH1 have identical host ranges, consistent with the high sequence similarity shown between their corresponding VP3 and VP6 proteins.

A previous study of SH1 could not detect transcripts across annotated ORFs 1–3 or 55-56 [30]. ORFs 1 and 56 occur in the terminal inverted repeat sequence and, in the present study, it was found that there were no ORFs corresponding to these in the PH1 genome. Given the close similarity of the two genomes, the comparative data are in agreement with the transcriptional data, indicating that these SH1 ORFs are not used. SH1 ORFs 2 and 3 are more problematic, as good homologs of these are also present in PH1. The conflict between the transcriptional data and the comparative genomic evidence requires further experimental work to resolve. The status of SH1 ORF55 has recently been confirmed by studies of the related virus, HHIV-2 (discussed above).

The pleolipovirus group has expanded dramatically in the last few years, and it now comprises a diverse group of viruses with different genome types and replication strategies. What is even more remarkable is that a similar expansion can now be seen with SH1-related viruses, which includes viruses with at least two replication modes and genome types (linear and circular). Evidence from haloarchaeal genome sequences show cases of not only provirus integrants or plasmids (e.g., pKH2 and pHH205) but also genomic loci (ViPREs) that contain cassettes of virus genes from different sources, and at least two cases (described in this study) have likely *att* sites that indicate circularisation and mobility. While there is a clear relationship between viruses with circular ds- and ssDNA genomes that replicate via the rolling circle method (i.e., the ssDNA form is a replication intermediate); it is more difficult to explain how capsid gene cassettes can move between these viruses and those with linear genomes and terminal proteins. Within the pleolipoviruses, His2 has a linear genome that contains a viral type B DNA polymerase and terminal proteins, while all the other described members have circular genomes that contain a *rep* homolog and probably replicate via the rolling circle method [14, 19]. Similarly among the halosphaeroviruses, PH1, SH1, and HHIV-2 have linear dsDNA with terminal proteins and probably replicate in the same way, but the related SNJ1 virus has a circular dsDNA genome. The clear relationships shown by the capsid genes of viruses within each group, plus the connections shown to haloviruses outside each group (e.g., the DNA polymerases of His1 and His2), all speak of a vigorous means of recombination; one that can readily switch capsid genes between viruses with radically different replication strategies. How is the process most likely to operate? One possibility is suggested by ViPREs, which appear to be mobile collections of capsid and replication genes from different sources. They offer fixed locations for recombination to occur, provide gene cassettes that can be reassorted to produce novel virus genomes, and some can probably recombine out as circular forms. For example, ViPRE$_{Hmuk1}$ carries genes related to pleolipoviruses, halosphaeroviruses, and other haloviruses. It is yet to be determined if any of the genes carried in these loci are expressed in the cell (such as the genes for capsid proteins) or if ViPREs provide any selective advantage to the host.

4. Materials and Methods

4.1. Water Sample. A water sample was collected in 1998 from Pink Lake (32°00′ S and 115°30′ E), a hypersaline lake on Rottnest Island, Western Australia, Australia. It was screened in the same year for haloviruses using the methods described in [8]. A single plaque on a *Har. hispanica* lawn plate was picked and replaque purified. The novel halovirus was designated PH1.

4.2. Media, Strains, and Plasmids. The media used in this study are described in the online resource, The Halohandbook (http://www.haloarchaea.com/resources/halohandbook/index.html). Artificial salt water, containing 30% (w/v) total salts, comprised of 4 M NaCl, 150 mM MgCl$_2$, 150 mM MgSO$_4$, 90 mM KCl, and 3.5 mM CaCl$_2$ and adjusted to pH 7.5 using ~2 mL 1 M Tris-HCl (pH 7.5) per litre. MGM containing 12%, 18%, or 23% (w/v) total salts and HVD (halovirus diluent) were prepared from the concentrated stock as previously described [46]. Bacto-agar (Difco Laboratories) was added to MGM for solid (15 g/L) or top-layer (7 g/L) media.

Table 1 lists the haloarchaea used in this study. All haloarchaea were grown aerobically at 37°C in either 18% or 23% (w/v) MGM (depending on the strain) and with agitation (except for "*Har. sinaiiensis*"). The plasmids used were pBluescript II KS+ (Stratagene Cloning Systems), pUBP2 [47], and pWL102 [48]. pUBP2 and pWL102 were first passaged through *E. coli* JM110 [49] to prevent *dam*-methylation of DNA, which has been shown to reduce transformation efficiency in some strains [45].

4.3. Negative-Stain Electron Microscopy. The method for negative-stain TEM was adapted from that of V. Tarasov, described in the online resource, The Halohandbook (http://www.haloarchaea.com/resources/halohandbook/index.html). A 20 μL drop of the sample was placed on a clean surface, and the virus particles were allowed to adsorb to a Formvar film 400-mesh copper grid (ProSciTech) for 1.5–2 min. They were then negatively stained with a 20 μL drop of 2% (w/v) uranyl acetate for 1-2 min. Excess liquid was absorbed with filter paper, and the grid was allowed to air dry. Grids were examined either on a Philips CM 120 BioTwin transmission electron microscope (Royal Philips Electronics), operating at an accelerating voltage of 120 kV, or on a Siemens Elmiskop 102 transmission electron microscope (Siemens AG), operating at an accelerating voltage of 120 kV.

4.4. Virus Host Range. Twelve haloarchaeal strains from the genera *Haloarcula*, *Halobacterium*, *Haloferax*, *Halorubrum*, and *Natrialba* (Table 1), Thirteen natural *Halorubrum* isolates (H. Camakaris, unpublished data), and five uncharacterized haloarchaeal isolates from Lake Hardy, Pink Lake (in Western Australia) and Serpentine Lake (D. Walker and M. K. Seah, unpublished data), were screened for PH1 susceptibility. Lysates from PH1-infected *Har. hispanica* cultures (1 × 10^{11} PFU/mL) were spotted onto lawns of each strain (using media with 12% or 18% (w/v) MGM, depending on the strain) and incubated for 2–5 days at 30 and at 37°C.

4.5. Large Scale Virus Growth and Purification. Liquid cultures of PH1 were grown by the infection (MOI, 0.05) of an early exponential *Har. hispanica* culture in 18% (w/v) MGM. Cultures were incubated aerobically at 37°C, with agitation, for 3 days. Clearing (i.e., complete cell lysis) of PH1-infected cultures did not occur, and consequently cultures were harvested when the absorbance at 550 nm reached the minimum, and the titre (determined by plaque assay) was at the maximum, usually ~10^{10}–10^{11} PFU/mL. Virus was purified using the method described previously for SH1 [8, 31], except that after the initial low speed spin (Sorvall GSA; 6,000 rpm, 30 min, 10°C); virus was concentrated from the infected culture by centrifugation at 26,000 rpm (13 hr, 10°C) onto a cushion of 30% (w/v) sucrose, in HVD. The pellet was resuspended in a small volume of HVD and loaded onto a preformed linear 5%–70% (w/v) sucrose gradient, followed by isopycnic centrifugation in 1.3 g/mL CsCl (Beckman 70Ti; 60,000 rpm, 20 hr, 10°C). The white virus band occurred at a density of 1.29 g/mL and was collected and diluted in halovirus diluent (HVD, see Section 4), and the virus pelleted (Beckman SW55; 35,000 rpm, 75 min, 10°C) and resuspended in a small volume of HVD and stored at 4°C. Virus recovery at the major stages of a typical purification is given in supplementary Table 1. The specific infectivity of pure virus solutions was determined as the ratio of the PFU/mL to the absorbance at 260 nm.

4.6. PH1 Single-Step Growth Curve. An early exponential phase culture of *Har. hispanica* grown in 18% (w/v) MGM was infected with PH1 (MOI, 50). Under these conditions the percentage of infected cells was approximately 30%. After an adsorption period of 1 hr at 37°C, the cells were washed three times with 18% (w/v) MGM (at room temperature), resuspended in 100 mL 18% (w/v) MGM (these methods ensured the removal of all residual virus), and incubated at 37°C, with shaking (100 rpm). Samples were removed at hourly intervals for measurements of absorbance at 550 nm and the number of infective centres. Immediately after sampling, titres were determined by plaque assay with an indicator lawn of *Har. hispanica*. Each experiment was performed in triplicate.

4.7. Halovirus Stability. After various treatments, samples were removed and diluted in HVD, and virus titres were determined by plaque assay on *Har. hispanica*. Each experiment was performed in triplicate, and representative data are shown. Chloroform sensitivity was examined by the exposure of PH1 lysates in 18% (w/v) MGM to chloroform in a volume ratio of 1 : 4 (chloroform to lysate). Incubation was at room temperature, with constant agitation. At appropriate time points, the mix was allowed to settle, and a sample was removed from the upper layer and diluted in HVD. The effect of a low ionic environment was examined by diluting PH1 lysates (in 18% (w/v) MGM) into double distilled H_2O in a volume ratio of 1 : 1,000 (lysate to double distilled H_2O). Incubation was at room temperature, with constant agitation. Samples were removed at various times and diluted in HVD. The pH stability of PH1 was determined by dilution of PH1 lysates in 18% (w/v) MGM in the appropriate pH buffer (HVD buffered with appropriate Tris-HCl) in a volume ratio of 1 : 100 (lysate to buffer). Incubation was at room temperature, with constant agitation. After 30 min, samples were removed and diluted in HVD. Thermal stability of PH1 was examined by a 1 hr incubation of a virus lysate in 18% (w/v) MGM at different temperatures, after which they were brought quickly to room temperature, diluted in HVD, and titrated.

4.8. Protein Procedures. To remove salts, purified virus preparations were mixed with trichloroacetic acid (10% (v/v) final concentration) and incubated on ice for 15 min to allow the proteins to precipitate. After centrifugation (16,000 g, 15 min, room temperature), the precipitate was washed three times in acetone, dried, and resuspended in double distilled H_2O. Proteins were dissolved in Laemmli sample buffer with 48 mM β-mercaptoethanol [50], heated in boiling water for 5 min, then separated on 12% (w/v) NuPAGE Novex Bis-Tris Gels using MES-SDS running buffer, according to the manufacturer's directions (Invitrogen). After electrophoresis, gels were rinsed in double distilled H_2O and stained with 0.1% (w/v) Brilliant Blue G in 40% (v/v) methanol and

10% (v/v) acetic acid. Gels were destained with several changes of 40% (v/v) methanol and 10% (v/v) acetic acid. Alternatively, gels were stained with GelCode Glycoprotein Staining Kit, Pro-Q Emerald 488 Glycoprotein Gel and Blot Stain Kit, SYPRO Ruby Protein Gel Stain, according to the manufacturer's directions (Invitrogen, Molecular Probes, Pierce Biotechnology).

Protein bands were cut from the gels and sent to the Australian Proteome Analysis Facility (Macquarie University) for trypsin digestion and analysis by matrix assisted laser desorption ionisation time of flight mass spectrometry (MALDI-TOF MS) on an Applied Biosystems 4700 Proteomics Analyser (Applied Biosystems).

4.9. DNA Procedures. Proteinase K-treated and nonproteinase K-treated DNA preparations were made from purified PH1 using the methods described previously for SH1 [8]. DNA was separated on 1% (w/v) agarose gels in Tris-acetate-EDTA electrophoresis buffer and was stained with ethidium bromide (Sigma-Aldrich).

λ DNA, DNase I (RNase-free), exonuclease III, mung bean nuclease, nuclease BAL-31, T4 DNA polymerase, T4 DNA ligase, T7 exonuclease, and type II restriction endonucleases were purchased from New England Biolabs. RNase A and yeast transfer RNA were purchased from Sigma-Aldrich. Proteinase K was purchased from Promega. Oligonucleotide primers were purchased from Geneworks, Australia.

To clone fragments of the PH1 genome, purified virus DNA was digested either with *Acc*I, *Eco*0109I, and *Mse*I or *Sma*I restriction endonucleases, blunted with DNA Polymerase I, Large (Klenow) Fragment where appropriate, and ligated into *Acc*I-, *Eco*0109I-, or *Sma*I-digested pBluescript II KS+ using T4 DNA ligase, according to the manufacturer's instructions (New England Biolabs). The DNA was introduced into *E. coli* XL1-Blue and transformants grown on Luria agar containing 15 μg/mL tetracycline and 100 μg/mL ampicillin. The resulting clones were sequenced, and the sequences were used to design specific oligonucleotide primers for PCR amplification and/or primer walking using the virus genome (or specific restriction fragments) as templates.

To amplify PH1 nucleic acid, approximately 10 ng virus DNA was combined with 500 nM primers, 100 μM of each dNTP, 1 U Deep Vent DNA Polymerase (New England Biolabs), and 1 × ThermoPol buffer (New England Biolabs), in a total reaction volume of 50 μL. Template was denatured at 95°C for 10 min, followed by 30 cycles of denaturation at 95°C for 30 sec, annealing at 56°C for 30 sec, extension at 75°C for 2 min, and a final extension at 75°C for 10 min. PCR reactions were performed on a PxE0.2 Thermo Cycler (Thermo Electro Corporation). Sequencing reactions using 3.2 pmol primer were performed by the dideoxy chain termination method using ABI PRISM Big Dye Terminator Mix version 3.1 on an ABI 3100 capillary sequencer (Applied Biosystems) by the Applied Genetic Diagnostics Sequencing Service at the Department of Pathology (The University of Melbourne).

To obtain terminal genomic fragments, PH1 DNA was digested by the enzyme PasI, and the head (~700 bp) and the tail (~1300 bp) fragments were purified by the QIAEX II gel extraction kits and treated with 0.5 M piperidine for

2 hr at 37°C to remove protein residues. The piperidine-treated fragments were sequenced by the Genomics Biotech Company (Taipei, Taiwan), using two primers: TGACCA-ATTAATTAGGCCGGTTCGCC (PH1 head-R) and GTGC-CATACTGCTACAATTCT (PH1 tail-F).

Four hundred fifty-four whole genome sequencing: PH1 DNA samples (~5 μg) were sequenced using parallel pyrosequencing on a Roche 454 Genome Sequencer System at Mission Biotech (Taipei, Taiwan). The largest contig was 27399 with 8803 reads (Newbler version 2.7), with 27387 positions being of Q40 quality. The average depth of coverage for each base was 82.9. The GenBank accession for the entire PH1 sequence is KC252997.

Acknowledgments

The authors thank Carolyn Bath, Mei Kwei (Joan) Seah, and Danielle Walker for their early work in PH1 characterization; Simon Crawford, Jocelyn Carpenter, and Anna Friedhuber for their assistance with the transmission electron microscopy; and Tiffany Cowie and Voula Kanellakis for their sequencing assistance. This paper has been facilitated by access to the Australian Proteome Analysis Facility established under the Australian Government's Major National Research Facilities program. SLT was supported by a grant from the Biodiversity Research Center, Academia Sinica, Taiwan.

References

[1] M. Krupovic, M. F. White, P. Forterre, and D. Prangishvili, "Postcards from the edge: structural genomics of archaeal viruses," *Advances in Virus Research*, vol. 82, pp. 33–62, 2012.

[2] P. Stolt and W. Zillig, "Gene regulation in halophage φH;—more than promoters," *Systematic and Applied Microbiology*, vol. 16, no. 4, pp. 591–596, 1994.

[3] R. Klein, U. Baranyi, N. Rössler, B. Greineder, H. Scholz, and A. Witte, "*Natrialba magadii* virus φCh1: first complete nucleotide sequence and functional organization of a virus infecting a haloalkaliphilic archaeon," *Molecular Microbiology*, vol. 45, no. 3, pp. 851–863, 2002.

[4] E. Pagaling, R. D. Haigh, W. D. Grant et al., "Sequence analysis of an Archaeal virus isolated from a hypersaline lake in Inner Mongolia, China," *BMC Genomics*, vol. 8, article 410, 2007.

[5] S. L. Tang, S. Nuttall, and M. Dyall-Smith, "Haloviruses HF1 and HF2: evidence for a recent and large recombination event," *Journal of Bacteriology*, vol. 186, no. 9, pp. 2810–2817, 2004.

[6] C. Bath and M. L. Dyall-smith, "His1, an archaeal virus of the Fuselloviridae family that infects *Haloarcula hispanica*," *Journal of Virology*, vol. 72, no. 11, pp. 9392–9395, 1998.

[7] C. Bath, T. Cukalac, K. Porter, and M. L. Dyall-Smith, "His1 and His2 are distantly related, spindle-shaped haloviruses belonging to the novel virus group, *Salterprovirus*," *Virology*, vol. 350, no. 1, pp. 228–239, 2006.

[8] K. Porter, P. Kukkaro, J. K. H. Bamford et al., "SH1: a novel, spherical halovirus isolated from an Australian hypersaline lake," *Virology*, vol. 335, no. 1, pp. 22–33, 2005.

[9] M. Dyall-Smith, S. L. Tang, and C. Bath, "Haloarchaeal viruses: how diverse are they?" *Research in Microbiology*, vol. 154, no. 4, pp. 309–313, 2003.

[10] A. Oren, G. Bratbak, and M. Heldal, "Occurrence of virus-like particles in the Dead Sea," *Extremophiles*, vol. 1, no. 3, pp. 143–149, 1997.

[11] T. Sime-Ngando, S. Lucas, A. Robin et al. et al., "Diversity of virus-host systems in hypersaline Lake Retba, Senegal," *Environmental Microbiology*, vol. 13, no. 8, pp. 1956–1972, 2010.

[12] N. Guixa-Boixareu, J. I. Calderón-Paz, M. Heldal, G. Bratbak, and C. Pedrós-Alió, "Viral lysis and bacterivory as prokaryotic loss factors along a salinity gradient," *Aquatic Microbial Ecology*, vol. 11, no. 3, pp. 215–227, 1996.

[13] S. T. Jaakkola, R. K. Penttinen, S. T. Vilen et al., "Closely related archaeal *Haloarcula hispanica* icosahedral viruses HHIV-2 and SH1 have nonhomologous genes encoding host recognition functions," *Journal of Virology*, vol. 86, no. 9, pp. 4734–4742, 2012.

[14] E. Roine, P. Kukkaro, L. Paulin et al., "New, closely related haloarchaeal viral elements with different nucleic acid types," *Journal of Virology*, vol. 84, no. 7, pp. 3682–3689, 2010.

[15] A. Sencilo, L. Paulin, S. Kellner, M. Helm, and E. Roine, "Related haloarchaeal pleomorphic viruses contain different genome types," *Nucleic Acids Research*, vol. 40, no. 12, pp. 5523–5534, 2012.

[16] H. T. Jäälinoja, E. Roine, P. Laurinmäki, H. M. Kivelä, D. H. Bamford, and S. J. Butcher, "Structure and host-cell interaction of SH1, a membrane-containing, halophilic euryarchaeal virus," *Proceedings of the National Academy of Sciences of the United States of America*, vol. 105, no. 23, pp. 8008–8013, 2008.

[17] M. K. Pietila, N. S. Atanasova, V. Manole et al., "Virion architecture unifies globally distributed pleolipoviruses infecting halophilic archaea," *Journal of Virology*, vol. 86, no. 9, pp. 5067–5079, 2012.

[18] M. L. Holmes and M. L. Dyall-Smith, "A plasmid vector with a selectable marker for halophilic archaebacteria," *Journal of Bacteriology*, vol. 172, no. 2, pp. 756–761, 1990.

[19] M. L. Holmes, F. Pfeifer, and M. L. Dyall-Smith, "Analysis of the halobacterial plasmid pHK2 minimal replicon," *Gene*, vol. 153, no. 1, pp. 117–121, 1995.

[20] M. L. Dyall-Smith, F. Pfeiffer, K. Klee et al., "*Haloquadratum walsbyi*: limited diversity in a Global Pond," *PLoS ONE*, vol. 6, no. 6, Article ID e20968, 2011.

[21] S. Casjens, "Prophages and bacterial genomics: what have we learned so far?" *Molecular Microbiology*, vol. 49, no. 2, pp. 277–300, 2003.

[22] R. W. Hendrix, M. C. M. Smith, R. N. Burns, M. E. Ford, and G. F. Hatfull, "Evolutionary relationships among diverse bacteriophages and prophages: all the world's a phage," *Proceedings of the National Academy of Sciences of the United States of America*, vol. 96, no. 5, pp. 2192–2197, 1999.

[23] D. Botstein, "A theory of modular evolution for bacteriophages," *Annals of the New York Academy of Sciences*, vol. 354, pp. 484–491, 1980.

[24] M. Krupovič, S. Gribaldo, D. H. Bamford, and P. Forterre, "The evolutionary history of archaeal MCM helicases: a case study of vertical evolution combined with Hitchhiking of mobile genetic elements," *Molecular Biology and Evolution*, vol. 27, no. 12, pp. 2716–2732, 2010.

[25] M. Krupovič, P. Forterre, and D. H. Bamford, "Comparative analysis of the mosaic genomes of tailed archaeal viruses and proviruses suggests common themes for virion architecture and assembly with tailed viruses of bacteria," *Journal of Molecular Biology*, vol. 397, no. 1, pp. 144–160, 2010.

[26] Z. Zhang, Y. Liu, S. Wang et al. et al., "Temperate membrane-containing halophilic archaeal virus SNJ1 has a circular dsDNA genome identical to that of plasmid pHH205," *Virology*, vol. 434, no. 2, pp. 233–241, 2012.

[27] D. H. Bamford, J. J. Ravantti, G. Rönnholm et al., "Constituents of SH1, a novel lipid-containing virus infecting the halophilic euryarchaeon *Haloarcula hispanica*," *Journal of Virology*, vol. 79, no. 14, pp. 9097–9107, 2005.

[28] H. M. Kivelä, E. Roine, P. Kukkaro, S. Laurinavičius, P. Somerharju, and D. H. Bamford, "Quantitative dissociation of archaeal virus SH1 reveals distinct capsid proteins and a lipid core," *Virology*, vol. 356, no. 1-2, pp. 4–11, 2006.

[29] H. Jäälinoja, *Electron Cryo-Microscopy Studies of Bacteriophage 8 and Archaeal Virus SH1*, University of Helsinki, Helsinki, Finland, 2007.

[30] K. Porter, B. E. Russ, J. Yang, and M. L. Dyall-Smith, "The transcription programme of the protein-primed halovirus SH1," *Microbiology*, vol. 154, no. 11, pp. 3599–3608, 2008.

[31] K. Porter and M. L. Dyall-Smith, "Transfection of haloarchaea by the DNAs of spindle and round haloviruses and the use of transposon mutagenesis to identify non-essential regions," *Molecular Microbiology*, vol. 70, no. 5, pp. 1236–1245, 2008.

[32] S. E. Luria, *General Virology*, John Wiley and Sons, New York, NY, USA, 1953.

[33] C. A. Thomas, K. Saigo, E. McLeod, and J. Ito, "The separation of DNA segments attached to proteins," *Analytical Biochemistry*, vol. 93, pp. 158–166, 1979.

[34] A. L. Delcher, D. Harmon, S. Kasif, O. White, and S. L. Salzberg, "Improved microbial gene identification with GLIMMER," *Nucleic Acids Research*, vol. 27, no. 23, pp. 4636–4641, 1999.

[35] R. Zhang, Y. Yang, P. Fang et al., "Diversity of telomere palindromic sequences and replication genes among *Streptomyces* linear plasmids," *Applied and Environmental Microbiology*, vol. 72, no. 9, pp. 5728–5733, 2006.

[36] X. Wang, H. S. Lee, F. J. Sugar, F. E. Jenney, M. W. W. Adams, and J. H. Prestegard, "PF0610, a novel winged helix-turn-helix variant possessing a rubredoxin-like Zn ribbon motif from the hyperthermophilic archaeon, *Pyrococcus furiosus*," *Biochemistry*, vol. 46, no. 3, pp. 752–761, 2007.

[37] S. W. Cline, L. C. Schalkwyk, and W. F. Doolittle, "Transformation of the archaebacterium *Halobacterium volcanii* with genomic DNA," *Journal of Bacteriology*, vol. 171, no. 9, pp. 4987–4991, 1989.

[38] S. Kamtekar, A. J. Berman, J. Wang et al., "The φ29 DNA polymerase: protein-primer structure suggests a model for the initiation to elongation transition," *EMBO Journal*, vol. 25, no. 6, pp. 1335–1343, 2006.

[39] X. Peng, T. Basta, M. Häring, R. A. Garrett, and D. Prangishvili, "Genome of the *Acidianus* bottle-shaped virus and insights into the replication and packaging mechanisms," *Virology*, vol. 364, no. 1, pp. 237–243, 2007.

[40] C. C. Yang, C. H. Huang, C. Y. Li, Y. G. Tsay, S. C. Lee, and C. W. Chen, "The terminal proteins of linear *Streptomyces* chromosomes and plasmids: a novel class of replication priming proteins," *Molecular Microbiology*, vol. 43, no. 2, pp. 297–305, 2002.

[41] C. C. Yang, Y. H. Chen, H. H. Tsai, C. H. Huang, T. W. Huang, and C. W. Chen, "In vitro deoxynucleotidylation of the terminal protein of *Streptomyces* linear chromosomes," *Applied and Environmental Microbiology*, vol. 72, no. 12, pp. 7959–7961, 2006.

[42] P. C. Chang and S. N. Cohen, "Bidirectional replication from an internal origin in a linear *Streptomyces* plasmid," *Science*, vol. 265, no. 5174, pp. 952–954, 1994.

[43] S. W. Wilhelm, W. H. Jeffrey, C. A. Suttle, and D. L. Mitchell, "Estimation of biologically damaging UV levels in marine surface waters with DNA and viral dosimeters," *Photochemistry and Photobiology*, vol. 76, no. 3, pp. 268–273, 2002.

[44] T. Allers and M. Mevarech, "Archaeal genetics—the third way," *Nature Reviews Genetics*, vol. 6, no. 1, pp. 58–73, 2005.

[45] M. L. Holmes, S. D. Nuttall, and M. L. Dyall-Smith, "Construction and use of halobacterial shuttle vectors and further studies on *Haloferax* DNA gyrase," *Journal of Bacteriology*, vol. 173, no. 12, pp. 3807–3813, 1991.

[46] S. D. Nuttall and M. L. Dyall-Smith, "HF1 and HF2: novel bacteriophages of halophilic archaea," *Virology*, vol. 197, no. 2, pp. 678–684, 1993.

[47] U. Blaseio and F. Pfeifer, "Transformation of *Halobacterium halobium*: development of vectors and investigation of gas vesicle synthesis," *Proceedings of the National Academy of Sciences of the United States of America*, vol. 87, no. 17, pp. 6772–6776, 1990.

[48] W. L. Lam and W. F. Doolittle, "Shuttle vectors for the archaebacterium *Halobacterium volcanii*," *Proceedings of the National Academy of Sciences of the United States of America*, vol. 86, no. 14, pp. 5478–5482, 1989.

[49] C. Yanisch-Perron, J. Vieira, and J. Messing, "Improved M13 phage cloning vectors and host strains: nucleotide sequences of the M13mp18 and pUC19 vectors," *Gene*, vol. 33, no. 1, pp. 103–119, 1985.

[50] U. K. Laemmli, "Cleavage of structural proteins during the assembly of the head of bacteriophage T4," *Nature*, vol. 227, no. 5259, pp. 680–685, 1970.

[51] D. G. Burns, H. M. Camakaris, P. H. Janssen, and M. L. Dyall-Smith, "Combined use of cultivation-dependent and cultivation-independent methods indicates that members of most haloarchaeal groups in an Australian crystallizer pond are cultivable," *Applied and Environmental Microbiology*, vol. 70, no. 9, pp. 5258–5265, 2004.

[52] G. Juez, F. Rodriguez-Valera, A. Ventosa, and D. J. Kushner, "*Haloarcula hispanica* spec. nov. and *Haloferax gibbonsii* spec. nov., two new species of extremely halophilic archaebacteria," *Systematic and Applied Microbiology*, vol. 8, pp. 75–79, 1986.

[53] A. Oren, M. Ginzburg, B. Z. Ginzburg, L. I. Hochstein, and B. E. Volcani, "*Haloarcula marismortui* (Volcani) sp. nov., nom. rev., an extremely halophilic bacterium from the Dead Sea," *International Journal of Systematic Bacteriology*, vol. 40, no. 2, pp. 209–210, 1990.

[54] M. Torreblanca, F. Rodriguez-Valera, G. Juez, A. Ventosa, M. Kamekura, and M. Kates, "Classification of non-alkaliphilic halobacteria based on numerical taxonomy and polar lipid composition, and description of *Haloarcula* gen. nov. and *Haloferax* gen. nov.," *Systematic and Applied Microbiology*, vol. 8, pp. 89–99, 1986.

[55] A. Ventosa and A. Oren, "*Halobacterium salinarum* nom. corrig., a name to replace *Halobacterium salinarium* (Elazari-Volcani) and to include *Halobacterium halobium* and *Halobacterium cutirubrum*," *International Journal of Systematic Bacteriology*, vol. 46, no. 1, p. 347, 1996.

[56] M. C. Gutierrez, M. Kamekura, M. L. Holmes, M. L. Dyall-Smith, and A. Ventosa, "Taxonomic characterization of *Haloferax* sp. ("*H. alicantei*") strain Aa 2.2: description of *Haloferax lucentensis* sp. nov.," *Extremophiles*, vol. 6, no. 6, pp. 479–483, 2002.

[57] M. F. Mullakhanbhai and H. Larsen, "*Halobacterium volcanii* spec. nov., a Dead Sea halobacterium with a moderate salt requirement," *Archives of Microbiology*, vol. 104, no. 1, pp. 207–214, 1975.

[58] S. D. Nuttall and M. L. Dyall-Smith, "Ch2, a novel halophilic archaeon from an Australian solar saltern," *International Journal of Systematic Bacteriology*, vol. 43, no. 4, pp. 729–734, 1993.

[59] P. D. Franzman, E. Stackebrandt, K. Sanderson et al., "*Halobacterium lacusprofundi* sp. nov., a halophilic bacterium isolated from Deep Lake, Antarctica," *Systematic and Applied Microbiology*, vol. 11, pp. 20–27, 1988.

[60] G. A. Tomlinson and L. I. Hochstein, "*Halobacterium saccharovorum* sp. nov., a carbohydrate metabolizing, extremely halophilic bacterium," *Canadian Journal of Microbiology*, vol. 22, no. 4, pp. 587–591, 1976.

[61] A. Ventosa, M. C. Gutiérrez, M. Kamekura, and M. L. Dyall-Smith, "Proposal to transfer *Halococcus turkmenicus, Halobacterium trapanicum* JCM 9743 and strain GSL-11 to *Haloterrigena turkmenica* gen. nov., comb. nov.," *International Journal of Systematic Bacteriology*, vol. 49, no. 1, pp. 131–136, 1999.

[62] M. Kamekura and M. L. Dyall-Smith, "Taxonomy of the family *Halobacteriaceae* and the description of two new genera *Halorubrobacterium* and *Natrialba*," *Journal of General and Applied Microbiology*, vol. 41, no. 4, pp. 333–350, 1995.

[63] T. J. Carver, K. M. Rutherford, M. Berriman, M. A. Rajandream, B. G. Barrell, and J. Parkhill, "ACT: the Artemis comparison tool," *Bioinformatics*, vol. 21, no. 16, pp. 3422–3423, 2005.

[64] S. W. Cline, W. L. Lam, R. L. Charlebois, L. C. Schalkwyk, and W. F. Doolittle, "Transformation methods for halophilic archaebacteria," *Canadian Journal of Microbiology*, vol. 35, no. 1, pp. 148–152, 1989.

17

Role of Motif III in Catalysis by Acetyl-CoA Synthetase

Cheryl Ingram-Smith, Jerry L. Thurman Jr., Karen Zimowski, and Kerry S. Smith

Department of Genetics and Biochemistry, Clemson University, Clemson, SC 29634-0318, USA

Correspondence should be addressed to Kerry S. Smith, kssmith@clemson.edu

Academic Editor: Herman van Tilbeurgh

The acyl-adenylate-forming enzyme superfamily, consisting of acyl- and aryl-CoA synthetases, the adenylation domain of the nonribosomal peptide synthetases, and luciferase, has three signature motifs (I–III) and ten conserved core motifs (A1–A10), some of which overlap the signature motifs. The consensus sequence for signature motif III (core motif A7) in acetyl-CoA synthetase is Y-X-S/T/A-G-D, with an invariant fifth position, highly conserved first and fourth positions, and variable second and third positions. Kinetic studies of enzyme variants revealed that an alteration at any position resulted in a strong decrease in the catalytic rate, although the most deleterious effects were observed when the first or fifth positions were changed. Structural modeling suggests that the highly conserved Tyr in the first position plays a key role in active site architecture through interaction with a highly conserved active-site Gln, and the invariant Asp in the fifth position plays a critical role in ATP binding and catalysis through interaction with the 2′- and 3′-OH groups of the ribose moiety. Interactions between these Asp and ATP are observed in all structures available for members of the superfamily, consistent with a critical role in substrate binding and catalysis for this invariant residue.

1. Introduction

AMP-forming acetyl-CoA synthetase (Acs, EC 6.2.1.1), which catalyzes the formation of acetyl-CoA from acetate, ATP, and CoASH (acetate + ATP + CoASH $\leftrightharpoons$ acetyl-CoA + AMP + PP$_i$), belongs to the acyl-adenylate-forming enzyme superfamily, which has newly been designated by Gulick [1] as the ANL superfamily of adenylating enzymes to reflect the three subfamilies, the acyl- and aryl-CoA synthetases, the adenylation domain of the nonribosomal peptide synthetases, and luciferase. Although distant members of this superfamily catalyze wholly unrelated reactions and employ different substrates, they share the property of formation of an enzyme-bound acyl-adenylate intermediate in the first step via activation with ATP with concurrent release of pyrophosphate.

Sequence alignment of members of this superfamily has revealed the presence of three signature motifs as defined by Chang et al. [2]:

motif I: **T**[S/G]-**S**[G]-[G]-[S/T]-**T**[S/E]-**G**[S]-[X]-**P**[M]-[K]-**G**[L/F],

motif II: **Y**[L/W/F]-**G**[S/M/W]-X-**T**[A]-**E**,

motif III: **Y**[F/L]-**R**[T/K/X]-**T**[S/V/A]-**G**-**D**,

(boldfaced residues are the predominant residue at each position, and alternative residues are indicated in bracket.)

Marahiel et al. [3] further identified ten conserved core motifs in the superfamily, in which the A3, A5, and A7 motifs overlap with or encompass motifs I, II, and III, respectively. All three of these motifs are located in or near the active site of each enzyme. Motifs I and II have been shown to play roles in formation of the adenylate based on evidence from enzymes altered at positions within these motifs [2, 4–13]. However, motif III is less well conserved and has received much less attention.

Here, we have investigated the role of motif III in acetyl-CoA synthetase (Acs) in the *Methanothermobacter thermautotrophicus* Acs1 (Acs1$_{Mt}$). This recombinant enzyme has been previously characterized and shows a strong preference for acetate as the acyl substrate and ATP as the nucleotide triphosphate [4, 14], typical of most Acs enzymes. Our results indicate an important role for motif III in catalysis as alteration of any position resulted in a strong decrease in the turnover rate. The highly conserved Tyr in the first position may play a key role in active-site architecture through interaction with a highly conserved active-site Gln. The invariant Asp in the fifth position plays a critical role in ATP binding and catalysis through interaction with the 2′- and 3′-OH groups of the ribose moiety of ATP. The role of this residue in Acs is discussed further in the context of its role in other members of the superfamily.

2. Experimental Procedures

2.1. Materials. Chemicals were purchased from VWR Scientific Products, Fisher Scientific, or Sigma Chemicals. Oligonucleotides for site-directed mutagenesis were purchased from Integrated DNA Technologies. IRD-700- and IRD-800-labeled oligonucleotides for DNA sequencing were purchased from Li-Cor Biosciences or MWG Biotech.

2.2. Sequence Alignment. Sequence alignments were performed using Clustal X [15] with a Gonnet PAM 250 weight matrix and the default parameters of 10.0 and 0.05 as the gap opening and gap extension penalties, respectively.

2.3. Site-Directed Mutagenesis. Site-directed mutagenesis of the gene-encoding *M. thermautotrophicus* Acs1 was performed using the QuikChange Site-Directed Mutagenesis Kit (Stratagene). Mutagenic primers were approximately 40 nucleotides, with the altered site located at the center. Mutations were confirmed by bidirectional DNA sequencing using the Thermo Sequenase Primer Cycle Sequencing Kit (GE Healthcare) at the Nucleic Acid Facility at Clemson University.

2.4. Heterologous Production and Purification of Acs1$_{Mt}$ Variants. Unaltered Acs1$_{Mt}$ and its variants were heterologously produced in *Escherichia coli* Rosetta Blue (DE3) (Novagen) as described previously [4, 14]. Cells harboring the Acs1$_{Mt}$ expression construct were grown at 37°C to an A_{600} of ~0.6, and enzyme production was induced by the addition of IPTG to a final concentration of 0.5 mM. Cell growth was continued overnight at ambient temperature, and cells were then harvested. Acs1$_{Mt}$ and variant enzymes were purified by a two-step procedure employing Q-sepharose anion exchange and phenyl sepharose hydrophobic interaction chromatography as previously described [4, 14]. An additional Source Q anion exchange chromatography step was added if the variant enzyme was not sufficiently pure after the first two steps. The purified enzymes were dialyzed and concentrated, and aliquots were stored at −20°C. Protein concentration was determined by the Bradford method [16].

2.5. Molecular Mass Determination of Enzyme Variants. The variants were subjected to gel filtration chromatography to determine subunit composition. A Superose 12 gel filtration column (GE Healthcare), preequilibrated with 50 mM Tris [pH 7.5] containing 150 mM KCl, was calibrated with chymotrypsinogen (25 kDa), ovalbumin (43 kDa), albumin (67 kDa), aldolase (158 kDa), catalase (232 kDa), ferritin (440 kDa), and blue dextran (2000 kDa).

2.6. Enzymatic Assay for ACS Activity. Enzymatic activity was determined by the hydroxamate assay, which monitors formation of activated acyl groups such as the acetyl-CoA product of the ACS reaction [17, 18]. The standard reaction contained 100 mM Tris [pH 7.5], 600 mM hydroxylamine-HCl [pH 7.0], and 2 mM glutathione (reduced form) in addition to the three substrates (HSCOA, MgATP, and acetate) in a 300 μL reaction volume. In all cases, the concentration of Mg^{2+} was the same as that for ATP. Reactions were performed at the optimal temperature for Acs1$_{Mt}$ of 65°C [4, 14], terminated by the addition of two volumes of stop solution (1 N HCl, 5% trichloroacetic acid, 1.25% FeCl$_3$), and the color change was measured by the change in absorbance at 540 nm.

For determination of apparent kinetic parameters, one substrate was varied, and the other two substrates were held at a saturating concentration, generally ten times the K_m value. The concentration of the variable substrate ranged from ~0.2 to 5–10 times the K_m value. The apparent kinetic parameters k_{cat} and k_{cat}/K_m were determined using nonlinear regression to fit the experimental data to the Michaelis-Menten equation. Each kinetic determination represents three replicates, and the standard errors are given. The enzyme variants followed Michaelis-Menten kinetics for all substrates.

2.7. Inhibition Assays. Inhibition of wild-type Acs1$_{Mt}$ by adenosine and its derivatives and ribose was determined using the hydroxamate assay. In these assays, all three substrates were held at saturating levels, and ribose or adenosine was added to the reaction mix to a final concentration ranging from 0 to 1000 mM or 0 to 100 mM, respectively. The K_i value for each inhibitor was determined by reciprocal plot of velocity versus inhibitor concentration.

2.8. Modeling Motif III Residues of Acs1$_{Mt}$. The structures of Acs1$_{Mt}$ and the variants were modeled on the *S. enterica* Acs (PDB ID: 2P2F) [19] and the *S. cerevisiae* Acs1 structures (PDB ID: 1RY2) [20] by DS Modeler (Accelrys) using the default parameters. Structures were compared to the *S. enterica* and *S. cerevisiae* Acs structures to ensure the modeling did not introduce major structural alterations.

3. Results and Discussion

Sequence alignment indicates considerable conservation of motif III among the Acs sequences, with the first, fourth, and fifth positions highly or completely conserved, but the second and third positions showing a higher level of variability and an overall consensus of Y-X-S/T/A-G-D (Table 1). An alignment of Acs with other members of the adenylate-forming superfamily confirms that these positions are highly conserved throughout (Table 1). The *Saccharomyces cerevisiae* ACS1 structure (Acs$_{Sc}$; PDB 1RY2) contains AMP [20] and is in a conformation thought to catalyze the first step of the Acs reaction in which acetate and ATP are bound and an enzyme-bound acetyl adenylate is formed with concomitant release of inorganic pyrophosphate. The *Salmonella enterica* Acs structure (Acs$_{Se}$; PDB 2P2F) [19] contains acetate, AMP,

TABLE 1: Alignment of motif III residues in Acs sequences.

Consensus		**YXXGD**	
M. thermoautotrophicus Acs1	498	**YTAGD**	503
Salmonella enterica Acs	496	**YFSGD**	501
Saccharomyces cerevisiae Acs1	555	**YFTGD**	560
Methanosaeta concilii Acs1	525	**YLAGD**	530
Methanosaeta concilii Acs2	545	**YLAGD**	550
Methanosaeta concilii Acs3	515	**YLAGD**	520
Methanosaeta concilii Acs4	509	**YFSGD**	514
Methanosaeta concilii Acs5	514	**YLAGD**	519
Homo sapiens ACSM2A	442	**WLLGD**	446
Alcaligenes sp. AL3007 CBL	442	**WLLGD**	446
Bacillus cereus DltA	379	**YRTGD**	383
		* *	

Positions within each sequence are shown, and conserved residues are indicated with an asterisk. M. thermoautotrophicus Acs1, gi:82541817; Salmonella enterica Acs, gi:16767525; Saccharomyces cerevisiae Acs1, gi:6319264; Methanosaeta concilii Acs1, gi: 330506788; Methanosaeta concilii Acs2, gi: 330506787; Methanosaeta concilii Acs3, gi:330506786; Methanosaeta concilii Acs4, gi:330506785; Methanosaeta concilii Acs5, gi:330508629; Homo sapiens ACSM2A, gi:58082049; Alcaligenes sp. AL3007 CBL, gi:197725159; Bacillus cereus DltA, gi:226887779.

TABLE 2: Kinetic parameters for $ACS1_{Mt}$ wild-type and variant enzymes.

Enzyme	k_{cat} (sec^{-1})	K_m acetate (mM)	K_m ATP (mM)	K_m CoA (mM)
$Acs1_{Mt}$	66.6 ± 0.9	3.5 ± 0.1	3.3 ± 0.2	0.19 ± 0.003
$Tyr^{498}Ala$	1.6 ± 0.04	7.5 ± 0.6	1.7 ± 0.3	0.10 ± 0.004
$Tyr^{498}Phe$	Inactive[a]			
$Thr^{499}Ala$	0.8 ± 0.01	3.0 ± 0.01	1.7 ± 0.1	0.24 ± 0.01
$Ala^{500}Thr$	1.5 ± 0.04	2.1 ± 0.2	3.6 ± 0.1	0.51 ± 0.04
$Gly^{501}Ala$	0.3 ± 0.01	1.4 ± 0.1	1.7 ± 0.07	0.08 ± 0.002
$Asp^{502}Ala$	Inactive[a]			
$Asp^{502}Glu$	Inactive[a]			
$Asp^{502}Asn$	Inactive[a]			

[a] Activity was tested over a wide range of concentrations for each substrate and at several enzyme concentrations, but no activity was observed.

and CoA [8] and is thought to be in the conformation for catalysis of the second step of the reaction, in which the C-terminal domain is repositioned near the active site to bring new residues into context for CoA binding and formation of the acetyl-CoA product with release of AMP [8, 19, 20].

Inspection of the Acs_{Se} and Acs_{Sc} structures places motif III in the active site, regardless of which conformation of the enzyme. The positioning of motif III residues near the adenylate moiety of the bound AMP ligand suggests that residues in this motif may play a role in ATP binding and/or catalysis. Modeling of $Acs1_{Mt}$ on the S. enterica and S. cerevisiae Acs structures places the motif III residues in a similar position to interact with substrates (Figure 1).

In $Acs1_{Mt}$, motif III has the sequence $^{498}YTAGD^{502}$. We individually altered each position of motif III in $Acs1_{Mt}$ and determined the kinetic parameters of the purified

enzyme variants. The residues in the highly conserved first, fourth, and fifth positions were changed to either Ala or a conservative amino acid replacement. The Thr residue at the more variable second position was changed to Ala, and the Ala residue in the third position was changed to Thr, as this is the residue found in many Acs sequences. The kinetic parameters for the purified enzyme variants were determined using the hydroxamate assay and are shown in Table 2.

The hydroxamate assay measures activated acyl groups including both acetyl-AMP and acetyl-CoA by their conversion to the acyl hydroxamate and subsequently to a ferric hydroxamate complex. Wilson and Aldrich [21] have shown that several stand-alone adenylating enzymes belonging to the same superfamily as Acs slowly release the acyl-adenylate intermediate in the absence of the native acceptor, and this released acyl-adenylate can react with hydroxylamine. Meng et al. [22] also witnessed this phenomenon with a medium chain acyl-CoA synthetase (Macs) that favors 2-methylbutyrate as the acyl substrate. In this case, Macs released the acyl adenylate to varying degrees in the absence of the CoA acceptor when a less favored acyl substrate such as propionate was used. However, little to no release of the acyl-adenylate intermediate was observed in the absence of CoA with the favored 2-methylbutyrate, suggesting that the acyl-adenylate intermediate is retained if the acyl moiety fits well in the active site but is more readily released in the absence of the native acceptor if the fit is suboptimal.

No activity was detected with $Acs1_{Mt}$ with acetate in the absence of HSCoA, indicating that the acetyl-AMP intermediate remains enzyme bound and that the bound intermediate is not reactive with hydroxylamine. Thus, the kinetic parameters shown in Table 2 are for the overall reaction, although the K_m values for ATP and acetate would likely be similar if just the first adenylation step of the reaction was measured.

Several of the variants were found to be inactive over a wide range of concentrations for each substrate and a range of enzyme concentrations. Enzymes that were inactive displayed similar behavior in both the ion exchange and hydrophobic interaction chromatography steps during purification, and gel filtration chromatography indicated the variants are dimeric as for the wild type enzyme, suggesting there are no gross structural alterations. Overall, alteration of any of the residues in motif III appeared to have a strong deleterious effect on catalysis, although substrate affinity was generally not impaired.

3.1. Positioning of Tyr^{498} Plays an Important Role in Active-Site Architecture.
Based on the two Acs structures, the highly conserved Tyr^{498} in the first position of motif III is part of a hydrogen bond network with Gln^{417} and through this hydrogen bond network may contribute to maintenance of the active-site architecture near the ATP binding site (Figures 1(a) and 1(b)). In the $Acs1_{Mt}$ model (Figure 1(c)), there is an additional interaction between Ala^{500} and Gln^{417}. To examine whether it is the hydrophobic and bulky nature of Tyr^{498} or its participation in this hydrogen bond network that plays the more important role in substrate binding and catalysis,

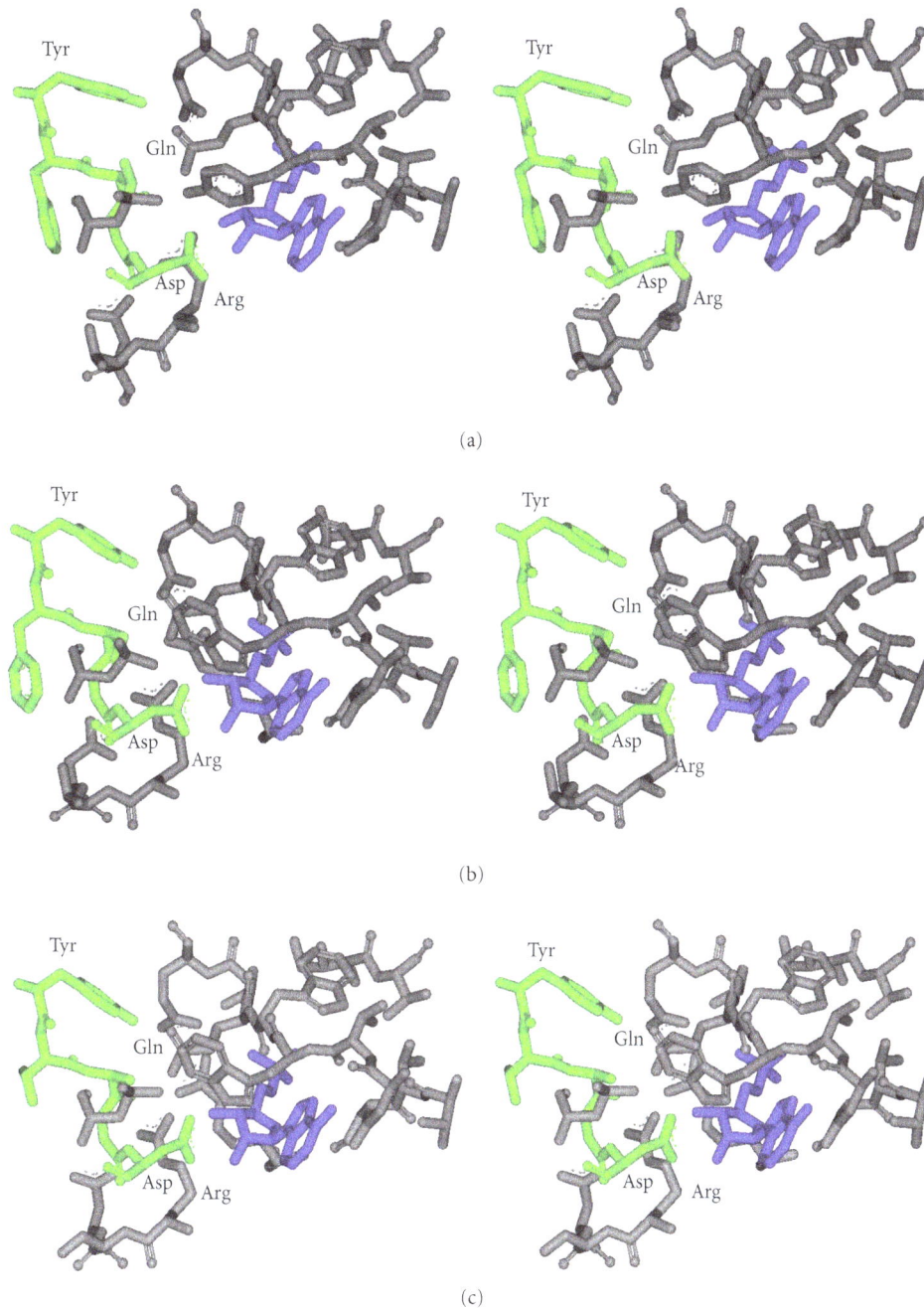

FIGURE 1: Position of motif III residues in the active sites of (a) Acs$_{Sc}$, (b) Acs$_{Se}$, and (c) the Acs1$_{Mt}$ structural model. Acs1$_{Mt}$ modeled on the Acs$_{Se}$ structure (PDB:2P2F) using Accelrys DS Modeler and the stereo images were created using Accelrys DS ViewerPro 5.0. For clarity in viewing, only some residues within the 10Å sphere of AMP are shown. AMP is shown in blue, and the motif III residues are shown in green. Residues discussed in the text are labeled.

this residue was altered to both Ala and Phe. The Tyr^{498}Ala alteration in Acs1$_{Mt}$ did not significantly affect the K_m for any substrate but reduced the turnover rate k_{cat} 41-fold (Table 2). However, the Tyr^{498} Phe variant was soluble but inactive at all substrate and enzyme concentrations tested. These results suggest that although the size of Tyr^{498} is important in maintaining active-site architecture, the hydroxyl moiety plays a critical role in properly positioning this large side chain through hydrogen bonding with Gln^{417}. Attempts at

chemical rescue of the Tyr^{498} Ala variant with phenol were unsuccessful.

3.2. Thr^{499} and Ala^{500} Are Less Well Conserved and Play Lesser Roles.

The second and third positions of motif III, represented by Thr^{499} and Ala^{500} in Acs1$_{Mt}$, are less well conserved than the other positions (Table 1). Thr^{499} is replaced by Phe in Acs$_{Se}$, Acs$_{Sc}$, and many other Acs sequences, but Leu is observed in that position in four of the five Acs sequences in

Methanosaeta concilii as well as all four of the Acs sequences in *Methanosaeta thermophila*. Ser and Thr are commonly observed at the third position in motif III, although Ala is present at this position in eight of the nine total Acs sequences in *M. concilii* and *M. thermophila*.

These positions were individually altered to Ala and Thr, respectively, in $Acs1_{Mt}$, and the purified variants were analyzed. The K_m values for substrates showed only minor changes (less than threefold increase or decrease) versus the unaltered enzyme. However, the k_{cat} value decreased 83-fold for the $Thr^{499}Ala$ variant and 44-fold for the Ala^{500} Thr variant (Table 2). Overall, these results suggest a less important role for these positions, which is consistent with the lower level of conservation observed.

3.3. Gly^{501} Is Highly Conserved and May Properly Position the Invariant Asp^{502}. Gly^{501} in the fourth position of motif III is almost completely conserved within the acyl-adenylate-forming enzyme superfamily except for a few members most distantly related to Acs. Replacement of this residue by Ala resulted in two- to threefold reduced K_m values for all three substrates; however, the turnover rate was over 200-fold reduced (Table 2). The strict conservation of this Gly and the reduced catalysis observed for the $Gly^{501}Ala$ variant are consistent with this residue playing a role in proper positioning of the critical Asp^{502} residue in the adjacent position.

3.4. Asp^{502} Plays a Critical Role in ATP Binding through Interaction with the $2'$-OH of the Ribose Moiety. To investigate the role of the invariant Asp residue of motif III, Asp^{502} of $Acs1_{Mt}$ was altered to Ala and the more conservative residues Glu and Asn. Although the enzyme variants were soluble, each of these alterations eliminated all detectable enzymatic activity, regardless of substrate concentrations or concentration of enzyme used. The fact that even the most conservative changes inactivated the enzyme indicates that this Asp is absolutely critical for activity, as might be expected since Asp^{502} is completely conserved among all ACSs and throughout the superfamily.

Inhibition assays were performed as an indirect approach to delineate the interaction between Asp^{502} and ATP. Ribose completely inhibited enzyme activity at concentrations above 600 mM, and the K_i was determined to be 53 mM (Figure 2(a)). The maximum adenosine concentration that could be reached in inhibition assays was 100 mM, which produced partial inactivation. However, extrapolation of the data indicated a K_i of ~121 mM for adenosine (Figure 2(b)). That the K_i for ribose was approximately half the estimated K_i for adenosine suggests that interaction between the enzyme and the ribose moiety plays an important role in ATP binding.

To determine more precisely the interaction between Asp^{502} and the $2'$- and $3'$-OH groups of the ribose sugar of adenosine, inhibition by $2'$- and $3'$-deoxyadenosine was examined. Although only partial inhibition was observed with either compound, extrapolation of the results gave apparent K_i values of 356 mM and 151 mM for $2'$- and

$3'$-deoxyadenosine, respectively (Figures 2(c) and 2(d)). These results suggest that interaction between Asp^{502} and adenosine is mediated primarily through the $2'$-OH group of the ribose sugar, as the absence of the $3'$-OH group had minimal effect.

The $Acs1_{Mt}$ model (Figure 1(c)) predicts hydrogen bonds between Asp^{502} and both the $2'$ and $3'$-OH groups of the ribose moiety. However, these hydrogen bonds are eliminated in the $Asp^{502}Ala$ variant. The inhibition results and the complete impairment of enzymatic activity by alterations at Asp^{502} suggest that the interaction of Asp^{502} of motif III with the $2'$-OH plays a key role although interaction with the $3'$-OH is also important for achieving optimal activity.

3.5. Interaction between the Invariant Asp and the Ribose Moiety of ATP in Other Members of the Enzyme Superfamily. The active-site architecture in Acs_{Sc} and Acs_{Se} is similar in the vicinity of the AMP ligand (Figures 1(a) and 1(b)). However, Asp^{559} of motif III in Acs_{Sc} interacts with both the $2'$- and $3'$-OH groups and with Arg^{574}, whereas Asp^{500} of motif III in Acs_{Se} hydrogen bonds with the $2'$-OH group and Trp^{413} and the $3'$-OH interacts with Gln^{415} and Arg^{515} [8, 20]. The Acs_{Sc} structure is proposed to be that for the enzyme poised to catalyze the first step of the reaction, whereas the Acs_{Se} structure has the C-terminal domain shifted inward toward the active-site to bring additional amino acid residues into context for substrate binding and catalysis of the second step of the reaction [8, 19, 20]. Whether these differences are due to slight changes in active-site architecture between two different enzymes or a movement of this Asp residue as the enzyme converts from one conformation to the other during catalysis of the two steps is unknown, as structures in both conformations are not available for either enzyme.

Structures have been determined for a number of enzymes spanning the adenylate-forming enzyme superfamily, including short, medium, and long-chain acyl-CoA synthetases, the aryl-CoA synthetases CBL and benzoyl-CoA ligase, several NRPS adenylation domains, and luciferase. These structures have revealed that domain alternation between the first and second steps of the reaction is universal among the superfamily [1]. Inspection of those structures with bound ligands indicates that in each case the invariant Asp in motif III/A7 interacts with one or both hydroxyl groups of the ribose moiety of ATP [6–8, 10, 19, 20, 23–29].

Three other members of the superfamily have structures in both the adenylate-forming and thioester-forming conformations. In 4-chlorobenzoate:CoA ligase (CBL), Asp^{385} hydrogen bonds with just the $2'$-OH, whereas the $3'$-OH interacts with Arg^{400} in the adenylate-forming conformation, whereas in the thioester-forming conformation, Asp^{385} maintains its hydrogen bond with the $2'$-OH and now also interacts with $3'$-OH group along with Arg^{400} [6, 7]. In DltA, the D-alanine:D-alanyl carrier protein ligase, Asp^{383}, interacts with both hydroxyl groups in both conformations. The $2'$-OH also interacts with Tyr^{394} in the adenylate-forming conformation, and the $3'$-OH also interacts with Arg^{396} in the thioester-forming conformation [25, 27]. Most recently, structures for the human medium-chain

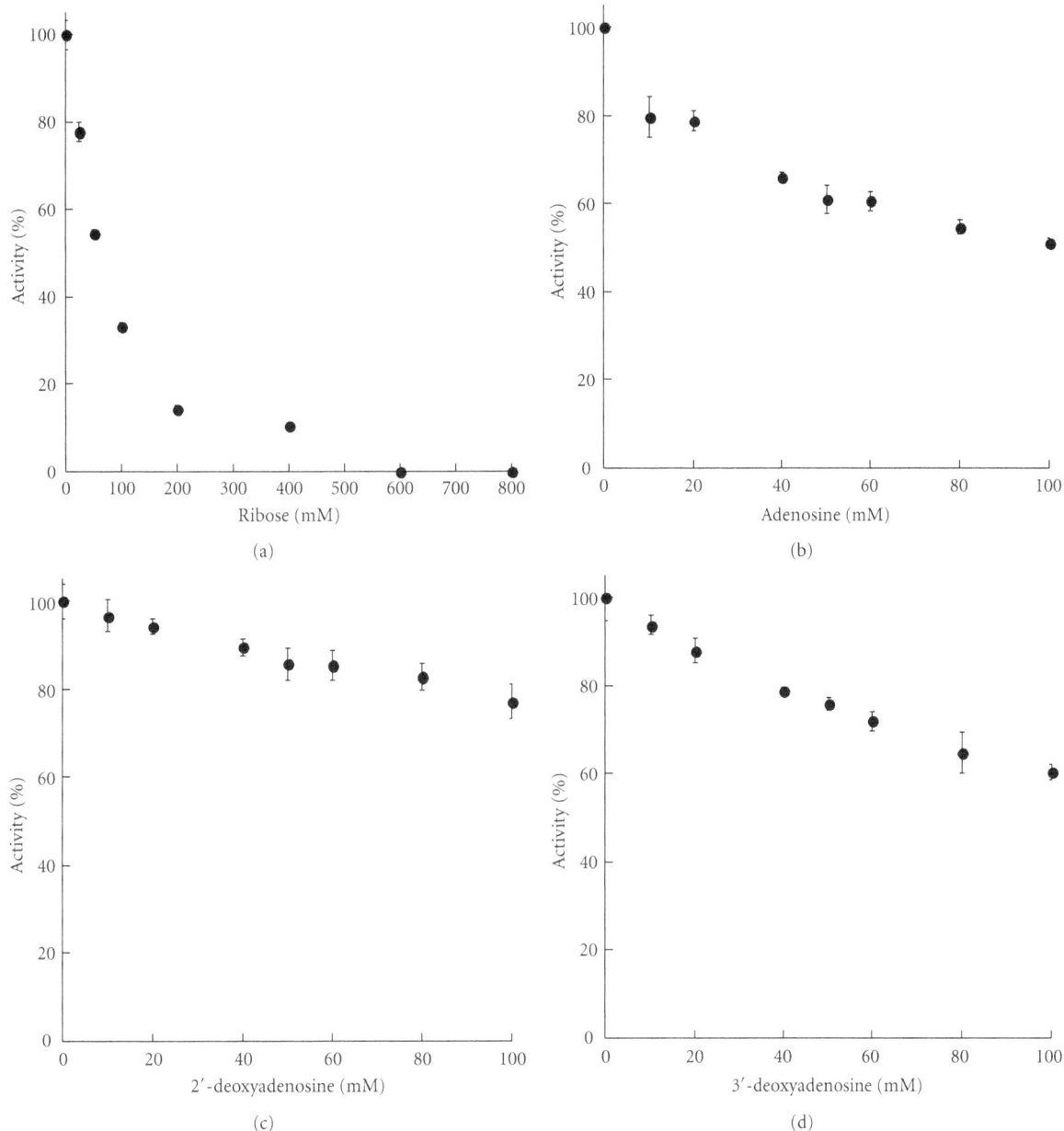

FIGURE 2: Inhibition of Acs1$_{Mt}$. (a) Ribose, (b) adenosine, (c) 2′-deoxyadenosine, and (d) 3′-deoxyadenosine. Assays were performed with the indicated concentrations of inhibitor in the reaction, and results are plotted as a percentage of the activity observed in the absence of inhibitor, with error bars as shown. The K_i value for each inhibitor was determined by extrapolation of the data.

acyl-CoA synthetase ACSM2A in both conformations have been reported [28]. Asp[446] interacts with both the 2′-OH and 3′-OH in both the adenylate-forming and thioester-forming conformations for this enzyme. Thus, different enzymes interact with the ribose moiety of ATP in different ways. In some cases, the interaction changes slightly after domain alternation. However, in all cases, the invariant Asp of motif III interacts with at least the 2′-OH, suggesting that this is the most important interaction.

3.6. Role of the Invariant Asp in Other Members of the Enzyme Superfamily. The role of this invariant Asp has been studied biochemically in only a few members of the superfamily.

In 3-chlorobenzoate-CoA ligase, alteration of this Asp to Val essentially eliminated all catalytic activity [30]. Gocht and Marahiel [31] reported that for gramicidin synthetase 1, replacement of this Asp residue with Asn or Ser reduced activity by 22% and 88%, respectively. Pavela-Vrancic et al. [32] observed with this same enzyme that replacement of ATP in the reaction with 2′-dATP resulted in a 20% reduction in activity versus a 74% reduction in activity when ATP was replaced with 3′-dATP. These results suggest that for gramicidin synthetase 1, as for Acs, interaction between the invariant Asp and the hydroxyl groups of the ribose moiety of ATP is important, with the interaction with the 2′-OH playing the most important role.

In CBL, the invariant Asp^{385} of motif III hydrogen bonds with the 2′-OH group. Alteration of this residue to Ala greatly reduced the overall rate of catalysis, primarily due to a reduced rate for the first step of the reaction, and resulted in increased K_m values for both ATP and 4-chlorobenzoate [7]. In D-alanyl carrier protein ligase, the invariant Asp hydrogen bonds with both the 2′- and 3′-OH groups of the ribose moiety [27]. Alteration of this residue to Asn reduced the rate of catalysis of the adenylation reaction only twofold but resulted in a 75-fold increased K_m value for ATP, leading the authors to conclude that this Asp plays a major role in tight binding of ATP and the adenylate intermediate [25].

4. Conclusions

The results from this investigation indicate that the motif III/A7 signature motif in Acs plays an important role in both active-site architecture and ATP binding and catalysis.

Although all positions of this motif appear to play an important role in catalysis, the Tyr at the first position that is highly conserved among Acs sequences helps maintain active site architecture through a hydrogen bond network with other active-site residues, particularly the well-conserved Gln in motif II/A5 (Gln^{417} of Acs1$_{Mt}$). Asp at the last position plays a critical role in active-site architecture through a hydrogen bond network and in ATP binding and catalysis through key interactions with the hydroxyl groups of the ribose moiety. This Asp is invariant across the entire superfamily, consistent with a critical role in ATP binding and catalysis of the adenylate-forming first step of the reaction in all members.

Authors' Contribution

C. I.-Smith and J. L. Thurman Jr. contributed equally to this work.

Acknowledgments

Financial support for this project was provided by NIH (Award GM69374-01A1), the South Carolina Experiment Station (Project SC-1700198), and Clemson University. This paper is Technical Contribution no. 6040 of the Clemson University Experiment Station.

References

[1] A. M. Gulick, "Conformational dynamics in the acyl-CoA synthetases, adenylation domains of non-ribosomal peptide synthetases, and firefly luciferase," *ACS Chemical Biology*, vol. 4, no. 10, pp. 811–827, 2009.

[2] K. H. Chang, H. Xiang, and D. Dunaway-Mariano, "Acyl-adenylate motif of the acyl-adenylate/thioester-forming enzyme superfamily: a site-directed mutagenesis study with the *Pseudomonas sp.* Strain CBS3 4-chlorobenzoate:coenzyme A ligase," *Biochemistry*, vol. 36, no. 50, pp. 15650–15659, 1997.

[3] M. A. Marahiel, T. Stachelhaus, and H. D. Mootz, "Modular peptide synthetases involved in nonribosomal peptide synthesis," *Chemical Reviews*, vol. 97, no. 7, pp. 2651–2673, 1997.

[4] C. Ingram-Smith, B. I. Woods, and K. S. Smith, "Characterization of the acyl substrate binding pocket of acetyl-CoA synthetase," *Biochemistry*, vol. 45, no. 38, pp. 11482–11490, 2006.

[5] M. A. Marahiel, "Multidomain enzymes involved in peptide synthesis," *FEBS Letters*, vol. 307, no. 1, pp. 40–43, 1992.

[6] A. S. Reger, R. Wu, D. Dunaway-Mariano, and A. M. Gulick, "Structural characterization of a 140 degrees domain movement in the two-step reaction catalyzed by 4-chlorobenzoate:CoA ligase," *Biochemistry*, vol. 47, no. 31, pp. 8016–8025, 2008.

[7] R. Wu, J. Cao, X. Lu, A. S. Reger, A. M. Gulick, and D. Dunaway-Mariano, "Mechanism of 4-chlorobenzoate:coenzyme A ligase catalysis," *Biochemistry*, vol. 47, no. 31, pp. 8026–8039, 2008.

[8] A. M. Gulick, V. J. Starai, A. R. Horswill, K. M. Homick, and J. C. Escalante-Semerena, "The 1.75 Å crystal structure of acetyl-CoA synthetase bound to adenosine-5'-propylphosphate and coenzyme A," *Biochemistry*, vol. 42, no. 10, pp. 2866–2873, 2003.

[9] A. R. Horswill and J. C. Escalante-Semerena, "Characterization of the propionyl-CoA synthetase (PrpE) enzyme of *Salmonella enterica*: residue lys^{592} is required for propionyl-AMP synthesis," *Biochemistry*, vol. 41, no. 7, pp. 2379–2387, 2002.

[10] J. J. May, N. Kessler, M. A. Marahiel, and M. T. Stubbs, "Crystal structure of DhbE, an archetype for aryl acid activating domains of modular nonribosomal peptide synthetases," *Proceedings of the National Academy of Sciences of the United States of America*, vol. 99, no. 19, pp. 12120–12125, 2002.

[11] H. P. Stuible, D. Büttner, J. Ehlting, K. Hahlbrock, and E. Kombrink, "Mutational analysis of 4-coumarate:CoA ligase identifies functionally important amino acids and verifies its close relationship to other adenylate-forming enzymes," *FEBS Letters*, vol. 467, no. 1, pp. 117–122, 2000.

[12] G. L. Challis, J. Ravel, and C. A. Townsend, "Predictive, structure-based model of amino acid recognition by nonribosomal peptide synthetase adenylation domains," *Chemistry and Biology*, vol. 7, no. 3, pp. 211–224, 2000.

[13] T. Stachelhaus, H. D. Mootz, and M. A. Marahiel, "The specificity-conferring code of adenylation domains in nonribosomal peptide synthetases," *Chemistry and Biology*, vol. 6, no. 8, pp. 493–505, 1999.

[14] C. Ingram-Smith and K. S. Smith, "AMP-forming acetyl-CoA synthetases in *Archaea* show unexpected diversity in substrate utilization," *Archaea*, vol. 2, no. 2, pp. 95–107, 2007.

[15] J. D. Thompson, T. J. Gibson, F. Plewniak, F. Jeanmougin, and D. G. Higgins, "The CLUSTAL X windows interface: flexible strategies for multiple sequence alignment aided by quality analysis tools," *Nucleic Acids Research*, vol. 25, no. 24, pp. 4876–4882, 1997.

[16] M. M. Bradford, "A rapid and sensitive method for the quantitation of microgram quantities of protein utilizing the principle of protein dye binding," *Analytical Biochemistry*, vol. 72, no. 1-2, pp. 248–254, 1976.

[17] F. Lipmann and L. C. Tuttle, "A specific micromethod for determination of acyl phosphates," *Journal of Biological Chemistry*, vol. 159, pp. 21–28, 1945.

[18] I. A. Rose, M. Grunberg-Manago, S. F. Korey, and S. Ochoa, "Enzymatic phosphorylation of acetate," *The Journal of Biological Chemistry*, vol. 211, no. 2, pp. 737–756, 1954.

[19] A. S. Reger, J. M. Carney, and A. M. Gulick, "Biochemical and crystallographic analysis of substrate binding and conformational changes in acetyl-CoA synthetase," *Biochemistry*, vol. 46, no. 22, pp. 6536–6546, 2007.

[20] G. Jogl and L. Tong, "Crystal structure of yeast acetyl-coenzyme A synthetase in complex with AMP," *Biochemistry*, vol. 43, no. 6, pp. 1425–1431, 2004.

[21] D. J. Wilson and C. C. Aldrich, "A continuous kinetic assay for adenylation enzyme activity and inhibition," *Analytical Biochemistry*, vol. 404, no. 1, pp. 56–63, 2010.

[22] Y. Meng, C. Ingram-Smith, L. L. Cooper, and K. S. Smith, "Characterization of an archaeal medium-chain acyl coenzyme A synthetase from *Methanosarcina acetivorans*," *Journal of Bacteriology*, vol. 192, no. 22, pp. 5982–5990, 2010.

[23] E. Conti, T. Stachelhaus, M. A. Marahiel, and P. Brick, "Structural basis for the activation of phenylalanine in the nonribosomal biosynthesis of gramicidin S," *The EMBO Journal*, vol. 16, no. 14, pp. 4174–4183, 1997.

[24] A. M. Gulick, X. Lu, and D. Dunaway-Mariano, "Crystal structure of 4-chlorobenzoate:CoA ligase/synthetase in the unliganded and aryl substrate-bound statest," *Biochemistry*, vol. 43, no. 27, pp. 8670–8679, 2004.

[25] K. T. Osman, L. Du, Y. He, and Y. Luo, "Crystal structure of *Bacillus cereus* D-alanyl carrier protein ligase (DltA) in complex with ATP," *Journal of Molecular Biology*, vol. 388, no. 2, pp. 345–355, 2009.

[26] Z. Zhang, R. Zhou, J. M. Sauder, P. J. Tonge, S. K. Burley, and S. Swaminathan, "Structural and functional studies of fatty acyl adenylate ligases from *E. coli* and *L. pneumophila*," *Journal of Molecular Biology*, vol. 406, no. 2, pp. 313–324, 2011.

[27] H. Yonus, P. Neumann, S. Zimmermann, J. J. May, M. A. Marahiel, and M. T. Stubbs, "Crystal structure of DltA: implications for the reaction mechanism of non-ribosomal peptide synthetase adenylation domains," *Journal of Biological Chemistry*, vol. 283, no. 47, pp. 32484–32491, 2008.

[28] G. Kochan, E. S. Pilka, F. von Delft, U. Oppermann, and W. W. Yue, "Structural snapshots for the conformation-dependent catalysis by human medium-chain acyl-coenzyme A synthetase ACSM2A," *Journal of Molecular Biology*, vol. 388, no. 5, pp. 997–1008, 2009.

[29] L. Du, Y. He, and Y. Luo, "Crystal structure and enantiomer selection by D-alanyl carrier protein ligase DltA from *Bacillus cereus*," *Biochemistry*, vol. 47, no. 44, pp. 11473–11480, 2008.

[30] S. K. Samanta and C. S. Harwood, "Use of the *Rhodopseudomonas palustris* genome sequence to identify a single amino acid that contributes to the activity of a coenzyme A ligase with chlorinated substrates," *Molecular Microbiology*, vol. 55, no. 4, pp. 1151–1159, 2005.

[31] M. Gocht and M. A. Marahiel, "Analysis of core sequences in the D-Phe activating domain of the multifunctional peptide synthetase TycA by site-directed mutagenesis," *Journal of Bacteriology*, vol. 176, no. 9, pp. 2654–2662, 1994.

[32] M. Pavela-Vrancic, H. Van Liempt, E. Pfeifer, W. Freist, and H. Von Dohren, "Nucleotide binding by multienzyme peptide synthetases," *European Journal of Biochemistry*, vol. 220, no. 2, pp. 535–542, 1994.

18

Synthetic Archaeosome Vaccines Containing Triglycosylarchaeols Can Provide Additive and Long-Lasting Immune Responses That Are Enhanced by Archaetidylserine

G. Dennis Sprott, Angela Yeung, Chantal J. Dicaire, Siu H. Yu, and Dennis M. Whitfield

Institute for Biological Sciences, National Research Council of Canada, 100 Sussex Drive, Ottawa, ON, Canada K1A 0R6

Correspondence should be addressed to G. Dennis Sprott, dennis.sprott@nrc-cnrc.gc.ca

Academic Editor: Angela Corcelli

The relation between archaeal lipid structures and their activity as adjuvants may be defined and explored by synthesizing novel head groups covalently linked to archaeol (2,3-diphytanyl-sn-glycerol). Saturated archaeol, that is suitably stable as a precursor for chemical synthesis, was obtained in high yield from *Halobacterium salinarum*. Archaeosomes consisting of the various combinations of synthesized lipids, with antigen entrapped, were used to immunize mice and subsequently determine CD8[+] and CD4[+]-T cell immune responses. Addition of 45 mol% of the glycolipids gentiotriosylarchaeol, mannotriosylarchaeol or maltotriosylarchaeol to an archaetidylglycerophosphate-O-methyl archaeosome, significantly enhanced the CD8[+] T cell response to antigen, but diminished the antibody titres in peripheral blood. Archaeosomes consisting of all three triglycosyl archaeols combined with archaetidylglycerophosphate-O-methyl (15/15/15/55 mol%) resulted in approximately additive CD8[+] T cell responses and also an antibody response not significantly different from the archaetidylglycerophosphate-O-methyl alone. Synthetic archaetidylserine played a role to further enhance the CD8[+] T cell response where the optimum content was 20–30 mol%. Vaccines giving best protection against solid tumor growth corresponded to the archaeosome adjuvant composition that gave highest immune activity in immunized mice.

1. Introduction

The total polar lipids extracted from various archaea hydrate to form liposomes (archaeosomes [1]), that were developed initially to improve the drug delivery application of conventional liposomes [2–4]. These total polar lipid archaeosomes were found subsequently to have an enhanced ability over conventional liposomes to serve as adjuvants, that promoted not only the antibody response to an entrapped protein antigen [5] but also the CD8[+] T cell response [6]. One mode of action could be correlated to an enhanced phagocytosis of archaeosomes compared to liposomes by various phagocytic cells [7]. This led to the observation that total polar lipids from various archaea, with their species-specific lipid structures, formed archaeosomes differing in receptor-mediated endocytosis and adjuvanticity [8].

Recently archaeol has been isolated from hydrolysed polar lipid extracts of *Halobacterium salinarum* to use as the lipid precursor to chemically synthesize various polar lipids, including glycolipids [9, 10]. The lipids so generated are described as synthetic or more precisely as semisynthetic, because the lipid moiety with specific archaeal sn-2,3 and R-methyl group stereochemistry is of biological origin, whereas a polar head group may be conjugated to the free sn-1 hydroxyl of the glycerol backbone to give a new polar lipid structure. In this way a chemically-defined, synthetic archaeosome could in theory be optimized for each application. Feasibility was demonstrated by synthesizing a series of diglycosylarchaeols and testing their interactions with antigen-presenting cells to produce immune responses *in vivo* [9].

The long-lasting CD8$^+$ T cell memory responses that are generally thought to be required for protection in intracellular pathogen and cancer vaccines are induced by certain total polar lipid archaeosomes and have been correlated to those archaeosomes having a high proportion of membrane-spanning caldarchaeol (tetraether) lipids [6, 11]. In this study we explore whether synthetic archaeosome adjuvants that are based on the archaeol lipids without caldarchaeols, can provide such long-term responses. Further, we explore if synthetic archaetidylserine, previously found to interact positively with the phosphatidylserine receptor of antigen-presenting cells [8, 12], can augment the adjuvant activity of synthetic glycolipid archaeosomes.

2. Materials and Methods

2.1. Growth of Archaea. *Halobacterium salinarum* (ATCC 33170) was grown aerobically at 37°C in a medium modified to be an all nonanimal origin medium consisting of: 15 g/L Phytone peptone UF (product 210931 from VWR International); 220 g/L NaCl; 6.5 g/L KCl; 10 g/L MgSO$_4$·7H$_2$O; 10 mL of 0.2 g/100 mL CaCl$_2$; 10 mL of 0.2 g/100 mL FeSO$_4$. Growth of *Haloferax volcanii* (ATCC 29605) was in medium ATCC 974 at 30°C with NaCl content of 12.5% [13]. The antifoam agent used was MAZU DF 204 (BASF Canada). Biomass was grown in 20 L medium in a 28 L New Brunswick Scientific fermentor and harvested after 72 h growth. Lipids were extracted from the biomass with chloroform/methanol/water and the total polar lipids precipitated from the lipid extract with cold acetone [14].

2.2. Purification of Archaeol. Typically, 3.5 g of total polar lipid from *H. salinarum* was dissolved in 45 mL of chloroform/methanol (2 : 1, v/v) and 190 mL methanol added. This mixture was cooled to 0°C in an ice bath, and 10 mL acetyl chloride added drop-wise while being stirred magnetically. Hydrolysis was accomplished by refluxing at 62°C for 3 h. The mixture was cooled and the volume reduced by rotary evaporation to 100 mL. Upon transfer to a separatory funnel, 12 mL water and 100 mL petroleum ether was added. The mixture was mixed and allowed to separate. The top ether phase containing lipid was pooled with a second ether extraction, and evaporated.

The archaeol oil obtained above was further purified by silica gel column chromatography. The oil dissolved in chloroform/methanol (2 : 1, v/v) was loaded on a Silica gel 60 (Merck) column and archaeol eluted with pressure using hexane/t-butylmethylether/acetic acid (80/20/0.5, v/v/v). Collected fractions were tested for archaeol by mini thin-layer chromatography using the eluting solvent, and fractions containing pure archaeol pooled and dried. The yield of archaeol from total polar lipid ranged from 43 to 53%. Structural identity and purity of archaeol was confirmed by both NMR spectroscopy and electrospray ionization mass spectrometry.

2.3. Chemical Synthesis. Archaetidylethanolamine was synthesized according to [15]. Mannotriosylarchaeol, maltotriosylarchaeol, gentiobiosylarchaeol, and gentiotriosylarchaeol

were synthesized according to our previous descriptions [10, 16] and structural details are shown in Figure 1. Synthesis methods for archaetidylserine can be found in Supplementary Material available online at doi:10.1155/2012/513231.

2.4. Purification of PGP. Archaetidylglycerolphosphate-O-CH$_3$ (PGP) was purified from the total polar lipids of *Haloferax volcanii* as described [13].

2.5. Archaeosome Vaccines. Archaeosomes were formed by hydrating 20–30 mg dried lipid at 40°C in 2 mL PBS buffer (10 mM sodium phosphate, 160 mM NaCl, pH 7.1) with ovalbumin Type VI (OVA, Sigma) as the test antigen dissolved at 10 mg/mL. Vesicle size was reduced to about 100–150 nm diameter by brief sonication in a sonic bath (Fisher Scientific), and OVA not entrapped was removed by centrifugation from 7 mL PBS followed by 2 washes (200,000 x g max for 90 min). Vesicle pellets were resuspended in 2–2.5 mL PBS and filter sterilized through 0.45 μm Millipore filters. Sterile conditions and pyrogen-free water were used throughout.

Quantification of antigen loading was conducted by separating OVA from lipids using SDS polyacrylamide gel electrophoresis and densitometry as described [14]. Loading was based on μg protein/mg salt corrected dry weight of lipid. Average diameters based on intensity were measured using a Malvern Nano Zetasizer with a He/Ne laser (Spectra Research Corp., Ontario, Canada).

2.6. Animal Trials. C57BL/6 female mice (6–8 weeks old) were immunized subcutaneously near the tail base with 0.1 mL vaccines containing the equivalent of 20 μg OVA, often entrapped in archaeosomes of various compositions. A booster consisting of the same vaccine and route was given on week 3. All protocols and SOPs were approved by the NRC Animal Care Committee and conducted within the guidelines of the Canadian Council on Animal Care.

2.7. Immune Responses. As a measure of the Th2 arm of CD4$^+$ T cell adjuvant activity, IgG antibody raised in response to the antigen in the vaccine and collected in the sera of mice (5-6 mice/group) was quantified by Elisa according to a previous description [17]. The CD8$^+$ T cell response was quantified by sacrificing duplicate mice/group to obtain their splenic cells. These splenic cells were assayed in triplicate for antigen-specific responses by standard Elispot and cytolytic T lymphocyte (CTL) methods [18].

2.8. Dendritic Cell (DC) Maturation Assay. Bone marrow was flushed from femurs and tibias of C57BL/6 mice to isolate DCs. Cells obtained were cultured in RPMI medium supplemented with 8% fetal calf serum (R8) (Thermo Scientific HyClone, UT, USA) and 5 ng/mL of granulocyte macrophage colony-stimulating factor (ID Labs, Inc., Ont., Canada) [19]. Nonadherent cells were removed on days 2 and 4 and supplied with fresh medium. Bone marrow DCs were harvested as the nonadherent cells on day 7. DC purity was greater than 90% based on flow cytometry

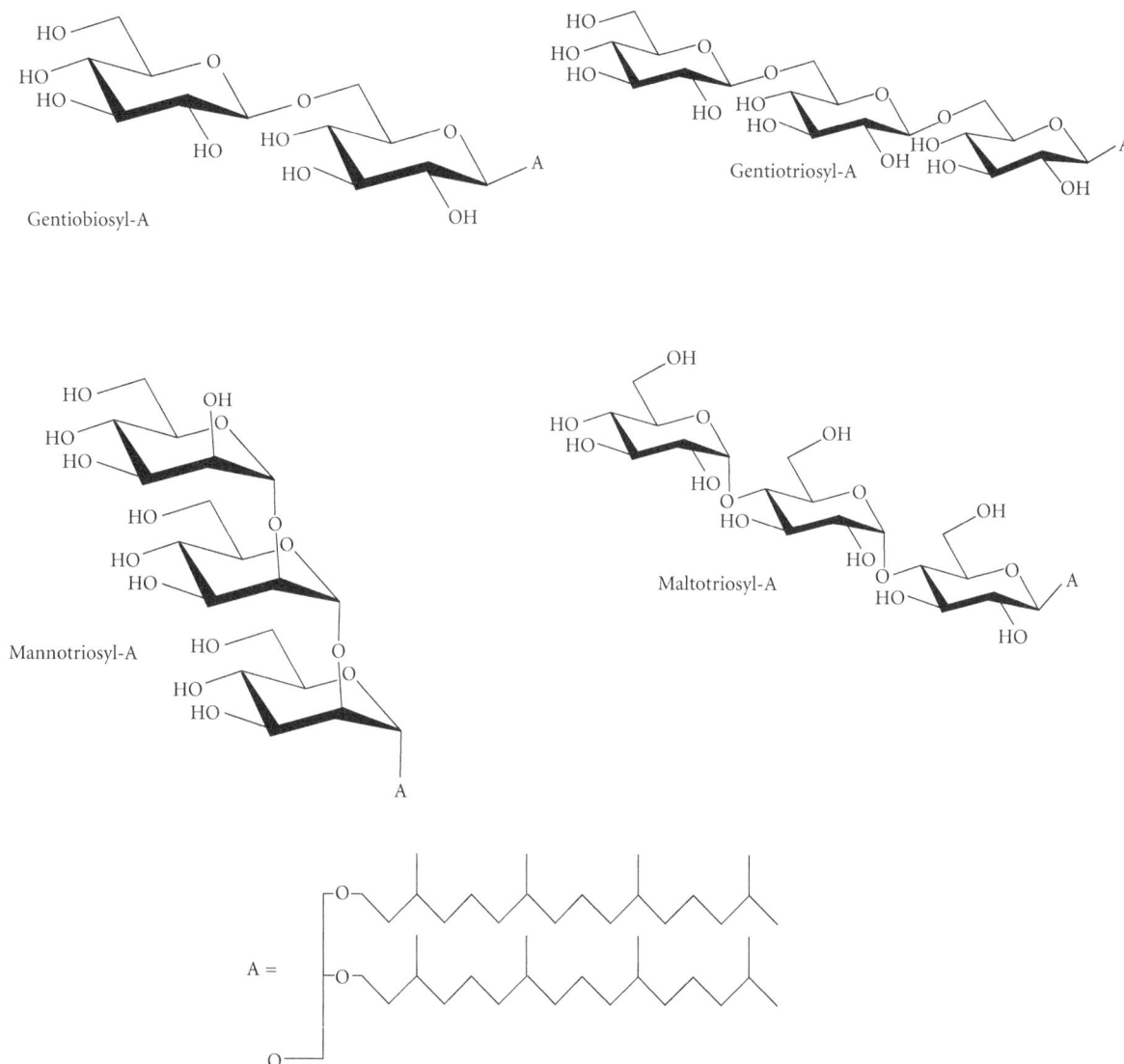

FIGURE 1: Semisynthetic glycoarchaeol structures showing head group details. "A" is the archaeol lipid precursor used for synthesis. gentiobiosyl-A (β-Glc$_p$-(1 → 6)-β-Glc$_p$-(1 → O)-archaeol); Gentiotriosyl-A (β-Glc$_p$-(1 → 6)-β-Glc$_p$-(1 → 6)-β-Glc$_p$-(1 → O)-archaeol); mannotriosyl-A (α-Man$_p$-(1 → 2)-α-Man$_p$-(1 → 2)-α-Man$_p$-(1 → O)-archaeol); maltotriosyl-A (α-Glc$_p$-(1 → 4)-α-Glc$_p$-(1 → 4)-β-Glc$_p$-(1 → O)-archaeol).

of cells labeled with PE-Cy7 conjugated anti-CD11c mAb (BD Biosciences, Ont., Canada). To activate, on day 7 DCs (3×10^5 cells/mL) were stimulated with 25 μg of various antigen-free archaeosomes or 1 μg E. coli lipopolysaccharide (LPS, Sigma-Aldrich, Ltd., Ont., Canada) per mL in 24-well plates for 24 h. Maturation was measured by the FITC-dextran (Sigma-Aldrich, Ltd., Ont., Canada) uptake assay using flow cytometry [20]. DCs were suspended in R8 medium and incubated with 1 mg/mL of FITC-dextran (Mr = 40 000) for 30 min at 4 or 37°C. After incubation, the cells were washed three times with ice cold 1% sodium azide in PBS. The quantitative uptake was calculated as the change in the Mean Fluorescence Index (MFI) between cell samples incubated at 37 and 4°C.

2.9. EG.7 Solid Tumour Model. C57BL/6 mice were immunized at 0 and 3 weeks subcutaneously with archaeosomes containing 20 μg OVA. A challenge consisting of 5×10^6 EG.7 cells was introduced subcutaneously in the shaved lower dorsal region at either 4.5 weeks, or 14 weeks from the second immunization. Tumour progression was measured in two dimensions with a digital calliper, and values multiplied to give tumour sizes. When a tumour mass of 300 mm^2 was reached the mouse was euthanized.

2.10. Statistics. A comparison of means for animal data was conducted using student's t-test to determine significance at 95% confidence, and two tailed P values calculated using GraphPad Prism 5.

Synthetic Archaeosome Vaccines Containing Triglycosylarchaeols Can Provide Additive and Long-Lasting
Immune Responses That Are Enhanced by Archaetidylserine

183

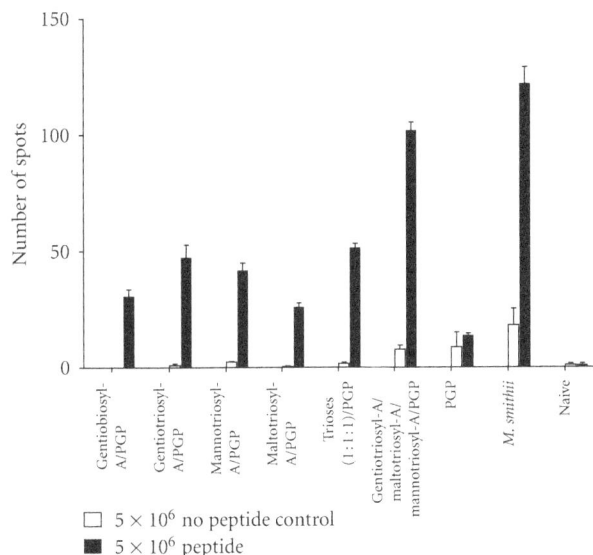

FIGURE 2: Antigen-specific CD8+ T cell activity in splenic cells of immunized mice as assayed by Elispot. Ratios of lipids in mol% for various compositions of archaeosomes were: di or triglycosylarchaeols/PGP (45/55), and gentiotriosyl-A/maltotriosyl-A/mannotriosyl-A/PGP (15/15/15/55), where "A" refers to archaeol. Trioses (1:1:1)/PGP refers to admixed triglycosyl-A/PGP vaccines, such that each contributed equal amounts of antigen. *M. smithii* represents OVA-loaded archaeosomes consisting of total polar lipids from *M. smithii*, as positive control. Mice were immunized subcutaneously at 0 and 3 weeks with OVA-loaded archaeosome adjuvants. Nonimmunized mice (naive) were included as negative controls. Spleens from duplicate mice were collected 5.5 weeks after first injection to determine the frequency (number of spots) of interferon-gamma (IFN-γ)-secreting splenic cells (spots) by enzyme-linked immunospot assay (Elispot). Omission of the major CD8 epitope of OVA (SIINFEKL) from the assay (no peptide control) was used to test for nonspecific responses. Means significantly different ($P < 0.05$) were gentiotriosyl-A/PGP versus maltotriosyl-A/PGP ($P = 0.0204$), mannotriosyl-A/PGP versus maltotriosyl-A/PGP ($P = 0.0135$), and maltotriosyl-A/PGP versus PGP ($P = 0.0032$). Those not significantly different were gentiotriosyl-A/PGP versus mannotriosyl-A/PGP ($P = 0.4238$), gentiotriosyl-A/PGP versus gentiobiosyl-A/PGP ($P = 0.0550$), mannotriosyl-A/PGP versus gentiobiosyl-A/PGP ($P = 0.0677$), and gentiotriosyl-A/maltotriosyl-A/mannotriosyl-A/PGP versus *M. smithii* ($P = 0.0657$).

3. Results

3.1. Synthetic Glycosylarchaeols as Adjuvants. To prepare stable glycolipid archaeosome adjuvants from neutrally charged glycosylarchaeols it was necessary to include a charged lipid. This function may be served by a conventional ester-phospholipid such as phosphatidylglycerol. Although mice vaccinated with archaeosomes consisting of synthetic diglycosylarchaeols mixed with dipalmitoyl phosphatidylglycerol and antigen developed short-term CD8+ T cell mediated immune responses [9], longer-term responses were lost [15]. Consequently, we avoided conventional lipids in this study designed to evaluate the potential for long-term immunity

from archaeol adjuvants, and chose instead an archaeol-based anionic lipid, PGP, purified from *H. volcanii*. The combination of glycosylarchaeols with PGP resulted in stable bilayers in the 100 nm average diameter range that entrapped the OVA antigen from 12–21 μg protein/mg dry weight (Table 1).

First, we tested CD8+ T cell responses in immunized mice using glycosylarchaeol/PGP adjuvants in short-term experiments assayed 2.5 weeks from the booster immunization (Figure 2). Elispot assays confirmed gentiotriosylarchaeol to be a better adjuvant than gentiobiosylarchaeol. Although not highly significant for the data shown here ($P = 0.055$), in other trials the difference in means was characteristically $P = 0.001$. Further, both gentiotriosylarchaeol and mannotriosylarchaeol were significantly better adjuvants with PGP than was maltotriosylarchaeol. When all three triosylarchaeol vaccines were admixed in equal proportion prior to immunization the CD8 response was not greatly improved. However, a strikingly improved adjuvant activity, approaching the *M. smithii* total polar lipid positive control, was observed when the triglycosylarchaeols were incorporated into the same archaeosome preparation during hydration.

CD8 responses in mice can also be measured by cytolytic T lymphocyte (CTL) assays that measure the ability of effector cells in the spleens of immunized mice to lyse an EG.7 target cell line expressing the dominant epitope (SIINFEKL) of OVA. In this assay (Figure 3) the same trends as found in Elispots occurred, although maltotriosylarchaeol/PGP was less effective as an adjuvant than pure PGP archaeosomes. The combined triosylarchaeols/PGP (45/55 mol%) again produced an adjuvant equivalent to the total polar lipid positive control. Because of these results, we omitted maltotriosylarchaeol from further studies, and continued with the combination of gentiotriosylarchaeol/mannotriosylarchaeol/PGP.

To evaluate the ability of archaeosome adjuvants to direct antigen via antigen-presenting cells through MHC class-II presentation to CD4+ T cells (see Figure 1 of [9]), we assayed anti OVA antibody titres in the peripheral blood of mice (Figure 4). Best titres were found for PGP archaeosomes, indicating that these archaeosomes favour an MHC-II route of antigen presentation versus MHC-I (as measured by CD8+ T cell responses). Antibody titres for PGP were significantly higher for all adjuvants except when compared to the combination of triosylarchaeols, which was not significantly different ($P = 0.056$).

3.2. Archaetidylserine (AS) and Archaetidylethanolamine (AE). The phosphatidylserine receptor is implicated in promoting phagocytosis of apoptotic cell debris [21] and archaeosomes [12]. Further, both archaetidylserine and archaetidylethanolamine are potentially fusogenic lipids, based on the assumption of similar activity to their ester analogs [22], and fusion of internalized archaeosomes with the phagolysosome membrane is the mechanism proposed to export antigen from archaeosomes to the MHC-I pathway

TABLE 1: Characterization of OVA-archaeosomes.

Archaeosome lipids	Average diameter (nm)	OVA content (µg/mg)
Gentiobiosyl-A/PGP	153 ± 54	19.6
Gentiotriosyl-A/PGP	90 ± 52	21.1
Mannotriosyl-A/PGP	76 ± 41	16.9
Maltotriosyl-A/PGP	97 ± 46	14.7
PGP	92 ± 54	14.2
M. smithii	88 ± 51	12.0

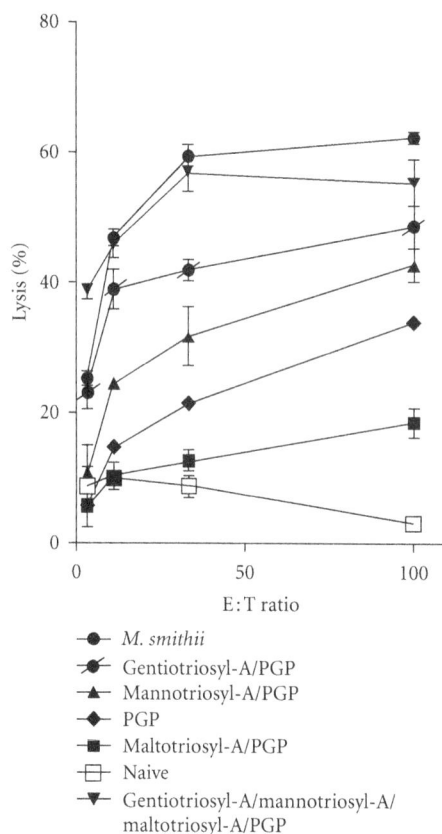

FIGURE 3: A cytotoxic T lymphocyte (CTL) lysis assay was used to assess the same populations of splenic cells as in Figure 1. The standard ^{51}Cr assay was conducted using specific and nonspecific target cells (EG.7 and EL-4, resp.). The ratio of effector splenic cells to target cells is shown as the E : T ratio in the graph. Results shown are for EG.7 targets. EL-4 targets produced only low nonspecific responses (not shown).

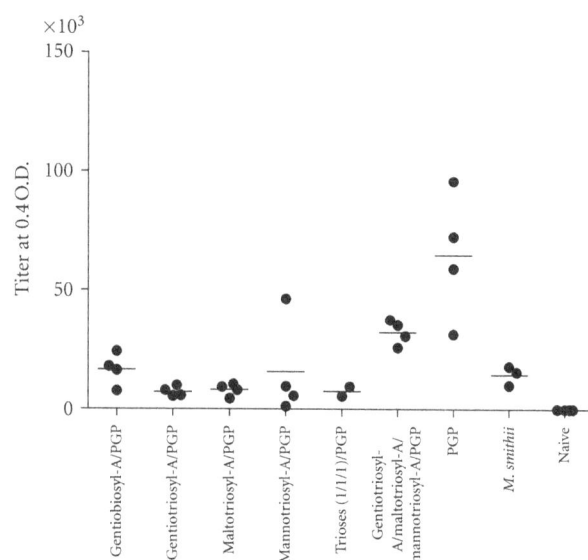

FIGURE 4: Antibody titres in sera of mice immunized with various archaeosome adjuvants. Peripheral blood was collected at 5.5 weeks, just prior to euthanizing mice for spleen removal (Figure 1). Each data point shown represents the titre in the serum from an individual mouse. Means were significantly higher for the OVA-PGP archaeosome vaccinated group ($P < 0.05$) compared to all groups except for Gentiotriosyl-A/Maltotriosyl-A/Mannotriosyl-A/PGP ($P = 0.0560$).

of antigen-presenting cells [12, 23]. Consequently, importance of AS or AE incorporated into the mannotriosylarchaeol/gentiotriosylarchaeol/PGP archaeosome was assessed in terms of adjuvanting CD8$^+$ T cell responses (Figure 5). Addition of 30 mol% AS to the glycotriosylarchaeol/PGP adjuvant resulted in a significantly higher CD8 response ($P = 0.0207$) that was not significantly different than the positive control (*M. smithii*). AE combined with AS had little further influence on adjuvanticity. As in other mouse trials, incorporation of glycoarchaeols to PGP archaeosomes produced a much improved CD8$^+$ T cell response. In contrast, anti OVA

antibody titres in peripheral blood were not significantly higher upon inclusion of AS (data not shown).

To quantify the optimal amount of AS to adjuvant the CD8$^+$ T cell mediated response, from 0 to 30 mol% AS was incorporated into the triglycosylarchaeol/PGP archaeosome. Elispot assays (Figure 6) showed little effect of 10% AS, with an optimal effect of >20–30 mol%. Archaeosomes could not be tested with >30 mol% AS because of instability. These findings were verified by CTL assays (Figure 7), that confirmed an adjuvant activity at 30 mol% AS to be somewhat higher than the positive control. As shown in Figure 5, the addition of AE to the adjuvant mix was rarely positive.

3.3. Maturation of DCs. Loss of ability to take up dextran was used to assess the extent of activation of DCs exposed *in vitro* to the various archaeosomes (lacking antigen) (Figure 8). LPS served as a positive control. Activation was similar for LPS, *M. smithii* archaeosomes, and the combination

Synthetic Archaeosome Vaccines Containing Triglycosylarchaeols Can Provide Additive and Long-Lasting
Immune Responses That Are Enhanced by Archaetidylserine

185

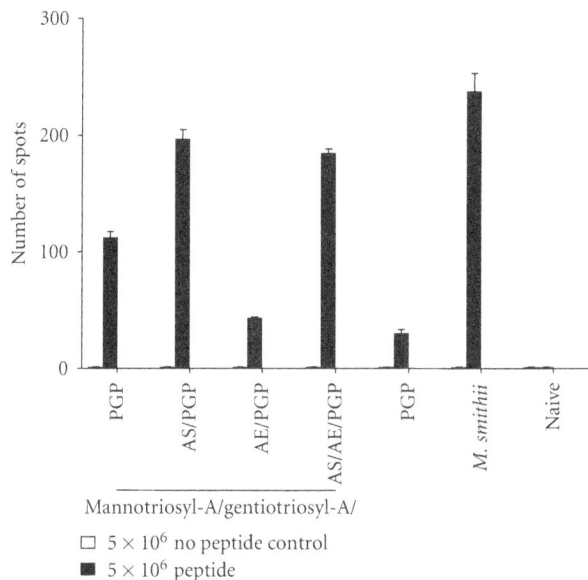

FIGURE 5: Elispot assay showing a relationship between adjuvant activity of glycosylarchaeol/PGP archaeosomes, archaetidylserine (AS), and archaetidylethanolamine (AE). Mol% compositions for OVA-archaeosome vaccines were mannotriosyl-A/gentiotriosyl-A/PGP (22.5/22.5/55), mannotriosyl-A/gentiotriosyl-A/AS/PGP (22.5/ 22.5/30/25), mannotriosyl-A/gentiotriosyl-A/AE/PGP (22.5/ 22.5/5/50), mannotriosyl-A/gentiotriosyl-A/AS/AE/PGP (22.5/ 22.5/30/5/20). Assays were conducted on splenic cells of mice 6 weeks post first immunization.

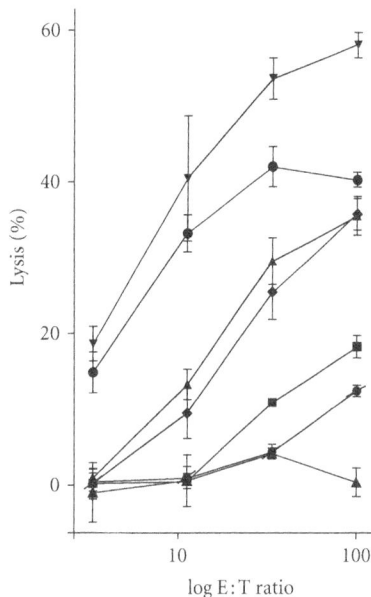

FIGURE 7: Cytotoxic T lymphocyte (CTL) lysis assay was used to assess the same populations of splenic cells as in Figure 6. Loadings of the OVA antigen are shown also in this figure. EL-4 control targets not expressing SIINFEKL gave <10% lysis in all cases (not shown).

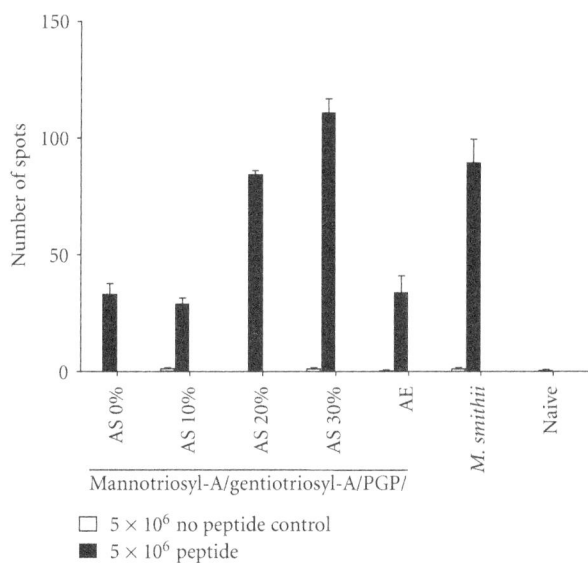

FIGURE 6: Elispot assay showing a relationship between mol% AS in a triglycosyl-A/PGP archaeosome and CD8 adjuvant activity. Mannotriosyl-A and gentiotriosyl-A were always 22.5 mol% each. AS was varied as shown at 0, 10, 20, and 30 mol%, with PGP making the remainder of each composition. For comparison, archaeosomes containing 5 mol% AE and *M. smithii* total polar lipid archaeosomes are included.

of gentiotriosylarchaeol/mannotriosylarchaeol/PGP with or without AS. Evidence for AS activation could be seen, however, by comparing AS/PGP (30/70 mol%) archaeosomes to either of AE/PGP (5/95 mol%), gentiotriosylarchaeol/PGP, or mannotriosylarchaeol/PGP archaeosomes.

3.4. Short and Long-Term Protective Immunity. M. smithii total polar lipid archaeosomes are capable of adjuvanting a $CD8^+$ T cell response that is long-lasting and provides protection in a solid tumour model in mice [11, 18]. Here we compare immune responses to protection in mice immunized with the various synthetic archaeosomes-OVA. Short-term immunity was assessed in animals ($n = 5$) by challenge with EG.7 tumour cells 4.5 weeks after the second immunization (Figure 9(a)). Protection could be correlated to the $CD8^+$ T cell immune responses achieved (see previous figures). Naive mice are considered unprotected and succumbed to tumour growth early. PGP-OVA archaeosomes showed only limited protection, with best protection achieved with mannotriosylarchaeol/gentiotriosylarchaeol/AS/PGP OVA-archaeosomes. Longer-term immunity was assessed by injection of EG.7 cells 14 weeks following the second immunization (Figure 9(b)). Nonimmunized naive mice and OVA immunizations (no adjuvant) showed no protection,

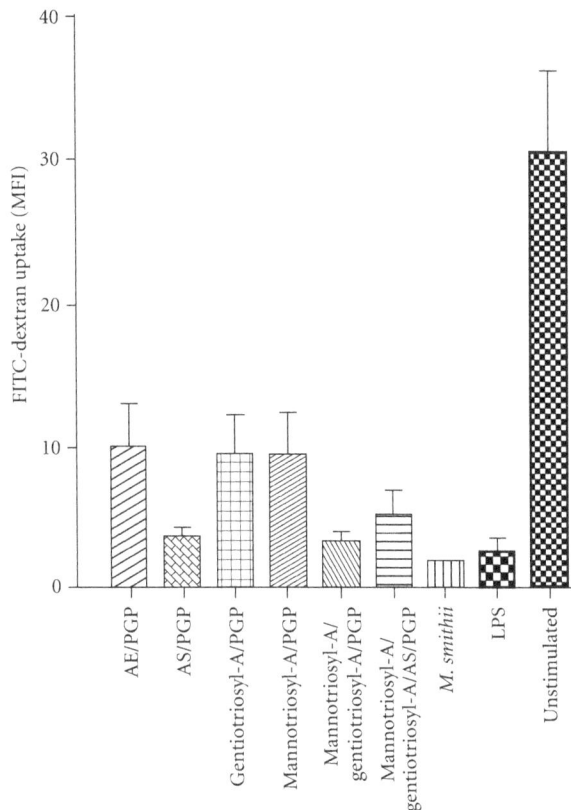

FIGURE 8: Maturation of dendritic cells (DCs) upon treatment with archaeosomes as measured by decrease of FITC-dextran uptake. Bone marrow DCs treated with archaeosomes *in vitro* were compared for their ability to take up FITC-dextran. The results depict the mean ΔMFI (37–4°C). Data represent means $\pm$ SD of triplicate cultures as indicated.

whereas the optimized archaeosome gave protection similar to the positive control.

4. Discussion

A goal of this study was to define an archaeosome adjuvant composition suitable for human application through use of synthetic archaeol-based lipids. Past studies on the mechanism of archaeosomes made from the total polar lipids of various archaea have shown that adjuvant activity occurs at the level of the antigen-presenting dendritic and macrophages cells [19, 24]. Glycolipids in these total polar lipid mixtures may presumably serve as effective adjuvant ingredients as they can target specific receptors on antigen-presenting cells [9, 25, 26].

As glycolipids are uncharged, a stable bilayer does not form when attempts are made to prepare pure glycolipid-liposome based vaccines. This can be achieved, as is the case for natural polar lipids consisting of both glyco and phospholipids, by including phospholipids in the glycolipid formulation [9]. Because inclusion of nonarchaeal lipids such as dipalmitoyl phosphatidylglycerol into archaeosomes results in decline in longer-term CD8$^+$ T cell mediated

immune responses [15], we used the diacidic extreme halophile lipid, PGP, in our synthetic glycoarchaeol formulations.

A series of mannosylarchaeols synthesized to have from 1 to 5 sugar units, hydrated best and gave best adjuvant activity at 3 or 4 linear sugar units [16]. Similarly, we found gentiotriosylarchaeol to be a better adjuvant than gentiobiosylarchaeol. Further, the additive adjuvant effect obtained by inclusion of both gentiotriosylarchaeol and mannotriosylarchaeol suggests multiple positive interactions with receptors, to account for an observed increased activation of antigen-presenting dendritic cells (Figure 8). This additive effect of glycosylarchaeols required that the archaeosome preparation be hydrated with all lipids present, suggesting that the various head groups on the archaeosome surface were presented simultaneously to multiple receptors *in vivo*.

Archaetidylserine (AS) as a component of gentiotriosylarchaeol/mannotriosylarchaeol/PGP archaeosomes increased the CD8$^+$ T cell immune response to entrapped antigen in a concentration dependent manner, without significantly enhancing the antibody response (Figures 6–7). *M. smithii* total polar lipid archaeosomes contain AS and their endocytosis has been linked to interaction with the phosphatidylserine receptor of antigen-presenting cells [12]. The pathway of cross-presentation of antigen carried in *M. smithii* archaeosomes occurs at the late phagolysosome stage [12] when calcium is internalized [27], suggesting that AS also contributes to membrane fusion promoted by calcium in analogy to phosphatidylserine [23]. Fusion of archaeosomes with the phagolysosome membrane would contribute to export of antigen to the cytosol and provide access to the MHC class-I presentation pathway.

The longevity of CD8$^+$ T cell memory induced by total polar lipid archaeosomes of *M. smithii* and *Thermoplasma acidophilum* is generally not found in archaeosomes prepared from total polar lipids of extreme halophiles, that lack caldarchaeols [6]. For this reason, it was proposed that long-term CD8$^+$ T cell memory may require the presence of high proportions of caldarchaeol membrane-stabilizing lipids. In this study we found that protective CD8$^+$ T cell memory responses could be induced in mice immunized with antigen-archaeosomes lacking caldarchaeols. This further indicated the importance of head group in lipid composition of an all archaeol-based adjuvant [9].

5. Conclusion

The immune response to antigen may be preferentially directed to either MHC-I (CD8) or MHC-II (CD4) presentations by selection of the head group(s) of an archaeol-based adjuvant. PGP archaeosomes direct antigen primarily to an antibody pathway of response as suggested previously [10]. Additions of glycoarchaeols to PGP archaeosomes enhance greatly the MHC class I pathway of antigen presentation producing the CD8$^+$ T cell response. Combination of gentiotriosyl- and mannotriosylarchaeols in the archaeosome adjuvant enhanced the CD8$^+$ T cell response over

Synthetic Archaeosome Vaccines Containing Triglycosylarchaeols Can Provide Additive and Long-Lasting
Immune Responses That Are Enhanced by Archaetidylserine

187

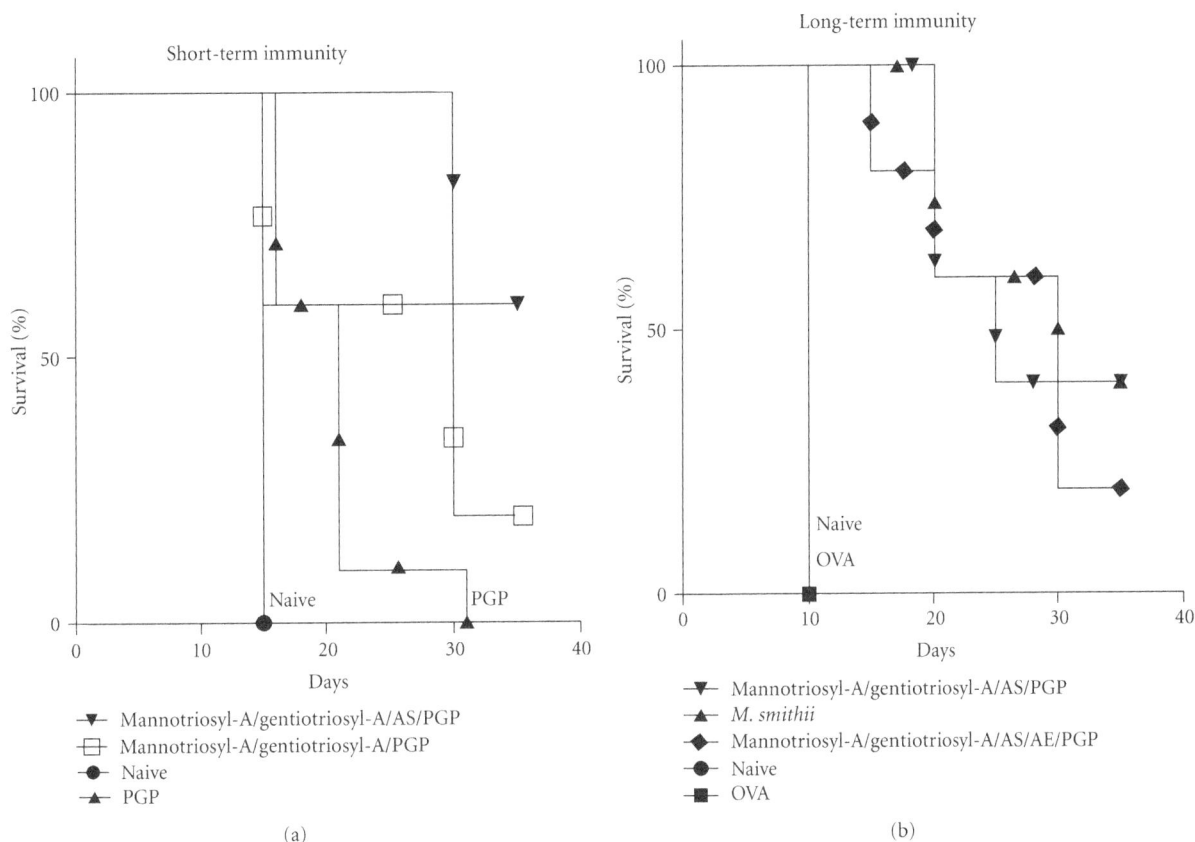

FIGURE 9: Protection of mice immunized with archaeosomes of various lipid compositions in a solid tumour model. Groups (n = 5) of unvaccinated mice (naive) or mice vaccinated with various OVA-archaeosomes were challenged with a subcutaneous injection of EG.7 cells (time zero) either 4.5 weeks (panel (a)) or 14 weeks (panel (b)) following their last vaccination.

either alone, and the additional presence of archaetidylserine was of further benefit. Finally, long-term immunity was obtained in an archaeol-based lipid archaeosome lacking caldarchaeols. We conclude that for a cancer or intracellular pathogen vaccine where a CD8$^+$ T cell response is needed, a favorable archaeosome composition is gentiotriosylar-chaeol, mannotriosylarchaeol, AS, and PGP in mol% ratio 22.5/22.5/30/25.

Acknowledgments

Halobacterium salinarum biomass was produced by Mr. John Shelvey in the NRC Bacterial Culture Facility. Animal Husbandry and the bleeding of mice were conducted in the NRC Animal Facility under the guidance of Dr. Craig Bihun. The final art work for figures was done by Mr. Tom Devecseri.

References

[1] G. D. Sprott, C. J. Dicaire, L. P. Fleming, and G. B. Patel, "Stability of liposomes prepared from archaeobacterial lipids and phosphatidylcholine mixtures," *Cells and Materials*, vol. 6, no. 1–3, pp. 143–155, 1996.

[2] C. G. Choquet, G. B. Patel, T. J. Beveridge, and G. D. Sprott, "Stability of pressure-extrudes liposomes made from archaeobacterial ether lipids," *Applied Microbiology and Biotechnology*, vol. 42, no. 2-3, pp. 375–384, 1994.

[3] A. Omri, B. Makabi-Panzu, B. J. Agnew, G. D. Sprott, and G. B. Patel, "Influence of coenzyme Q10 on tissue distribution of archaeosomes, and pegylated archaeosomes, administered to mice by oral and intravenous routes," *Journal of Drug Targeting*, vol. 7, no. 5, pp. 383–392, 2000.

[4] K. Ring, B. Henkel, A. Valenteijn, and R. Gutermann, "Studies on the permeability and stability of liposomes derived from a membrane spanning bipolar archaebacterial tetraetherlipid," in *Liposomes as Drug Carriers*, K. H. Schmidt, Ed., pp. 101–123, Georg Thieme, Stuttgart, Germany, 1986.

[5] G. D. Sprott, D. L. Tolson, and G. B. Patel, "Archaeosomes as novel antigen delivery systems," *FEMS Microbiology Letters*, vol. 154, no. 1, pp. 17–22, 1997.

[6] L. Krishnan and G. D. Sprott, "Archaeosomes as self-adjuvanting delivery systems for cancer vaccines," *Journal of Drug Targeting*, vol. 11, no. 8–10, pp. 515–524, 2003.

[7] D. L. Tolson, R. K. Latta, G. B. Patel, and G. D. Sprott, "Uptake of archaeobacterial liposomes and conventional liposomes by phagocytic cells," *Journal of Liposome Research*, vol. 6, no. 4, pp. 755–776, 1996.

[8] G. D. Sprott, S. Sad, L. P. Fleming, C. J. Dicaire, G. B. Patel, and L. Krishnan, "Archaeosomes varying in lipid composition differ in receptor-mediated endocytosis and differentially adjuvant immune responses to entrapped antigen," *Archaea*, vol. 1, no. 3, pp. 151–164, 2003.

[9] D. G. Sprott, C. J. Dicaire, J. P. Côté, and D. M. Whitfield, "Adjuvant potential of archaeal synthetic glycolipid mimetics critically depends on the glyco head group structure," *Glycobiology*, vol. 18, no. 7, pp. 559–565, 2008.

[10] D. M. Whitfield, S. H. Yu, C. J. Dicaire, and G. D. Sprott, "Development of new glycosylation methodologies for the synthesis of archaeal-derived glycolipid adjuvants," *Carbohydrate Research*, vol. 345, no. 2, pp. 214–229, 2010.

[11] L. Krishnan, S. Sad, G. B. Patel, and G. D. Sprott, "Archaeosomes induce long-term CD8+ cytotoxic T cell response to entrapped soluble protein by the exogenous cytosolic pathway, in the absence of CD4+ T cell help," *Journal of Immunology*, vol. 165, no. 9, pp. 5177–5185, 2000.

[12] K. Gurnani, J. Kennedy, S. Sad, G. D. Sprott, and L. Krishnan, "Phosphatidylserine receptor-mediated recognition of archaeosome adjuvant promotes endocytosis and MHC class I cross-presentation of the entrapped antigen by phagosome-to-cytosol transport and classical processing," *Journal of Immunology*, vol. 173, no. 1, pp. 566–578, 2004.

[13] G. D. Sprott, S. Larocque, N. Cadotte, C. J. Dicaire, M. McGee, and J. R. Brisson, "Novel polar lipids of halophilic eubacterium Planococcus H8 and archaeon Haloferax volcanii," *Biochimica et Biophysica Acta*, vol. 1633, no. 3, pp. 179–188, 2003.

[14] G. D. Sprott, G. B. Patel, and L. Krishnan, "Archaeobacterial ether lipid liposomes as vaccine adjuvants," *Methods in Enzymology*, vol. 373, article 11, pp. 155–172, 2003.

[15] C. J. Dicaire, S. H. Yu, D. M. Whitfield, and G. D. Sprott, "Isopranoid- and dipalmitoyl-aminophospholipid adjuvants impact differently on longevity of CTL immune responses," *Journal of Liposome Research*, vol. 20, no. 4, pp. 304–314, 2010.

[16] D. M. Whitfield, E. E. Eichler, and G. D. Sprott, "Synthesis of archaeal glycolipid adjuvants-what is the optimum number of sugars?" *Carbohydrate Research*, vol. 343, no. 14, pp. 2349–2360, 2008.

[17] L. Krishnan, C. J. Dicaire, G. B. Patel, and G. D. Sprott, "Archaeosome vaccine adjuvants induce strong humoral, cell-mediated, and memory responses: comparison to conventional liposomes and alum," *Infection and Immunity*, vol. 68, no. 1, pp. 54–63, 2000.

[18] L. Krishnan, S. Sad, G. B. Patel, and G. D. Sprott, "Archaeosomes induce enhanced cytotoxic T lymphocyte responses to entrapped soluble protein in the absence of interleukin 12 and protect against tumor challenge," *Cancer Research*, vol. 63, no. 10, pp. 2526–2534, 2003.

[19] L. Krishnan, S. Sad, G. B. Patel, and G. D. Sprott, "The potent adjuvant activity of archaeosomes correlates to the recruitment and activation of macrophages and dendritic cells in vivo," *Journal of Immunology*, vol. 166, no. 3, pp. 1885–1893, 2001.

[20] M. Kato, T. K. Nell, D. B. Fearnley, A. D. McLellan, S. Vuckovic, and D. N. J. Hart, "Expression of multilectin receptors and comparative FITC-dextran uptake by human dendritic cells," *International Immunology*, vol. 12, no. 11, pp. 1511–1519, 2000.

[21] V. A. Fadok, D. L. Bratton, D. M. Rose, A. Pearson, R. A. B. Ezekewitz, and P. M. Henson, "A receptor for phosphatidylserine-specific clearance of apoptotic cells," *Nature*, vol. 405, no. 6782, pp. 85–90, 2000.

[22] S. Martens and H. T. McMahon, "Mechanisms of membrane fusion: disparate players and common principles," *Nature Reviews Molecular Cell Biology*, vol. 9, no. 7, pp. 543–556, 2008.

[23] G. D. Sprott, J. P. Côté, and H. C. Jarrell, "Glycosidase-induced fusion of isoprenoid gentiobiosyl lipid membranes at acidic pH," *Glycobiology*, vol. 19, no. 3, pp. 267–276, 2009.

[24] L. Krishnan and G. D. Sprott, "Archaeosome adjuvants: immunological capabilities and mechanism(s) of action," *Vaccine*, vol. 26, no. 17, pp. 2043–2055, 2008.

[25] R. N. Coler, S. Bertholet, M. Moutaftsi et al., "Development and characterization of synthetic glucopyranosyl lipid adjuvant system as a vaccine adjuvant," *PLoS ONE*, vol. 6, no. 1, article e16333, 2011.

[26] I. Matsunaga and D. B. Moody, "Mincle is a long sought receptor for mycobacterial cord factor," *Journal of Experimental Medicine*, vol. 206, no. 13, pp. 2865–2868, 2009.

[27] K. A. Christensen, J. T. Myers, and J. A. Swanson, "pH-dependent regulation of lysosomal calcium in macrophages," *Journal of Cell Science*, vol. 115, Part 3, pp. 599–607, 2002.

19

Crystal Structure of PAV1-137: A Protein from the Virus PAV1 That Infects *Pyrococcus abyssi*

N. Leulliot,[1,2] S. Quevillon-Cheruel,[1] M. Graille,[1,3] C. Geslin,[4] D. Flament,[4] M. Le Romancer,[4] and H. van Tilbeurgh[1]

[1] *Institut de Biochimie et de Biophysique Moléculaire et Cellulaire, CNRS-UMR 8619, IFR115, Université Paris-Sud, Bâtiment 430, 91405 Orsay, France*

[2] *Laboratoire de Cristallographie et RMN Biologiques-CNRS UMR-8015, Université Paris Descartes, Faculté des Sciences Pharmaceutiques et Biologiques, 4, av de l'Observatoire, 75270 Paris CEDEX 06, France*

[3] *Laboratoire de Biochimie (BIOC), CNRS UMR 7654, Ecole Polytechnique, Route de Saclay, 91128 Palaiseau, France*

[4] *Université de Brest, CNRS, IFREMER, UMR 6197, Laboratoire de Microbiologie des Environnements Extrêmes, OSU-IUEM, Technopôle Brest-Iroise, Avenue Dumont D'Urville, 29280 Plouzané, France*

Correspondence should be addressed to H. van Tilbeurgh; herman.van-tilbeurgh@u-psud.fr

Academic Editor: Martin Lawrence

Pyrococcus abyssi virus 1 (PAV1) was the first virus particle infecting a hyperthermophilic Euryarchaeota (*Pyrococcus abyssi* strain GE23) that has been isolated and characterized. It is lemon shaped and is decorated with a short fibered tail. PAV1 morphologically resembles the fusiform members of the family Fuselloviridae or the genus *Salterprovirus*. The 18 kb dsDNA genome of PAV1 contains 25 predicted genes, most of them of unknown function. To help assigning functions to these proteins, we have initiated structural studies of the PAV1 proteome. We determined the crystal structure of a putative protein of 137 residues (PAV1-137) at a resolution of 2.2 Å. The protein forms dimers both in solution and in the crystal. The fold of PAV1-137 is a four-α-helical bundle analogous to those found in some eukaryotic adhesion proteins such as focal adhesion kinase, suggesting that PAV1-137 is involved in protein-protein interactions.

1. Introduction

The archaea domain is organized into two major phyla, the Crenarchaeota and the Euryarchaeota. The first phylum contains mainly the extremely thermophilic Sulfolobales, Desulfurococcales, and Thermoproteales. The vast majority of hyperthermophilic viruses were isolated from the Crenarchaeota infecting in particular the genera *Sulfolobus*, *Thermoproteus*, *Acidianus*, *Pyrobaculum*, *Stygiolobus*, and *Aeropyrum* [1, 2]. Their shapes are characterized by unusual morphologies very different from bacterialviruses and eukaryotic viruses. Genomic sequences were determined for some of these archaeal viruses and revealed a very high portion of ORFan genes [3]. Due to their exceptional morphological and genomic properties, they were assigned to eight novel viral families [1].

The Euryarchaeota phylum includes extreme halophiles, methanogens, and hyperthermophilic sulfur reducers (*Thermococcales*). Most of the viruses infecting this phylum are isolated from mesophilic hosts and are tailed viruses, whereas pleomorphic types are relatively rare [4]. The knowledge about archaeal viruses is still very limited and this is even more poignant for viruses that infect hyperthermophilic Euryarchaeota [5, 6]. To date, PAV1 and TPV1 (*Thermococcus prieurii* virus 1) are the only viruses isolated from cultivated marine hyperthermophilic euryarchaea. These spindle-shaped viruses are morphologically similar to the haloviruses of the genus *Salterprovirus* [7, 8] that infect extreme halophiles, and to crenarchaeal viruses assigned to the fusiform family *Fuselloviridae* [9], but they do not share any genomic properties. PAV1, isolated from *Pyrococcus abyssi*, was the first virus isolated and described in *Thermococcales* [10]. PAV1 virions display a lemon-shaped morphology (120 nm long and 80 nm wide) with a short tail (15 nm) terminated by fibers. Very recently a novel fusiform virus, TPV1, was isolated and characterized from the hyperthermophilic euryarchaeal genus *Thermococcus* [11].

PAV1 and TPV1 are released during all phases of host growth without causing host lysis. A simple procedure to spot viruses on cellular lawns and directly observe their impact has been specially designed. This allows determination of the host range and infectivity of viruses isolated from anaerobic hyper/thermophile sulfur-reducing microorganisms. We used this approach to prove the infectivity of PAV1 and to confirm the host range of TPV1, both of them being genus specific [12].

The genome of PAV1 is composed of a double stranded circular DNA. It was shown that the free viral genome exists as a multicopy plasmid in the host strain, but no integrated prophage could be detected. The complete genome of PAV1 contains 18,098 bp [13]. A number of 25 ORFs (open reading frames) encoding at least 50 amino acids were identified and almost all are located on the same strand. The shape of the viruses is perturbed upon treatment with organic solvents or detergents, suggesting that their envelopes contain lipids. This observation is supported by the fact that half of the PAV1 genome has predicted transmembrane helices. Sixty-five percent of the predicted proteins have no homologues in the sequence databases. Functions could only be vaguely suggested for three proteins. A 59 amino acid protein (PAV1-59) shares similarities to the CopG transcriptional regulators. Two other ORFs (PAV1-676 and PAV1-678) that are the only ORFs shared between the hyperthermophilic euryarchaeal viruses PAV1 and TPV1 contain one or two copies of the laminin G-like jelly roll fold and may hence be involved in adhesion to the host. Polycistronic mRNA analysis showed that all predicted genes are transcribed in six mRNAs.

In contrast with other lemon-shaped viruses isolated either from hypersaline waters (*Salterprovirus*) or from extreme geothermal terrestrial environments (*Fuselloviridae*), PAV1 was isolated from a remote deep-sea hydrothermal vent. So, the uniqueness of the PAV1 genome compared to those of other archaeal viruses may be a consequence of its evolutionary history [14]. Since no function could be proposed for most of the predicted ORFs, we set out to analyze the structures of the proteins encoded by the PAV1 genome. The assignment of function to archaeal proteins suffers from the absence of genetic data and sequence analogs in better-characterized organisms [15–18]. 3D structure is better conserved than sequence and may reveal similarities that remain undiscovered by sequence analysis [19]. Therefore, structure determination offers a valuable alternative for investigating protein function. Indeed, the determination of the crystal structure of the AvtR protein from a hyperthermophilic archaeal lipothrixvirusallowed us to establish a role for this protein in the transcriptional regulation of viral genes [20]. We want *in fine* to find out if PAV1 protein structures are related to those of other archaeal virus proteins. We present here the X-ray crystal structure of a putative ORFan protein PAV1-137, to our knowledge the first for a euryarchaeal viral protein.

2. Results and Discussion

We purified the C-terminal His-tagged protein from a genetic construct deleted for the 14 first residues because sequence analysis predicted an unstructured conformation for this N-terminal region [21]. MALDI-TOF mass spectrometry analysis of the purified recombinant protein shows that the N-terminal methionine was cleaved off during the production in *E. coli*. Gel filtration analysis suggests that the protein forms dimers in solution (not shown). Crystals were obtained in 35% PEG400, 0.5 M NH_4Cl, 0.1 M Na citrate at pH 4, at a concentration of 15 mg/mL for the protein. Details of data collection and refinement are found in Table 1. All residues of the construct, except for the affinity tag, are visible in the electron density. Two copies of PAV1-137 are present in the asymmetric unit and their structures are almost identical (root mean square deviation = 0.5 Å).

PAV1-137 contains four amphipathic helices that are organized as a helical bundle. The three N-terminal helices form a parallel up and down configuration while the C-terminal helix is shorter and packs with an angle of about 45° against helices 1 and 3. The core of the 3 helices is very hydrophobic consisting mainly of branched aliphatic amino acids. The connections between helices 1, 2, and 3 are short. The connection between helices 3 and 4 is a longer stretch, resulting in a less tight packing of helix 4 compared to the other helices.

The A and B monomers in the asymmetrical unit form a two-fold symmetrical dimer and the symmetry axis runs perpendicular to the direction of the long helices (Figure 1). The helices of both monomers associate to form an antiparallel super helical bundle. The dimer interface involves helices 1, 2, and 4. The accessible surface of each subunit buried by dimer formation is substantial. The accessible surface area for the monomer is 7300 Å^2. Dimerisation buries 1278 Å^2 per monomer corresponding to 17% of the solvent accessible surface area. The majority of interactions between the monomers are conferred by helix 4 that lies against the extremities of helices 1 and 2. The interface is more hydrophilic than the core of the helical bundle and is stabilized by 9 hydrogen bonds, mainly between side chain and side or main chain atoms.

Orthologs of PAV1-137 were recently reported in genomic sequences of new thermococcus plasmids [14]. The gene coding for the ortholog of PAV1-137 is found in tandem with an ortholog of gene PAV1-375 in three of these plasmids. Surprisingly, it also formed a three-gene cluster, which is conserved within the provirus A3 VLP of the euryarchaeal methanogens *Methanococcus voltae* A3 [22]. PAV1-375 likely encodes for a P-loop ATPase, but its association with PAV1-137 remains unclear. PAV1-137 has presently no structural analogues in the Protein Data Bank that could help defining its function. The lack of sequence analogues at the start of this study suggested that PAV1-137 might adopt a new fold. Helical bundles are extremely common in protein structures. Therefore we found substantial structural similarities with proteins sharing helical bundle architecture. Examination of the function of these proteins clearly shows that the structural similarity does not indicate functional relationships. For example, significant overlaps are found between PAV1-137

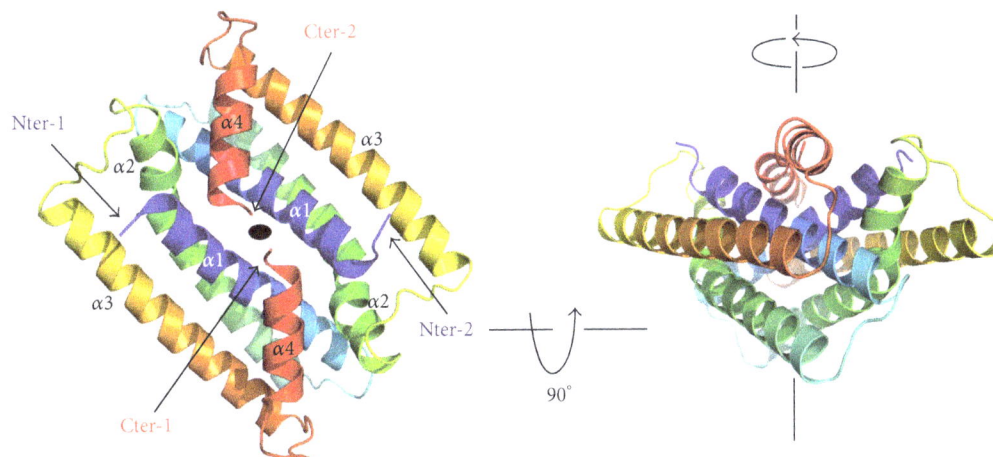

FIGURE 1: Two perpendicular views of the X-ray structure of the PAV1-137 dimer. The two subunits are represented in rainbow colouring going from blue (N-terminal) to red (C-terminal). The N and C terminus and the helices of the two subunits are labelled. The two-fold axis is indicated.

and fragments of Talin and Focal adhesion kinase, both from eukaryotic origin (Z-score 5 and root mean square difference of 3.2 Å for 100 aligned residues) [23, 24]. Helices 1, 2, and 3 of PAV1-137 superpose well onto helices 2, 3, and 4 of the eukaryotic helical bundles. No equivalent is present in these eukaryotic analogues for the fourth helix found in PAV1-137. The C-terminal helix adopts a different orientation and has no equivalent in Talin or Focal adhesion kinase.

Since PAV1-137 does not seem to carry any active site, its biological function is probably connected with protein-protein or protein-nucleic acid interactions. Helical bundle proteins are frequently involved in this type of interactions, exemplified by Talin. Virtually nothing is known about the life cycle of PAV1. Its genome sequence was a first step towards a better understanding of this new type of viruses which may have evolved from a recombination event between different mobile genetic elements harbored by both hyperthermophilic and methanogenic euryarchaeota. We report here on the first results of the investigation on structure and function of proteins encoded by this virus.

3. Materials and Methods

3.1. Protein Production and Purification. The PAV1-137 ORF lacking the region encoding for the 14 N-terminal residues was amplified by PCR using genomic DNA of PAV1 virus as a template. An additional sequence coding for a 6-histidine tag was introduced at the $3'$ end of the ORF during amplification. The PCR product was then cloned into pET28 vector. Expression was done at 37°C using the *E. coli* BL21 (Gold)DE3 strain. The His-tagged protein was purified on a Ni-NTA column (Qiagen Inc.) followed by gel filtration using a buffer composed of 20 mM Tris-HCl pH 7.5, 200 mM NaCl, and 10 mM b-mercaptoethanol. Selenomethionine-substituted PAV1-137 was produced and purified as the native protein. The peak fractions were concentrated to 15 mg/mL and used for crystallization.

3.2. Crystallization and Data Collection. Crystallization was performed using a Cartesian crystallization robot in 200 × 200 μl sitting drops (volume for protein and liquid mother) and reproduced manually in 1 × 1 μL drops. Crystals were obtained from the following crystallization conditions: 35% PEG400, 0.5 M NH_4Cl, 0.1 M Na citrate at pH 4 at 18°C. Crystals were transferred in the mother liquor containing 30% glycerol prior to flash freezing in liquid nitrogen. X-ray diffraction data of SeMet substituted protein crystals were collected on the ID14-4 ESRF beamline and were processed using MOSFLM and SCALA [25]. The crystals belong to the $P6_122$ space group with two molecules per asymmetric unit. The cell parameters and data collection statistics are reported in Table 1.

3.3. Structure Solution and Refinement. The structure was solved at a resolution of 2.2 Å by single anomalous diffraction (SAD) using data collected at the Selenium peak wavelength. The Hyss module of Phenix program [26] was used to find the Selenium sites using the entire resolution range. The sites were refined with SHARP [27], and solvent flattening was performed with DM. Arp/Warp [28] built 95% of the visible residues. The model was fully refined and completed from the native data using Buster and the graphics programme O [29, 30]. Refinement was carried out using noncrystallographic symmetry restraints and one TLS group per chain (statistics are shown in Table 1) with the program Buster. All the residues fall in favourable regions of the Ramachandran plot.

For homology search, we did a Psi-BLAST analysis using standard procedures as provided by the NIH blastserver (http://blast.ncbi.nlm.nih.gov.gate1.inist.fr/Blast.cgi).

Accession Number

The structure factor amplitudes and the refined coordinates of PAV1-137 have been deposited in the Protein Data Bank as entry 4HR1.

TABLE 1: Data collection and refinement statistics.

Data collection	
Space group	P6$_1$22
Cell dimensions a, b, c (Å)	73.55, 73.55, 191.55
Wavelength (Å)	0.9793
Resolution (Å)	20.0–2.20
Outer resolution shell (Å)	2.32–2.20
Number of observed reflections/unique	94031/15911
Completeness (%) (outer shell)	97.4 (85.3)
Multiplicity (outer shell)	5.9 (2.8)
$I/\sigma(I)$ (outer shell)	19.0 (2.4)
R_{merge} (%)[1] (outer shell)	5.8 (45.9)
Refinement	
Resolution (Å)	20.0–2.20
Reflections (working/test)	15873/800
R/R_{free} (%)[2]	21.5/24.7
RMSD	
Bond lengths (Å)	0.010
Bond angles (°)	1.15
B-factors (Å^2)	
Protein	63
Ramachandran statistics (%)	
Most favored	98.71
Allowed	1.29

[1]$R_{merge} = \sum_h \sum_i |I_{hi} - \langle I_h \rangle|/ \sum_h \sum_i I_{hi}$, were I_{hi} is the ith observation of the reflection h, while $\langle I_h \rangle$ is the mean intensity of reflection h.
[2]$R = \sum ||F_o| - |F_c||/|F_o|$. R_{free}: was calculated with a set of randomly selected reflections (5%).

Acknowledgments

This work was supported by a grant from the Agence Nationale de Recherche (no. ANR-BLAN-0408-03). The authors are indebted to Benjamin Dray and Nathalie Ulryck for the technical assistance.

References

[1] D. Prangishvili and R. A. Garrett, "Exceptionally diverse morphotypes and genomes of crenarchaeal hyperthermophilic viruses," *Biochemical Society Transactions*, vol. 32, no. 2, pp. 204–208, 2004.

[2] T. Mochizuki, Y. Sako, and D. Prangishvili, "Provirus induction in hyperthermophilic archaea: characterization of Aeropyrum pernix spindle-shaped virus 1 and Aeropyrum pernix ovoid virus 1," *Journal of Bacteriology*, vol. 193, no. 19, pp. 5412–5419, 2011.

[3] D. Prangishvili, R. A. Garrett, and E. V. Koonin, "Evolutionary genomics of archaeal viruses: unique viral genomes in the third domain of life," *Virus Research*, vol. 117, no. 1, pp. 52–67, 2006.

[4] F. Eiserling, A. Pushkin, M. Gingery, and G. Bertani, "Bacteriophage-like particles associated with the gene transfer agent of Methanococcus voltae PS," *Journal of General Virology*, vol. 80, no. 12, pp. 3305–3308, 1999.

[5] H. W. Ackermann and D. Prangishvili, "Prokaryote viruses studied by electron microscopy," *Archives of Virology*, vol. 157, no. 10, pp. 1843–1849, 2012.

[6] C. Geslin, M. Le Romancer, M. Gaillard, G. Erauso, and D. Prieur, "Observation of virus-like particles in high temperature enrichment cultures from deep-sea hydrothermal vents," *Research in Microbiology*, vol. 154, no. 4, pp. 303–307, 2003.

[7] C. Bath, T. Cukalac, K. Porter, and M. L. Dyall-Smith, "His1 and His2 are distantly related, spindle-shaped haloviruses belonging to the novel virus group, *Salterprovirus*," *Virology*, vol. 350, no. 1, pp. 228–239, 2006.

[8] C. Bath and M. L. Dyall-smith, "His1, an archaeal virus of the *Fuselloviridae* family that infects Haloarcula hispanica," *Journal of Virology*, vol. 72, no. 11, pp. 9392–9395, 1998.

[9] P. Redder, X. Peng, K. Brügger et al., "Four newly isolated fuselloviruses from extreme geothermal environments reveal unusual morphologies and a possible interviral recombination mechanism," *Environmental Microbiology*, vol. 11, no. 11, pp. 2849–2862, 2009.

[10] C. Geslin, M. Le Romancer, G. Erauso, M. Gaillard, G. Perrot, and D. Prieur, "PAV1, the first virus-like particle isolated from a hyperthermophilic euryarchaeote, 'Pyrococcus abyssi'," *Journal of Bacteriology*, vol. 185, no. 13, pp. 3888–3894, 2003.

[11] A. Gorlas, E. V. Koonin, N. Bienvenu, D. Prieur, and C. Geslin, "TPV1, the first virus isolated from the hyperthermophilic genus Thermococcus," *Environmental Microbiology*, vol. 14, no. 2, pp. 503–516, 2012.

[12] A. Gorlas and C. Geslin, "A simple procedure to determine the infectivity and host range of viruses infecting anaerobic and hyperthermophilic microorganisms," *Extremophiles*, 2013.

[13] C. Geslin, M. Gaillard, D. Flament et al., "Analysis of the first genome of a hyperthermophilic marine virus-like particle, PAV1, isolated from *Pyrococcus abyssi*," *Journal of Bacteriology*, vol. 189, no. 12, pp. 4510–4519, 2007.

[14] M. Krupovic, M. Gonnet, W. Ben Hania, R. Forterre, and G. Erauso, "Insights into dynamics of mobile genetic elements in hyperthermophilic environments from five new thermococcus plasmids," *PLoS One*, vol. 8, no. 1, p. e49044, 2013.

[15] R. Khayat, L. Tang, E. T. Larson, C. M. Lawrence, M. Young, and J. E. Johnson, "Structure of an archaeal virus capsid protein reveals a common ancestry to eukaryotic and bacterial viruses," *Proceedings of the National Academy of Sciences of the United States of America*, vol. 102, no. 52, pp. 18944–18949, 2005.

[16] E. T. Larson, D. Reiter, M. Young, and C. M. Lawrence, "Structure of A197 from Sulfolobus turreted icosahedral virus: a crenarchaeal viral glycosyltransferase exhibiting the GT-A fold," *Journal of Virology*, vol. 80, no. 15, pp. 7636–7644, 2006.

[17] J. Keller, N. Leulliot, B. Collinet et al., "Crystal structure of AFV1-102, a protein from the acidianus filamentous virus 1," *Protein Science*, vol. 18, no. 4, pp. 845–849, 2009.

[18] A. Goulet, M. Pina, P. Redder et al., "ORF157 from the archaeal virus Acidianus filamentous virus 1 defines a new class of nuclease," *Journal of Virology*, vol. 84, no. 10, pp. 5025–5031, 2010.

[19] J. C. Whisstock and A. M. Lesk, "Prediction of protein function from protein sequence and structure," *Quarterly Reviews of Biophysics*, vol. 36, no. 3, pp. 307–340, 2003.

[20] N. Peixeiro, J. Keller, B. Collinet et al., "Structure and function of AvtR, a novel transcriptional regulator from a hyperthermophilic archaeal lipothrixvirus," *Journal of Virology*, vol. 87, no. 1, pp. 124–136, 2013.

[21] S. Hirose, K. Shimizu, S. Kanai, Y. Kuroda, and T. Noguchi, "POODLE-L: a two-level SVM prediction system for reliably predicting long disordered regions," *Bioinformatics*, vol. 23, no. 16, pp. 2046–2053, 2007.

[22] M. Krupovič and D. H. Bamford, "Archaeal proviruses TKV4 and MVV extend the PRD1-adenovirus lineage to the phylum Euryarchaeota," *Virology*, vol. 375, no. 1, pp. 292–300, 2008.

[23] S. T. Arold, M. K. Hoellerer, and M. E. M. Noble, "The structural basis of localization and signaling by the focal adhesion targeting domain," *Structure*, vol. 10, no. 3, pp. 319–327, 2002.

[24] E. Papagrigoriou, A. R. Gingras, I. L. Barsukov et al., "Activation of a vinculin-binding site in the talin rod involves rearrangement of a five-helix bundle," *EMBO Journal*, vol. 23, no. 15, pp. 2942–2951, 2004.

[25] A. Leslie, *Joint CCP4 and EACMB Newsletter Protein Crystallography*, Daresbury Laboratory, Warrington, UK, 1992.

[26] P. D. Adams, K. Gopal, R. W. Grosse-Kunstleve et al., "Recent developments in the PHENIX software for automated crystallographic structure determination," *Journal of Synchrotron Radiation*, vol. 11, no. 1, pp. 53–55, 2004.

[27] C. Vonrhein, E. Blanc, P. Roversi, and G. Bricogne, "Automated structure solution with autoSHARP," *Methods in Molecular Biology*, vol. 364, pp. 215–230, 2007.

[28] R. J. Morris, A. Perrakis, and V. S. Lamzin, "ARP/wARP and automatic interpretation of protein electron density maps," *Methods in Enzymology*, vol. 374, pp. 229–244, 2003.

[29] E. Blanc, P. Roversi, C. Vonrhein, C. Flensburg, S. M. Lea, and G. Bricogne, "Refinement of severely incomplete structures with maximum likelihood in BUSTER-TNT," *Acta Crystallographica Section D*, vol. 60, no. 12, pp. 2210–2221, 2004.

[30] T. A. Jones, "Interactive electron-density map interpretation: from INTER to O," *Acta Crystallographica Section D*, vol. 60, no. 12, pp. 2115–2125, 2004.

20

Dynamics of the Methanogenic Archaea in Tropical Estuarine Sediments

María del Rocío Torres-Alvarado,[1] **Francisco José Fernández,**[2]
Florina Ramírez Vives,[2] **and Francisco Varona-Cordero**[1]

[1] Department of Hydrobiology, Universidad Autónoma Metropolitana-Iztapalapa, Avenida San Rafael Atlixco No. 86,
Colonia Vicentina, 09340 Mexico City, DF, Mexico
[2] Department of Biotechnology, Universidad Autónoma Metropolitana-Iztapalapa, Avenida San Rafael Atlixco No. 86,
Colonia Vicentina, 09340 Mexico City, DF, Mexico

Correspondence should be addressed to María del Rocío Torres-Alvarado; rta@xanum.uam.mx

Academic Editor: Martin Krüger

Methanogenesis may represent a key process in the terminal phases of anaerobic organic matter mineralization in sediments of coastal lagoons. The aim of the present work was to study the temporal and spatial dynamics of methanogenic archaea in sediments of tropical coastal lagoons and their relationship with environmental changes in order to determine how these influence methanogenic community. Sediment samples were collected during the dry (February, May, and early June) and rainy seasons (July, October, and November). Microbiological analysis included the quantification of viable methanogenic archaea (MA) with three substrates and the evaluation of kinetic activity from acetate in the presence and absence of sulfate. The environmental variables assessed were temperature, pH, Eh, salinity, sulfate, solids content, organic carbon, and carbohydrates. MA abundance was significantly higher in the rainy season (10^6–10^7 cells/g) compared with the dry season (10^4–10^6 cells/g), with methanol as an important substrate. At spatial level, MA were detected in the two layers analyzed, and no important variations were observed either in MA abundance or activity. Salinity, sulfate, solids, organic carbon, and Eh were the environmental variables related to methanogenic community. A conceptual model is proposed to explain the dynamics of the MA.

1. Introduction

Coastal and marine environments, including estuaries and coastal lagoons, are characterized by large amounts of organic matter, which is mineralized primarily in sediments through anaerobic processes, sulfate reduction being the dominant metabolic pathway [1, 2]. However, although these ecosystems are the typical habitat of sulfate-reducing prokaryotes (SRP), methanogenic archaea (MA) and methane production have also been detected [3, 4].

MA are strict anaerobes that produce methane as endproduct of their metabolism. These organisms are common in anoxic environments in which electron acceptors such as nitrate and sulfate are either absents or present at low concentrations and are usually dominant in freshwater environments. In the presence of these electron acceptors, methanogenesis is outcompeted by anaerobic respiration, mainly for thermodynamic reasons [5]. MA distribution patterns and its number, as well as physical, chemical, and nutritional parameters controlling their abundance and distribution have been studied in lacustrine sediments [6] and in coastal environments [7, 8].

Most of the ecological studies assessing the structure of methanogenic communities in estuarine systems have been performed in temperate latitudes where temperature is one of the major factors regulating ecosystem function. These investigations have included an evaluation of the MA in the intertidal zone of marshes with the presence of *Spartina alterniflora*, whose roots provide organic carbon and contribute to create aerobic microhabitats [9, 10]. MA abundance has been quantified with two or three substrates, of which acetate and hydrogen have been reported as the two most important ones

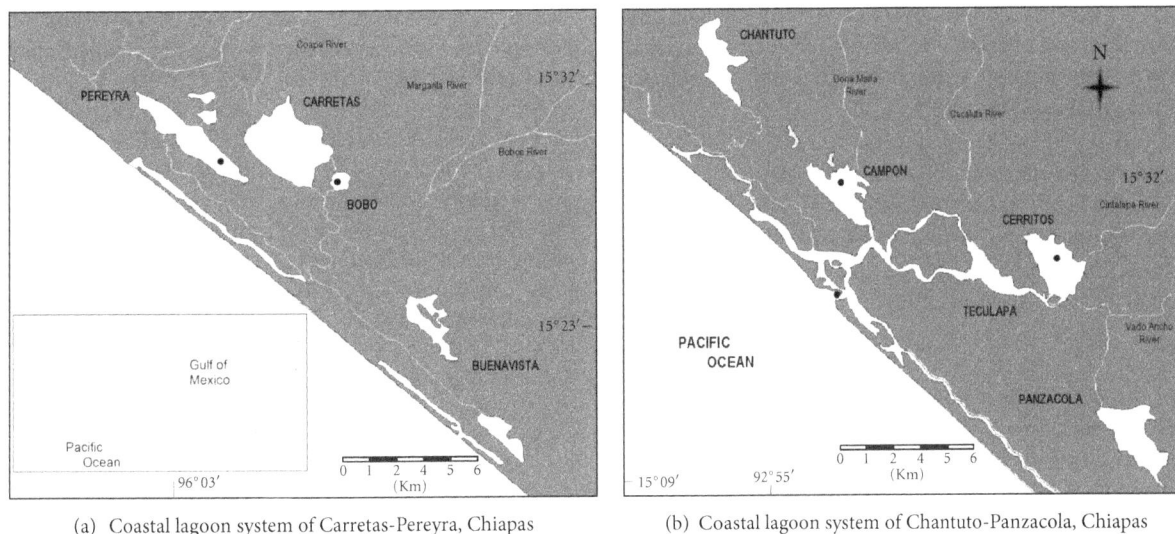

(a) Coastal lagoon system of Carretas-Pereyra, Chiapas

(b) Coastal lagoon system of Chantuto-Panzacola, Chiapas

FIGURE 1: Study area and sampling sites (•).

[4, 11]. Additionally, it has been established that in estuaries, where a salinity gradient exists from the marine zone to a river entrance, MA are prevalent upstream in the freshwater region and decrease towards the brackish and marine ends; sulfate reduction has been identified as the key factor related to the MA distribution [7, 10, 12, 13]. Depth profiles of MA distribution have been observed, their abundance increase in deeper layers of the sediment column, because the MA are dependent on heterotrophs and fermenters during the organic matter decomposition, its decline is also related to a decrease in both sulfate concentration and redox potential [8].

In contrast to estuaries, coastal lagoons generally have restricted communication with the sea and in tropical lagoons, as a result of strong seasonal precipitation patterns, there are significant fluctuations in river discharge, and associated hydrological conditions (salinity). These variations might affect the structure of microbial communities involved in the terminal phases of the anaerobic organic matter mineralization, as well as to the biogeochemical processes related to it. In spite of its importance, studies focused on these ecosystems to assess the dynamics of anaerobic microbiota, especially MA, are scarce. It has been reported that MA using methylamines are the primary microbial components in sediments of coastal lagoons associated to mangroves, with higher densities during the summer and premonsoon [14, 15]. In another study, a peak of methane production in mangrove sediments has been recorded in the postmonsoon season [16]. In Mexico, where coastal lagoons are abundant, investigations on methanogenic communities are virtually absent; hence, the aim of the present study was to explore the spatial and temporal dynamics of the methanogenic community in sediments from two tropical coastal systems: Chantuto-Panzacola and Carretas-Pereyra, located in the Mexican southern Pacific and to propose a conceptual model on MA dynamics in sediments for the tropical coastal lagoons studied.

2. Materials and Methods

2.1. Study Site. The Chantuto-Panzacola and Carretas-Pereyra lagoon systems are located in the State of Chiapas, Mexican Pacific coast (Figure 1); they are part of the International Biosphere Reserve "La Encrucijada". The climate of the region is warm ($28°C$) and humid (89%) with abundant summer rainfall; annual rainfall ranges between 1,300 and 3,000 mm. The rainy season begins between May and June and continues through November; the dry season occurs from December to May [17]. Lagoon systems are characterized by high temperatures in the water column ($29-35.5°C$), with a variable salinity ranging from 0 to 34.5‰ in Chantuto-Panzacola and from 0 to 22.7‰ in Carretas-Pereyra, depending on the season. There is a limited exchange with the sea and a significant phosphorus supply from rivers, which favors high chlorophyll-*a* levels. Systems are bordered by mangrove forests and freshwater wetlands. Mangrove detritus results in high humic substance levels (>150 mg/L) in the rainy season [18] also recording high ammonium concentrations derived from mineralization [19].

The Chantuto-Panzacola lagoon has an area of 18,000 ha and comprises five lagoons: Chantuto, Campón, Teculapa, Cerritos, and Panzacola. In this system, samples were collected from the Cerritos and Campón lagoons (Figure 1). The Cerritos lagoon ($15°09'54.4''$ N, $92°45'34.0''$ W) has a mean depth of 1.1 m in the dry season and 1.3 m during the rainy season. The Cintalapa River flows into this lagoon, contributing a volume between $66.2 \, m^3/s$ in October and $0.4 \, m^3/s$ in May (dates proportionated by the National Water Commission in Mexico). The Campon lagoon ($15°12'30.0''$ N, $92°51'24.2''$ W) has a mean depth of 0.8 m in the dry season and of 0.9 m in the rainy season. The Cacaluta River flows into this lagoon, with a maximum inflow in October ($144.2 \, m^3/s$) and a minimum inflow in May ($0.5 \, m^3/s$). Sediments are a mixture of silt and sand in both lagoons.

The Carretas-Pereyra system covers an area of 3,696 ha and comprises four water bodies: Pereyra, Carretas, Bobo, and Buenavista, sampling took place in Pereyra and Bobo (Figure 1). The Pereyra lagoon ($15°31'26.1''$ N, $92°51'24.2''$ W) has a mean depth of 0.7 m in the dry season and 1.0 m in the rainy season. Sediment is silt-sand. The Margaritas River drains into the Pereyra lagoon (discharge volume unknown). The Bobo lagoon ($15°29'22.0''$ N, $93°08'44.6''$ W) has a mean depth of 0.5 m and 0.7 m in the dry and rainy seasons, respectively. It lacks freshwater inputs and sediment is silt-sand.

2.2. Sample Collection and Preparation Procedures. Sediment cores were collected with a 45 cm long and 4.5 cm wide plex-iglass coring device during the dry (February, May, and early June) and rainy seasons (July, October, and November). Temperature, Eh, and pH were simultaneously measured when sampling the cores at two sediment depths (6 and 12 cm) using standard electrodes and an Ionanalizer (Conductronic pH 120). pH was measured with a glass electrode and the sediment redox potential was measured using a platinum electrode and a saturated KCl calomel reference electrode (Instrulab, Mexico). The standard potential of the reference (+198) was added to the mean value to obtain the Eh of the sediment medium. Electrodes were routinely standardized in the field using a ZoBell Solution [20]. Subsequently, samples were transported to the laboratory.

Cores obtained in each sampling station were segmented in two sections (0–6 cm and 6–12 cm) under a nitrogen atmosphere. After each section was homogenized in a plastic bag using steady shaking, subsamples were immediately taken to quantify MA. The remaining sediment was maintained under low temperature to perform physical-chemical analyses.

2.3. Microbiological Analyses. Enumeration of viable MA was performed using the Most Probable Number (MPN) method by a ten-fold dilution series (10^{-1} to 10^{-10}) for each sample using four tubes per dilution. The MPN analyses included the quantification with substrates commonly used by the different groups of MA: acetate, $CO_2 + H_2$, and methanol, with the basic medium by Balch et al. [21]. Salinity in the culture medium was adjusted with a NaCl (330 g/L) solution to obtain similar values to those measured in the original sediment sample; the pH was adjusted to 7.2 with a bicarbonate (10%) solution. Cultures were incubated at $32°C$ for one month. Methane was detected with a GOW-MAC Series 580 GC with a thermal conductivity detector (TCD) under the following operation conditions: column, detector, and injector temperatures of 140, 190, and $170°C$, respectively; $25°C$/min rate; column packed with carbosphere 80/100, helium as carrier gas at 25 mL/min; polarity of 120 mA.

In order to determine the effect of sulfate on MA for a competitive substrate, methanogenic activity was determined in a medium without sulfates (sulfate-free), using 125-mL serum bottles, with 42 mL of the Balch et al. [21] and acetate as substrate to a final concentration of 20 mM. Experiments were conducted in parallel in which the culture medium was supplemented with sulfate (final concentration 20 mM). Bottles were inoculated with 8 mL of moist sediment and incubated at $32°C$ in the dark for 42 days; the incubations were shaken three times per week. Each experiment was run by duplicate for each sample, including the respective controls (without acetate), with and without sulfates in the medium. Mineralization was evaluated by determining changes in acetate concentration and percent methane production in bottles. For acetate analysis, 1.5 mL samples were centrifuged at 1,120 gf for 10 min. The supernatant was filtered. A 950 μL aliquot was acidified with 50 μL of HCl (2.2 M). The acetate concentration was measured by flame ionization gas chromatography (Agilent Series 6890 Plus) using an Agilent crosslinked FFAP capillary column (15 m × 0.530 mm × 1.00 μm). Column, injection port, and FID temperatures were 120, 130, and $150°C$, respectively. The temperature of the column, detector, and injector were 120, 150, and $130°C$, respectively. The carrier gas was N_2 (4.5 mL/min).

2.4. Physicochemical Analyses. Sediment samples were centrifuged at 1,602.76 gf at low temperature (4-5°C) for 20 minutes to separate porewater from sediments [22]. Porewater was filtered through 0.45 μm Millipore membranes and the following parameters were determined: salinity, with an optical refractometer (American Optical); sulfate [23] and total dissolved carbohydrates, with the phenol-sulfuric acid technique [24]. Total solids and volatile solids were quantified in moist sediments [25], porosity was determinate by measuring the weight loss by drying sediment samples of know volumes and weights. Organic carbon content was measured through the method by Gaudette et al. [26] in a sediment sample dried at $60°C$.

2.5. Statistical Analyses. The data matrix included MA abundances and physicochemical variables. To meet the normality assumptions, data for variables were transformed through $\log x + 1$ [27]. For the temporal analysis, variables were grouped into two climate seasons (dry and rainy); for the spatial analysis, data were grouped into two depth categories (0–6 cm and 6–12 cm). An analysis of variance (ANOVA) was conducted to test for significant differences between seasons in each system, on the one hand, and between depth categories, on the other. The significance of specific differences was assessed through the Tukey-Kramer multiple comparison test [27]. A Canonical Correspondence Analysis (CCA) was used to investigate the relationship between microbial abundance and environmental variables [28]. These analyses were conducted with the Statistica 10 (Academic) and MVSP 3.12b Software.

3. Results and Discussion

The aim of this study was to analyze the changes in the abundance and activity of MA and relate these community characteristics with some physicochemical variables to propose a conceptual model of methanogenic community dynamics in coastal lagoon sediments.

TABLE 1: Environmental variables in the coastal lagoon sediments of Chantuto-Panzacola and Carretas-Pereyra, Chiapas. Mean ± Standard deviation.

Depth	Dry season		Rainy season	
	6	12	6	12
Chantuto-Panzacola				
Temperature (°C)	29.2 ± 1.1	28.3 ± 1.3	28.1 ± 1.5	26.7 ± 1.5
Salinity (‰)	21.3 ± 6.1	18.6 ± 5.1	2.5 ± 2.5	2.8 ± 3.1
Sulphate (mM)	11.0 ± 1.8	9.8 ± 1.2	3.8 ± 1.4	2.9 ± 0.9
pH	7.1 ± 0.1	7.0 ± 0.1	6.7 ± 0.2	6.8 ± 0.1
Eh (mV)	−206 ± 76	−356 ± 34	−104 ± 4	−286 ± 53
Total Solids (TS, g/L)	445.50 ± 120.65	338.12 ± 79.11	320.79 ± 153.2	303.50 ± 151.07
Volatile Solids (VS, g/L)	42.61 ± 20.19	47.40 ± 34.86	75.82 ± 41.0	68.76 ± 51.98
Porosity (g/cm^3)	0.3 ± 0.1	0.4 ± 0.08	0.4 ± 0.1	0.4 ± 0.1
Organic matter (%)	7.2 ± 3.4	5.9 ± 3.8	9.8 ± 5.5	5.8 ± 3.2
Organic carbon (%)	4.1 ± 2.0	3.4 ± 2.2	5.7 ± 3.1	3.3 ± 1.8
Carbohydrates (mg/L)	5.6 ± 4.0	6.5 ± 4.0	5.0 ± 1.0	5.9 ± 4.1
Carretas-Pereyra				
Temperature (°C)	29.4 ± 0.8	28.5 ± 0.7	28.5 ± 1.9	28.3 ± 0.9
Salinity (‰)	27.3 ± 5.3	23.5 ± 3.3	4.3 ± 4.08	3.2 ± 3.8
Sulphate (mM)	13.0 ± 1.4	11.7 ± 1.3	3.5 ± 1.98	1.9 ± 1.4
pH	6.9 ± 0.1	6.8 ± 0.1	6.8 ± 0.1	6.7 ± 0.1
Eh (mV)	−296 ± 83	−411 ± 66	−152 ± 46	−369 ± 99
Total Solids (TS, g/L)	261.70 ± 135.49	229.49 ± 134.29	211.56 ± 123.36	188.41 ± 97.24
Volatile Solids (VS, g/L)	75.22 ± 35.27	85.05 ± 71.58	28.40 ± 12.90	40.39 ± 22.98
Porosity (g/cm^3)	0.2 ± 0.07	0.2 ± 0.1	0.4 ± 0.1	0.5 ± 0.1
Organic matter (%)	12.5 ± 4.5	25.4 ± 19.2	10.0 ± 4.5	15.7 ± 8.8
Organic carbon (%)	7.2 ± 4.5	14.5 ± 11.02	6.1 ± 2.6	9.03 ± 5.04
Carbohydrates (mg/L)	6.8 ± 3.6	6.0 ± 3.5	3.8 ± 1.2	5.3 ± 3.5

3.1. *Environmental Variables.* Conditions in the sedimentary habitat in the Chantuto-Panzacola and Carretas-Pereyra lagoon systems resulted from seasonal variations between the dry and rainy seasons. Temperature in the sediment was higher in the dry season in comparison with rainy season (Table 1); the temporal variations were significant in Chantuto-Panzacola (Table 2). Significant differences in pH were observed (Table 2). In the dry season, a greater marine influence favors neutral conditions; by contrast, in the rainy season the higher fluvial inflow decreased marine influence, and acid conditions were registered (Table 1). The redox conditions were similar to those reported for sediments from mangroves [29] and were significantly less reductive in the rainy season (Table 1) when the freshwater inflow favored sediment suspension in the water column (turbidity = 126–224 NTU), with an increase in porosity and less reduced conditions at the sediments. In the dry season redox potential decreased as a result of sediment deposition (turbidity = 31–107 NTU).

The major changes were determined in salinity and sulfate content (Tables 1 and 2). Maximum values were recorded in the dry season and minimum in the rainy season; even totally freshwater conditions existed in both systems in October (0‰). The decrease in salinity and sulfates was due to an increase in fluvial inflow and precipitation. Salinity in coastal lagoons varies according to annual cycles, which depend on the local climate, continental freshwater runoff, connection with the sea, and influence of tides. Knoppers and Kjerfve [30] point out that seasonal pulses in freshwater inflow exert a marked impact on the ecology of coastal lagoons, besides controlling salinity, increasing the water level, and holding open communication to the sea.

No significant temporal variations were observed in the concentration of total solids and organic fractions (volatile solids, organic matter, organic carbon, and carbohydrates) ($P > 0.05$); and their supply was constant through rivers and wetlands. The high rate of freshwater inflow with organic debris from land and run-off as well as from adjacent mangroves is a key factor related to the contribution of organic matter in coastal zones [31].

Spatially there was no pattern of physicochemical conditions in the sedimentary habitat as evidenced by the null significance observed for the temperature, pH, salinity and sulfates ($P > 0.05$). An exception was the Eh, which decreased significantly with depth (Tables 1 and 2). The vertical fluctuations in Eh may be attributed to a reduction in the oxygen diffusion rate in porewater as the depth of the sediment column increases [32]. There were no significant variations in solids content and organic fractions ($P > 0.05$) (Table 1). However the organic carbon content was higher in the sediment layer of 12 cm, dos Santos Fonseca et al. [33] point out that this behavior seems to result from the fact

TABLE 2: Results of the ANOVA (F) and multiple comparisons analysis (MCA) (Tukey test) of environmental and microbiological variables between seasons and sediment depth in Chantuto-Panzacola and Carretas-Pereyra. P: significance. Seasons: D: dry and R: rainy. Depth: 6 cm and 12 cm.

Variables	Season			Depth		
	F	P	MCA	F	P	MCA
Chantuto-Panzacola						
Temperature (°C)	4.66	0.0421	D > R	3.75	0.0684	—
Salinity (‰)	62.03	0.0000	D > R	0.08	0.9311	—
Sulphate (mM)	109.00	0.0000	D > R	0.45	0.5349	—
pH	28.81	0.0000	D > R	0.01	0.9427	—
Eh (mV)	4.54	0.0446	D < R	38.04	0.0000	6 < 12
MA-Acetate (cells/g)	112.38	0.0000	D < R	0.01	0.7842	—
MA-Hydrogen (cells/g)	15.10	0.0008	D < R	0.05	0.8195	—
MA-Methanol (cells/g)	5.92	0.0236	D < R	3.36	0.0528	—
Activity + SO_4 (mM acetate/g VS/day)	14.71	0.0009	D > R	1.50	0.2321	—
$CH_4 + SO_4$	66.12	0.0000	D < R	0.85	0.3085	—
$CH_4 - SO_4$	4.96	0.0364	D < R	2.17	0.0831	—
Carretas-Pereyra						
Temperature (°C)	0.97	0.3344		1.28	0.2705	—
Salinity (‰)	154.47	0.0000	D > R	0.26	0.6156	—
Sulphate (mM)	210.03	0.0000	D > R	0.45	0.5101	—
pH	10.47	0.0038	D > R	1.09	0.3088	—
Eh (mV)	3.80	0.0641	—	19.80	0.0002	6 < 12
MA-Acetate (cells/g)	4.82	0.0390	D < R	0.13	0.7193	—
MA-Hydrogen (cells/g)	9.39	0.0057	D < R	0.48	0.4952	—
MA-Methanol (cells/g)	2.71	0.1142	—	1.06	0.3142	—
Activity + SO_4 (mM acetate/g VS/day)	12.62	0.0018	D > R	0.46	0.5042	—
$CH_4 + SO_4$	15.39	0.0007	D < R	6.24	0.0204	6 < 12
$CH_4 - SO_4$	7.21	0.0135	D < R	7.88	0.0103	6 < 12

that the most labile substrate is readily used by the microbial community in the top centimeters of sediment, and the refractory fraction builds up in deeper layers, where it will be degraded slowly. The presence of refractory material (wood and phytoplankton debris identified with a light microscope Zeiss Axioscop) concentrated largely in the 6–12 cm-deep layer in Pereyra and Campón lagoons seem to support this hypothesis.

3.2. Abundance and Distribution of MA. Viable MA in the sediments of Chantuto-Panzacola and Carretas-Pereyra systems were evaluated with MPN, obtaining a range of abundance between 10^4 and 10^7 cells/g. MA density reached peak levels in the rainy season, with a significant decrease of as much as two orders of magnitude during the dry season ($P < 0.05$) (Figures 2(a)–2(c)). In the rainy season, increased freshwater input created favorable conditions for MA proliferation. In this season highest levels of MA were recorded with acetate and methanol in Chantuto-Panzacola and with methanol and H_2-CO_2 in Carretas-Pereyra. During the dry season, high MA levels were obtained with methanol in both lagoon systems; the second substrate in importance

was H_2-CO_2 and the lowest levels correspond to acetate (Table 3).

The constant occurrence of MA was probably the result of their ability to use different electron donors in an ecosystem with a constant supply of organic matter provided by the rivers and run-off from adjacent mangroves. Verma et al. [34] mentioned that the continued presence of MA in coastal lagoons is possible by the presence of "noncompetitive" substrates, (methanol and methylamines), that are used exclusively by the MA, as well as the constant availability of "competitive" substrates (acetate and hydrogen), used by methanogen and other anaerobic microorganisms.

Methanol was an important substrate in both seasons, may be released from methoxy groups during degradation of lignin. Methanol-utilizing MA have a broad substrate spectrum, can also grow on acetate, growth on H_2-CO_2 is restricted to some *Methanosarcina* species [5]. There is evidence supporting the hypothesis that cometabolism of a broad range of substrates by generalist microorganisms may confer competitive advantages [35]. Purdy et al. [13] mention that, within the methanogenic community, the presence of generalist groups implies that these are better adapted to the variations in the estuarine conditions. Additionally methanol

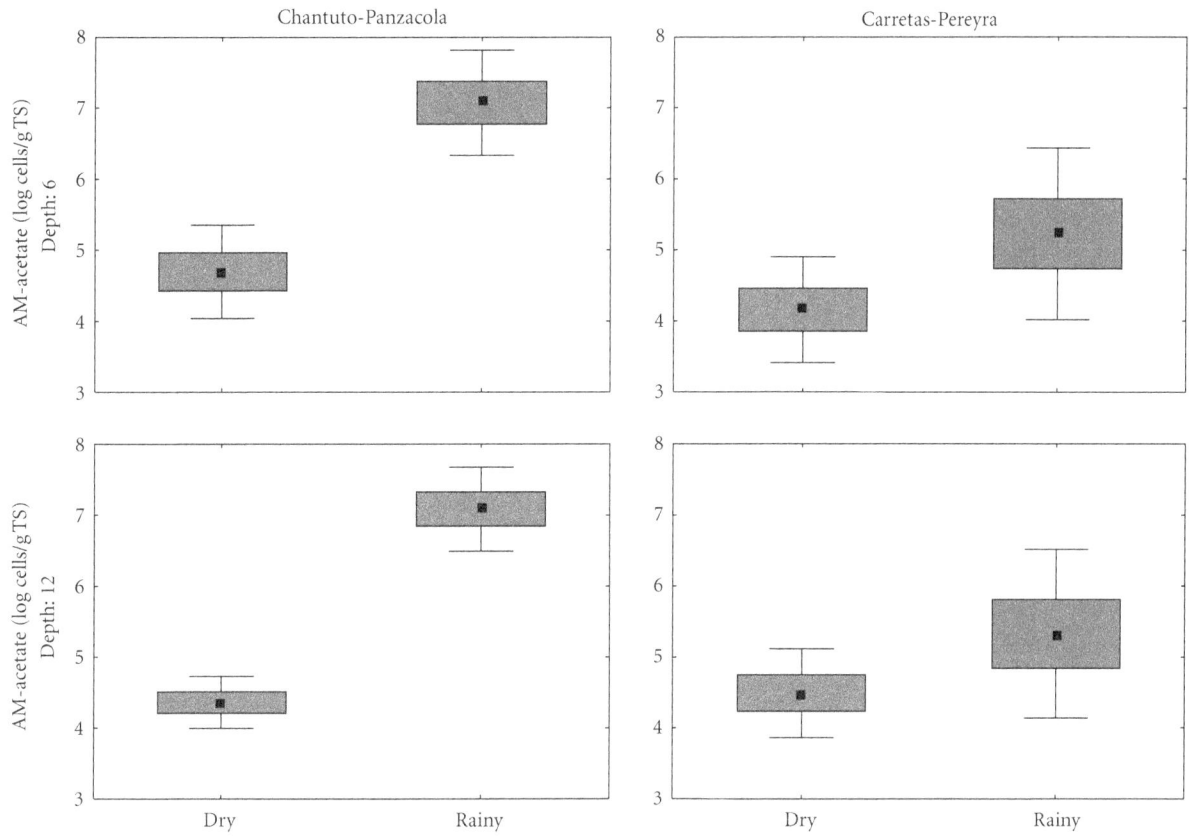

(a)

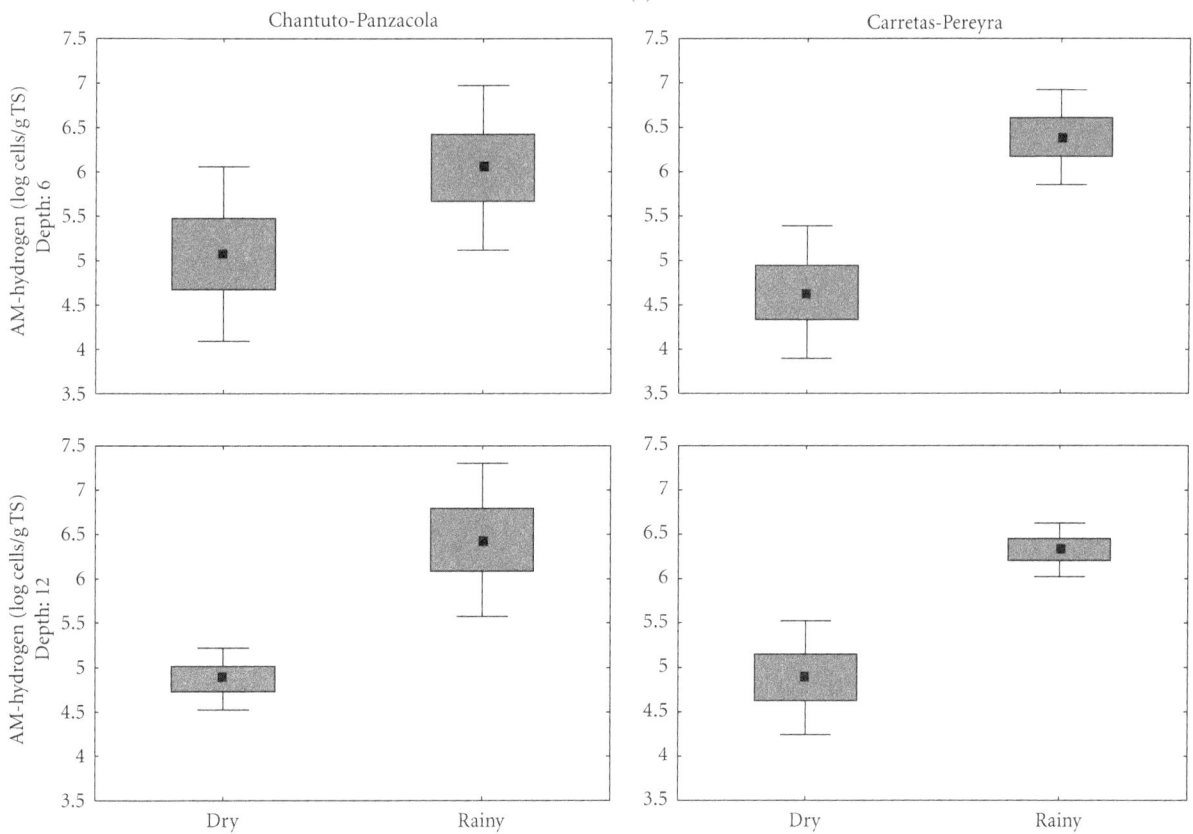

(b)

FIGURE 2: Continued.

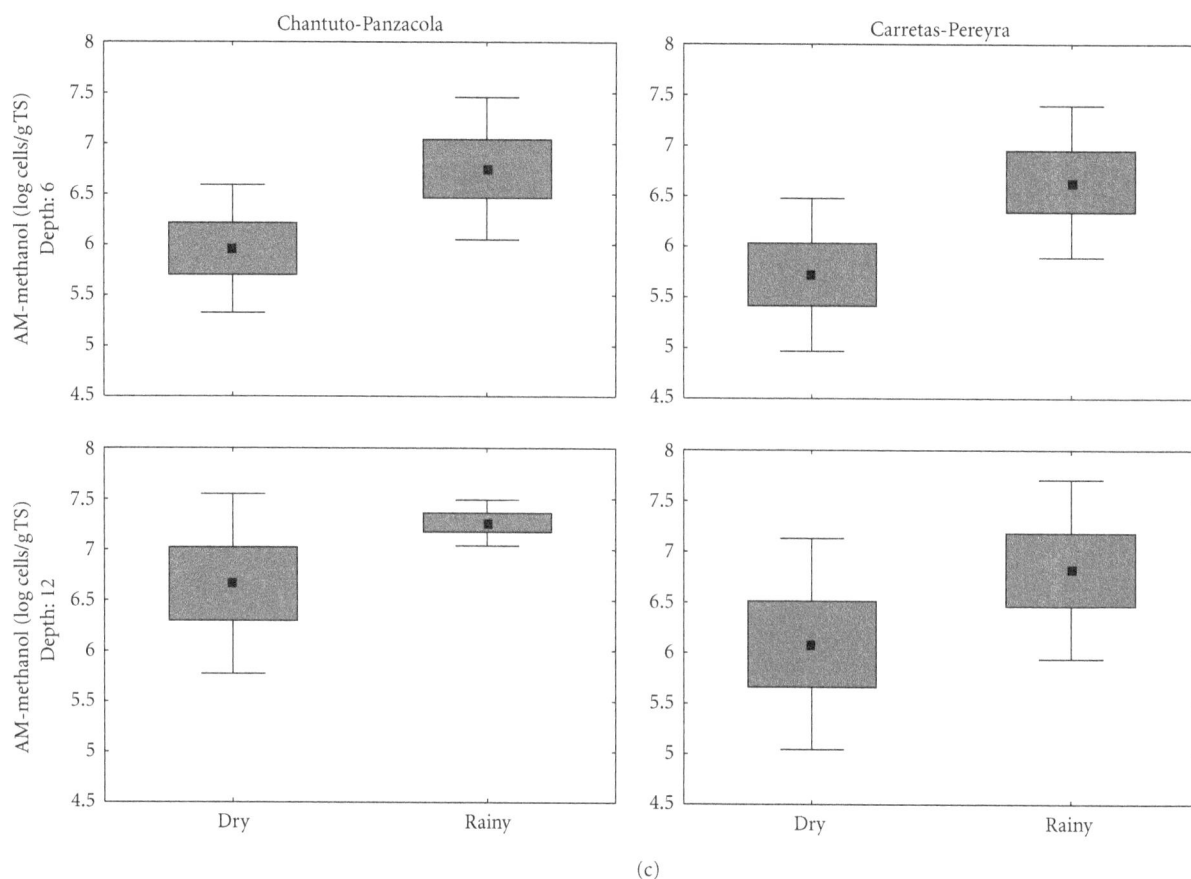

(c)

FIGURE 2: Temporal and spatial variation in the abundance of MA (log cells/g TS).

TABLE 3: Abundance of MA, acetoclastic activity, and methane production in sediments of Chantuto-Panzacola and Carretas-Pereyra, Chiapas. Mean values.

Depth	Dry season		Rainy season	
	6	12	6	12
Chantuto-Panzacola				
MA-acetate (cells/g)	1.30×10^5	2.99×10^4	2.20×10^7	2.09×10^7
MA-Hydrogen (cells/g)	1.63×10^6	9.55×10^4	9.37×10^6	8.63×10^6
MA-methanol (cells/g)	1.79×10^6	1.97×10^7	1.17×10^7	2.06×10^7
Acetate activity without SO_4^{-2} (mM acetate/g VS/day)	0.03	0.03	0.02	0.01
Acetate activity with SO_4^{-2} (mM acetate/g VS/day)	0.05	0.03	0.01	0.01
% CH_4 without SO_4^{-2}	4.81	7.91	23.50	29.73
% CH_4 with SO_4^{-2}	2.78	4.63	5.64	8.77
Carretas-Pereyra				
MA-acetate (cells/g)	4.52×10^4	6.17×10^4	1.90×10^6	1.32×10^6
MA-Hydrogen (cells/g)	1.34×10^5	1.51×10^5	4.49×10^6	2.64×10^6
MA-methanol (cells/g)	1.34×10^6	8.24×10^6	1.27×10^7	2.21×10^7
Acetate activity without SO_4^{-2} (mM acetate/g VS/day)	0.03	0.02	0.01	0.01
Acetate activity with SO_4^{-2} (mM acetate/g VS/day)	0.04	0.03	0.01	0.01
% CH_4 without SO_4^{-2}	7.83	13.01	15.42	23.47
% CH_4 with SO_4^{-2}	4.02	6.55	6.41	13.02

allows MA to maintain their populations in the presence of sulfate, which act favoring sulfate reduction. The key role of other methylated compounds was demonstrated in mangrove areas in India, where MA were quantified from methylamines [14, 15].

In the rainy season, methanol remained important, but the abundance of MA from hydrogen and acetate increased under low sulfate concentrations, hydrogen theoretically contributes 33% to total methanogenesis when carbohydrates or similar organic matter are degraded, being important in environments with high sedimentation rates (≈ 10 cm/year) and organic carbon supplementation [36]. In the coastal lagoons studied, a high concentration of organic carbon (3.4–14.5%) was quantified, and a sedimentation rate of 6 cm/year was observed in Carretas-Pereyra. Acetate can produce approximately two thirds of total methane in freshwater sediments; however, its contribution to methane formation decreases when is consumed in other anaerobic processes as the sulfate reduction [4]. The effect of sulfate on methanogenesis was demonstrated in temperate estuaries, where the contribution of acetate for this process has been found to increase when sulfate concentration is low in freshwater zone, and the sulfate reduction decreased [7, 13]. The acetate and hydrogen are also important substrates for methanogenesis in salt marshes areas [10].

This study has revealed that acetate-utilizing and hydrogen-utilizing MA does not have a distinct vertical distribution pattern in Chantuto-Panzacola and Carretas-Pereyra sediments, whereas the methanol-based group apparently being more abundant in the 6–12 cm layer ($P = 0.05$). The presence of MA along 12 cm of sediment column seems to be a result of the availability of substrates for these microorganisms; the constant supply of different substrates favors the presence of MA at different sediment layers as also has been demonstrated in sediments of tidal flats, coastal marshes, and mangroves [8, 10, 14].

3.3. Acetoclastic Metabolic Activity. In all kinetic experiments, there was an increase in the concentration of acetate in the first days, along with other volatile fatty acids (propionate and butyrate); this pattern reveals the presence of fermentation processes in sediments. The continued presence of acetate along with other intermediaries (butyrate and propionate) is similar to that reported in other studies where methanogenesis has been assessed [37]. Acetate is an important intermediate produced during the anaerobic mineralization of organic matter, followed by propionate and other volatile fatty acids [38]. The fermentation activity is important because it releases organic substrates, such as acetate, that can be used by the MA, which cannot directly use complex organic compounds. Subsequent to the production of volatile fatty acids, acetate consumption started on day 7 in sulfate-enriched media and between days 14 and 21 in sulfate-free media. Methane production was recorded on day 21.

Acetoclastic activity in sulfate-free experiments had no significant temporary differences ($P > 0.05$) (Figures 3(a)–3(c)). The experiments with sulfate showed significant temporal fluctuations, with high values in the dry season

(Table 3; Figures 3(a)–3(c)). Vertical variations did not reach statistical significance ($P > 0.05$).

Methane formation was observed in all experiments, with differences depending on the specific conditions of each medium. The addition of acetate results in an increase in methane production in relation to the amount observed in controls (no carbon supplementation).

Methane production was higher in sulfate-free media compared with sulfate-enriched media (Table 3; Figures 3(b)–3(d)). Temporal differences ($P < 0.05$) in methane production from acetate were observed in both systems. Methane levels were higher in the rainy season than in the dry season (Table 3). Significant vertical changes ($P < 0.05$) were observed only in Carretas-Pereyra: a lower production in the upper 6 cm and a higher methane production in the 6–12 cm layer (Table 3, Figures 3(b)–3(d)).

The presence of sulfate in the culture media influenced methanogenic activity. In the sulfate-free experiments a peak of acetoclastic activity was observed coupled with a rise in methane production in sediments during the rainy season and in the deep layer, suggesting that methanogenesis was favored. Studies demonstrated that potential methanogenesis from acetate was higher in the absence of sulfates [37]. By contrast, the addition of sulfate resulted in an increase of acetoclastic activity in the dry months and in the upper sediment layer, and methane production declined. In sediments of coastal lagoons and mangrove areas in India, an increase in the production and emission of methane was determined in freshwater areas compared to brackish regions. Also, methane emissions were higher in the postmonsoon season, when salinity and sulfate concentration were lower [16, 34].

3.4. Environmental Variables and MA. The correlation coefficients between environmental variables and ordination axes (interset correlation) obtained by CCA denote the relative importance of each environmental variable in the distribution of the methanogenic community. For Chantuto-Panzacola, the MA-environment correlation was 0.92 corresponded to a salinity-sulfate gradient and 0.60 for pH. CCA results for Carretas-Pereyra showed a correlation of 0.74 for pH and volatile solids, and 0.43 for volatile solids. The ordination diagram obtained by CCA showed a change in the structure of the methanogenic community with regard to certain environmental variables (Figure 4). The first axis accounted for 65.62% of total variance in Chantuto-Panzacola, corresponding to a salinity-sulfate gradient (Figure 4(a)). In the right side of the diagram, those sites with the highest sulfate concentration, temperature, and pH (dry season) were grouped, in these conditions methanol-utilizing MA were abundant. The left side of the plot-grouped sites with highest total solids content where hydrogen-utilizing MA prospered, whereas acetate-utilizing MA abound in sites with a higher porosity and less reduced conditions (Figure 4(a)). In Carretas-Pereyra, to the plot's upper left side, the first axis accounted for 29.08% of variance and salinity-sulfate, Eh and organic carbon concentration were all correlated with hydrogen-utilizing MA abundance, mainly during the rainy season. Abundance of methanol-utilizing

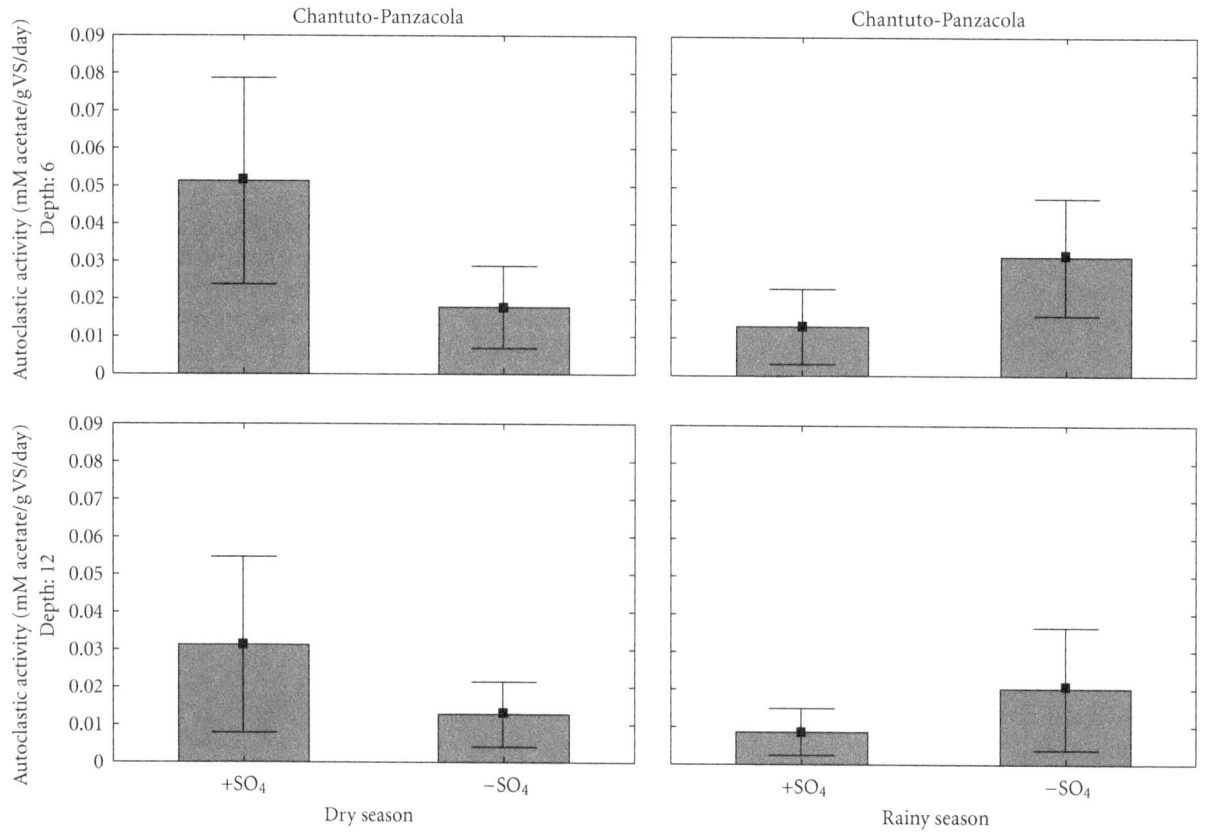

(a)

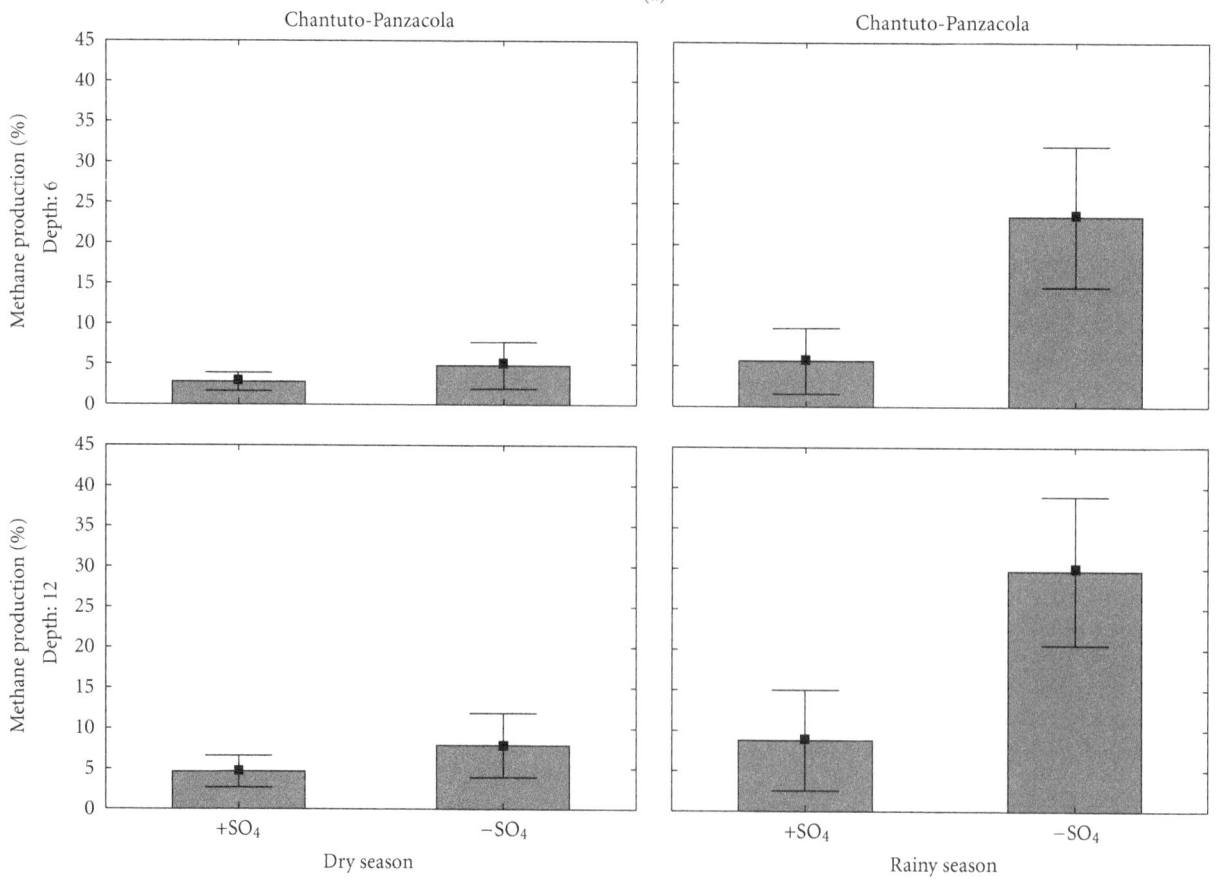

(b)

FIGURE 3: Continued.

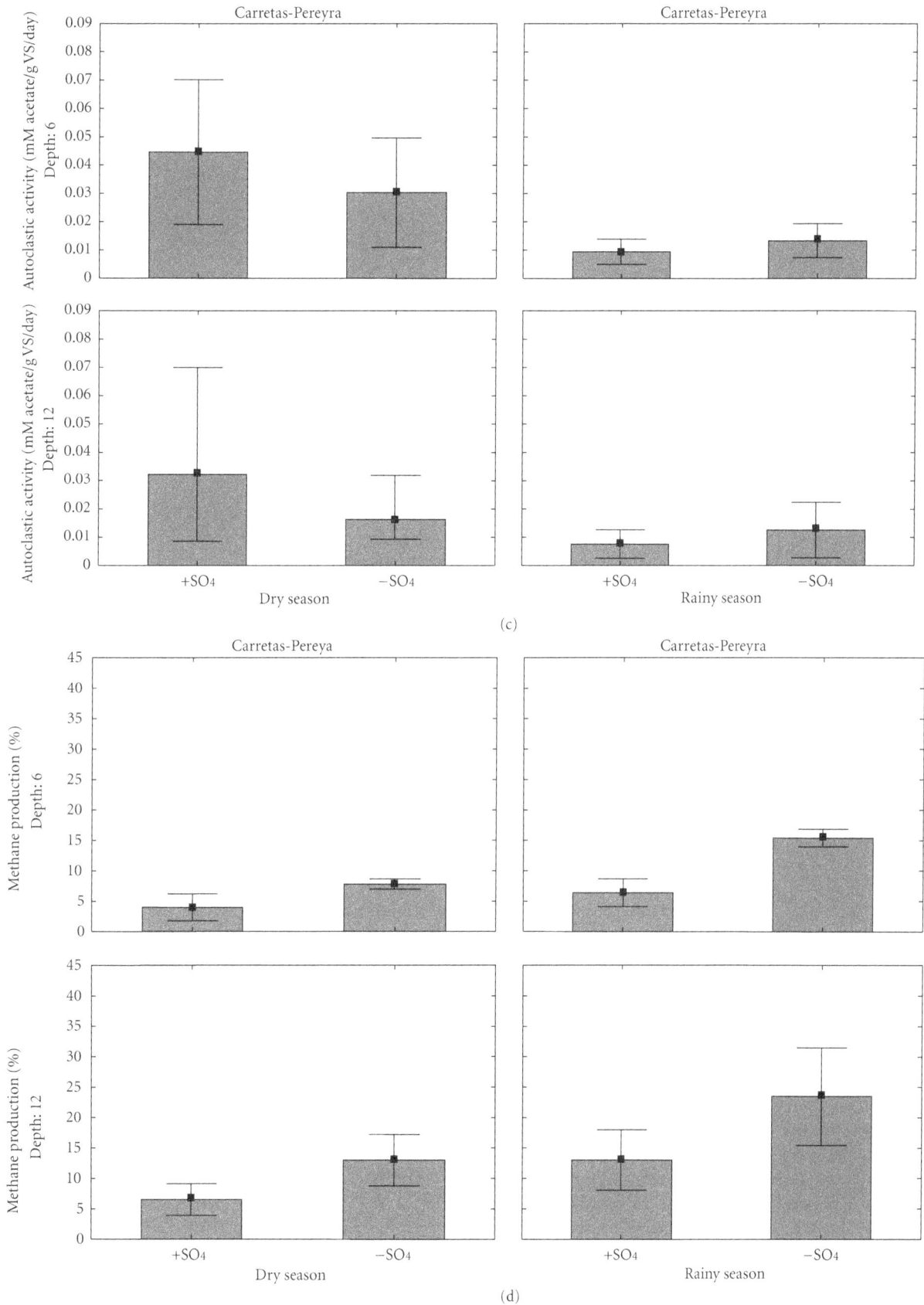

FIGURE 3: Temporal and spatial variations of acetoclastic activity and methane production in Chantuto-Panzacola (a, b) and Carretas-Pereyra (c, d).

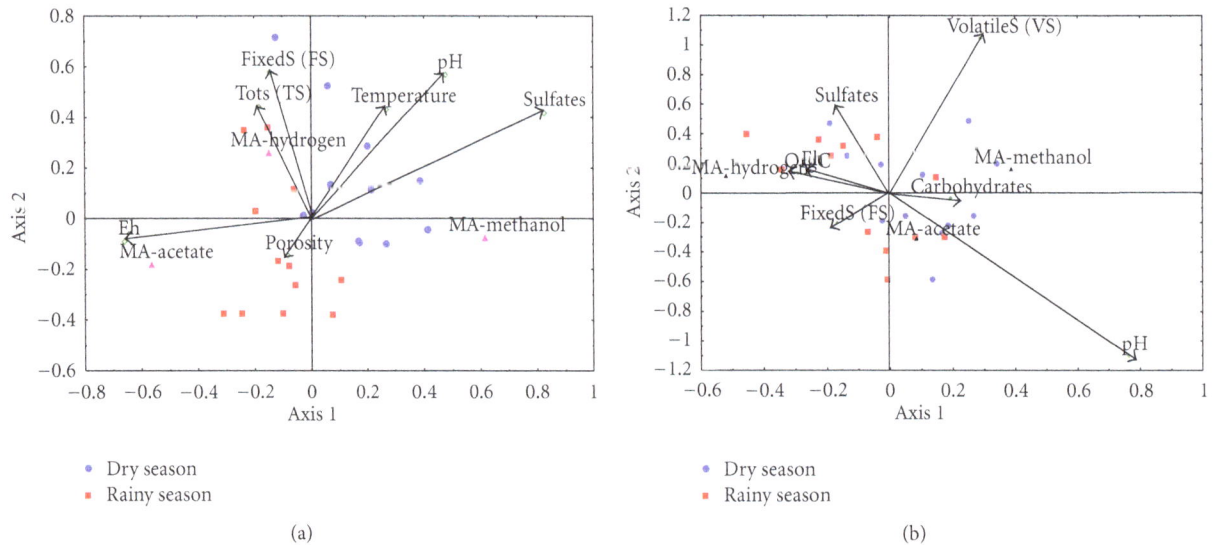

FIGURE 4: Relationship between environmental variables and methanogenic community in Chantuto-Panzacola (a) and Carretas-Pereyra (b).

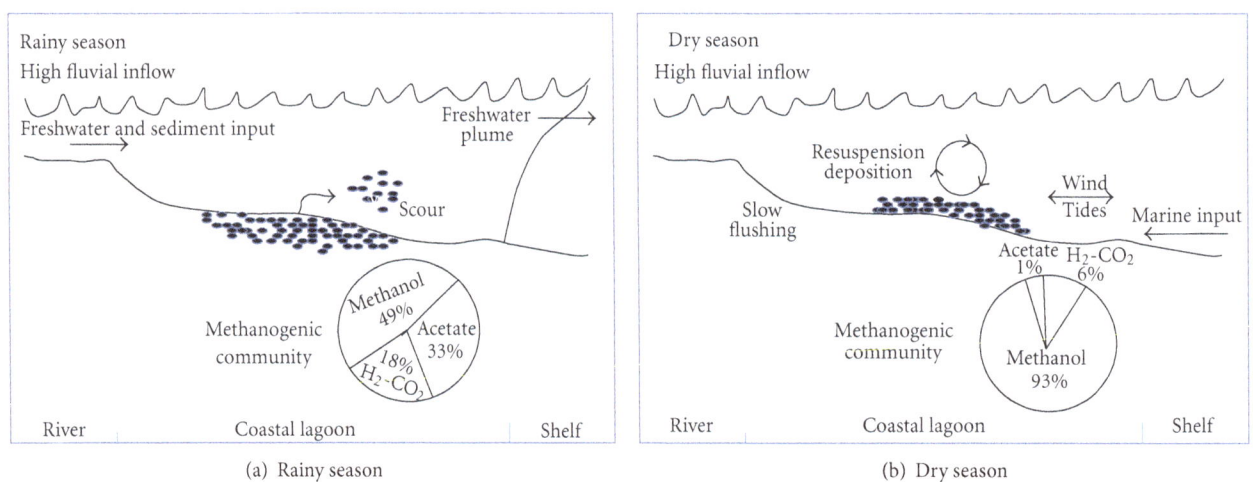

(a) Rainy season   (b) Dry season

FIGURE 5: Conceptual model on MA dynamics in sediments for tropical coastal lagoons.

MA was correlated with volatile solids during the dry season and acetate-utilizing MA prospered in a pH gradient in both seasons.

The presence of sulfate significantly influenced the abundance of MA in both systems. In sulfate-rich conditions the anaerobic process that is most favored is the sulfate reduction. Elevated levels of SRP in the dry season (10^8–10^{10} cells/g) and decrease in the rainy season (10^5–10^7 cells/g) quantified in coastal lagoons studied support this hypothesis [39]. The relationship between MA and sulfate content is consistent with knowledge on these communities in sediments of temperate estuaries, tropical coastal lagoons, coastal marshes, and mangroves [7–10, 13, 14, 16].

The temperature, pH, Eh, and organic fractions were other variables contributing to the presence of MA. In this study the bacteria grew in a temperature range of 26.7–29.4°C, the optimum temperature reported for the development of methanogens is 30–32°C in tropical wetlands, whereas in mangrove sediments MA have been observed at temperature ranges between 26 and 30°C [14, 15]. The pH conditions (6.7–7.1) were favorable for methanogenic community. Mohanraju and Natarajan [15] associated the presence of MA with pH values of 6.6–7.2 in mangrove sediments, whereas in coastal marshes sediments MA were recorded in pH ranges of 6.1–7.5 [10]. The reducing (−100 to +100 mV) and highly reducing (−300 to −100 mV) characteristics of sediment also contributed to MA development, these have been reported at levels from Eh below −150 mV in coastal marshes [40].

4. Conclusions

The MA was a constant component involved in anaerobic mineralization of organic matter in the sediments of the coastal systems Chantuto-Panzacola and Carretas-Pereyra. Their populations were active by its ability to utilize different substrates, especially methanol. In these ecosystems, changes in precipitation and its influence on fluvial inputs significantly impacted salinity and sulfate content, which was the main factor regulating the temporal dynamics of methanogenic community. In the rainy season, the increase in river inflow to lagoons produces freshwater conditions, low sulfate concentrations, sediment resuspension, and less reducing redox potentials. The environmental characteristics that prevail in this season lead to an increase in MA abundance, with the following decreasing rank by substrate: methanol > H_2-CO_2 > acetate (Figure 5(a)). Methanogenic activity from acetate is higher and results in a rise in methane production. The peak of methanogenic activity in the rainy season suggests that these ecosystems may be an important source of atmospheric CH_4 and CO_2 in this season. In the dry season, the freshwater inflow declines and salinity, sulfate, and inorganic solids content increases, along with more negative redox conditions. In this conditions a lower MA density was observed (Figure 5(b)), with the following order by substrate: methanol > acetate > H_2-CO_2. Our hypothesis is that this mechanism is cyclic and is controlled by changes associated to the seasonal fluctuations in fluvial inflow and precipitation. We considered that this model could be applied to other coastal lagoons and wetlands (mangroves) in tropical latitudes, characterized by a high organic matter concentration and a permanent river discharge, with significant seasonal variations in discharge volume. These characteristics favor the continued presence of the MA in tropical coastal systems and control their temporal dynamics. In estuaries, the absence of barriers that restrict communication with the sea difficult to apply this model because the tidal influence decreases the impact of freshwater input. In these systems the freshwater influence is more important at spatial level.

Conflict of Interests

Authors of the paper have no conflict of interests with Millipore, Statistica 10 (Academic) and MVSP (3.12b Software). The authors do not have any association with Millipore or companies who designed Statistica 10 (Academic) and MVSP (3.12b Software).

Acknowledgment

This study was funded by the project "Ecological study of estuarine systems Chantuto-Panzacola and Carretas-Pereyra, Chiapas" DCBS, Universidad Autónoma Metropolitana-Iztapalapa.

References

[1] B. B. Jørgensen, "Mineralization of organic matter in the sea bed—the role of sulphate reduction," *Nature*, vol. 296, no. 5858, pp. 643–645, 1982.

[2] M. Fukui, J. Suh, Y. Yonezawa, and Y. Urushigawa, "Major substrates for microbial sulfate reduction in the sediments of Ise Bay, Japan," *Ecological Research*, vol. 12, no. 2, pp. 201–209, 1997.

[3] D. Marty, P. Bonin, V. Michotey, and M. Bianchi, "Bacterial biogas production in coastal systems affected by freshwater inputs," *Continental Shelf Research*, vol. 21, no. 18-19, pp. 2105–2115, 2001.

[4] D. E. Canfield, E. Kristensen, and B. Thamdrup, "The methane cycle," in *Advances in Marine Biology. Aquatic Geomicrobiology*, A. Southward, P. A. Tyler, C. M. Young, and L. A. Fuiman, Eds., vol. 48, chapter 10, pp. 383–418, Elsevier, London, UK, 2005.

[5] R. K. Thauer, A. K. Kaster, H. Seedorf, W. Buckel, and R. Hedderich, "Methanogenic archaea: ecologically relevant differences in energy conservation," *Nature Reviews Microbiology*, vol. 6, no. 8, pp. 579–591, 2008.

[6] K. Zepp Falz, C. Holliger, R. Großkopf et al., "Vertical distribution of methanogens in the anoxic sediment of Rotsee (Switzerland)," *Applied and Environmental Microbiology*, vol. 65, no. 6, pp. 2402–2408, 1999.

[7] K. J. Purdy, M. A. Munson, D. B. Nedwell, and T. M. Embley, "Comparison of the molecular diversity of the methanogenic community at the brackish and marine ends of a UK estuary," *FEMS Microbiology Ecology*, vol. 39, no. 1, pp. 17–21, 2002.

[8] R. Wilms, H. Sass, B. Köpke, J. Köster, H. Cypionka, and B. Engelen, "Specific Bacterial, Archaeal, and Eukaryotic communities in tidal-flat sediments along a vertical profile of several meters," *Applied and Environmental Microbiology*, vol. 72, no. 4, pp. 2756–2764, 2006.

[9] M. J. Franklin, J. W. William, and W. B. Whitman, "Populations of methanogenic bacteria in a Georgia salt marsh," *Applied and Environmental Microbiology*, vol. 54, no. 5, pp. 1151–1157, 1988.

[10] M. A. Munson, D. B. Nedwell, and T. M. Embley, "Phylogenetic diversity of Archaea in sediment samples from a coastal salt marsh," *Applied and Environmental Microbiology*, vol. 63, no. 12, pp. 4729–4733, 1997.

[11] R. Segers and S. W. M. Kengen, "Methane production as a function of anaerobic carbon mineralization: a process model," *Soil Biology and Biochemistry*, vol. 30, no. 8-9, pp. 1107–1117, 1998.

[12] S. Takii and M. Fukui, "Relative importance of methanogenesis, sulfate reduction and denitrification in sediments of the lower Tama river," *Bulletin of Japanese Society of Microbial Ecology*, vol. 6, no. 1, pp. 9–17, 1991.

[13] K. J. Purdy, M. A. Munson, T. Cresswell-Maynard, D. B. Nedwell, and T. M. Embley, "Use of 16S rRNA-targeted oligonucleotide probes to investigate function and phylogeny of sulphate-reducing bacteria and methanogenic archaea in a UK estuary," *FEMS Microbiology Ecology*, vol. 44, no. 3, pp. 361–371, 2003.

[14] T. Ramamurthy, R. Mohanraju, and R. Natarajan, "Distribution and ecology of methanogenic bacteria in mangrove sediments of Pitchavaram, east coast of India," *Indian Journal of Marine Sciences*, vol. 19, no. 4, pp. 269–273, 1990.

[15] R. Mohanraju and R. Natarajan, "Methanogenic bacteria in mangrove sediments," *Hydrobiologia*, vol. 247, no. 1–3, pp. 187–193, 1992.

[16] H. Biswas, S. K. Mukhopadhyay, S. Sen, and T. K. Jana, "Spatial and temporal patterns of methane dynamics in the tropical mangrove dominated estuary, NE coast of Bay of Bengal, India," *Journal of Marine Systems*, vol. 68, no. 1-2, pp. 55–64, 2007.

[17] INE-SEMARNAP, *Programa de Manejo Reserva de la Biósfera La Encrucijada*, Instituto Nacional de Ecología-Secretaría del Medio Ambiente, Recursos Naturales y, Pesca, Mexico, 1999.

[18] F. J. Flores-Verdugo, F. González-Farías, D. S. Zamorano, and P. Ramírez-García, "Mangrove ecosystems of the Pacific Coast of Mexico: distribution, structure, litter fall, and detritus dynamics," in *Coastal Plant Communities of Latin America*, U. Seliger, Ed., vol. 17, pp. 269–288, Academic Press, New York, NY, USA, 1992.

[19] F. J. Flores-Verdugo, G. de la Lanza-Espino, F. Contreras-Espinosa, and C. M. Agraz-Hernández, "The tropical Pacific Coast of Mexico," in *Coastal Marine Ecosystems of Latin American, Ecological Studies*, U. Seliger and B. Kjerve, Eds., vol. 144, pp. 307–314, Springer, Berlin, Germany, 2001.

[20] D. Langmuir, "Eh-pH determination," in *Proceedings Sediments and Petrology*, R. E. Conver, Ed., p. 653, Wiley-Interscience, New York, NY, USA, 1971.

[21] W. E. Balch, G. E. Fox, L. J. Magrum, C. R. Woese, and R. S. Wolfe, "Methanogens: reevaluation of a unique biological group," *Microbiological Reviews*, vol. 43, no. 2, pp. 260–296, 1979.

[22] B. L. Howes, "Effects of sampling technique on measurements of porewater constituents in salt marsh sediments," *Limnology and Oceanography*, vol. 30, no. 1, pp. 221–227, 1985.

[23] R. W. Howarth, "A rapid and precise method for determining sulfate in seawater, estuarine waters, and sediment pore waters," *Limnology and Oceanography*, vol. 23, no. 5, pp. 1066–1069, 1978.

[24] M. Dubois, K. A. Gilles, J. K. Hamilton, P. A. Rebers, and F. Smith, "Colorimetric method for determination of sugars and related substances," *Analytical Chemistry*, vol. 28, no. 3, pp. 350–356, 1956.

[25] APHA, AWWA, and WPCF, Eds., *Standard Methods For the Examination of Water and Wastewater*, American Public Health Association, American Water Works Association and Water Pollution Control Federation, Washington, DC, USA, 2005.

[26] H. Gaudette, W. Fligh, L. Toner, and D. Folger, "An inexpensive titration method for the determination of organic carbon in recent sediments," *Journal of Sediments and Petrology*, vol. 44, no. 1, pp. 249–253, 1974.

[27] J. H. Zar, *Bioestatistical Analysis*, Prentice Hall, New York, NY, USA, 1999.

[28] C. J. F. Ter Braak, "Canonical correspondence analysis: a new eigenvector technique for multivariate direct gradient analysis," *Ecology*, vol. 67, no. 5, pp. 1167–1179, 1986.

[29] E. Lallier-Vergès, B. P. Perrussel, J. R. Disnar, and F. Baltzer, "Relationships between environmental conditions and the diagenetic evolution of organic matter derived from higher plants in a modern mangrove swamp system (Guadeloupe, French West Indies)," *Organic Geochemistry*, vol. 29, no. 5–7, pp. 1663–1686, 1998.

[30] B. Knoppers and B. Kjerfve, "Coastal lagoons of Southeastern Brazil: physical and biogeochemical characteristics," in *Estuaries of South America*, G. M. E. Perillo, M. C. Piccolo, and M. Pino-Quivira, Eds., pp. 35–66, Springer, New York, NY, USA, 1997.

[31] M. R. Preston and P. Prodduturu, "Tidal variations of particulate carbohydrates in the Mersey estuary," *Estuarine, Coastal and Shelf Science*, vol. 34, no. 1, pp. 37–48, 1992.

[32] R. W. Howarth, "Microbial processes in salt-marsh sediments," in *Aquatic Microbiology*, T. E. Ford, Ed., pp. 239–260, Blackwell Scientific Publications, Boston, Mass, USA, 1993.

[33] A. L. dos Santos Fonseca, M. Minello, C. Cardoso Marinho, and F. de Assis Esteves, "Methane concentration in water column and in pore water of a coastal lagoon (Cabiúnas Lagoon, Macaé,

RJ, Brazil)," *Brazilian Archives of Biology and Technology*, vol. 47, no. 2, pp. 301–308, 2004.

[34] A. Verma, V. Subramanian, and R. Ramesh, "Methane emissions from a coastal lagoon: vembanad Lake, West Coast, India," *Chemosphere*, vol. 47, no. 8, pp. 883–889, 2002.

[35] T. Egli, "The ecological and physiological significance of the growth of heterotrophic microorganisms with mixtures of substrates," in *Advances in Microbial Ecology*, J. Gwynfryn, Ed., vol. 14, pp. 305–386, Plenum Press, New York, NY, USA, 1995.

[36] R. Conrad, "Contribution of hydrogen to methane production and control of hydrogen concentrations in methanogenic soils and sediments," *FEMS Microbiology Ecology*, vol. 28, no. 3, pp. 193–202, 1999.

[37] M. Holmer and E. Kristensen, "Coexistence of sulfate reduction and methane production in an organic-rich sediment," *Marine Ecology Progress Series*, vol. 107, no. 1-2, pp. 177–184, 1994.

[38] H. T. S. Boschker, W. de Graaf, M. Köster, L. A. Meyer-Reil, and T. E. Cappenberg, "Bacterial populations and processes involved in acetate and propionate consumption in anoxic brackish sediment," *FEMS Microbiology Ecology*, vol. 35, no. 1, pp. 97–103, 2001.

[39] M. R. Torres-Alvarado, *Determinación de la diversidad y actividad bacteriana sulfatorreductora y metanogénica en los sedimentos de dos ecosistemas estuarino-lagunares del Estado de Chiapas [Ph.D. thesis]*, Universidad Autónoma Metropolitana-Iztapalapa, México, Mexico, 2009.

[40] H. K. Kludze and R. D. DeLaune, "Methane emissions and growth of Spartina patens in response to soil redox intensity," *Soil Science Society of America Journal*, vol. 58, no. 6, pp. 1838–1845, 1994.

Permissions

The contributors of this book come from diverse backgrounds, making this book a truly international effort. This book will bring forth new frontiers with its revolutionizing research information and detailed analysis of the nascent developments around the world.

We would like to thank all the contributing authors for lending their expertise to make the book truly unique. They have played a crucial role in the development of this book. Without their invaluable contributions this book wouldn't have been possible. They have made vital efforts to compile up to date information on the varied aspects of this subject to make this book a valuable addition to the collection of many professionals and students.

This book was conceptualized with the vision of imparting up-to-date information and advanced data in this field. To ensure the same, a matchless editorial board was set up. Every individual on the board went through rigorous rounds of assessment to prove their worth. After which they invested a large part of their time researching and compiling the most relevant data for our readers. Conferences and sessions were held from time to time between the editorial board and the contributing authors to present the data in the most comprehensible form. The editorial team has worked tirelessly to provide valuable and valid information to help people across the globe.

Every chapter published in this book has been scrutinized by our experts. Their significance has been extensively debated. The topics covered herein carry significant findings which will fuel the growth of the discipline. They may even be implemented as practical applications or may be referred to as a beginning point for another development. Chapters in this book were first published by Hindawi Publishing Corporation; hereby published with permission under the Creative Commons Attribution License or equivalent.

The editorial board has been involved in producing this book since its inception. They have spent rigorous hours researching and exploring the diverse topics which have resulted in the successful publishing of this book. They have passed on their knowledge of decades through this book. To expedite this challenging task, the publisher supported the team at every step. A small team of assistant editors was also appointed to further simplify the editing procedure and attain best results for the readers.

Our editorial team has been hand-picked from every corner of the world. Their multi-ethnicity adds dynamic inputs to the discussions which result in innovative outcomes. These outcomes are then further discussed with the researchers and contributors who give their valuable feedback and opinion regarding the same. The feedback is then collaborated with the researches and they are edited in a comprehensive manner to aid the understanding of the subject.

Apart from the editorial board, the designing team has also invested a significant amount of their time in understanding the subject and creating the most relevant covers. They scrutinized every image to scout for the most suitable representation of the subject and create an appropriate cover for the book.

The publishing team has been involved in this book since its early stages. They were actively engaged in every process, be it collecting the data, connecting with the contributors or procuring relevant information. The team has been an ardent support to the editorial, designing and production team. Their endless efforts to recruit the best for this project, has resulted in the accomplishment of this book. They are a veteran in the field of academics and their pool of knowledge is as vast as their experience in printing. Their expertise and guidance has proved useful at every step. Their uncompromising quality standards have made this book an exceptional effort. Their encouragement from time to time has been an inspiration for everyone.

The publisher and the editorial board hope that this book will prove to be a valuable piece of knowledge for researchers, students, practitioners and scholars across the globe.

List of Contributors

Christoph Wrede
Institute of Microbiology and Genetics, Georg-August-University, Grisebachstr. 8, 37077 Göttingen, Germany
Hannover Medical School, Institute of Functional and Applied Anatomy, Carl-Neuberg-Str. 1, 30625 Hannover, Germany

Sebastian Kokoschka
Institute of Microbiology and Genetics, Georg-August-University, Grisebachstr. 8, 37077 Göttingen, Germany

Anne Dreier
Institute of Microbiology and Genetics, Georg-August-University, Grisebachstr. 8, 37077 Göttingen, Germany
Courant Centre Geobiology, Georg-August-University, Goldschmidtstr. 3, 37077 G"ottingen, Germany

Michael Hoppert
Institute of Microbiology and Genetics, Georg-August-University, Grisebachstr. 8, 37077 Göttingen, Germany
Geoscience Centre Göttingen, Georg-August-University, Goldschmidtstr. 3, 37077 Göttingen, Germany

Joachim Reitner
Courant Centre Geobiology, Georg-August-University, Goldschmidtstr. 3, 37077 Göttingen, Germany
Geoscience Centre Göttingen, Georg-August-University, Goldschmidtstr. 3, 37077 Göttingen, Germany

Christina Heller
Geoscience Centre Göttingen, Georg-August-University, Goldschmidtstr. 3, 37077 Göttingen, Germany
Federal Institute for Geosciences and Natural Resources, Stilleweg 2, 30655 Hannover, Germany

Kian-Hong Ng
Cell Division Laboratory, Temasek Life Sciences Laboratory, 1 Research Link, National University of Singapore, Singapore 117604

Vinayaka Srinivas
Cell Division Laboratory, Temasek Life Sciences Laboratory, 1 Research Link, National University of Singapore, Singapore 117604
Department of Biological Sciences, National University of Singapore, 14 Science Drive 4, Singapore 117543

Ramanujam Srinivasan
Mechanobiology Institute, National University of Singapore, 5A Engineering Drive 1, Singapore 117411

Mohan Balasubramanian
Cell Division Laboratory, Temasek Life Sciences Laboratory, 1 Research Link, National University of Singapore, Singapore 117604
Department of Biological Sciences, National University of Singapore, 14 Science Drive 4, Singapore 117543
Mechanobiology Institute, National University of Singapore, 5A Engineering Drive 1, Singapore 117411

Xuan Zhou, Michael Kreuzer and Johanna O. Zeitz
ETH Zurich, Institute of Agricultural Sciences, Universitaetstrasse 2, 8092 Zurich, Switzerland

Leo Meile
ETH Zurich, Institute of Food, Nutrition and Health, Schmelzbergstrasse 7, 8092 Zurich, Switzerland

Bernd Wemheuer, Robert Taube, Pinar Akyol and Rolf Daniel
Department of Genomic and Applied Microbiology and Goettingen Genomics Laboratory, Institute of Microbiology and Genetics, Georg-August-University Goettingen, Grisebachstraße 8, 37077 Goettingen, Germany

Franziska Wemheuer
Section of Agricultural Entomology, Department for Crop Sciences, Georg-August-University Goettingen, Grisebachstraße 6, 37077 Goettingen, Germany

Parkson Lee-Gau Chong, Umme Ayesa, Varsha Prakash Daswani and Ellah Chay Hur
Department of Biochemistry, Temple University School of Medicine, 3420 North Broad Street, Philadelphia, PA 19140, USA

Stephen T. Abedon and Kelly L. Murray
Department of Microbiology, The Ohio State University, 1680 University Drive, Mansfield, OH 44906, USA

Jennifer Gebetsberger
Department of Chemistry and Biochemistry, University of Bern, Freiestraße 3, 3012 Bern, Switzerland
Graduate School for Cellular and Biomedical Sciences, University of Bern, 3012 Bern, Switzerland

Andrea Künzi
Department of Chemistry and Biochemistry, University of Bern, Freiestraße 3, 3012 Bern, Switzerland

Norbert Polacek
Department of Chemistry and Biochemistry, University of Bern, Freiestraße 3, 3012 Bern, Switzerland
Division of Genomics and RNomics, Innsbruck Biocenter, Innsbruck Medical University, Innrain 80/82, 6020 Innsbruck, Austria

Marek Zywicki
Division of Genomics and RNomics, Innsbruck Biocenter, Innsbruck Medical University, Innrain 80/82, 6020 Innsbruck, Austria
Laboratory of Computational Genomics, Institute of Molecular Biology and Biotechnology, Adam Mickiewicz University, 61-712 Poznan, Poland

Ajda Ota and Dejan Gmajner
Department of Food Science and Technology, Biotechnical Faculty, University of Ljubljana, Jamnikarjeva 101, 1000 Ljubljana, Slovenia

Marjeta Šentjurc
EPR Center, Institute Jo˘zef Stefan, Jamova 39, 1000 Ljubljana, Slovenia

Nataša Poklar Ulrih
Department of Food Science and Technology, Biotechnical Faculty, University of Ljubljana, Jamnikarjeva 101, 1000 Ljubljana, Slovenia
Centre of Excellence for Integrated Approaches in Chemistry and Biology of Proteins (CipKeBiP), Jamova 39, 1000 Ljubljana, Slovenia

Alvaro Orell, Rafaela Arancibia and Carlos A. Jerez
Laboratory of Molecular Microbiology and Biotechnology, Department of Biology and Millennium Institute for Cell Dynamics and Biotechnology, Faculty of Sciences, University of Chile, Santiago, Chile

Francisco Remonsellez
Department of Chemical Engineering, North Catholic University, Antofagasta, Chile

Elisabeth W. Vissers and Paul L. E. Bodelier
Department of Microbial Ecology, Netherlands Institute of Ecology (NIOO-KNAW), Droevendaalsesteeg 10, 6708 PBWageningen, The Netherlands

Flavio S. Anselmetti
Swiss Federal Institute of Aquatic Science and Technology (Eawag), ¨Uberlandstrasse 133, 8600 D¨ubendorf, Switzerland
Institute of Geological Sciences and Oeschger Centre for Climate Change Research, University of Bern, Z¨ahringerstrasse 25, 3012 Bern, Switzerland

Gerard Muyzer
Department of Microbial Ecology, Netherlands Institute of Ecology (NIOO-KNAW), Droevendaalsesteeg 10, 6708 PBWageningen, The Netherlands
Department of Aquatic Microbiology, University of Amsterdam, Science Park 904, 1098 XH Amsterdam, The Netherlands

Christa Schleper
Department of Genetics in Ecology, University of Vienna, Althanstrasse 14, 1090 Vienna, Austria

Maria Tourna
AgResearch Ltd., Ruakura Centre, East Street Private Bag 3115, Hamilton 3240, New Zealand

Hendrikus J. Laanbroek
Department of Microbial Ecology, Netherlands Institute of Ecology (NIOO-KNAW), Droevendaalsesteeg 10, 6708 PBWageningen, The Netherlands
Institute of Environmental Biology, Utrecht University, Padualaan 8, 3584 CH Utrecht, The Netherlands

Stefanie Berger, Cornelia Welte and Uwe Deppenmeier
Institute for Microbiology and Biotechnology, University of Bonn, Meckenheimer Allee 168, 53115 Bonn, Germany

Brian C. Thomas
Department of Earth and Planetary Science, University of California, Berkeley, 307 McCone Hall, Berkeley, CA 94720-4767, USA

Karen Andrade
Department of Environmental Science, Policy, and Management, University of California, Berkeley, 54 Mulford Hall, Berkeley, CA 94720, USA

Joanne B. Emerson
Department of Earth and Planetary Science, University of California, Berkeley, 307 McCone Hall, Berkeley, CA 94720-4767, USA
Cooperative Institute for Research in Environmental Sciences, University of Colorado, Boulder, CO, USA

Jillian F. Banfield
Department of Earth and Planetary Science, University of California, Berkeley, 307 McCone Hall, Berkeley, CA 94720-4767, USA
Department of Environmental Science, Policy, and Management, University of California, Berkeley, 54 Mulford Hall, Berkeley, CA 94720, USA

Anders Norman
Department of Earth and Planetary Science, University of California, Berkeley, 307 McCone Hall, Berkeley, CA 94720-4767, USA
Department of Biology, University of Copenhagen, Copenhagen, Denmark

Eric E. Allen
Marine Biology Research Division, Scripps Institution of Oceanography, La Jolla, CA, USA
Division of Biological Sciences, University of California, San Diego, La Jolla, CA 92093-0202, USA

Karla B. Heidelberg
Department of Biological Sciences, University of Southern California, Los Angeles, CA 90089, USA

Elina Roine and Dennis H. Bamford
Department of Biosciences and Institute of Biotechnology, University of Helsinki, P.O. Box 56, Viikinkaari 5, 00014 Helsinki, Finland

L. A. Fernández-Güelfo, C. J. Á lvarez-Gallego and L. I. Romero García
Department of Chemical Engineering and Food Technology, Faculty of Science, University of Cadiz, Cadiz, 11510 Puerto Real, Spain

D. Sales Márquez
Department of Environmental Technologies, Faculty of Marine and Environmental Sciences, University of Cadiz, Cadiz, 11510 Puerto Real, Spain

Takeru Yokoi, Keisuke Isobe, Tohru Yoshimura and Hisashi Hemmi
Department of Applied Molecular Bioscience, Graduate School of Bioagricultural Sciences, Nagoya University, Furo-cho, Chikusa-ku, Nagoya, Aichi 460-8601, Japan

Kate Porter
Biota Holdings Limited, 10/585 Blackburn Road, Notting Hill, VIC 3168, Australia

Sen-Lin Tang, Chung-Pin Chen, Pei-Wen Chiang and Mei-Jhu Hong
Biodiversity Research Center, Academia Sinica, Nankang, Taipei 115, Taiwan

Mike Dyall-Smith
School of Biomedical Sciences, Charles Sturt University, Locked Bag 588, WaggaWagga, NSW2678, Australia

Cheryl Ingram-Smith, Jerry L. Thurman Jr., Karen Zimowski and Kerry S. Smith
Department of Genetics and Biochemistry, Clemson University, Clemson, SC 29634-0318, USA

G. Dennis Sprott, Angela Yeung, Chantal J. Dicaire, Siu H. Yu and Dennis M. Whitfield
Institute for Biological Sciences, National Research Council of Canada, 100 Sussex Drive, Ottawa, ON, Canada K1A 0R6

N. Leulliot
Institut de Biochimie et de Biophysique Moléculaire et Cellulaire, CNRS-UMR 8619, IFR115, Université Paris-Sud, Bˆatiment 430, 91405 Orsay, France
Laboratoire de Cristallographie et RMNBiologiques-CNRSUMR-8015, Université Paris Descartes, Faculté des Sciences Pharmaceutiques et Biologiques, 4, av de l'Observatoire, 75270 Paris CEDEX 06, France

S. Quevillon-Cheruel and H. van Tilbeurgh
Institut de Biochimie et de Biophysique Moléculaire et Cellulaire, CNRS-UMR 8619, IFR115, Université Paris-Sud, Bˆatiment 430, 91405 Orsay, France

M. Graille
Institut de Biochimie et de Biophysique Moléculaire et Cellulaire, CNRS-UMR 8619, IFR115, Université Paris-Sud, Bˆatiment 430, 91405 Orsay, France
Laboratoire de Biochimie (BIOC), CNRS UMR 7654, Ecole Polytechnique, Route de Saclay, 91128 Palaiseau, France

C. Geslin, D. Flament and M. Le Romancer
Université de Brest, CNRS, IFREMER,UMR 6197, Laboratoire de Microbiologie des Environnements Extrêmes, OSU-IUEM, Technopôle Brest-Iroise, Avenue Dumont D'Urville, 29280 Plouzané, France

María del Rocío Torres-Alvarado and Francisco Varona-Cordero
Department of Hydrobiology, Universidad Autónoma Metropolitana-Iztapalapa, Avenida San Rafael Atlixco No. 86, Colonia Vicentina, 09340 Mexico City, DF, Mexico

Francisco José Fernández and Florina Ramírez Vives
Department of Biotechnology, Universidad Autónoma Metropolitana-Iztapalapa, Avenida San Rafael Atlixco No. 86, Colonia Vicentina, 09340 Mexico City, DF, Mexico